Fire and Emergency Services Instructor

Principles and Practice

THIRD EDITION

Forest F. Reeder

Alan E. Joos

JONES & BARTLETT LEARNING

Jones & Bartlett Learning
World Headquarters
25 Mall Road
Burlington, MA 01803
978-443-5000
info@jblearning.com
www.jblearning.com

International Society of Fire Service Instructors
14001C Saint Germain Drive
Suite 128
Centreville, Virginia 20121
www.isfsi.org

National Fire Protection Association
1 Batterymarch Park
Quincy, MA 02169-7471
www.NFPA.org

International Association of Fire Chiefs
4025 Fair Ridge Drive
Fairfax, VA 22033
www.IAFC.org

Jones & Bartlett Learning books and products are available through most bookstores and online booksellers. To contact Jones & Bartlett Learning directly, call 800-832-0034, fax 978-443-8000, or visit our website, www.jblearning.com.

Substantial discounts on bulk quantities of Jones & Bartlett Learning publications are available to corporations, professional associations, and other qualified organizations. For details and specific discount information, contact the special sales department at Jones & Bartlett Learning via the above contact information or send an email to specialsales@jblearning.com.

Copyright © 2020 by Jones & Bartlett Learning, LLC, an Ascend Learning Company, and the National Fire Protection Association®

All rights reserved. No part of the material protected by this copyright may be reproduced or utilized in any form, electronic or mechanical, including photocopying, recording, or by any information storage and retrieval system, without written permission from the copyright owner.

The content, statements, views, and opinions herein are the sole expression of the respective authors and not that of Jones & Bartlett Learning, LLC. Reference herein to any specific commercial product, process, or service by trade name, trademark, manufacturer, or otherwise does not constitute or imply its endorsement or recommendation by Jones & Bartlett Learning, LLC and such reference shall not be used for advertising or product endorsement purposes. All trademarks displayed are the trademarks of the parties noted herein. *Fire and Emergency Services Instructor: Principles and Practice, Third Edition* is an independent publication and has not been authorized, sponsored, or otherwise approved by the owners of the trademarks or service marks referenced in this product.

There may be images in this book that feature models; these models do not necessarily endorse, represent, or participate in the activities represented in the images. Any screenshots in this product are for educational and instructive purposes only. Any individuals and scenarios featured in the case studies throughout this product may be real or fictitious, but are used for instructional purposes only.

The International Association of Fire Chiefs, International Society of Fire Service Instructors, National Fire Protection Association, and the publisher have made every effort to ensure that contributors to *Fire and Emergency Services Instructor: Principles and Practice, Third Edition* materials are knowledgeable authorities in their fields. Readers are nevertheless advised that the statements and opinions are provided as guidelines and should not be construed as official International Association of Fire Chiefs, International Society of Fire Service Instructors, National Fire Protection Association policy. The recommendations in this publication or the accompanying resource manual do not indicate an exclusive course of action. Variations taking into account the individual circumstances and local protocols may be appropriate. The International Association of Fire Chiefs, International Society of Fire Service Instructors, National Fire Protection Association, and the publisher disclaim any liability or responsibility for the consequences of any action taken in reliance on these statements or opinions.

17602-5

Production Credits
General Manager and Executive Publisher: Kimberly Brophy
VP, Product Development: Christine Emerton
Senior Managing Editor: Donna Gridley
Executive Editor: Bill Larkin
Senior Development Editor: Janet Morris
VP, Sales: Phil Charland
Director of Marketing Operations: Brian Rooney
Project Manager: Kristen Rogers
Digital Products Manager: Jordan McKenzie
Digital Project Specialist: Rachel Reyes
Production Services Manager: Colleen Lamy
VP, Manufacturing and Inventory Control: Therese Connell
Composition: S4Carlisle Publishing Services
Project Management: S4Carlisle Publishing Services
Cover Design: Kristin E. Parker
Text Design: Scott Moden
Director, Content Services and Licensing: Joanna Gallant
Rights & Media Manager: Shannon Sheehan
Rights & Media Specialist: John Rusk
Cover Image and Title Page: © Jones & Bartlett Learning. Photographed by Steven Sulewski.
Printing and Binding: LSC Communications
Cover Printing: LSC Communications

Library of Congress Cataloging-in-Publication Data
Names: Reeder, Forest, author. | Joos, Alan, author.
Title: Fire and emergency services instructor : principles and practice / Forest F. Reeder, Alan E. Joos.
Other titles: Fire service instructor
Description: Third edition. | Burlington, MA : Jones & Bartlett Learning, [2020] | Revised edition of: Fire service instructor / International Association of Fire Chiefs, International Society of Fire Service Instructors, National Fire Protection Association, Forest Reeder, Alan Joos. Second edition. 2014. | Includes bibliographical references and index.
Identifiers: LCCN 2018059562 | ISBN 9781284172331 (pbk. : alk. paper)
Subjects: LCSH: Fire prevention--Study and teaching. | Fire extinction--Study and teaching.
Classification: LCC TH9120 .F583 2020 | DDC 628.9/2071--dc23 LC record available at https://lccn.loc.gov/2018059562

6048

Printed in the United States of America
25 24 23   10 9

Brief Contents

SECTION 1
Fire and Emergency Services Instructor I

CHAPTER 1 Today's Fire and Emergency Services Instructor — 3

CHAPTER 2 The Learning Process — 37

CHAPTER 3 Methods of Instruction — 73

CHAPTER 4 Communication Skills — 101

CHAPTER 5 Using Lesson Plans — 121

CHAPTER 6 Technology in Training — 141

CHAPTER 7 Training Safety — 159

CHAPTER 8 Evaluating the Learning Process — 183

SECTION 2
Fire and Emergency Services Instructor II

CHAPTER 9 Instructional Development — 201

CHAPTER 10 Instructional Delivery — 221

CHAPTER 11 Evaluation and Testing — 239

CHAPTER 12 Program Management and Training Resources — 265

SECTION 3
Fire and Emergency Services Instructor III

CHAPTER 13 Program Development — 295

CHAPTER 14 Program Evaluation — 323

CHAPTER 15 Program Administration — 339

SECTION 4
Live Fire Instructor and Live Fire Instructor-in-Charge

CHAPTER 16 Live Fire Instructor Introduction — 371

CHAPTER 17 Live Fire Instructor — 381

CHAPTER 18 Preburn Inspection and Planning — 403

CHAPTER 19 Conducting Burn Evolutions — 429

CHAPTER 20 Post Live Fire Evolution — 445

APPENDIX A: Resources for Fire and Emergency Services Instructors — 455

APPENDIX B: An Extract from NFPA 1041, Standard for Fire and Emergency Services Instructor Professional Qualifications, 2019 Edition — 474

APPENDIX **C: Correlation to** *NFPA 1041:*
Standard for Fire and Emergency Services
Instructor Professional Qualifications,
2019 Edition **480**

APPENDIX **D: An Extract from**
NFPA 1401, Recommended Practice for
Fire Service Training Reports and Records,
2017 Edition **488**

APPENDIX **E: An Extract from:**
NFPA 1500, Standard on Fire Department
Occupational Safety, Health, and Wellness
Program, **2018 Edition** **491**

APPENDIX **F: An Extract from**
NFPA 1403, Standard on Live Fire
Training Evolutions, **2018 Edition** **493**

GLOSSARY **496**

INDEX **501**

Contents

SECTION 1
Fire and Emergency Services Instructor I

CHAPTER 1
Today's Fire and Emergency Services Instructor — 3

Introduction	4
Qualities of an Instructor	4
Levels of Fire and Emergency Services Instructors	5
Roles and Responsibilities of the Fire and Emergency Services Instructor I	6
Where Do I Fit In?	7
The Roles of an Effective Instructor	8
The Fire and Emergency Services Instructor as a Leader	8
The Fire and Emergency Services Instructor as a Mentor	9
The Fire and Emergency Services Instructor as a Coach	10
The Fire and Emergency Services Instructor as an Evaluator	10
The Fire and Emergency Services Instructor as a Teacher	11
Setting Up the Learning Environment	11
Physical Elements Affecting the Learning Environment	11
Emotional Elements Affecting the Learning Environment	12
The Instructor's Role in the Future of the Department	13
The Instructor's Role in Succession Planning	13
Instructor Credentials and Qualifications	14
Laying the Groundwork	14
Meeting Standards	14
Continuing Education	15
Building Confidence	15
Issues of Ethics in the Training Environment	15
Leading by Example: What You Do, You Teach	16
Accountability in Training	16
Recordkeeping	16
Trust and Confidentiality	17
The Law as It Applies to Fire and Emergency Services Instructors	17
Federal Employment Laws	18
Records, Reports, and Confidentiality	20
Copyright and Public Domain	21
Managing Multiple Priorities as an Instructor	21
Time Management	21
Training Priorities	22
Planning a Program	22
Training Through Delegation	23
Professional Development	24
Staying Current	24
Higher Education	25
Conferences	25
Lifelong Learning	26
Professional Organizations	27
The Next Generation	28
Identifying	28
Mentoring	29
Coaching	29
Sharing	30
Summary of Instructor I Duties	30

CHAPTER 2
The Learning Process — 37

Introduction	38
The Interactive Process of Learning	39
What Is Adult Learning?	39
Similarities and Differences Among Adult Learners	40
Influences on Adult Learners	40
Today's Adult Learners: Generational Characteristics	41
The Laws and Principles of Learning	45
Use of Senses	45
The Behaviorist and Cognitive Perspectives	45
Competency-Based Learning Principles	46
Forced Learning	46
Learning Skills for Adult Learners	47
Classroom Study Tips	47
Personal Study Time	47
Test Preparation Skills	48
Maslow's Hierarchy of Needs	48
Level One: Physiological Needs	49
Level Two: Safety, Security, and Order	49
Level Three: Social Needs and Affection	49
Level Four: Esteem and Status	50
Level Five: Self-Actualization	50
Learning Domains	50
Cognitive Learning	51
Psychomotor Learning	51
Affective Learning	52
Learner Characteristics	53
How Does an Individual Learn?	53
Effective Teaching and Learning for All Students	54
Student-Centered Learning	55
Student-Centered Teaching Methods and Strategies	56
Motivation and Learning	60
Motivation of Adult Learners	61
Motivation as a Factor in Class Design	62
Learning Disabilities	62
Instructing Students with Disabilities	63
Disruptive Students	64

CHAPTER 3
Methods of Instruction — 73

Introduction	75
Methods of Instruction	75
Lecture	75
Demonstration or Skill Drill	75
Discussion	76
The Four-Step Method of Instruction	77
Pre-Course Survey and Prerequisites	78
Enhanced Instructional Methods	78
Lesson Presentation Skills and Techniques	79
Effective Communication	81
Questioning Techniques	82
Managing Disruptive Behavior	82
Transitioning Between Methods of Instruction	84
Audience and Department Culture	85
Audience	85
Department Culture	85
The Learning Environment	85
The Physical Environment	86
The Indoor Classroom	87
The Outdoor Classroom	89
Teamwork and Self-Actualization	92
Demographics in the Learning Environment	93
Gender Considerations	93
Offensive Language, Gestures, and Dress	93
Adapting the Lesson Plan Based on Demographics	94

CHAPTER 4
Communication Skills — 101

Introduction	102
The Basic Communication Process	103
The Sender	103
The Message	103
The Medium	103
The Receiver	103
Feedback	103
The Environment	104

Nonverbal Communication	104
Active and Passive Listening	104
Verbal Communication	105
Language	106
Audience Analysis	107
Written Communications	108
Reading Level	109
Writing Format	109
The Rules of Writing	110
Style	111
Reports	112
Writing a Decision Document	113

CHAPTER 5
Using Lesson Plans 121

Introduction	123
Why Use a Lesson Plan?	123
Understanding Learning Objectives	124
The Components of Learning Objectives	124
Parts of a Lesson Plan	126
Lesson Title	126
Level of Instruction	126
Method of Delivery	128
Learning Objectives	128
References	129
Materials Needed	130
Lesson Outline	130
Lesson Summary	130
Assignment	130
Instructional Preparation	130
Student Preparation	130
Organizational Techniques	130
Procuring Instructional Materials and Equipment	131
Preparing to Instruct	131
Scheduling	131
Adapting Versus Creating a Lesson Plan	131
Reviewing Instructional Materials for Adaptation	132
Evaluating Local Conditions	133
Evaluating Facilities for Appropriateness	133
Meeting Local SOPs	133
Evaluating Limitations of Students	134
Adapting the Method of Instruction	134
Accommodating Instructor Style	134
Meeting the Needs of the Students	134

CHAPTER 6
Technology in Training 141

Introduction	142
Technology-Based Instruction	142
Distance Learning	143
Computer-Based Training and Learning Management Systems	143
Multimedia Applications in Instruction	145
Understanding Multimedia Tools	145
When to Use Multimedia Presentations	145
Multimedia Presentations: Best Practices	146
Technology and Software	147
Maintaining Technology	151
Troubleshooting Common Multimedia Problems	151

CHAPTER 7
Training Safety 159

Introduction	160
The 16 Fire Fighter Life Safety Initiatives	161
Leading by Example	162
Safety in the Learning Environment	162
Hands-on Training Safety	163
Personal Protective Equipment	163
Rehabilitation Practices and Hands-on Training	163
Safety Policies and Procedures for Training	164
Influencing Safety Through Training	164
Planning Safe Training	167
Hidden Hazards During Training	169
Overcoming Obstacles	169
Live Fire Training	170
Other Training Considerations	170
Developing a Safety Culture	171
Anticipating Problems	171
Accident and Injury Investigation	172
Student Responsibilities for Safety During Training	172
Legal Considerations	172
Negligence, Misfeasance, and Malfeasance	173

CHAPTER 8
Evaluating the Learning Process — 183

Introduction	184
Legal Considerations for Testing	185
Purposes and Types of Tests	185
Written Tests	185
Performance Tests	185
Oral Tests	185
Standard Testing Procedures	186
Proctoring Tests	186
Proctoring Written Tests	186
Proctoring Oral Tests	187
Proctoring Performance Tests	187
Written Test Items	187
Components of a Test Item	187
Multiple Choice Test Items	189
Matching Test Items	189
Arrangement Test Items	189
Identification Test Items	190
Completion Test Items	190
True/False Test Items	191
Essay Test Items	191
Cheating During an Exam	192
Grading Student Oral, Written, and Performance Tests	192
Reporting Test Results	192
Web-Based Training and Testing	193

SECTION 2
Fire and Emergency Services Instructor II

CHAPTER 9
Instructional Development — 201

Introduction	202
The Lesson Plan Components	202
Creating a Lesson Plan	203
Instructor- and Student-Centered Learning	203
Determining the Learning Outcomes	204
Developing the Learning Objectives	204
Developing the Lesson Outline	211
Writing the Evaluation Plan	215

CHAPTER 10
Instructional Delivery — 221

Introduction	222
Conducting a Training Session: Methods of Instruction	223
Instructional Techniques and Transitioning	224
Supervising Other Instructors	225
Evaluating the Instructor	226
Formative Evaluation	226
Summative Evaluation	227
Student Evaluation of the Instructor	227
The Evaluation Process	229
Preparation	229
Observation	229
Lesson Plan	230
Evaluation Forms and Tools	230
Instructor Feedback	231
Evaluation Review	232
Supervision During High-Risk Training	233
Live Fire Training	233
Safety Briefings	233

CHAPTER 11
Evaluation and Testing — 239

Introduction	240
Types and Purpose of Testing	240
Uniform Guidelines for Employee Selection	242
Test-Item Validity	242
Currency of Information	242
Test Item and Test Analysis	244
The Role of Testing in the Systems Approach to Training Process	244
Test-Item Development	245
Test Specifications	245
Written Test	246

Complete a Test-Item Development and Documentation Form	246
Selection-Type Objective Test Items	247
Multiple Choice Test Items	247
Matching Test Items	247
Arrangement Test Items	248
Identification Test Items	249
True/False Test Items	250
Essay Test Items	250
Performance Testing	253
Test Generation Strategies and Tactics	254
Computer and Web-Based Testing	254
Developing Class Evaluation Forms	256

CHAPTER 12
Program Management and Training Resources — 265

Introduction	266
Scheduling of Instruction	267
Developing a Training Schedule	267
Types of Training Schedules	268
Scheduling for Success	271
Master Training Schedule	274
Selection of Instructors	276
Record Management	276
Confidentiality	277
Other Considerations for Records	277
Budget Development and Administration	278
Introduction to Budgeting	278
Budget Terminology	278
Budget Preparation	278
The Budget Cycle	279
Capital Expenditures and Training	279
Training-Related Expenses	281
Acquiring and Evaluating Training Resources	282
Training Resources	282
Hand-Me-Downs	282
Purchasing Training Resources	283
Evaluating Resources	284

SECTION 3
Fire and Emergency Services Instructor III

CHAPTER 13
Program Development — 295

Introduction	297
Training Program Development	297
Training Needs Analysis	297
Designing Training Programs	301
Program Goals and Program Structure	301
Program Outcomes	302
Program Evaluation	302
Designing Courses	303
Writing Course Objectives	303
Course Content Outlines	306
Course Evaluation Plans	310
Interpreting Evaluation Results	316

CHAPTER 14
Program Evaluation — 323

Introduction	324
Acquiring Evaluation Results	324
Storing Evaluation Results	325
Dissemination of Evaluation Results	325
Analyzing Evaluation Tools	327
Analyzing Student Evaluation Instruments	327
Item Analysis	327
Data Collection	328
Computing the P+ Value	331
Computing the Discrimination Index	331
Computing the Reliability Index	333
Identifying Items as Acceptable or Needing Review	333
Revising Test Items Based on Posttest Analysis	334
Psychomotor Skills Evaluations	334

CHAPTER 15
Program Administration — 339

Introduction	341
Training Record Systems	341
Using Training Records	341

Types of Training Records	341
A Typical Training File	343
Types of Training Records	346
Reports for the Training Program	347
Format for Training Reports	348
Legal Aspects of Training Records	350
Records Storage and Retention	350
Training Program Policies	350
Writing Policy	353
Adoption and Implementation	353
Personnel Management: Selecting Staff	354
Candidate Evaluation	354
Selection Policy and Procedures	355
Performance-Based Evaluation Plans	356
Budgets	357
Budget Categories	359
Budget Organization	359
Budget Process	359
Budget Skills Application	361
Budgetary Justifications	361
Purchasing Guidelines and Policies	361
Purchasing Process	362

SECTION 4
Live Fire Instructor and Live Fire Instructor-in-Charge

CHAPTER 16
Live Fire Instructor Introduction — 371

Introduction	372
The Importance of NFPA 1403	373
The History of Live Fire Training Evolutions	373
The Impact of NFPA 1403 on Live Fire Training	374
Referenced Standards	374
Using NFPA 1403	376
Establishing Clear Objectives	376
The Importance of the 1001 Prerequisite JPRs	377
Safety	377

CHAPTER 17
Live Fire Instructor — 381

Introduction	383
Preparing for a Live Fire Evolution	383
Protective Clothing and Self-Contained Breathing Apparatus	383
Live Fire Evolutions	385
Personal Protective Equipment Use	385
Accountability	385
Fire Behavior and Structural Fire Dynamics	386
Heat Release Rate	386
Smoke	387
Students in the Live Fire Training Environment	388
Recruit Students	388
Student Psychology: Fire Fighter Style	389
Physiological Aspect of Fire Training	390
Cardiovascular and Thermal Strain of Firefighting	390
Factors Affecting Cardiovascular and Thermal Strain	393
Heat Emergencies	397
Cardiac Emergencies	398
Incident Scene Rehabilitation	400
Crew Resource Management	400

CHAPTER 18
Preburn Inspection and Planning — 403

Introduction	405
Initial Evaluation of the Site	405
Permits	405
Developing the Preburn Plan	406
Learning Objectives	406
Participants	406
Water Supply	406
Apparatus Needs	409
Building Plan	409
Fuel Needs	409
Site Plan	411
Parking, Staging, and Areas of Operations	412
Visitors and Spectators	412
Emergency Plans	412
Weather	414

List of Training Evolutions	414
Order of Operations	414
Emergency Medical Plan	415
Communications Plan	415
Staffing and Organization	416
Protective Clothing and Self-Contained Breathing Apparatus	416
Rehabilitation Plan	416
Agency Notification Checklist	416
Demobilization Plan	416
Preburn Inspection	416
Inspection and Preparation of Acquired Structures	416
Exterior Preparation	418
Interior Preparation	420
Preparation and Inspection of Props and Facilities	423

CHAPTER 19
Conducting Burn Evolutions — 429

Introduction	430
Instructor-in-Charge Responsibilities	431
Learning Objectives	432
New Recruits and Live Fire Training Evolutions	432
Experienced Students	433
Preburn Briefing	433
Selecting Instructors for Live Burn Evolutions	434
Staffing and Organization	437
Safety Officer	438
Fire Control Team	438
Staff and Participant Rotation	440
Evolution Safety	440
Emergency Plans	441
Developing the Preburn Plan	441

CHAPTER 20
Post Live Fire Evolution — 445

Introduction	446
Reports and Documentation	447
Documentation Before the Event	447
General Reports and Documentation	447
Documentation of an Emergency Incident	448
Photographs and Videos	448
Documentation of Unusual Events	449
The Conclusion of the Training Exercise	450
Acquired Structures	450
Gas-Fired and Non-Gas-Fired Structures	450
Exterior Training Props	450
Postburn Inspection	451
Overhaul of Acquired Structures	451
Postburn Inspection of Structures and Props	451
The Postevolution Debriefing	451
Postevolution Evaluation Forms	452

APPENDIX A: Resources for Fire and Emergency Services Instructors	455
APPENDIX B: An Extract from NFPA 1041, Standard for Fire and Emergency Services Instructor Professional Qualifications, 2019 Edition	474
APPENDIX C: Correlation to NFPA 1041: Standard for Fire and Emergency Services Instructor Professional Qualifications, 2019 Edition	480
APPENDIX D: An Extract from NFPA 1401, Recommended Practice for Fire Service Training Reports and Records, 2017 Edition	488
APPENDIX E: An Extract from: NFPA 1500, Standard on Fire Department Occupational Safety, Health, and Wellness Program, 2018 Edition	491
APPENDIX F: An Extract from NFPA 1403, Standard on Live Fire Training Evolutions, 2018 Edition	493
GLOSSARY	496
INDEX	501

Acknowledgments

Authors

Forest F. Reeder, MS

Forest Reeder began his fire service career in 1978. He serves as fire chief of the Tinley Park (IL) Fire Department. Forest is the fire officer training coordinator for the Illinois Fire Chiefs Association Educational and Research Foundation and has served as the chairman of many Illinois State Fire Marshal certification committees. He is the author of the weekly drill features at www.firefighterclosecalls.com and at www.fireengineering.com. He has instructed at Fire Department Instructor's Conference (FDIC) for nearly 20 years and trains both locally and nationally on fire service training, safety, and officer development related topics. Several articles on these programs have been published in major trade publications.

Forest holds many Illinois fire service certifications including Fire Officer 3, Training Program Manager, and Fire Department Safety Officer. He has an associate of applied science degree in fire science technology, a bachelor's degree in fire science management, and a master's degree in public safety administration from Lewis University. He was awarded the George D. Post Instructor of the Year by the International Society of Fire Service Instructors at FDIC in 2008.

Alan E. Joos, MS, EFO, CFO, FIFireE

Alan E. Joos currently serves as the chief of firefighter training within the State Fire Marshal's Office of Nebraska. Prior to his appointment in Nebraska he was an assistant director at Louisiana State University Fire and Emergency Training Institute (FETI). Prior to FETI, he worked at the Utah Fire and Rescue Academy for 12 years as assistant director of the certification system and assistant director of the training division.

Alan received associate's and bachelor's degrees from Utah Valley University in fire science and business technology management and a master's degree from Grand Canyon University in Phoenix. He is currently a doctoral candidate in human resource education/workforce development at Louisiana State University. He is also a graduate of the Executive Fire Officer's Program at the National Fire Academy, has received Chief Fire Officer Designation from the Commission on Professional Credentialing, and is a fellow with the Institution of Fire Engineers.

Alan's fire service background began in 1985 as a career fire fighter in a combination fire department and then as an on-call fire fighter/EMT-I in Utah. He has continued his fire service involvement with the City of Gonzales, Louisiana, as a contract fire fighter. During his fire service career, he has served as a training officer, driver, shift officer, and EMT.

Alan has been involved with several national organizations, including the International Fire Service Accreditation Congress (IFSAC) and the ProBoard, and has previously served on the NFPA Technical Committee for the Professional Qualification Standards.

Alan is married to Carla Joos, and their family includes Nathan and his wife Shauntelle, Jordan (deceased), and Dallan, and granddaughter Chloe and grandson Gavin.

Section 4 Authors

Dave Casey, MPA, EFO, CFO
Susan Schell, MBA
Mike Kemp, Retired Battalion Chief
Brian P. Kazmierzak, EFO, CTO
Denise L. Smith, PhD

Contributors and Reviewers

John P. Alexander
Sr. Instructor
Connecticut Fire Academy
Windsor Locks, Connecticut

Marcus Allen
Captain
Turkey Creek Fire and Rescue
Holly Ridge, North Carolina

Robert Jay Alley
Dean
Blue Ridge Community College
Flat Rock, North Carolina

Roderick Armstrong
WVPST (West Virginia Public Service Training) and WVFSE (West Virginia Fire Service Extension)
WVPST, Region 5, Currently Dunbar, West Virginia
WVUFSE, WV State Fire Academy, Jackson Mills, West Virginia

Angela L. Bennett
Manager, Institute Development Section
Maryland Fire and Rescue Institute
University of Maryland
College Park, Maryland

Ryan Benson
Assistant Chief
Cleveland Fire Department
Cleveland, Ohio

Kendal E. Bortisser
Miramar College Fire Technology Program
San Diego, California

Dave Burgess
Program Manager
Justice Institute of British Columbia
New Westminster, British Columbia

Rodney Burris
Assistant Chief
Spout Springs Emergency Services
Cameron, North Carolina

Troy Cailler
Fire Fighter/Co-Owner
Tri-County Training Association, LLC
Litchfield, Maine

Frank Chapman
Firefighter/Instructor
Leon Volunteer Fire Department
Leon, West Virginia

Shea Chwialkowski
Richfield Fire Department
Richfield, Minnesota

Andrew Clark
Chief
Albion Fire-Rescue
Albion, Maine

Mike Consie
Captain
Duluth Fire Department
Duluth, Minnesota

Nathaniel Contreras
Operations Captain
Scarborough Fire Department
Scarborough, Maine

Brent Cowx
Instructor
Justice Institute of British Columbia
New Westminster, British Columbia

Mike Crews
Fire Chief
Taylorville Fire Department
Taylorville, Illinois

Michael Dade
Instructor
Illinois Fire Service Institute
Champaign, Illinois

Leo DeBobes
Associate Professor and Fire Protection Technology Program Coordinator
Suffolk County Community College, Fire Protection Technology
Selden, New York

Jason Decremer
Director of Certification
Commission on Fire Prevention and Control
Windsor Locks, Connecticut

Blake J. Deiber
District Coordinator
Western Technical College
Sparta, Wisconsin

Dan Diehl
Assistant Director
Valencia College
Orlando, Florida

E. Brene Duggins
Fire Prevention Coordinator/Technology Specialist
Holly Grove Fire Department
Lexington, North Carolina

Michael F. Dunlap
South Carolina State Fire Academy
Columbia, South Carolina

Doug Eggiman
Fire Chief
Midway Fire Rescue
Pawleys Island, South Carolina

Robert Fash
National Fire Protection Association
Quincy, Massachusetts

J. Michael Freeman
Director
WV Public Service Training
Clarksburg, West Virginia

Christopher L. Gilbert
Paramedic/Hazmat Specialist
Alachua County
Gainesville, Florida

John Glass
Professor
Valencia College
Orlando, Florida

Casey Hall
Kentucky Fire Commission
Greenville, Kentucky

Scot Hughes
Training Captain
Williamson EMA/Fire
Williamson County, Tennessee

Jim Huyser
Captain
Jackson Fire Department
Jackson, Wisconsin

Jennifer Johnson
Kansas City Kansas Fire Department and Kansas City Kansas Community College
Kansas City, Kansas

Joel E. Jones
Senior Captain, Training Officer
Kingsport Fire Department
Kingsport, Tennessee

Shawn Kelley
International Association of Fire Chiefs
Fairfax, Virginia

Jeremy Kircher
Captain
Training and Safety
Boise Fire Department
Boise, Idaho

Steve Knight
Coordinator / Professor
Lambton College
Sarnia, Ontario

Daniel Krakora
Manager
Fire Science and EMS
College of DuPage
Glen Ellyn, Illinois

Joe Lachowski
Lead Instructor
National Park Service
Empire, Michigan

Chad Landis
Acting Captain
City of Pineville Fire Department
Pineville, Louisiana

Dawn Landry
Firefighter
Hahnville Volunteer Fire Department
Hahnville, Louisiana

Chris Lau
Battalion Chief
Alaska Department of Public Safety
Department of Fire and Life Safety
Anchorage, Alaska

Marc M. Lavoie
Training Officer
Professional Development, Training and Safety
Halifax Regional Fire and Emergency
Halifax, Nova Scotia

Jason Loeb
Hoffman Estates Fire Department
Hoffman Estates, Illinois

Frank Marcinkiewicz
Lieutenant
Cat Spring Volunteer Fire Department
Cat Spring, Texas

Johnny Mason
Lieutenant/Assistant Academy Commander
Great Oaks
Cincinnati, Ohio

Patrick McArdle
Chief of Training
Fayetteville Technical Community College
Fayetteville, North Carolina

Steve Metz
Fire Instructor
Moraine Park Technical College
Fond Du Lac, Wisconsin

Andrew Mihans
Captain
Arlington Fire Department
Poughkeepsie, New York

Michael C. Mire
Assistant Fire Chief
Houston Fire Department
Houston, Texas

Nick Morgan
Captain
St. Louis Fire Department
St. Louis, Missouri

Keith Padgett
Director, College of Safety and Emergency Services
Columbia Southern University
Orange Beach, Alabama

Ian Pleet
Senior Exercise Planner, Leidos
Life Member, Manassas Volunteer Fire Company
Reston, Virginia

Christopher Poremby
Maine Fire Service Institute
Brunswick, Maine

Jeff Pricher
Division Chief
Columbia River Fire and Rescue
Scappoose Fire District
Scappose, Oregon

Joseph Ramsey
Battalion Chief
Winston-Salem Fire Department
Winston-Salem, North Carolina

Acknowledgments

Steven Riley
Lieutenant
Columbus Fire Department
Columbus, Ohio

Peter Rines
Certification Program Manager
Maine Fire Service Institute
Brunswick, Maine

Lt. John W. Ross, Jr.
City of Aurora Fire Department
Aurora, Illinois

Mike Rutkowski
Battalion Chief
Grayslake Fire Protection District
Grayslake, Illinois

Kelli J. Scarlett, JD
Adjunct Professor
University of Maryland University College
Adelphi, Maryland

Richard Siebel
Captain
Raleigh Fire Department
Raleigh, North Carolina

Timothy Smith
District Fire Chief
Boston Fire Department
Boston, Massachusetts

Jeremy Souza
National Fire Protection Association
Quincy, Massachusetts

Larry Straffin
Deputy Chief/Training Officer
North Berwick Fire Department
Fire Instructor
Maine Fire Service Institute and New Hampshire
    Fire Academy and EMS
Brunswick, Maine

Bobby Strozier
LSU Fire and Emergency Training Institute
Baton Rouge, Louisiana

Robert Taggart
Training Officer
Shreveport Fire Department
Shreveport, Louisiana

Luis F. Videgaray
Captain/Training Officer
Natchez Fire Department Training Center
Natchez, Mississippi

Chris Walker
Captain
Fort Wayne Fire Department
Fort Wayne, Indiana

Laura Walker
Associate Professor
Montezuma Fire Department
Sinclair Community College Fire Academy
Dayton, Ohio

Devon Wells
International Society of Fire Service Instructors
Hood River Fire and EMS
Hood River, Oregon

John West
Oregon Department of Public Safety Standards and Training
Salem, Oregon

Angela White
Education Director, Fire Service Training
Wisconsin Technical College System
Madison, Wisconsin

Becki White
Eden Prairie Fire Department
Eden Prairie, Minnesota

Kim Williams
Deputy Director
Fire and Rescue Commission
Raleigh, North Carolina

Ronnie L. Willis Sr.
Fire Chief
Grenada Fire Department
Grenada, Mississippi

Christopher Young
Deputy Fire Chief
Crowley Fire Department
Crowley, Texas

Gray Young
Assistant Manager
Louisiana State University, Fire and Emergency
    Training Institute
Baton Rouge, Louisiana

Daniel J. Zimmer Sr.
Safety and Training Officer
Arlington (VT) Fire Department
Arlington, Vermont

Alan J. Zygmunt
Department of Emergency Services and Public Protection
Connecticut Fire Academy
Windsor Locks, Connecticut

SECTION 1

Fire and Emergency Services Instructor I

CHAPTER 1 **Today's Fire and Emergency Services Instructor**

CHAPTER 2 **The Learning Process**

CHAPTER 3 **Methods of Instruction**

CHAPTER 4 **Communication Skills**

CHAPTER 5 **Using Lesson Plans**

CHAPTER 6 **Technology in Training**

CHAPTER 7 **Training Safety**

CHAPTER 8 **Evaluating the Learning Process**

CHAPTER 1

Today's Fire and Emergency Services Instructor

NFPA 1041 JOB PERFORMANCE REQUIREMENTS

4.1.1 The Fire and Emergency Services Instructor I shall meet the JPRs defined in Sections 4.2 through 4.5 of this standard.

4.2.5 Complete training records and reports, given policies and procedures, so that required reports are accurate and submitted in accordance with the procedures.

(A) Requisite Knowledge.

Types of records and reports required, and policies and procedures for processing records and reports.

(B) Requisite Skills.

Report writing and record completion.

4.4.2 Organize the learning environment, given a facility and an assignment, so that lighting, distractions, climate control or weather, noise control, seating, audiovisual equipment, teaching aids, and safety are addressed.

(A) Requisite Knowledge.

Learning environment management and safety, advantages and limitations of audiovisual equipment and teaching aids, classroom arrangement, and methods and techniques of instruction.

(B) Requisite Skills.

Use of instructional media and teaching aids.

KNOWLEDGE OBJECTIVES

After studying this chapter, participating in a structured learning environment, and completing assigned assessments, you will be able to:

- Define the roles and responsibilities of the Fire and Emergency Services Instructor I. (**NFPA 1041: 4.1.1**, pp 6–7)
- Identify the physical elements of the learning environment. (**NFPA 1041: 4.4.2**, pp 11–12)
- Explain the importance of succession planning for the instructor. (pp 13–14)
- Identify issues of ethics for the instructor. (p 15)
- Explain the importance of continuing learning for the instructor. (p 15)
- Describe how laws and standards apply to the instructor. (pp 17–21)
- Explain the importance of proper recordkeeping. (**NFPA 1041: 4.2.5**, p 16)
- Identify three methods instructors can use to manage multiple priorities. (pp 21–23)
- Identify professional organizations that will help in the professional development of the instructor. (p 27)
- Identify and discuss the value and importance of coaching and mentoring the next generation of instructors. (pp 29–30)

SKILLS OBJECTIVES

After studying this chapter, participating in a structured learning environment, and completing assigned assessments, you will be able to:

- Complete a record of a training event. (**NFPA 1041: 4.2.5**, pp 16; 20–21)
- Prepare a report following a training event. (**NFPA 1041: 4.2.5**, pp 20–21)

You Are the Fire and Emergency Services Instructor

As the only Fire Instructor I assigned to your shift, your company officer has asked you to lead a training session with your company. Because this is your first assignment of this nature, ask yourself the following questions:

1. What are the roles and responsibilities of the Fire Instructor I?
2. What are four ethical issues an instructor must be aware of?
3. Which laws and standards, particularly NFPA standards, apply to the instructor in your jurisdiction?

 *Access Navigate for more practice activities.*

Introduction

Remember when you began your career in the fire service? Somewhere in your memory is probably an outstanding fire and emergency services instructor. He or she even may be the person who introduced you to your career. Instructors are the guardians of knowledge, skills, and ability in the fire service; their knowledge is wrapped up in a long tradition that has served many generations of fire fighters well. As protectors of that tradition, instructors look to and rely on innovation and creativity while constantly striving for perfect training.

In years past, firefighting training was often left to on-the-job education. New recruits would be handed coats, boots, helmets, and gloves and told to jump on the back of the rig as it rushed to the scene of the emergency. In those days, the fire ground was the recruit's training ground—and the title of "fire and emergency services instructor" may have been handed out too generously. Formal instructors and, in some cases, formal instruction were left to the larger departments that had budgets large enough to hire specialized personnel and build training facilities.

Clearly, times have changed. The fire service is now being pushed to the limit by communities that expect more out of the types of services provided. Retired Phoenix Fire Chief Alan Brunacini often referred to today's recipient of that service as "Mrs. Smith." As a pioneer of customer service, Chief Brunacini knew the value of the instructor in preparing the troops to meet Mrs. Smith's demands. Today, fire departments rely on their instructors both to train new recruits and to maintain the skill levels of existing fire fighters.

Qualities of an Instructor

What does it take to join this elite group of fire fighters on whose shoulders rest the success and safety of emergency operations? Although excellent instructors possess many attributes, a few important qualities come to mind and rise to the top of the list. First, there

FIGURE 1-1 The instructor brings a blend of experience and education to every training opportunity.
© Jones & Bartlett Learning. Photographed by Glen E. Ellman.

is desire—a desire to be of assistance to those in need. In this case, we are not referring to those persons who require our assistance during times of trouble or emergency, but rather to those individuals who would require the knowledge needed to assist those in trouble. These fire fighters must be trained so that they will gain the knowledge, skills, and abilities necessary to serve and respond to those in need of assistance. The instructor gives freely without hesitation in the pursuit of excellence (**FIGURE 1-1**). He or she should certainly have experience in the subject matter taught. Students have more confidence in instructors who have demonstrated competence in the subject matter through experience; in turn, experience opens the door for the instructor to reach the minds of the students. Even so, good instructors understand that they cannot rely on experience alone. They must remain vigilant to their dynamic environment and gather the newest information, technology, and skills to remain out front and stay current with an ever-changing work and instructional environment.

Good instructors also demonstrate flexibility. They are able to work in a variety of environments while offering instruction on a variety of topics to a class full of students with a variety of talents, backgrounds, and experience levels. Instructing the adult learner can be challenging, and the instructor must be willing to alter the approaches to education used to reach all students. In some cases, the fire department's schedule provides the greatest challenge, as instruction time is often interrupted by calls for assistance. In other fire departments, it is the fire fighters' personal schedules that present the challenge, as training time competes with family time and full-time work schedules.

Motivation is one of the keys to bringing excitement to the training environment. Nothing breeds fire fighter motivation like a motivated instructor. The right kind of motivation leads to the attitude, "I can't wait to teach," and, just as important, "I can't wait to learn." Motivation is contagious; it can spread from the learning environment to the station floor, and ultimately it can drive the quality of the service provided to the community. Motivated instructors bring creativity and ingenuity to the learning environment as a means of creating excitement about the learning process.

The skills to be an effective instructor can be learned by a person who has the desire to develop and perfect them. The ability to use these skills at the correct time and in a manner that helps others learn a topic is developed by teaching courses and being mentored, coached, and supervised by experienced trainers. The learning process never ends for those who desire to improve their knowledge and skills. All instructors start out at the Instructor I level; with additional training and professional development, they may reach the top of the qualification set owing to their desire to learn and develop their skills. Desire matched with knowledge creates great instructors.

In addition to being motivated and skilled, instructors will benefit from a sense of ethics. Ethics can be broken down into personal beliefs, interpersonal actions, and professional ethics. Instructors should recognize the interrelated aspects of personal, interpersonal, and professional ethics.

Are you up for the challenge? Can you accept the responsibility? Can you be a steward of tradition while remaining open to teaching new ideas? Can you strive to maintain desire, experience, flexibility, and motivation as you pursue operational excellence? If your answers to these questions are all "yes," then you are on your way to joining the ranks of the greatest profession. Ideally, your curriculum will serve as the launching pad for excellent fire fighters and serve them well as a critical resource for years to come.

This text provides information to meet the job performance requirements (JPRs) outlined in National Fire Protection Association (NFPA) NFPA 1041: *Standard for Fire and Emergency Service Instructor Professional Qualifications*. These definitions resulted from a task analysis intended to validate these levels and to create specific requirements that would apply to each level. Although the levels are identified as instructor qualifications, in many states they may become certification levels, with candidate prerequisites to be completed before a certification is granted.

Levels of Fire and Emergency Services Instructors

According to NFPA 1041, the duties performed by specific instructors are broken down into three distinct levels. These three classifications build on one another and progressively give the instructor additional skills, duties, and responsibilities.

Instructor I is defined as follows (**NFPA 1041, Section 3.3.2.1**):

> A fire and emergency services instructor who has demonstrated the knowledge and ability to deliver instruction effectively from a prepared lesson plan, including instructional aids and evaluation instruments; adapt lesson plans to the unique requirements of the students and authority having jurisdiction; organize the learning environment so that learning and safety are maximized; and meet the record-keeping requirements of the authority having jurisdiction.

Stated in the most basic terms, the Instructor I delivers instruction from prepared materials at the direction, and often under the supervision, of an Instructor II or higher. Emphasis of this level of instructor is on the ability to communicate effectively and to use various methods of instruction, including hands-on training and lecture. In many cases, the Instructor I will deliver the majority of the fire department company training.

The Instructor II is an instructor who, in addition to meeting the Instructor I qualifications, satisfies the following criteria (**NFPA 1041, Section 3.3.2.2**):

> Has demonstrated the knowledge and ability to develop individual lesson plans for a specific topic including learning objectives, instructional aids, and evaluation instruments; schedule training sessions based on overall training plan of authority having jurisdiction; and supervise and coordinate the activities of other instructors.

The Instructor II functions at a higher level of authority and responsibility than the Instructor I; he or she is responsible for all duty areas of the Instructor I, and is also able to create the training materials. In the purest sense, the Instructor II will create the training materials for distribution to the Instructor I to present to the students. In reality, both tasks may be completed by the same person.

An Instructor III is defined as follows (**NFPA 1041, Section 3.3.2.3**):

> A fire and emergency services instructor who, in addition to meeting Instructor II qualifications, has demonstrated the knowledge and ability to develop comprehensive training curricula and programs for use by single or multiple organizations; conduct organization needs analysis; design record keeping and scheduling systems; and develop training goals and implementation strategies.

The Instructor III typically works as an overall training program manager and oversees the entire spectrum of a comprehensive training program. This includes the planning, development, and implementation of curricula and development of an evaluation plan and training program budget.

Roles and Responsibilities of the Fire and Emergency Services Instructor I

Throughout our lives, we are asked to conform to someone else's idea of how we should act or behave. Child, adolescent, adult, parent, spouse, employee, supervisor, owner, citizen—all are examples of roles that we fill while negotiating through life. With each of these roles, expectations direct our actions and allow us to evaluate our success or failure at that role.

The fire service mirrors life in that it also contains various roles, each with its own expectations and responsibilities. At every step along the fire service path, from recruit to fire chief, we strive to meet these expectations and responsibilities. The instructor is one of those important roles in the fire service that requires dedicated and competent individuals who can positively influence the entire fire department.

The roles and responsibilities of a fire instructor vary greatly by fire department according to the size, make-up, and delivery system used. Understanding the responsibilities for the Instructors I, II, and III is essential for success in these key positions.

The role of the Instructor I includes the following:

- Manage the basic resources and the records and reports essential to the instructional process.
- Assemble course materials.
- Prepare training records and reports.
- Prepare requests for training resources.
- Schedule instructional sessions.
- Review and adapt prepared instructional materials.
- Deliver instructional sessions using prepared course materials.

- Organize the learning environment, laboratory, or outdoor learning environments.
- Use instructional media and materials to present prepared lesson plans.
- Adjust presentations to students' different learner characteristics, abilities, and behaviors.
- Operate and use audiovisual equipment and demonstration devices.
- Administer and grade student evaluation instruments.
- Deliver oral, written, or performance tests.
- Grade students' oral, written, or performance tests.
- Report test results.
- Provide examination feedback to students.

The relationship between training and education is often confusing. Education is the process of imparting knowledge or skill through systematic instruction. Education programs are conducted through academic institutions and are primarily directed toward an individual's comprehension of the subject matter. Training is directed toward the practical application of education to produce an action, which can be an individual or a group activity. There is an important distinction between these two types of learning.

Within the fire service, training has been considered essential for many years. In contrast, the emphasis on fire fighter education is a much more recent development. The first Wingspread Conference on Fire Service Administration, Education and Research was sponsored by the Johnson Foundation and held in Racine, Wisconsin, in 1966. This conference brought together a group of leaders from the fire service to identify needs and priorities. They agreed that a broad knowledge base was needed and that an educational program was necessary to deliver that knowledge base. Their conclusions became the blueprint for the development of community college fire science and fire administration programs, as well as the degrees-at-a-distance program. The transition from training to education had begun.

In 1998, the U.S. Fire Administration hosted the first Fire and Emergency Services Higher Education (FESHE) conference. That conference produced a document, "The Fire Service and Education: A Blueprint for the 21st Century," that initiated a national effort to address and update the academic needs of the fire service. Participants at the 2000 FESHE conference began work to develop a model fire science curriculum that would span from the community college through graduate school levels. At the 2002 conference, U.S. Fire Administration Education Specialist Edward Kaplan compared the results of the FESHE effort with the Wingspread higher education curriculum. The FESHE work affirmed the soundness of the original Wingspread model, with information technology being the only new knowledge item added to the curriculum.

Where Do I Fit In?

Much has been written about the responsibilities of positions within various types of organizations, and the fire service is no different. Fire chiefs have their role, rooted in visions of leadership, as the commander-in-chief of the organization. They have a defined responsibility that not only allows them to assume this role but also gives them the authority to shape their own destiny.

Within the fire service's remaining command structure lie the positions of supervision, including frontline lieutenants, captains, and various chief officers. These ranks have had their roles defined through a history of developed policies and job descriptions created by necessity, the chief's vision, and, yes, tradition. These positions are often referred to as middle management, as they are created and reside in the middle of the chain of command. They are pushed and tested by fire fighters at the bottom and by fire chiefs at the top.

As tough as these positions are, there is one job that still finds itself in a somewhat more precarious spot: the instructor. All too often, an instructor may ask, "Where do I fit in?" Is the instructor a line officer, a training officer, a fire fighter, or someone with a specific specialty? In some fire departments, line officers are assigned to the training division as part of their duties. In volunteer organizations, the training officer often volunteers for the job; alternatively, he or she may be assigned to it. Regardless of how you come to be assigned to training, you may feel challenged if you are not given the proper authority to carry out that job.

In examining fire departments' **organizational charts**, you will find positions identified for operational activities, clearly outlining the chain of command and the corresponding responsibilities for firefighting personnel and their fire officer counterparts. Instructors, if shown at all, are usually found off to the side on the organizational chart, preferably within a training division. This disconnect may create a sense of isolation from the fire fighters whom instructors are expected to train. It may leave the instructor with a void, feeling unfulfilled in terms of a

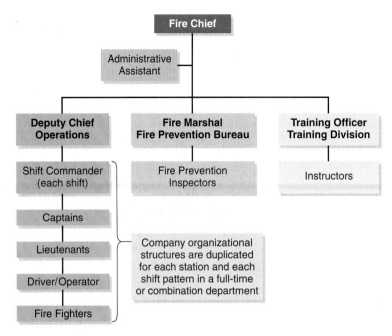

FIGURE 1-2 A sample organizational chart for a fire department.
© Jones & Bartlett Learning.

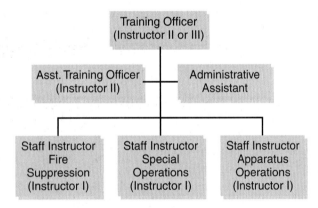

FIGURE 1-3 A training division organizational chart for a typical fire department. The size and make-up of the department will dictate the instructor certification levels for the various positions.
© Jones & Bartlett Learning.

management role that demonstrates the fire department's lack of respect for the role (**FIGURE 1-2**). Fire fighters and even some fire officers are often seen as resisting or even obstructing the training process, because all too often those within the operations staff do not view training as a high priority.

As an instructor, you must remain focused, upbeat, positive, proactive, and true to your role (**FIGURE 1-3**). You must remind yourself that you hold the power to shape the future fire service. Every member of the fire service—from fire fighter through fire officer—has had an instructor affect their careers. That fact demonstrates the importance of your position within the fire department.

The Roles of an Effective Instructor

As a fire and emergency services instructor, you must look beyond charts and titles and instead focus on those important, yet sometimes invisible, roles that produce lasting contributions to overall fire department organizational health and success. Today's instructor is asked to fulfill these roles. These roles are not unique to the fire service; indeed, examples of each can be found in many different professions. Each role is important, however, and will help you develop the talents found within the organization. Understanding each of these roles can assist you in creating and building your own tradition within the fire department.

Fire instructors at all levels must learn to manage the following roles to be effective in carrying out their duties in a department (**FIGURE 1-4**):

- Leader
- Mentor
- Coach
- Evaluator
- Teacher

The Fire and Emergency Services Instructor as a Leader

Instructors are often asked to lead fire departments into the future by preparing fire fighters for new missions. Instructors set the example for all fire fighters to follow in terms of performance excellence. As such,

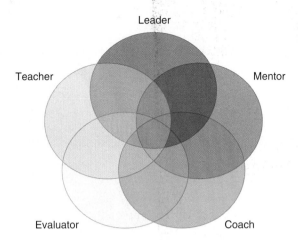

FIGURE 1-4 The five major roles filled by fire instructors.
© Jones & Bartlett Learning.

FIGURE 1-5 Mentoring and preparing your successors are both key attributes in quality instruction.
© Jones & Bartlett Learning. Photographed by Glen E. Ellman.

they must remain true to the direction set by upper management and supportive of the mission as defined by the organization.

Your place in front of the classroom puts you in a visible—and sometimes vulnerable—position where you may have to endure the anxieties of an impatient rank and file. Do not waver from the training mission by letting the instructional environment be turned into a debate on issues of management within the fire department. The most powerful leadership tool is the one over which you have the most control: Lead by example. Leading others is a major responsibility. Part of that responsibility is leading in a positive manner. Belittling a new recruit over a mistake may cause the recruit to hide future mistakes, which places both the recruit and others at risk. Instructors should foster an environment that allows for errors and turns mistakes during training into a teaching opportunity.

The Fire and Emergency Services Instructor as a Mentor

Good leaders are also good at identifying future talent. Instructors are in the best position to observe first-hand both the raw talent of recruits and the ongoing growth of department personnel. They can support and enhance the careers of fire fighters by identifying their talents and mentoring them appropriately. Instructors with mentorship abilities enhance the fire department's succession planning by evaluating and developing the talent pool of future instructors and officers. For example, established instructors assist future instructors by showing confidence in their abilities and recommending additional training and opportunities. In short, good instructors mentor future good instructors and leaders (**FIGURE 1-5**).

As an instructor, you must have the respect of the trainees. Respect may come from a title or position, but it can also come from the individual, based on his or her expertise and knowledge. If you find yourself charged with the task of teaching those of higher rank, be prepared to work for their respect.

Earning someone's respect can be done in several ways. Do not wait to begin building credibility until you are asked to teach—be a professional at your job from the first day you join the fire department. Read books, trade magazines, and other material so that you will become well versed in the trade. Take additional courses and training, and learn how to use the internet to research the latest information on topics that you are assigned to teach or that are of interest to you. Get out and practice what you know: For example, organize the medical kit. Demonstrate that you are an avid learner and practitioner of your craft. Other fire fighters will notice and respect your thirst for knowledge, which builds your credibility when you are teaching any topic.

Credibility is a direct result of your knowledge of the subject. Rank does not always equate with knowledge—but hard work does. Go over the material so you can determine the flow of the course. Read the activities carefully so you have the right materials on hand and you understand the intended result. Write notes in the margins of the instructor's guide about comments you want to bring up and examples of the practical application of the material. An ill-prepared instructor is doomed to fall short when it comes to gaining credibility.

In addition to your before-class preparation, be ready to build credibility in the learning environment. Respect the knowledge that the students possess. Take advantage of those individuals with specialized knowledge to add value to the course. Do not allow yourself to be chastised. Be confident in your skills—obviously whoever asked you to teach believes in you. Take control of the class and be a professional. In the long run, you will be respected for your teaching skills if you have the credibility to support them.

Building respect and credibility with students is essential for those who do not have a formal position in the fire department or when teaching those with greater rank. Although it can be more challenging, success can be achieved with proper preparation and implementation.

The Fire and Emergency Services Instructor as a Coach

In their book *A Passion for Excellence*, Tom Peters and Nancy Austin define coaching as follows (Austin and Peters, 1986, 325–326):

> Face-to-face leadership that pulls together people with diverse backgrounds, talents, experiences and interest; encourages them to step up to responsibility and continued achievement; and treats them as full-scale partners and contributors.

In the sports world, the coach receives most of the attention and pressure because he or she is cast in the role of determining the success or failure of the team. In the fire service, we often stress the need for teamwork in all aspects of the job. Teamwork is an important ingredient on the fire ground as the incident commander orchestrates the company's actions through various tactical assignments. Even in the fire station, much of the nonemergency activity is accomplished not by an individual, but rather by a dedicated team of professionals. Whether on the fire ground or in the station, the incident commander acts much like the quarterback of a football team, calling out the formations and plays. In the training environment, the instructor plays a role analogous to that of the football coaching staff: Whereas the coaching staff works on drills to prepare the players for the next game, the instructor works to prepare the fire fighters for the next incident response.

The instructor's job within the fire service corresponds to that of a coach in the sports world in many ways. It is the instructor who must prepare the team for battle. An exterior attack can be as challenging as an interior one; students must be prepared for both equally. The instructor studies the team's opposition (the fire) and prepares the team for what they may face on game day. He or she builds team skills

FIGURE 1-6 The instructor builds team skills through practice and repetition on the drill ground.
© Jones & Bartlett Learning.

through practice and repetition on the drill ground (**FIGURE 1-6**). The instructor, much like a coach, must enthusiastically encourage fire fighters as they struggle with new ideas and techniques. Finally, the instructor must remember to remain positive, even in tough times, and to support the direction in which the team is headed as determined by the general manager (fire chief) and the owners (residents). Coaches and instructors lead, train, and drive their team's performance.

The Fire and Emergency Services Instructor as an Evaluator

Who is in the best position to evaluate the capabilities of a fire department? The answer is found standing in the front of the classroom: the instructor. As obvious as that answer is, how many fire chiefs have taken advantage of the knowledge fire instructors have assembled through their daily interactions with the troops? Instructors must evaluate students' learning and competencies at many points during the training process. To do so, they must sharpen their skills in evaluating the proficiency and knowledge retained by their students. The ability of the instructor to assess the comprehension of the class or student accurately can determine the pace of the training. Determining the right time to introduce new techniques or to review previously taught topics cannot be done without the ability to evaluate students honestly. Proper evaluations done in the course of training save the department from the heartache of critical evaluations performed by the public at the scene of a mistake.

Instructors are in a unique position to evaluate the capabilities and limitations of fire fighters. In many fire departments, input from the training division and instructors is regularly sought by fire department leadership to help them develop policies and procedures. The instructor can evaluate fire fighters, suggest operational directives, and field-test new standard operating procedures (SOPs).

Instructors should also be evaluators of the learning environment. That is, they must always be able to evaluate their effectiveness, recognizing what the students need and what the students are learning. They must evaluate themselves as well, looking for ways to improve their own teaching techniques. Learning environments and self-evaluations are essential in keeping the training program proactive and on the road to continuous improvement.

The Fire and Emergency Services Instructor as a Teacher

As simple as it sounds, the instructor must be a teacher. What comes to mind as you hear the word *teacher*? Perhaps it is someone you remember from your elementary school days. Why does that name come to mind? Teachers affect students in many ways—many times positively, but unfortunately sometimes negatively. Instructors must teach students new skills and abilities, thereby shaping the abilities of the team. They must remember the potential influence they have with their students. It is not just what the students learn that is important; rather, the shaping of a fire fighter's demeanor, attitudes, and desires ultimately determine how the instructor will be remembered.

Being an effective teacher should be the goal of each instructor. Mastering the interpersonal side of the learning environment goes a long way toward reaching that goal. The best and most remembered teachers shape and mold the total person to become part of the team and contribute the best they have to offer. A true teacher helps others see their potential and develop a path to accomplish those goals.

Setting Up the Learning Environment

Every instructor should understand the effect that the learning environment has on the ability of students to grasp the material being presented. You need to take command of that environment and use it to your advantage. The learning environment includes those environmental factors that influence the learning process and can include multiple elements, both physical (e.g., lighting, temperature, furniture configuration) and emotional (e.g., attitudes, comments, learning).

Physically, the classroom and drill ground must provide a safe, comfortable, and distraction-free environment. Emotionally, students' minds must be kept focused on the instructional material in front of them and away from the sometimes-contentious issues found in many fire departments. To make matters even more challenging, some of the factors affecting the learning environment are outside the instructor's control. If you are not careful, you may bring some frustrations into the learning environment, thereby impeding the learning process. To avoid these mistakes, familiarize yourself with the elements that affect the learning environment, including the audience.

COMPANY-LEVEL INSTRUCTOR TIP

One foundational component of ensuring a quality program is paying attention to the learning environment details and arrangement. You can facilitate better learning by eliminating unnecessary distractions and providing a comfortable setting.

Physical Elements Affecting the Learning Environment

Imagine yourself as you walk into a classroom. What do you notice first? Lighting? Cleanliness? Temperature? Or perhaps how far from the front of the room you can sit? Every student who has ever taken a class has evaluated the learning environment. The simple arrangement of tables and chairs can affect how we interact and ultimately learn. The best arrangements

FIGURE 1-7 Tables arrangements such as a U-shape **(A)** or a square **(B)** allow face-to-face interactions.
A: Courtesy of Forest Reeder; **B:** Courtesy of Bill Larkin.

allow for the free exchange of information, both from student to student and from student to instructor.

Traditional school classroom setups can lead to the formation of groups in distinct areas of the room, as students tend to sit next to those with whom they are most comfortable. Both the quiet and the uninterested students vie for the seats farthest from the front, not wanting to become involved in the learning discussion. Tables arrangements such as a U-shape or a square or tables arranged in groups, which allow face-to-face interactions, can sometimes neutralize the hierarchy of traditional settings (**FIGURE 1-7**). These arrangements also allow you to move among the students freely, improving the exchange of information and increasing the attentiveness of students.

In setting up the learning environment, you must also consider both the natural light and the installed lighting. Lighting affects many aspects of the learning environment. If projectors are used in the presentation, lighting becomes critical to the ability of the students to view the projected information. If natural light is a problem, a simple rearrangement of the room may allow the movement of a screen to a better location. If that is not possible, perhaps the addition of window blinds can correct the problem.

You must also evaluate the installed lighting in the room. Improperly installed lights may make it difficult to dim the lights enough to ensure quality projection. Conversely, dimming the lights too much may create reading or note-taking difficulties for students. If installed lighting is causing problems, your supervisor might be able to recommend improvements to the system through the annual department budget process.

Room temperature can also create problems for the learning environment. Take the entire class into account when setting the room temperature. Finding a comfortable compromise for all may turn you into a negotiator. Of course, some classrooms have environmental controls that cannot be changed by those using the facility. In those cases, you may have to alter lecture and break times if temperatures are uncomfortable. You may also have to become a student advocate and seek improvements to the environmental controls. More information on the physical learning environment can be found in Chapter 3, *Methods of Instruction*.

Emotional Elements Affecting the Learning Environment

The instructor has a responsibility to help protect the individual student from the emotional letdown that comes with the inability or difficulty to learn a new task or subject. The fire service training environment includes both rookies and senior personnel, adult learners and students fresh out of school. In addition, it is populated by both students who know what to do and those who think they know what to do. Fast learners and those requiring more individual efforts will challenge even the most seasoned instructors.

The best instructors can present material effectively for all types of learners. Instructors who protect those who struggle by placing them in positions where they can make progress without the embarrassment of failure in front of the greater group will be successful. Good instructors also learn how to use the more talented department members to coach and teach those with lesser skills.

Learning to work within the environment challenges you to be flexible, loyal, confident, and fair. Building a bond of trust with your students will enable you to present new ideas and make needed revisions to old traditions. As the instructor, you are the visionary of the fire service.

The Instructor's Role in the Future of the Department

Have you ever thought about what drives an organization? Why do some survive and others struggle as time forces change?

In examining the structure of a typical fire department, it is easy to point out the formal leader: It is the individual at the top of the pyramid, the chief who provides the formal direction for the troops to follow. The fire chief is responsible for the ultimate success or failure of the fire department. Of course, organizational charts are filled with many other positions as well, each of which has its own responsibility for providing direction in support of organizational objectives. It is within these ranks that you find assistant, deputy, battalion, and division chiefs; line officer positions, including captains and lieutenants; and fire marshals and inspectors.

Where does the instructor fit within this scheme? In many fire departments, the instructor is treated as an operational support assignment more than an official rank or position. The instructor position is often viewed from one of two extremes: as vitally important to the overall operation or as unnecessary by the administration. They may also have a rank, be scene safety officers, or hold other responsibilities in the department. Wise leaders understand that instructors are important members of the team. They use their instructors to maintain their department's state of readiness and prepare their fire fighters for future missions.

It is this role—preparation for future activities—that requires you to keep ahead of the curve. To fulfill it, you must become a visionary force within the fire department. It is vision that drives instructors to keep abreast of the ever-changing world of firefighting. Changes and advancements in firefighting tactics must be reviewed and implemented through revisions in training curriculum. If the fire department desires a change in operations, the instructor and training division will be charged with educating and training fire fighters to make that change. New techniques and advancements must be tested and made applicable to each organization, because many changes are not "one size fits all" measures. You will be challenged as you introduce new ideas that seem to conflict with established traditions.

As the mission of the fire service continues to evolve, you must be able to prepare the troops to carry out the new mission. The history of the fire service is full of pertinent examples, as the instructor has had to evolve to provide training for medical, hazardous materials, and technical rescue services. Today the threats of terrorism, active shooters, and scenes of violence, along with the use of weapons of mass destruction (WMDs), present their own unique challenges for the fire service, with instructors once again being called upon to lead the troops into these new areas of service.

Instructors must monitor the ever-changing learning environment as well. Struggles with budgets and staffing present ongoing problems for instructors, who must continually seek to keep training at the forefront of nonemergency operations. In the instructor role, you must find new and creative ways to reach students and present training. Today, visionary instructors are turning to the cyberworld as online training programs gain a foothold in the instructional world. Technology will continue to advance and, consequently, affect the delivery of training. Instructors with the vision to see how this new technology can be used for training purposes are establishing virtual learning environments and using satellite and video classrooms as effective learning media.

Given the unique challenges apparent in the modern-day fire service, the fire chief would be wise to select the very best personnel for the position of instructor. An instructor with vision can greatly assist the fire chief in meeting the fire department's future challenges.

The Instructor's Role in Succession Planning

Preparing fire fighters for battle is not the only job of the instructor. For any organization to survive and grow, a continuity plan must be in place. Continuity of the organization provides security for the community that the organization serves. The instructor can assist in that regard by becoming involved in **succession planning**.

As trainers of fire fighters, instructors are often in the best position to recognize potential talent and leadership qualities in the fire fighters and officers they train. The instructor's role is to nurture and challenge these fire fighters through the training program. During the course of training, the instructor may use some fire fighters to assist in training other members. By placing fire fighters in leadership positions within the training environment, you allow them to refine, enhance, and demonstrate their leadership qualities.

The instructor may also be in the position of providing input to ranking officers on the performance of fire fighters in training. By recommending high achievers to fire department administrators, you assist in the identification of future fire officers and instructors. You may also be in the best position to identify those fire fighters who do not fit the model fire officer that the fire department has established.

Succession planning also means looking for future instructors who someday will take your place as the lead instructor in your department. As difficult or as uncomfortable as it might seem, identifying, nurturing, and training their replacements is something successful instructors need to learn to do. Another way to approach this goal is to identify someone who is capable of building upon the foundation that you, as an instructor, have established, and can move it to the next level. Because you have worked hard to develop a successful training program, the last thing you would like to see happen is to have the program fail. Herein lies the opportunity to find another fire fighter who is passionate, skilled, and has the desire to train the other members of the department. Succession planning is discussed later in this chapter.

The instructor walks a fine line between operations and training. In organizations with a weak operations leadership, the instructor may become absorbed with setting the operational direction and standards simply because he or she has expertise in specific areas. In other cases, the instructor has a formal role in establishing policy. Of course, you should be careful to ensure that those policies exist to support the direction of the operations division—not the other way around. Although you may not always agree with the standards chosen by the operations chief, you must accept that direction and train fire fighters accordingly.

Instructor Credentials and Qualifications

You've heard it before: "Walk the talk." It's a phrase that can certainly be applied to the instructor. Your best friend is the confidence that your students have in your abilities and knowledge. This is a quality that cannot be learned from a book. Often, instructors are born in the learning environment from energetic students. Others vow to develop their instructional skills after attending a highly charged training session or a lecture delivered by an impassioned instructor. As an instructor, you must be aware that your success or failure might rest with the degree of effort and preparation you put forth at the beginning of your career in the fire service. It is here that dividends are paid.

Laying the Groundwork

Good instructors are born from good students—students who thrive on the knowledge gained through training, and those who actively participate in training activities and are not afraid to learn from their initial failures. These individuals are the fire fighters who understand the value of education and refinement of skills. They continually place themselves in learning situations, looking to upgrade their skills and knowledge.

Think for a moment about the instructors who have influenced you. Students attending any training program immediately focus their attention on the instructor as they begin looking for clues about the quality of the program. They might ask what experience the instructor has in the subject area. What is the instructor's firefighting experience level? If the training is being held in-house, your students might remember the days when you were a student. Your credibility in the learning environment depends on your past behavior. The past always has a way of finding the future, so it is always wise to protect your future by engaging in proper behavior in the present. Laying the groundwork for becoming a good instructor begins the day you join the profession. While some mistakes will be made, you must always guard your credibility and integrity.

Meeting Standards

Standards dictate many things within the fire service. Training programs are not immune from national standards. Standards set the bar for instructors' proficiency and knowledge. They seek to establish uniformity for instructors across the entire profession. NFPA 1041 outlines the instructional levels in the fire service and guides instructor trainer programs; this book is written to be in compliance with this standard. Fire fighters seeking local, state, provincial, or national instructor certifications may be asked to demonstrate compliance with NFPA 1041 as well, though it may not be the only standard that the instructor must meet.

A **degree** is awarded by an institution of higher learning after a person has completed acquisition of the required knowledge in a particular field. A **certificate** is given after attendance at a course. A **certification** is awarded after a person has passed an examination process that is based on a set standard such as an NFPA Professional Qualifications standard.

Fire chiefs and fire departments are free to establish their own set of qualifications for instructors. If the instructor is asked to teach only one class in one department, he or she may not be asked to obtain a formal certification from a certifying agency. Instead, the instructor may simply receive the training that the fire department deems necessary from other in-house instructors. In some cases, instructors may be asked to have a certain number of years of experience prior to taking on an instructional assignment. In other cases, instructor positions might be reserved for those with command authority, holding a line officer's or chief officer's position. Some fire departments set few or no additional requirements in an effort to find a fire fighter willing to take on the extra responsibilities of the instructor.

Whatever the case, it is wise to meet these challenges head on. Becoming an instructor may open doors for future promotions. Meeting the qualifications of this position will also prepare you for future leadership assignments; for example, Fire and Emergency Services Instructor I is a prerequisite for Fire Officer I in the NFPA standards system.

Continuing Education

Meeting requirements, whether set by a national standard or through fire department policy, is just the beginning for the instructor. Working within the dynamic world of the fire service means you need to continue your own professional growth and development. While some individuals dread the idea of **continuing education**, you need to understand the need to improve your knowledge, as it allows you to provide the very best and up-to-date information in the learning environment. The idea of requiring continuing education is not a new one. Many professions, including health care, education, and inspections, require their members to participate in continuing education to remain licensed or certified.

For the instructor, the organization that issues the initial certification decides what, if any, continuing education is necessary. Some states may require only proof of continued instructional activity. Do not rely on the requirements of outside entities as a motivational force to professional growth. You should always strive to be on the cutting edge of the fire service. Attend outside trainings, seminars, and instructional conferences to stay up-to-date with the most current fire-ground tactics, management practices, and instructional techniques available.

Building Confidence

By keeping abreast of the latest information, you demonstrate the very value of education. For learning to occur, students must believe in your knowledge of the subject—which is not to say that you will always know everything about a particular topic.

There is a saying that in the learning environment: "The instructor is always right." Believing in this old adage could be a fatal mistake and result in the eventual loss of your credibility. It does not take many times of being proven wrong in the learning environment to lose the confidence of your students.

Building and maintaining student confidence in both the instructor and the training program go hand-in-hand. Instructors can build high levels of confidence in several ways. For example, you can demonstrate your own commitment to lifelong learning through continuing education. Be willing to admit that you are in doubt about a particular question and seek the answers for the inquisitive student instead of trying to make something up on the fly in an attempt to impress your pupils. Be open to the suggestions and ideas of the students. Because instructors often serve in operational roles as well, be very careful about following all teachings when working the emergency scene or when in the supervision role. One sure-fire way to lose credibility is to project a "do as I say, not as I do" attitude.

Issues of Ethics in the Training Environment

Firefighting is perhaps one of the most respected professions in today's society. The task of maintaining the public's trust in its fire service is held in high regard by all those who wear the uniform.

Ethics reaches beyond laws and standards to define behavior. Codes of ethics spell out what is acceptable and what is unacceptable within a fire department. For many individuals, ethics is rooted in their own personal assessment of right versus wrong (or values), which begins during their early upbringing. Ethics can reflect the attributes of the people we respect and interact with in our own profession. In the end, it is perhaps that feeling in the pit of your stomach—your "gut" instinct—that gives you the best clue about whether a certain behavior is ethical.

The issue of ethical behavior also affects the training program. Instructors are often the ones who are asked to judge which candidates are ready to provide public service. Failing to hold students accountable for their learning objectives or simply passing fire fighters through the training program in the interest of moving on is unacceptable by any standard. This failure to hold to agency standards will affect not only the fire fighter who "slips" through but also those with whom this fire fighter works, and ultimately the instructor who did not maintain the standards of the agency.

Another aspect of ethical conduct from an instructor's point of view is to ensure that your teaching methods and learning environment behavior are acceptable to all students. The fire service has evolved

FIGURE 1-8 An effective instructor is aware of who their students are and have a desire for all of them to be successful.
© Jones & Bartlett Learning. Photographed by Glen E. Ellman.

over the years to include members from across all demographics and socioeconomic levels of our society; as an instructor, you have a responsibility to be aware of and sensitive to the diverse backgrounds of your students. This might sound challenging or difficult, but the solution is actually quite simple—treat everyone equally and fairly. It is important to acknowledge the diversity that is found in our learning environments and fire departments. Effective instructors are aware of who their students are and strive to teach all of them with compassion, fairness, and a desire for all of them to be successful in whatever subject is being taught (**FIGURE 1-8**).

Leading by Example: What You Do, You Teach

Ethics in training must begin with the instructor. Instruction must extend beyond the learning environment and into everyday operations. Instructors demonstrate ethical behavior through their "do as I say and as I do" lifestyles. Leading by good example lays the cornerstone for training ethics. If you are responsible for teaching the confidentiality of disciplinary measures in a leadership class but are later seen discussing an employee's behavior around the coffee table, you will lose credibility. Instructors must recognize early on that they are always being watched and their actions are often mimicked by their students.

> **COMPANY-LEVEL TRAINING TIP**
>
> Increasing diversity in the fire service is a major initiative. Instructors at all levels must possess the cultural competencies to provide training to a variety of genders, ethnicities, and generations.

Accountability in Training

Another ethical issue related to the training program is the need to maintain student accountability for training. As an example, if 10 fire fighters attend training on pump operations but only 4 fire fighters actually operate the pump while the remaining 6 chat away, who should receive credit for pump training?

Never put yourself in a position of simply passing a student by rote. It is important to document students who fail to meet training objectives. Because the training program is tasked with preparing fire fighters for critical life-safety operations, you must be dedicated to protecting the program's integrity. The classroom or drill ground is no place for favors and friendship to affect a student's achievement record. At the end of the day, the fire fighter's performance ultimately judges your ability as an instructor.

Recordkeeping

Good records are important to any training program, but only if those records are thorough and accurate. Just as you must maintain accountability in individual training accomplishments, you must also maintain accurate records of program achievements. Training that is not fully completed owing to interruptions should never be recorded as accomplished.

Some instructors have been known to falsify records by simply recording trainings or skill attainments that did not occur for one reason or another. This act of falsification, also known as "pencil whipping an exercise," might look good on paper, but can lead to disastrous consequences once fire fighters have to perform the task under the stress of a true emergency or during a legal issue.

All training sessions must be documented, and accuracy is critical in those records. The records must contain a factual listing of the accomplished objectives. Whenever training records are found to be inaccurate, whether due to negligence or simply because of an error, your credibility will be called into question. Accurate recordkeeping builds confidence in the training program and sets ethical expectations for all to follow.

The accuracy of training records is not only an ethical issue but also a legal issue. In recent court cases dealing with fire fighter fatalities, training records have been used as evidence both for and against the authority having jurisdiction (AHJ). Clearly, accurate and correct training records will have a tremendous influence in the outcome of such lawsuits. Recordkeeping as it pertains to the law is discussed in more detail later in this chapter.

Trust and Confidentiality

Trust and confidentiality go hand-in-hand: Lose one and you risk losing both. Both of these attributes have tremendous influence on how well you are able to lead in the learning environment and maintain an effective teaching ability.

Instructors are often one of the first points of contact fire fighters make when joining the fire department. As such, you are placed in a position where your trust and confidentiality are critical to students. Struggling students may relay personal information referencing problems in their personal lives to explain why they are having difficulty in completing a task. A student might also share information about learning disabilities that require special considerations in the learning environment. Conversations such as these require that you make every effort to maintain confidentiality. If circumstances require that fire department management become involved in the issue, then you should make that fact known first to the student. You may also provide guidance to students about where to go for assistance within the fire department's command structure.

Another area of concern is the need to maintain the confidentiality of student performance in the program. Information dealing with test scores, student evaluations, attendance, and behavioral issues should always be protected. Guidelines for the recording and sharing of this type of information should be written into policy to protect students' rights. Instructors are often required to handle sensitive student information and must consistently demonstrate the ability to do so with professionalism.

The Law as It Applies to Fire and Emergency Services Instructors

In the fire service, training is designed both to keep existing skills sharp and to introduce new skills. Every training session exposes the instructor and the fire department to potential liability. For example, inappropriate comments voiced in a training session based on religion, nationality, gender, or sex—even if made in jest—can subject the instructor and the fire department to liability. A good working knowledge of the law, standard operating guidelines, and the department's own rules and regulations will reduce the potential exposure to liability. An understanding of the legal issues pertaining to training is critical for instructors. You will likely encounter situations during training evolutions, classroom participation, and your daily duties when issues pertaining to federal law will arise. You may face questions about compliance with the Americans with Disabilities Act (ADA), hostile work environment claims, and complaints of sexual harassment. A strong familiarity with these federal laws will prepare you to fulfill your obligations properly. State law could come into play on issues of privacy, indemnification, injuries to training participants, and equipment failures.

It is only through familiarity with the appropriate laws that you can insulate yourself and the fire department from potential liability. Directly associated with familiarity is proper recordkeeping. Maintaining an appropriate documentation system affords you some level of protection; conversely, inadequate or incomplete records may expose you to increased liability.

As an instructor, your conduct is governed by three sources of law:

- Federal law, which is made up of statutory laws
- State law, which is established by each state's legislature
- Policies and procedures crafted by each individual fire department

All three sources set expectations for and restrictions on the conduct of instructors (**FIGURE 1-9**). Nevertheless, federal and state laws differ from a fire department's policies and procedures in an important way. Federal and state laws apply evenly to all citizens and are the standard by which all instructors are measured. The ADA and the Age Discrimination in Employment Act (ADEA) are two examples of federal laws that govern all citizens across the nation. By contrast, policies and procedures are specific to a particular

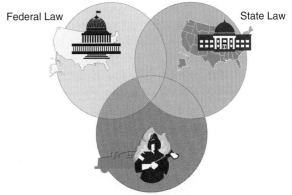

FIGURE 1-9 All three sources of law set expectations for and restrictions on the instructor's conduct.
© Jones & Bartlett Learning.

fire department and vary from region to region within a state, because each fire department must take into consideration numerous variables, such as the size of the community it serves, its locale, and its resources.

A fire department's policies and procedures must not violate any state or federal laws. Indeed, they do not have the same status in the law as state and federal statutes. A fire department's policies and procedures represent the minimum expectations of the fire department and do not necessarily impose a legal duty, as does a state or federal law. The only true benchmark against which the departmental policies and procedures are measured is the national standards or guidelines set within the fire service. Departmental policies and procedures can place obligations upon an instructor that are not otherwise required by federal and state laws. For instance, a fire department's policies and procedures may require the department's instructor to keep all department licenses and permits up-to-date, even though there is no legal requirement for the instructor to do so. The policies and procedures may also specify additional recordkeeping requirements not set forth by state or federal laws.

Federal law is the broadest, in that it applies to all persons and requires the same expectations across the board regardless of your chosen profession or the state in which you live or work. For example, the ADA, the ADEA, and Title VII govern not only fire departments but private employers as well. State laws are somewhat narrower, in that they apply only to the state in which they are enacted. State laws set some expectations for that state, but are not necessarily specific to a profession. The policies and procedures of a fire department have the narrowest scope, being designed specifically for an individual department and the professionals who work in it.

Although the law is designed to set certain standards and expectations, it also can be used to protect instructors from lawsuits. Instructors must have a working knowledge of federal and state laws. In addition, they must be well versed in the policies and procedures of their particular fire department, which requires that they possess a strong working knowledge of national guidelines. Although national guidelines are not legally binding on a fire department, they clearly set out the consensus of the fire service across the United States. The NFPA standards, for example, are the national standards upon which most fire departments model some of their own standards.

Should an instructor's conduct be called into question, the national guidelines serve as the backdrop against which that conduct will be judged. For example, if a live fire training session gets out of control, the instructor's methods will be compared to NFPA 1403: *Standard on Live Fire Training Evolutions* to determine whether the instructor's conduct was reasonable. NFPA 1403 outlines a checklist of safety activities when performing a live fire training session, which includes activities such as ensuring that all utilities are disconnected and that proper permits have been obtained. If an instructor does not follow this checklist, his or her conduct may be considered unreasonable and result in liability.

Instructors have an immense responsibility to conduct, manage, and administer training programs. With that responsibility comes the potential for liability if an injury, accident, fatality, or lawsuit occurs. In our increasingly litigious society, the instructor must be aware of the laws that pertain to each level of instructor responsibility. All certification levels of instructor will have different applications of laws and statutes in the performance of their duty areas, and professional development as an instructor will be necessary throughout your career to review trends and keep up-to-date on cases that have occurred and judgments for and against a fire department and its training program.

> **COMPANY-LEVEL INSTRUCTOR TIP**
>
> Instructors who teach to the national standards, document their instruction, and provide a safe learning environment are not high-risk targets for lawsuits. It is when an instructor does not teach the correct information or does not maintain safety as a top priority that potential liabilities increase.

Federal Employment Laws

As an instructor, you are not expected to know all the laws of the land, but to perform your job correctly you do need to understand three laws: the Americans with Disabilities Act, Title VII, and Section 1983 of the Civil Rights Act of 1964. In all likelihood, you will be called upon to teach or arrange instruction on each of these laws. You will also encounter situations that will require you to act or react to situations involving the practical application of these three laws. Here is a brief summary of their content:

- Americans with Disabilities Act: Prohibits discrimination against a qualified person because of a disability, where discrimination may include hiring, firing, promotions, and compensation
- Civil Rights Act of 1964: Prohibits discrimination based on race, sex, or national origin
- Title VII of the Civil Rights Act: Prohibits discrimination that creates a hostile work environment

Americans with Disabilities Act

The Americans with Disabilities Act of 1990 is designed to protect individuals with disabilities from discrimination in the workplace. A **disability** is a physical or mental impairment that substantially limits one or more major life activities. In 2008, the ADA was amended with the primary purpose of expanding what constitutes a disability. Determining whether an impairment substantially limits a major life activity should not take into account possible mitigating measures. Mitigating measures include medication, medical supplies, prosthetics, hearing aids, mobility devices, and the like. The mitigating measures of ordinary eyeglasses and contact lenses, however, are considered in determining whether an impairment substantially limits a major life activity. While the definition itself did not change with the 2008 revision of the ADA, the amendment added instructions to follow when determining what constitutes a disability. **Major life activities** include functions such as caring for oneself, performing manual tasks, walking, seeing, hearing, speaking, breathing, learning, and working. Conditions that previously would not have qualified as a disability—such as compulsive gambling, kleptomania, pyromania, and current illegal use of drugs—may now be considered a disability. It should also be noted that an impairment that is episodic or on remission is considered a disability if it would limit a major life activity when active.

In the fire service, it is unlikely that you will encounter physical disabilities such as blindness or the inability to walk. Nevertheless, you may encounter many other types of disabilities, including psychological disorders, neurological disorders, cardiac issues, respiratory issues, diabetes, epilepsy, dyslexia, drug and alcohol addiction, and prosthetic devices.

If you are approached by an individual alleging a disability, you should undertake a two-step analysis. First, you must determine whether an actual disability exists. If the answer to that question is yes, then you must determine whether the disability can be reasonably accommodated. The ADA defines **reasonable accommodation** as a modification or adjustment to a job or work environment that will enable a qualified applicant or employee with a disability to perform the essential functions of the job or enjoy the benefits and privileges of employment, and that does not create an undue hardship for the employer. For example, if a fire fighter has diabetes, a reasonable accommodation may be to allow for periodic blood glucose testing during the day or during training (**FIGURE 1-10**).

Examples of reasonable accommodations found to be in compliance with the ADA include modifying

FIGURE 1-10 A reasonable accommodation may include allowing for periodic blood glucose testing during the day or during training.
© Jones & Bartlett Learning.

facilities to make them accessible to employees with disabilities, restructuring a job, modifying the work schedule, acquiring or modifying equipment, and reassigning a disabled employee to a vacant position for which the individual is qualified. Accommodation is not automatic—that is, you are not required to make an accommodation if it would impose an undue hardship on the operation. An undue hardship exists if the accommodation would require significant difficulty or expense relative to the size, resources, and nature of the employer.

Further, a reasonable accommodation does not have to be provided if doing so would present a **direct threat** to the health or security of others. Four factors are considered when assessing whether a direct threat exists: the duration of the risk, the nature and severity of the potential harm, the likelihood that potential harm will occur, and the imminence of the potential harm.

Like many legal situations, ADA issues are fact driven. There is no "bright line" test; each case must be analyzed based on the particular facts at hand. Courts have ruled on many cases involving fire fighters and the ADA, and those rulings can provide guidance for you in similar situations. For example, the U.S. Court of Appeals for the Fifth Circuit has held that a fire department is not required to accommodate a fire fighter who is unable to perform the essential duties of firefighting owing to a back injury when no light-duty position exists in a small fire department (*Burch v. City of Nacogdoches*, 5th Cir.). In Pennsylvania, a fire department was not required to provide a fire fighter with a second chance at the academy course after he failed due to depression. In Maryland, the court held that it was proper for a fire department to prohibit a fire fighter from using an inhaler when the brand used was highly flammable (*Huber v. Howard County Maryland*, 4th Cir.).

It is important for the instructor and managers up the chain of command to be aware of both the requirements of the ADA and those personnel under their command who qualify for reasonable accommodation under the ADA. In situations when an accommodation is requested during a training evolution, you should take great care to evaluate the request and, if needed, contact a higher authority for advice. Once a decision is made, it should be documented immediately to protect both you and the fire department. Not all medical and psychological issues qualify as disabilities under the ADA. Each issue must be examined separately. Also, any accommodation request must be reasonable.

Freedom of Information Act

Each state has freedom of information laws that govern public access to government records, documents, and other information kept by government bodies. The purpose of the Freedom of Information Act (FOIA) is to promote government accountability and the public's right to know. Although each state has its own variation on FOIA, generally the laws require the following records and information to be made available to the general public upon request:

- Administrative manuals and procedures
- Instructions to staff
- Substantive rules
- Statements and interpretations of policy
- Final planning policies
- Factual reports (such as inspection reports)
- Information about accounts, vouchers, and contracts involving the expenditure of public funds
- Names, salaries, titles, and dates of employment for all employees and public officers
- Meeting notes or recordings
- All records, reports, forms, writings, letters, memoranda, books, and papers (even if in electronic form). This may include training records and reports, such as training rosters.

Generally, members of the public can make a FOIA request through designated public officials. Depending on the applicable state law, there may be exceptions to the disclosure of certain documents or the government body releasing the documents may be required to redact personal information from certain documents. In addition, government entities may be required to retain public records for a specific period of time. In most cases, state record retention laws will dictate the minimum length of time public records must be retained by government entities. Please review your state law and consult with your attorney for more specific information about the requirements of FOIA, the exceptions to FOIA disclosure, and the record retention requirements in your state.

Records, Reports, and Confidentiality

Once you understand the laws that apply to you as an instructor, the next logical question is this: How do you perform your job so that you abide by the law and have protection should a lawsuit be filed? The first line of defense is always proper and thorough recordkeeping. Creating and maintaining records, reports, memoranda, e-mails, and other records in the course of your duties places you in a good position. These records should be constructed contemporaneously and maintained for a reasonable period of time and in accordance with any local law.

The recordkeeping process starts at the outset of each activity. For example, when you are organizing a training evolution, the materials you prepare should set forth the purpose of the training, the activities to be performed, the equipment to be used, and the goals of the training. You should have available any sources that were relied upon in creating the evolution and should confirm that the training complies with all policies and procedures of your fire department.

Lawsuits evolve over long periods of time, which means that the accuracy of your records is vital in mounting a defense against such litigation. Proper recordkeeping is as important as the training you provide to your department. In cases where an injury occurs to a participant in or an observer of a training session, it is crucial to document what occurred, who was present, and what each person observed. The same holds true for claims of sexual harassment: It is necessary for the person who receives the report to document what was said and which actions were taken once the complaint was received.

Proper documentation is essential to the performance of your job, as is storage of the records in an organized and secure setting. Both state and federal laws (Privacy Act of 1974) prohibit the disclosure of certain information. Most likely, your department has a policy or procedure that requires the protection of personal information that is kept in the control of the instructor. Failure to abide by the **confidentiality** provisions required by law or your department could lead to liability or adverse action against you.

Records that should be kept secure and confidential include the following:

- Personnel files: Usually include information such as date of birth, Social Security number, dependent information, and medical information
- Hiring files: Include test scores, pre-employment physical reports, psychological reports, and personal opinions about the candidate
- Disciplinary files: Any report or document about an individual's disciplinary history and related reports

Regardless of the type of training being provided, recordkeeping should be a top priority as part of risk management. Records with sign-in sheets and the topic presented should be maintained for all training sessions for a period of at least 5 years, or longer as required by law. You should also keep copies of the handouts, training outline, and other training materials presented with the sign-in sheet. Records, reports, or memoranda related to any unusual occurrences that happen during training or that are reported to you should be kept for at least 5 years. Because many lawsuits are filed long after the event occurred, the only real protection from this risk for the fire department may be the records kept. Incomplete or missing records often make defending suits more difficult.

Copyright and Public Domain

Many materials you may want to use in training are protected by copyright. More than 3 decades ago, Congress passed the Copyright Act of 1976, which governs copyright issues. The act divides work into two general categories: those protected by the law and those considered to be part of the public domain. Works include written words, photographs, and some artwork. The Copyright Act sets out some specific time frames pertaining to copyrighted material. Any work published before 1923 is considered to be in the public domain. Any works published between 1923 and 1978 have a 95-year copyright protection from the date of actual publication. For any works published after 1978, the author, artist, or photographer has a copyright for life plus 70 years. Violation of the copyright law can result in liability, with the creator of the work being eligible to recover any damages suffered or lost profits.

When providing training materials for fire service employees, many of the handouts used are photocopied from copyrighted materials. The Copyright Act does not prohibit their use outright or require that permission be sought before they are used. Indeed, the fact that some material is protected by a copyright does not necessarily prohibit you from using it in your training, because a "fair use" exception exists relative to the protection afforded by copyrights. Examples of fair use include copying materials for the purpose of criticizing, commenting, news reporting, teaching, and research. Copies for the class in a training session generally fall into this category. Ultimately, fair use depends on the nature of the copyrighted work and how much of the work is copied. For example, it would not be fair use to purchase one copy of a textbook and make copies of the entire book for the class. By contrast, if you want to use an article from a magazine or newspaper for learning environment purposes, that would likely fall within the fair use exception.

If a particular use does not constitute fair use, you may still be able to use the material in a learning environment by contacting the copyright holder and seeking permission in advance to use that material. You should obtain permission in writing prior to using the material.

Whether a work is copied under the fair use exception or with permission, it should include a notice indicating who holds the copyright to the work.

Managing Multiple Priorities as an Instructor

In today's fast-paced world, we are bombarded with many priorities, each demanding our attention and time. As a consequence, success or failure is often determined by the ability to understand and prioritize tasks. Recognition, planning, and delegation are skills that can assist those facing multiple tasks while maneuvering through sometimes hectic and complicated assignments.

Time Management

As an instructor, one of the first skills you need to learn and develop is time management. Several methods or techniques are available to help you make the best use of your time. You should select those techniques that suit your personality and are easy and convenient to use. For example, setting a goal to read one magazine article per shift, or per week, is a good start. Attending one fire service conference per year and earning a new certification are goals that will help in your development.

One helpful method is to set time aside each day to create a "to do" list. Prioritize the items on this list, and then work your way through them, checking off what you have accomplished. Included on this list might be reading an article or reviewing a department SOP. Another method is to set goals for the year and break the goals down into months and weeks. Use whatever method works for you: The key is to develop a system that fits your needs. When creating your personal "to do" list, be mindful of any recertification requirements for your certifications or job position. As a fire fighter and instructor, you might have required hours of continuing education or "recert" hours that must be earned in a set period of time. As you are setting your goals and priorities for the year, make sure that your own credentials do not expire! Budget the time and resources needed to obtain new or renewed certifications, and do not wait until the last moment to complete these requirements.

Regardless of which system you use or develop, perhaps the most important part of successful time management is learning to prioritize your "to do" list. Too often we spend a great deal of time on low-priority items that are easy to handle but distract us from the more important items on our list. Learning how to prioritize your list takes self-discipline and practice. Another challenge to accomplishing the items on your "to do" list is balancing your list against that of your organization. When developing this list, make sure every item supports the mission and goals of your agency. Finding this balance will help you develop both as an individual and as a professional in your organization.

Training Priorities

Managing today's modern fire and emergency services creates multiple priorities for the training program. Deciding what and when to teach is just one of the challenges. The direction that the program takes may be decided at different levels depending on the fire department's organization. In some cases, upper management may choose to lay out the training schedule for the period and then leave it to the instructor to decide how to accomplish it. If the fire chief takes a hands-off approach to training, then the instructor may be asked to develop both the schedule and the training program's objectives.

Training priorities may also reflect specific community characteristics. For example, a community with a heavy chemical industrial presence may place additional emphasis on hazardous materials training. Large cities with high-rise construction may require extra training on high-angle rescue and high-rise fire operations. Rural communities may require training on water supply shuttle operations and farm rescue. Training priorities are driven by mandatory training requirements to maintain certifications or Occupational Safety and Health Administration (OSHA) regulations. Clearly, the task of setting training priorities is not a simple one.

Decisions about how precious training time should be spent must include input from all levels of the fire department. Fire company officers may be able to provide valuable insight into fire-ground performance issues affecting operations; perhaps some problematic issues could be corrected through additional training. Fire chiefs may have knowledge of changes in future missions that will require fire fighters to learn new skills. Training committees may need to be established to provide a broad-based perspective regarding the fire department's training needs, tackling the question, "Should we train more on our most serious hazards or on those hazards we respond to most often?" Both elements deserve consideration in the planning process (**FIGURE 1-11**).

Planning a Program

The level of success of any training program is directly proportional to the planning efforts. In career settings, the schedules of on-duty personnel must be considered when setting up training programs. By contrast, instructors who are dealing with volunteer and part-time fire fighters must take into account the availability of personnel to leave their full-time jobs and to balance family time with the many department requirements. Holidays, vacations, multiple shifts, injuries, and illnesses all affect the need to reschedule or make up lost training opportunities.

The program must also plan for the use of training facilities and equipment. In some cases, outside instructors with special expertise may need to be scheduled.

No employee likes surprises, so consideration of the fire fighter's efforts to be put forth during training and other personal commitments require that training program schedules and completion requirements be established well in advance and communicated clearly to all. Once established, training schedules need to be followed as closely as possible. Although changes will

FIGURE 1-11 A refinery or industry fire is an example of a low-frequency-high-risk incident **(A)**. Training committees can identify high-frequency incidents such as a motor vehicle accident and ensure that risk management practices, such as proper apparatus positioning and PPE use, are in place **(B)**.
A: Courtesy of University of Nevada, Reno Fire Science Academy; B: © Jones & Bartlett Learning. Photographed by Glen E. Ellman.

inevitably need to be made, any changes should be implemented with as much advance notice as possible and with sufficient time planned for rescheduling of the training to allow fire fighters to readjust their own schedules.

In some locations, a learning environment may be shared by the city government, community access groups, and fire department training programs. Nothing is more frustrating to both fire fighters and instructors than to have an overlap of scheduling or other unforeseen event force a last-minute cancellation or location change of a training session. Given the very nature of emergency services, it is always difficult to schedule training and execute it according to schedule without interruption. In a combination or volunteer organization, conflicting priorities of personal and professional life outside the fire station make the program management portion of fire service training all the more important.

Program management aspects of running a training division and training program include many basic management skills learned in fire officer training, albeit focused in this case on the goals and objectives of the training division. Administration of training policy and procedures, training recordkeeping systems, selection of instructional staff, review of curricula, and design of programs and courses that meet the current and future needs of the organization require a training team to accomplish many tasks.

Instructors also need to plan the content of individual training sessions. Without planning, conflicts among competing groups can arise in the use of facilities and lead to a delay or cancellation of a scheduled training session, further eroding confidence in the instructor's abilities. Contingencies for the use of specialized equipment needed in training evolutions must be built in if dedicated training equipment is not available. The instructor should also plan how a missed training session will be made up and ensure that make-up sessions accomplish the same learning objectives as the original training. Training time should be planned to eliminate distractions and interruptions. This may not always be possible when students attend training while on duty and as part of in-service training, as the need to respond to an emergency can interrupt even the best-laid plans.

Training Through Delegation

Instructors are faced with many tasks in managing the fire department's training program. Planning and scheduling for a multitude of priorities, developing and reviewing curricula, evaluating program effectiveness, coaching and mentoring students, and instructing all place enormous demands on the instructor's time. Without assistance, it is easy to become burned out, even for the most dedicated instructor.

One way to minimize the potential for burnout is through **delegation**. Be alert to those fire fighters who have shown through recognized efforts the ability to handle additional training assignments. Be observant in the learning environment for students who show an interest in instructional activities. It is often said that there is no better way to learn a subject than to have to teach it. Learning can be achieved through delegation, and you can obtain valuable assistance through the assignment of certain instructional tasks to competent fire fighters. Delegate only those tasks where a student has shown competency and where such a hand-off of responsibilities will not jeopardize the safety of other students.

Professional Development

Recall the instructors who shaped and influenced your development (both good and bad). What is the common thread among these individuals who had a positive impact on you? Most likely, they always had a magazine or an article in their hand or their nose in a book. The individuals who actively sought opportunities to learn were often the ones called upon to teach at a moment's notice because they had a wealth of knowledge that others wanted to hear. The same individuals attended conferences to learn and focused on networking with their peers. Eventually, they became presenters after obtaining a thorough knowledge base and developing their own ideas and theories to share with the fire service.

Another way to look at this idea is through an analogy. The speaker who is often heard at various fire service conferences did not start out as the keynote speaker. At one time he or she was just a fire fighter sitting in the audience, taking notes, and listening to every word. This person took what was learned at the conference, returned to the firehouse, and looked for ways to put that new knowledge into practice. As he or she learned more and advanced into higher positions within the department, the instructor had opportunities to teach new concepts and ideas to the next generation. The ideas, concepts, or theories that this individual shared stood the test of time and eventually became the basis for new policies and procedures followed within the department. Soon, word of mouth spread about these new ideas. The fire fighter who once sat in the audience taking notes is now on the stage leading the discussion and is respected by his or her peers as a leader in the profession. All of this came about because of the individual's desire to learn and to develop—because the fire fighter was a true lifelong learner.

An important part of becoming a quality instructor is having the basic desire to always improve and to become a better fire service professional. Who benefits when you develop into a respected instructor? Both you and the fire department benefit, because a good instructor supports the department's goals and, in turn, the department recognizes the instructor's critical value to the fire service.

As grand and interesting as being an instructor sounds, it also carries a great deal of responsibility. A department is often judged by the performance of its fire fighters. When a fire fighter is injured or killed in the line of duty, some of the first items reviewed and scrutinized are the training records and the qualifications of the instructors who taught that fire fighter during recruit school and during his or her time in the department. Keeping up-to-date on all changes and improvements in the fire service is a difficult but critical task. How do you stay informed, do your job effectively, and still maintain your personal life? When teaching a class, title an easel board "Resources." Over the course of the class, add resource information to this list and encourage students to do the same. Typically, this information will include website addresses, books, and names of other instructors.

Staying Current

Subscribe to trade magazines to stay up-to-date on trends and new developments in the fire service. Read other professional magazines in related fields, such as homeland security or business, as well. The benefit from reading trade journals is that this practice helps you stay current in the fire service. The benefit from reading material outside the fire service profession is that it helps you develop into a well-rounded individual and professional.

Another tool for you as an instructor is the internet. The web has opened the door to resources that were difficult to obtain in earlier years, but now are readily available to anyone who has a computer and internet access. The numerous websites hosted by colleges, professional journals, professional organizations, the National Fire Academy (NFA) and state fire academies, and government agencies offer volumes of information at minimal or no cost. Another online resource is the NFPA, which provides a variety of reports, online training, conferences, and workshops. This organization is also the source for the Fire Service Professional Qualifications Standards. Some material on the NFPA website is free, but other access requires a subscription.

One drawback to information found on the internet is the sometimes questionable accuracy of the content presented. If you use material from the internet, you need to verify the authenticity of the source and the accuracy of the information presented. More than one instructor has taken information from the internet and used it in a lesson plan, only to find that it was misleading or completely wrong. This, in turn, hurts the reputation of the instructor and may put students in harm's way. Another challenge when researching content on the internet is that a great deal of information is available at the click of a mouse, and you can easily spend too much time sifting through all you find. The internet is a tool and, just like with any other tool, you need to know how to use it properly.

Higher Education

Another method you can use for personal and professional development is to pursue a degree from an institution of higher education. Most departments require fire fighters to have a high school diploma or a GED as a condition of employment. In addition, many community or state colleges now offer degrees in fire science, emergency management, homeland security, or communication. Although this coursework is within your discipline, colleges also require you to take general education courses such as history and English literature as part of the degree program so that you will become a well-rounded thinker.

Higher education has become more accessible today than ever before, with degrees being offered by traditional colleges as well as online programs offered on the internet. Both of these delivery methods (traditional or online) provide the opportunity to earn an advanced degree and leave you the choice of which methodology you wish to pursue. One big advantage of pursuing an online degree is the convenience and flexibility it offers the student. You can "attend" class on your own schedule, while at the station or at home. Some higher education programs also offer a blended learning format, which combines the use of the internet for much of the classroom work but requires students to meet face-to-face with the instructor during certain points of the course. Check the institute's accreditation before obtaining an online degree. Some programs are not accredited, so your degree will not be recognized by other schools or your employer. Online courses and blended courses are discussed further in Chapter 6, *Technology in Training*.

An increasing number of departments have begun requiring an advanced degree for promotion, such as an associate's degree for the captain's position and a bachelor's degree for the battalion chief's position. Other departments offer financial incentives for engaging in continuing education, paying for college or even giving fire fighters a raise for each college-level course they complete. Attending college can be a time-consuming and expensive proposition. You may be willing to make the time commitment, but your finances may not be flexible enough to handle the entire cost. If so, take some time to investigate the many possible funding sources for which you may qualify. For example, your fire department or city may have a tuition reimbursement program. The federal government provides grants that are awarded to thousands of students every year. If you served in the armed forces, you may qualify for funding as a veteran. Many local charitable organizations also provide scholarships to students, including nontraditional students. Do not automatically assume that you cannot attend college because of money concerns. If you are willing to make the commitment and expend the effort, funding sources are available to support your continuing education.

A major incentive for obtaining a bachelor's degree is the requirement in place at the National Fire Academy that requires an advanced degree to be eligible to enter the Executive Fire Officer Program (EFOP). Beginning October 1, 2009, a bachelor's degree was required for entrance into the EFOP. The EFOP is considered by many to be a key to promotion as a chief officer and is a desirable qualification to add to your toolbox.

Conferences

Attendance at conferences, workshops, or seminars also offers you a chance to increase your skills and knowledge as an instructor (**FIGURE 1-12**). The many professional conferences put on every year provide opportunities for growth and development at all levels. Some conferences provide hands-on training opportunities, whereas others are limited to classroom

FIGURE 1-12 Attending conferences is an excellent self-improvement activity.
Courtesy of Brian Rooney.

lectures or presentations. One of the largest annual conferences that all instructors should attend if given the opportunity is the Fire Department Instructors Conference (FDIC), which is held each spring in Indianapolis, Indiana. FDIC offers a wide range of courses for everyone in the fire service, from the chief officer to the newest fire fighter. Many courses are directed at the instructor to improve presentation skills and research techniques, and provide new learning tools for instruction and a variety of other skills.

Other instructor conferences are held at the local and regional levels and provide opportunities to learn new skills, enhance your existing teaching skills, and learn from others. Regardless of which type of conference you choose to attend, take advantage of each and every opportunity to learn from your peers. Be willing to look outside the fire service and look for instructional learning opportunities in clubs, community organizations, and community college courses.

Take advantage of the opportunities at these conferences and workshops to network with other instructors from both within and outside the fire service community. Such conferences are great ways to learn new presentation skills that you can add to your own toolbox. **Networking** is a fundamental strategy used by great instructors to learn from others and to bounce an idea off a peer to determine whether it is valid. Networking is a powerful addition to your toolbox that can open the door to sharing or collaborating on an idea and developing a new methodology or course.

A major factor affecting today's fire service is the limited budget that every type of department faces. When an instructor asks the chief permission to attend FDIC or to pursue an advanced degree, the chief is thinking, "How does the department benefit from this investment?" The real benefit comes from instructors who return from the conferences or workshops and improve the way they teach or share new information with their fellow fire fighters and instructors. The relationship between the chief and the instructor is one of respect and trust, and one of investment and return on investment.

Although networking is often associated with politics and "schmoozing," what you are really doing when you network is creating a web of friends. The bigger the web, the more likely you are to have access to the precise resource you need when the time comes. Networking is designed as a give-and-take system.

The best way to create a large web is to be a giver. When someone needs something that you can provide, offer it. If someone has a question, answer it. Helping others be successful creates a powerful network that can work to your benefit. By giving, you become the person who is sought out: The more sought out you are, the more contacts you will have. Develop your network not by schmoozing but by being genuinely interested in helping others succeed.

Lifelong Learning

Lifelong learning is an endless path marked by endless opportunities. For the one person who says he or she has seen it all, there are a hundred other individuals who ask for more. A major career crusher is adopting the attitude that you cannot teach students anything new. Your greatest challenge is to create the desire to learn in all your students.

In your pursuit of becoming a lifelong learner, you need to remember the basics. Your teaching skills got you where you are, so continually brush up on your existing presentation skills and learn new ones. Remember the teacher in high school whom no one liked because he taught as if he was still locked in the past? Remember the high school teacher who stayed current and fresh to students? Those teachers who worked at truly "teaching" a subject and encouraging students to learn were the same teachers whose classes were the first to fill during registration. Learning can be enjoyable. Instructors who are willing to learn new methodologies to present information become valuable assets to the organization because of their willingness to learn and develop themselves into strong instructors.

Keeping your teaching skills sharp and current is just as important as reading magazines, networking, or going to college. If you are the senior instructor in the training division, make a point of teaching one subject once a month so that you keep your teaching skills up-to-date. As your knowledge base about the fire service has grown, have you stayed current with the types of students entering the fire service today? Are you still an effective instructor for the new generation of fire fighters? Test yourself by teaching a class. Be honest with yourself about how effective you were. Great instructors are lifelong teachers who love to teach.

An important skill to add to your toolbox is the acknowledgment and acceptance of the changes we are seeing in the fire service. The new generation of fire fighters is more diverse and has a strong understanding of technology. Effective instructors should recognize that today's students are different and learn how to teach them by using methods that will allow these students to learn the important skills that will save their lives and the lives of others. Perhaps instructional methodology needs to change to assist the new and diverse learners, but what we teach has not changed.

Firefighting is a dangerous profession, and fire will kill either a young or old fire fighter without regard to how they learn. Our job as instructors is to learn whatever methods will enable us to teach life-saving skills to those students whom we are responsible for teaching and training. Part of being an effective instructor is learning to scan the horizon for industry trends and to anticipate which will pan out. Part of this scanning process involves looking for new ways to enhance your skills and abilities. Attending a 1-day skills development program and reading a new book are two ways to stay on the cutting edge of your profession.

Professional Organizations

There are many professional organizations that you can join to help in your professional development. Some organizations operate on the national level, but most are found at the state, local, regional, provincial, or county level. (Among these are state training organizations.) Regardless of the level, these organizations provide many opportunities to network, share resources, and develop contacts. Some are very formal and highly structured, whereas others are more informal and relaxed.

The International Society of Fire Service Instructors (ISFSI) is an organization that is committed to the development of instructors. ISFSI actively supports the professional development of the fire service through its push for qualified instructors teaching quality material The ISFSI has partnered with UL Firefighter Safety Research Institute (FSRI) and other agencies to develop model curriculums and instructional resources. ISFSI participates in state, local, and national instructor training events, including the FDIC, along with hosting its own national conference. As a member-driven organization, you can participate in many ways by becoming a member. Visit the ISFSI website for information.

The National Fire Academy, which is part of the National Emergency Training Center in Emmitsburg, Maryland, is the nation's "fire university." At the National Fire Academy, you can attend 6-day or 2-week courses that teach you how to develop lesson plans and build courses. In addition, other courses are available to help in all aspects of your professional and educational development. The National Fire Academy also supports two other programs that will help in your professional development: the FESHE initiative and the EFOP.

A key part of what the National Fire Academy is doing to professionalize the fire service can be seen in the FESHE initiative. Across the United States, many different community and state colleges and universities offer degrees in different aspects of fire service management or engineering, but there are no standardized model degree programs. The FESHE initiative is intended to bring together representatives from colleges and universities to develop a standardized degree format and model curricula. If college degree programs follow the same set of rules, students will be able to transfer coursework from one school to another more readily. In addition, courses would present a standard curriculum, which increases the ability to transfer college credits and standardizes the information taught. Another purpose of the FESHE initiative is to develop a career path for those seeking to be chief officers.

Part of this development process has resulted in the National Professional Development Model. This model summarizes the combination of education, training, certification, and experience that leads to the development of a well-rounded and fully developed fire officer who has the skills and ability to lead the fire service into the future.

The EFOP is another program developed by the National Fire Academy that has helped to professionalize the fire service. Its purpose is to provide senior officers with the tools they need to become leaders of the future. Acceptance into the EFOP is by application process only; that criteria can be found on the Federal Emergency Management Agency (FEMA) website. One of the key criteria for application in the EFOP is the individual's "potential for future impact on the fire service." Of the many positions in the fire service, instructors have one of the most crucial roles in influencing the future of their organizations and the fire service in their communities.

The EFOP requires a major time commitment from both the fire officer and the fire department. The program takes 4 years to complete, with attendance at the National Fire Academy for four 2-week courses. At the end of each course, the student is required to submit an applied research project (ARP) within 6 months of the course's ending date. This ARP is evaluated by peer reviewers against a set of criteria established by the National Fire Academy staff. Once the ARP is approved, the student can register for the next course in the series.

Each spring, the National Fire Academy hosts the annual EFOP Graduate Symposium on its campus in Emmitsburg, Maryland. Each year the symposium features a topic of interest to the fire service, and EFOP graduates give presentations on the symposium topic. This symposium serves as an opportunity for EFOP graduates to renew friendships, network, and "charge their batteries" as leaders in their organizations.

The Next Generation

A true mark of a great leader is his or her willingness to plan to be replaced. Succession planning is the process of identifying others within the department who could be mentored and coached to replace you and to carry out the functions you perform.

The concept of training your replacement can be very difficult and challenging to handle on an emotional level. It may be easier if you understand that the process of training another person to take your place usually means that you are also moving up and onward to something new. Another benefit of training your replacement is that you can select someone who shares your own ideals and goals, thereby ensuring that your training program will continue after you leave. Through this legacy, your influence could be felt for many decades in your department (**FIGURE 1-13**).

How do you train your replacement and help him or her to be successful? Four basic concepts will aid you in teaching your replacement:

1. Identifying
2. Mentoring
3. Coaching
4. Sharing

Identifying

Identifying is the process of selecting those whom you would like to mentor, coach, and develop. The identification process almost requires a "crystal ball" approach of looking at individuals—that is, you need to see the person today and forecast who the person could be tomorrow. Look for these qualities:

- The quick learner. Look for those who truly understand how, why, and when (and when not).
- Individuals who expect 110 percent and inspire others by leading by example.
- Individuals who are competent and knowledgeable as fire fighters. This type of fire fighter stands out and is respected by his or her peers.
- Individuals who possess organizational and preparation skills.
- Individuals who understand and exercise time management skills.
- Fire fighters with a passion for the fire service. Passion is contagious and can push a crew or a training division to move to the next level.

There are many important questions to ask the potential instructor. Oral interview questions may include the following:

1. Assume that you are hired to be the training officer and have been approached by members of a fire company who complain that their officer treats training as an unimportant issue and spends as little time as possible on drills. The members respect the officer as a good fire-ground commander, but claim that he has no appreciation for training and does little to support the training program. Which action would you take in this situation?
2. Why are you interested in becoming the training officer? What have you done to prepare for this position?
3. Which personal attributes that you possess will be of the greatest benefit to you in fulfilling the duties of this position?
4. Do you have any plans for personal development that will benefit you in this position?
5. Throughout the course of the year, members of the department may need motivation, direction, and guidance regarding professional training and education needs. Do you see this as part of your role as training officer? Why or why not? If so, briefly describe your thoughts on how you can best provide this need.
6. What is your top priority or priorities that would have to be accomplished in the first 3 months?
7. Describe specifically how you will gain the support, backing, and assistance of those company officers who do not consider training to be a high priority.
8. One of the responsibilities of this position will be to function as the on-scene safety officer at

FIGURE 1-13 Succession planning ensures that the goals and objectives for the department or organization continue.

incidents. Have you had prior experience in this type of position? Also, explain your philosophy and approach to the position of scene safety officer.
9. How would you structure, document, and provide follow-up to ensure that all personnel complete required training?
10. How would you structure training sessions to accomplish training and maintain district coverage?
11. Assuming this department is a combination department of full-time and part-time personnel, which additional challenges do you anticipate in developing and conducting training for this type of structure?
12. What is your philosophy regarding the role of the company officer in training?

Written/practical evaluation questions may include the following items:

1. Prepare a written plan (not to exceed three pages—single-sided) describing how you will structure, conduct, and document the self-contained breathing apparatus (SCBA) training program.
2. Full-time personnel hired by the department are required to have prior certification as a Fire Fighter II. These individuals, once hired, are placed directly on shift. A current training issue is to develop a 40-hour program to provide an orientation to the department and complete a basic assessment of skill levels. Prepare a written plan (not to exceed three pages—single-sided) of how you would structure, conduct, and document a new fire fighter orientation/assessment program.
3. An important aspect of department training programs will be the addition of certification courses that are in compliance with the state's certification or training requirements. Prepare a written plan (not to exceed three pages—single-sided) detailing the steps that will be needed to implement an in-house fire apparatus engineer program.

Mentoring

Mentoring is a tool you can use to nurture and develop a fire fighter who has the basic skills and abilities to become a great instructor. As a mentor, you select an individual to begin the process as either a formal or an informal protégé. Your role as a mentor is to demonstrate how to perform functions or skills and to explain and answer questions. Most importantly, you act as a sounding board if your protégé has a problem.

The following ideas form the foundation for a solid mentoring program:

- The mentor's job is to promote intentional learning and to provide opportunities for the protégé to learn and grow. This does not mean that you throw your protégé to the wolves and walk away; rather, it means that you give your protégé opportunities to teach or demonstrate a skill while you watch.
- As a mentor, you need to share your experiences, both good and bad. Learning comes from both positive and negative experiences.
- The mentoring process takes time, and patience needs to be shown on the part of both the mentor and the protégé.
- Mentoring is a joint venture and requires both parties to be actively involved in the process.

Sometimes the identification and selection process does not work as anticipated. For example, you might select a person whom you feel is ready for mentoring, but he or she does not respond to the process or fails to perform as desired due to lack of skill or ability. In this scenario, there is still an opportunity to learn and grow from the experience by understanding why the process failed and then learning from it. This allows everyone to part on good terms and leaves the door open to future contact.

Coaching

Coaching is a process usually associated with sports. The coach is a person who has experience and assists players in developing and improving their skills. Coaching is most often tied to manipulative skills development and physical training. The difference between mentoring and coaching is subtle. Coaching is the process of helping an individual see his or her abilities and come up with solutions independently. A coach encourages through words but allows the player to develop his or her own skills. A good coach helps a player see his or her potential and then gives the player the freedom and support to achieve it.

As an instructor, you must have expertise in numerous skills, such as firefighting skills for all levels you could be expected to teach. In addition to your technical skills, you have another set of skills at your disposal—namely, the skills used in the presentation of lessons, lesson and curriculum development, and use and development of evaluation tools. Part of the coaching process consists of allowing the fire fighter you are mentoring the opportunity to develop these skills as well.

Coaching also requires that you evaluate those persons whom you are grooming to replace you. These evaluations may take the form of either formal or informal evaluations or critiques. Just as when teaching a child to ride a bike, you must help your protégé up to a point, and then allow him or her to go on alone. You might run alongside the bike for a while, but as the child gets the hang of it, you back off. The same process applies when coaching a new instructor: You teach your protégé how to do a presentation during training, and then the fire fighter gets to do it for real in a recruit class while you stand in the back of the room and watch. Afterward, you give some advice, highlighting good points and offering suggestions for rough areas. This feedback helps your protégé grow and improve. Coaching involves taking a "big brother" or "big sister" approach toward helping a person use the skills that he or she has and then coaching/teaching the person to do even better.

Sharing

The last skill in succession planning focuses on the basic concept of **sharing**. Sharing is a key part of mentoring and coaching. To be successful in succession planning, you must mentor and coach with a giving attitude. Not only do you need to take your protégé under your wing, you also need to do so with a sharing attitude. You must be willing to share everything that has made you such a successful instructor. This arsenal of experiences includes your skills and knowledge, as well as all the little nuances that have made your job easier and fun. All too often, those persons charged with the task of training their replacements hold back the little tricks of the trade. Some have the attitude, "I had to learn it myself; so can you!" This attitude hurts both the department and your reputation. Share your knowledge and insight with your protégé as he or she takes on the responsibilities of training others. Give your protégé the extra step up rather than making him or her repeat all of your mistakes and waste all of that extra time and effort. When you share what you know with others and help others excel, you benefit both personally and professionally.

When you approach succession planning with this positive attitude and mentor the next generation of instructors by teaching and coaching with a caring and sharing attitude, both you and your department move forward. Leave a legacy of professionalism that will outlive your time on the job.

Summary of Instructor I Duties

At the Instructor I level, the instructor is beginning to develop the basic skills to present a lesson plan to a room of students or a crew of fire fighters sitting around a break room table. Even at this basic level, the instructor must have the basic skills to be effective in delivering the assigned material. The instructor at this level exercises leadership in presenting the material in a way that instills confidence in students and helps build the team. In the process of presenting lesson material, the instructor uses mentoring and coaching skills that will help students obtain new knowledge and build the necessary skills to perform their jobs as part of the organization. As part of the teaching role, the Instructor I evaluates student performance to ensure competency in learning and demonstrating the new knowledge and skill sets.

A critical aspect of the Instructor I role is accurate and thorough recordkeeping, along with reporting and recording test scores and other evaluative information from the course. This type of documentation might not seem important at the time, but future events might require documentation of course material presented, students' scores and evaluation sheets, and the instructor who presented the material.

Training BULLETIN

JONES & BARTLETT FIRE DISTRICT
TRAINING DIVISION
5 Wall Street, Burlington, MA, 01803
Phone 978-443-5000   Fax 978-443-8000

Instant Applications: Today's Emergency Services Instructor

Drill Assignment

Apply the chapter content to your department's operation, training division, and your personal experiences to complete the following questions and activities.

Objective

Upon completion of the instant applications, fire and emergency services instructor students will exhibit decision making and application of job performance requirements of the instructor using the text, class discussion, and their own personal experiences.

Suggested Drill Applications

1. Identify the changes in training you have seen or experienced since you began your career in the fire service.
2. Check industry websites and reference materials to identify at least three hot topics facing instructors today.
3. Identify a new piece of equipment purchased by your department. Discuss how the initial training was accomplished on that equipment before it was placed in service. How will your department ensure that the members remain proficient in the equipment's operation after the initial training?
4. Ask the newest or youngest member of your organization about his or her training. Identify similarities and differences between current fire service training and the traditional education of a fire fighter.
5. Review the Incident Report in this chapter and be prepared to discuss your analysis of the incident from a training perspective and as an instructor who wishes to use the report as a training tool.

Incident Report: NIOSH Report F2000-27

Drill Assignment

Review the information in this incident report and prepare a practice presentation for class delivery at the direction of your instructor. Your presentation should include a summary of the incident facts and a review of the NFPA 1403 compliance and noncompliance findings. Use an outline to organize your thoughts. You may be evaluated on your communication skills during your presentation.

Greenwood, Delaware—2000

A volunteer fire department was conducting a live fire training exercise in a 100+-year-old house (**FIGURE 1-A**). Upon completing interior operations, personnel were preparing for the "final burn down" of the 2½-story house.

Three officers were in the attic. One officer was in full protective clothing and SCBA. He was using a sprayer can of diesel fuel to spray fuel. Two other officers, without SCBA, lit multiple fires. The fire spread rapidly and as conditions worsened, the two ignition officers exited to the top of the staircase and told the remaining officer to follow. He agreed to follow, but heat and fire conditions in the attic became untenable very rapidly. The others escaped down the stairs, but fire blocked their colleague's escape.

Multiple rescue attempts were unsuccessful, and the roof quickly collapsed into the attic.

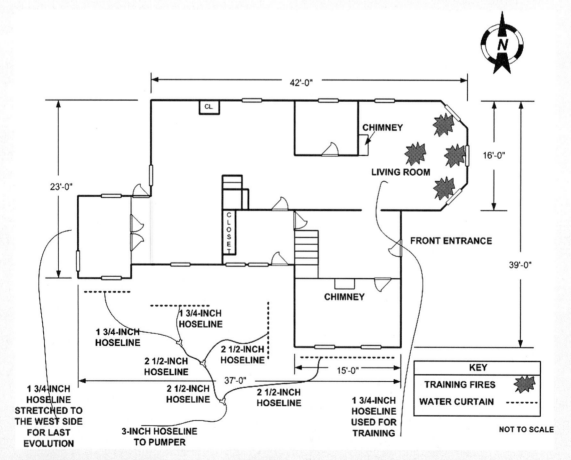

FIGURE 1-A Floor plan of Greenwood, Delaware.
From the NIOSH Fire Fighter Fatality Investigation and Prevention Program, Investigation Report F2000-27.

Postincident Analysis Greenwood, Delaware

NFPA 1403 Noncompliant

- Interior personnel without hose line
- Flashover and fire spread unexpected
- Flammable/combustible liquids used in large quantities
- Multiple fire sets
- Interior stairs only exit other than windows without ladders in place
- Participants not wearing PPE or SCBA

After-Action REVIEW

IN SUMMARY

- NFPA 1041 categorizes instructor duties into three classifications:
  - The Fire and Emergency Services Instructor I is able to deliver presentations from a prepared lesson plan, adapt lesson plans to meet the needs of a jurisdiction, organize the learning environment, and provide appropriate recordkeeping.
  - The Fire and Emergency Services Instructor II meets the requirements of Instructor I and is able to develop lesson plans, schedule training sessions, and supervise other instructors.
  - The Fire and Emergency Services Instructor III meets the requirements of Instructor II and is able to develop training curricula for single or multiple organizations, conduct needs analysis, develop training goals and implement strategies, and develop a comprehensive program and instructor evaluation plan.
- Instructors can have a wide range of ranks, ranging from fire fighter to chief of department.
- Instructors will act in a variety of roles, including leader, mentor, coach, evaluator, and teacher.
- Both physical considerations and emotional issues influence the learning environment.
- Instructors have a responsibility to stay current on topics and aid in succession planning.
- Instructors regularly face ethical dilemmas that require them to make value-based judgments.
- Learning is a lifelong process. An important part of becoming a quality instructor is having the basic desire to always improve and to become a better fire service professional.
- Professional development can take the form of training, certifications, and formal education.
- Professional development can be achieved by participating in local, state, and national organizations and by attending training conferences.
- Networking is a powerful tool that can open the door to sharing or collaborating on an idea and developing a new methodology or course.
- It is important to keep your teaching skills sharp and current. As you grow in your knowledge, stay current with the types of students entering the fire service. Test yourself by teaching a class and be honest with yourself about how effective you were.
- Succession planning is the process of developing individuals who can be mentored and coached to take the lead in the future of the fire department.
- One benefit of training your replacement is that you can select someone who shares your own ideals and goals, thereby ensuring that your training program will continue after you leave. Through this legacy, your influence could be felt for many decades in the department.

KEY TERMS

Access Navigate for flashcards to test your key term knowledge.

Certificate Document given for the completion of a training course or event.

Certification Document awarded for the successful completion of a testing process based on a standard.

Coaching The process of helping individuals develop skills and talents.

Confidentiality The requirement that, with very limited exceptions, employers must keep medical and other personal information about employees and applicants private.

Continuing education Education or training obtained to maintain skills, proficiency, or certification in a specific position.

Degree Document awarded by an institution for the completion of required coursework.

Delegation Transfer of authority and responsibility to another person for the purpose of teaching new job skills or as a means of time management.

Direct threat A situation in which an individual's disability presents a serious risk to his or her own safety or the safety of his or her co-workers.

Disability A physical or mental condition that interferes with a major life activity.

Ethics Principles used to define behavior that is not specifically governed by rules of law but rather in many cases by public perceptions of right and wrong. They are often defined on a regional or local level within the community.

Identifying The process of selecting those persons whom the instructor would like to mentor, coach, and develop.

Learning environment A combination of a physical location (classroom or training ground) and the proper emotional elements of both an instructor and a student.

Major life activity Basic functions of an individual's daily life, including, but not limited to, caring for oneself, performing manual tasks, breathing, walking, learning, seeing, working, and hearing.

Mentoring A relationship of trust between an experienced person and a person with less experience for the purpose of growth and career development.

Networking An activity or process of like-minded individuals meeting and developing relationships for professional and personal growth through sharing of ideas and beliefs.

Organizational chart A graphic display of the fire department's chain of command and operational functions.

Reasonable accommodation An employer's attempt to make its facilities, programs, policies, and other aspects of the work environment more accessible and usable for a person with a disability.

Sharing The basic concept of giving to others with nothing expected in return.

Standards A set of guidelines outlining behaviors or qualifications of positions or specifications for equipment or processes. Often developed by individuals within the regulated profession, they may be applied voluntarily or referenced within a rule or law.

Succession planning The act of ensuring the continuity of the organization by preparing its future leaders.

REFERENCES

Austin, Nancy, and Tom Peters. 1989. *A Passion for Excellence: The Leadership Difference.* New York: Grand Central Publishing.

National Fire Protection Association. 2019. *NFPA 1041: Standard for Fire and Emergency Services Instructor Professional Qualifications.* Quincy, MA: National Fire Protection Association.

National Fire Protection Association. 2018. *NFPA 1403: Standard on Live Fire Training Evolutions.* Quincy, MA: National Fire Protection Association.

CHAPTER PRESENTATION EXERCISE

Instructions: Your course instructor will assign individual or group discussions on the key points and teaching tips of this chapter. You or your group will review the chapter teachings and identify the major learning points from the chapter. You should be able to discuss the points, why they are important, and how/where they apply to your responsibilities at your level of instructor training.

REVIEW QUESTIONS

1. Describe some of the main distinctions among the three levels of Fire and Emergency Services Instructors.
2. What are several important factors that create a good learning environment?
3. List some of the initiatives created by the National Fire Academy and explain how they support instructors.

CHAPTER 2

The Learning Process

NFPA 1041 JOB PERFORMANCE REQUIREMENTS

Note: An asterisk denotes that the 1041 standard contains further information in its annex section.

4.3.2 Review instructional materials, given the materials for a specific topic, target audience, learner characteristics, and learning environment, so that elements of the lesson plan, learning environment, and resources that need adaptation are identified.

(A) Requisite Knowledge.

Recognition of student learner characteristics and diversity, methods of instruction, types of resource materials, organization of the learning environment, and policies and procedures.

(B) Requisite Skills.

Analysis of resources, facilities, and materials.

4.4.3 Present and adjust prepared lessons, given a prepared lesson plan that specifies the presentation method(s), so that the method(s) indicated in the plan are used and the stated objectives or learning outcomes are achieved, applicable safety standards and practices are followed, and risks are addressed.

(A)* Requisite Knowledge.

The laws and principles of learning, methods and techniques of instruction, lesson plan components and elements of the communication process, and lesson plan terminology and definitions; learner characteristics; student-centered learning principles; instructional technology tools; the impact of cultural differences on instructional delivery; safety rules, regulations, and practices; identification of training hazards; elements and limitations of distance learning; distance learning delivery methods; and the instructor's role in distance learning.

(B) Requisite Skills.

Oral communication techniques, methods and techniques of instruction, ability to adapt to changing circumstances, and utilization of lesson plans in an instructional setting.

4.4.4* Adjust to differences in learner characteristics, abilities, cultures, and behaviors, given the instructional environment, so that lesson objectives are accomplished, disruptive behavior is addressed, and a safe and positive learning environment is maintained.

(A)* Requisite Knowledge.

Motivation techniques, learner characteristics, types of learning disabilities and methods for dealing with them, and methods of dealing with disruptive and unsafe behavior.

(B) Requisite Skills.

Basic coaching and motivational techniques, correction of disruptive behaviors, and adaptation of lesson plans or materials to specific instructional situations.

KNOWLEDGE OBJECTIVES

After studying this chapter, participating in a structured learning environment, and completing assigned assessments, you will be able to:

- Define learner characteristics and explain their significance for instructors. (**NFPA 1041: 4.3.2**, p 39)
- Identity generational groups and explain general characteristics of each group. (**NFPA 1041: 4.4.3**, pp 41–44)
- Describe the six basic principles or "laws" of learning. (**NFPA 1041: 4.4.3**, p 45)

- Identify methods individuals can use for classroom learning, personal study, and test preparation. (pp 47–48)
- Relate human needs to the instructional environment and process. (pp 48–50)
- Identify the three domains of learning. (pp 39; 50–53)
- Define student-centered learning. (p 53)
- Describe common categories of learning disabilities and methods to enhance and adapt instructional techniques to meet the needs of individuals with learning disabilities. (**NFPA 1041: 4.4.4**, pp 62–64)

SKILLS OBJECTIVES

After studying this chapter, participating in a structured learning environment, and completing assigned assessments, you will be able to:

- Perform an audience analysis to determine learner characteristics. (**NFPA 1041: 4.3.2**, pp 41–42)

You Are the Fire and Emergency Services Instructor

Recently, several members of your shift returned from a joint, countywide training program that involved instructors from outside your area, as well as members of neighboring departments who serve as their own department training officers. When asked how the training went, several of those who attended immediately started to remark on how the instructors spent too much time in the classroom telling "war stories" instead of letting those in attendance "get their hands on the equipment on the training grounds." They further criticized the instructors for not understanding that fire fighters are "hands-on" people, contending that the training session was destined to fail because there wasn't enough practical, hands-on involvement. In an effort to learn from this lesson, do the following:

1. Identify the three types of learning domains, and provide examples of different fire service topics that would be appropriate for each.

2. Define the different learner characteristics, and discuss the effects they would have on the success or failure of a specific training session.

3. Compare and contrast the similarities and differences among adult learners.

 Access Navigate for more practice activities.

Introduction

Learning is a change in a person's ability to behave in certain ways. This change can be traced to two key factors—experience with the subject (e.g., in the field) and practice (e.g., in the classroom). Learning can occur both formally (inside the classroom) and informally (around the dinner table) (Connick, 1997). Formal learning does not occur by accident—it is the direct result of a program designed by an instructor (Butler and McManus, 1998). An adult learner may intentionally set out to learn by taking classes or by reading about a subject. He or she may also gather information through the experience of living that changes the learner's behavior. Informal learning occurs spontaneously and continually changes the adult learner's behavior. Ideally, learning is created through the blending of individual curiosity, reflection, and adaptation (Stewart, 2003).

FIGURE 2-1 As a fire and emergency services instructor, you have the opportunity to positively influence fire fighters at all levels of the fire department.
© Jones & Bartlett Learning. Photographed by Glen E. Ellman.

Learning takes time and patience. As a fire and emergency services instructor, you have the opportunity to influence fire fighters positively at all levels of the fire department (**FIGURE 2-1**). As a leader in the fire department, you have a responsibility to teach information, to hone skills, and to promote positive values and **motivation** in your students. A wise fire service leader once said, "Classroom learning plus street experience equals wisdom." As an instructor, you have a major impact on the first part of this equation.

All levels of instructors will learn that each student in your class has different ways that they understand information. As you participate in this course, you are experiencing this phenomenon by reading text and reviewing charts, graphs, photographs, and different text fonts, sizes, and colors. Each of these elements allows different learner characteristics to be experienced. Other characteristics are demonstrated as your instructor presents course materials.

The Interactive Process of Learning

Learning is an interactive process in which the adult learner encounters and reacts to a specific learning environment. That learning environment encompasses all aspects of the adult learner's surroundings—sights, sounds, temperature, other students, and so on. In addition, each adult learner has personal, academic, social, or cognitive attributes that may influence how they learn—that is, **learner characteristics**. Some adults learn by listening to and sharing ideas with others, others by thinking through ideas themselves, and still others by testing theories or synthesizing content and context.

Learning takes place in three domains: cognitive (knowledge), psychomotor (physical use of knowledge), and affective (attitudes, emotions, or values). These domains are discussed later in this chapter.

Effective learning follows a natural process:

1. Adult learners begin with what they already know, feel, or need, based on the groundwork that was laid prior to what is currently taking place. In other words, learning does not take place in a vacuum.
2. Seeing that "newly" learned information has real-life connections prepares the adult learner for the next step—learning something new.
3. Adult learners use the new content, practicing how it can be applied to real life.
4. Adult learners take the material learned in the classroom and apply it to the real world, including in situations that were not always covered in the original lesson.

What Is Adult Learning?

Fire service training and educational programs are designed with the adult learner in mind. An adult learner is defined as someone who has passed adolescence and is out of secondary school. Self-motivation is rarely a problem for these individuals, who typically see the courses as a way both to advance their careers and to better themselves as people (**FIGURE 2-2**).

According to Bender (1999), adult learning is the integration of new information into existing values, beliefs, and behaviors. New information is learned and synthesized so that students will be able to consider alternative answers to never-before-asked questions,

FIGURE 2-2 Fire service training and educational programs are designed with the adult learner in mind.
© Jones & Bartlett Learning. Photographed by Glen E. Ellman.

new applications of knowledge, and creative ways of thinking. Adult students need realistic, practical, and knowledgeable lessons that deliver information in a variety of ways to accommodate their differing learner characteristics. In particular, adult learners may have difficulty imagining a piece of equipment, responding to a written scenario, or performing a skill if they do not have a realistic application for the learning. The more realistic the lesson, the more likely the adult learner will generalize the information and be able to apply it in the field.

Similarities and Differences Among Adult Learners

There are two general lines of thought in the realm of the learning process: pedagogy and andragogy. **Pedagogy** has been defined as the art and science of teaching children. In the pedagogical model, the teacher has full responsibility for making decisions about what will be learned, how it will be learned, when it will be learned, and whether the material has been learned. Pedagogy—or teacher-directed instruction, as it is commonly known—places the student in a submissive role requiring obedience to the teacher's instructions (Knowles, 1984). This process may occur in young adults who are members of an entry-level fire academy or young volunteer fire fighters who have not entered the workforce or higher learning environments. **Andragogy** (adult learning), in contrast, attempts to identify the way in which adults learn and help them in the process of learning. According to this theory's originator—professor Malcolm Knowles—although adult learners differ, some commonalities can be found (Knowles, 1990). For example, adults typically need to know why they need to know something and how they can use it. Adult learners also tend to focus on life-centered (practical) tasks and activities. They like to learn at their own pace and use their preferred learning style. Successful learning also takes advantage of adult learners' rich history of life experiences.

Not all adult learners are alike. Two early theorists in behavioral science, Erikson and Levinson, focused on age to help understand how adult learners differ. According to these researchers, young adults seem to be primarily focused on issues surrounding the development of a sense of identity, establishing a career, pursuing a life dream, having an intimate relationship, getting married, and achieving parenthood. By comparison, middle-aged adults focus more on caregiving for children and perhaps older family members, career changes, physical and mental changes, role changes, and planning for retirement. Older adults, by contrast, often face a series of upheavals—loss of their jobs (via retirement); the transition from parent to grandparent; age-related physical and mental changes; and the deaths of friends, spouses, and eventually themselves. Stage of life, in turn, directly affects the adult learner in the classroom. Many training sessions in the fire station will involve experienced (older) and less experienced (younger) fire fighters sitting side-by-side or working together on the same hose line. Youthful enthusiasm can be tempered by wisdom and naturalization of skills in this setting; when the two are combined, an effective firefighting team will emerge. A young fire fighter in an experienced company may be a motivation for the older fire fighters to engage at a faster and more educated pace than before the young member was introduced.

Houle (1961) identified three types of adult learners:

1. For learner-oriented adults, the content is less important than the act of being involved in learning.
2. Goal-oriented learners acquire knowledge with the intention of improving their job prospects or learning a new skill. Members of this group view the instructor as being responsible for disseminating knowledge that they need.
3. Activity-oriented learners learn by doing. To succeed in learning, they require personal productive time, with control over content and learning style, with the goal of improving social contact rather than the acquisition of knowledge.

With all of these orientations, the adult learners set goals, identify objectives, select relevant resources, and use the instructor as a facilitator for learning.

Influences on Adult Learners

Adult learning theories are based on the premise that adults have had different experiences than children and adolescents, and that these differences are relevant to learning. Although age is certainly a factor that can affect learning, other factors—such as motivation, prior knowledge, the learning context, and influences exerted by situational and social conditions—are influential as well. For example, in the classroom an adult learner might not pay close attention to the topic because a previous experience left the impression that the subject is not important. This view is merely reinforced for the adult learner when others in the class share the same feelings about the subject.

Adult learners are more apt to learn and retain information when it is meaningful to their lives. For instance, information directly related to a new piece of equipment will capture the interest of fire fighters working on that apparatus. In contrast, material that is outside the realm of the adult learner's experience and knowledge is viewed as less meaningful. The prior knowledge and experiences of the adult learner are often more diverse than those of a child or adolescent, yet you must always be aware that some adult learners may lack the expected prerequisite knowledge needed to learn new information.

What influences adult learning? Here is one answer to that question (Merriam & Caffarela, 1999, 107):

> Adults often engage in learning as one way to cope with the life events they encounter, whether that learning is related to or just precipitated by a life event. Learning within these times of transition is most often linked to work and family.

Fire fighters work in a very intense environment. What is learned during times of intense stress inevitably influences learning in the classroom. If a fire fighter was just on an intense call, he or she may have trouble concentrating or sitting still in the classroom. Coping mechanisms (e.g., deflection of emotions, exaggeration of dark humor, being talkative or withdrawn) may affect behavior in the classroom. Always be on alert for signs of critical incident stress. Fire fighters need to discuss incident events in a proper environment.

Other influences on adult learners include social and cultural perspectives, such as social roles, race, and gender. Some adult learners, for example, may feel uncomfortable with classmates who are more educated or who hold a higher rank. You must ensure that in the classroom everyone is equal.

A unique feature of adult learning is that adults often desire to be self-directed learners—that is, they want to take their own direction within the learning process. Self-directed learning is related to humanism, the concept that the adult learner can make "significant personal choices within the constraints imposed by heredity, personal history, and environment" (Elias & Merriam, 1980). Although adults may have the desire to be self-directed in their learning, their prior experience with didactic classroom-based instruction may not have prepared them sufficiently for self-management in the learning environment. As the instructor, it is your responsibility to provide the necessary guidance that keeps your students on track.

Critical reflection—that is, the ability to reflect back on prior learning to determine whether what has been learned is justified under present circumstances—is often a unique advantage attributed to adult learners, and a key asset that you should utilize when teaching. This kind of reflection involves "meta-thinking" about the strategies required to achieve a goal. Such reflection often occurs when adult learners compare their ideas to those of their mentors or those of their classmates.

Over time, an adult acquires basic cultural beliefs and ways of interpreting the world. Often such learning and beliefs are assimilated uncritically. In the classroom, however, adult learners may reflect critically on the assumptions and attitudes that underlie their knowledge. Such reflection helps them become deeper thinkers and can enhance their processes of thinking.

The benefits of this evolution in thinking are readily evident in the fire service. How many times have fire fighters gone to training with an attitude of "I really don't want to be here"? These students would rather be doing something else, but they go because the education is mandated. This disdain for training may represent an assimilated attitude—perhaps others in the station feel the same way. It may not indicate the way that a particular fire fighter truly feels about training or education; fire fighters naturally want to fit in socially, so they tend to adopt the attitudes that others around them have. As the instructor, part of your job is to help all students become motivated to learn the information presented.

Factors such as participation in formal educational programs or training programs since high school will also affect the adult learner's ability to learn. For many adult learners, traditional learning skills (e.g., reading and understanding textbook material, taking notes during lectures, and studying for tests) may have been forgotten or, in some cases, were never developed properly. Many adult learners have developed a learning preference that is limited to "doing" (McClincy, 1995). Adult learners with this learning characteristic learn best when they can perform the skill. Difficulty with developing study skills, and problems with unlearning and relearning, present challenges to both the adult learner and you, as the instructor.

Today's Adult Learners: Generational Characteristics

Many types of learners exist, and many groups of students emerge with similar traits and skill sets that the instructor must take into account when developing or presenting a lesson plan. This type of assessment, known as an **audience analysis**, is part of all levels of instructor responsibility when planning instruction. Audience analysis is the determination of characteristics common to a group of people—in this case, students. An audience analysis can be used to choose the best instructional approach. Factors considered in an audience analysis include the following:

- Age of the students and their experience in the subject matter
- Who the lesson is intended for and how students will use the material

- Number of students in the class
- Level of previous knowledge of the class material

Public speaking expert Lenny Laskowski devised the following use of the term *audience* as an acronym to help identify this essential element in presentation preparation and delivery:

- Analysis: Who is the audience?
- Understanding: What is the audience's knowledge of the subject?
- Demographics: What are the ages, genders, educational backgrounds, and other distinguishing characteristics?
- Interest: Why are the students attending your class?
- Environment: Where will this presentation be delivered, and which possible barriers to effective learning will exist?
- Needs: What are the audience's needs associated with your presentation, and how will they use the material in the future or as part of their job?
- Customization: Which specific needs and interests should you, as the instructor, address relating to the specific audience? Are appropriate department SOPs and terminology being used?
- Expectations: What does the audience expect to learn from your presentation? The audience members should walk away having their initial questions answered and explained, or they should be able to perform a task.

The following are some generalized categories of learners divided by the generation in which they were born. Most share common characteristics of behavior, knowledge, and skills. Again, understanding your students is a key behavior for any instructor.

Baby Boomers

The **baby boomers** were born between 1946 and 1964. At the end of World War II, soldiers returned home, married, and began families in the midst of an economic boom. The baby boom ended when the increased use of birth control caused a sharp decline in the birth rate in 1964.

The baby-boomer generation was raised on television, which had a tremendous impact on them and caused many of them to see world events from a single source (**FIGURE 2-3**). The following statements tend to apply to this generation:

- Rock and roll strongly influenced their formative years.
- They have high standards concerning work and place great value on education, and they expect the same of others.

FIGURE 2-3 The baby boomer generation of fire fighters are the backbone of many departments, and their trust must be earned by new instructors.
Courtesy of Fire Chief Gordon J. Nord Jr.

- They are independent, confident, and self-reliant.
- Speaking out against social injustice was a milestone period in their lives.

How do you teach to baby boomers? Today, this generation of individuals may largely compose the upper level management of the department or training division. Look at the senior members of your fire department and observe how you interact with them. As with all students, you must first gain baby boomers' trust and show them respect. Draw on their experience to assist you in teaching your course. Baby boomers will want to share their experiences and knowledge. This generation is known for both accepting change and resisting it because it pushes them out of their comfort zones. An effective method for teaching baby boomers is using group activities that will allow them to take a lead role and share their knowledge.

Generation X

As B. L. Brown points out in *New Learning Strategies for Generation X* the "gap between Generation X and earlier generations represents much more than age and technological differences. It reflects the effect of a changing society on a generation" (Brown, 1997). The term

Generation X has been used to describe the generation consisting of those people whose teenaged years were touched by the 1980s, although this definition excludes the oldest and youngest Gen Xers. In broad terms, members of Generation X are those persons born in the late 1960s to late 1970s. This was the first generation for whom it was common for both parents to work outside the home. Their parents' absence required children to learn to take care of themselves and their siblings until Mom and Dad returned home. While their physical development was the same as earlier generations, the psychological and learning development of Generation X was not. This generation was the first to abandon the traditional family models, to consider divorce to be the norm, and to promote single-parent families. Many Gen Xers prefer to stay childless (**FIGURE 2-4**).

Many members of this generation were raised in single-parent homes, growing up with "fast food, remote controls, entertainment and quick response devices such as automatic teller machines and microwave ovens, all of which provide instant gratification" (Brown, 1997). These life experiences shaped the way they learned and resulted in certain learning characteristics that require some modifications of traditional classroom tactics and new teaching strategies. Members of Generation X are typically familiar with being educated and trained by people from earlier generations. The cultural gap between the generations reflects the diverse life experiences of the individuals in each generation. One of the major challenges facing instructors is the need to bridge the generation gap.

The following statements tend to define the majority of Generation X learners:

- They are independent problem solvers and self-starters. These individuals want support and feedback, but they do not want to be controlled.

FIGURE 2-4 Instructors must learn techniques to bridge the generation gaps between their students. Generation X firefighters were the first generation of technology literate students to enter the classroom.
© Jones & Bartlett Learning. Photographed by Glen E. Ellman.

- They are technologically literate. Gen Xers are familiar and comfortable with computer technology and using the internet for sourcing information.
- They are conditioned to expect immediate gratification. Gen Xers crave stimulation and expect immediate answers and feedback.
- Members of this generation want their work to be meaningful to them. They want to know why they must learn something before they will take the time to learn the subject.
- They do not expect to grow old working for the same company, so they view their job environments as places to grow. They seek continuing education and training opportunities; if they do not find them, they will look for new jobs where they can get what they want.
- Members of Generation X often seek employment at technology start-ups, start their own small businesses, and even adopt worthy causes.
- Adversity—far from discouraging members of this generation—has given them a harder, even ruthless edge. Many believe, "I have to take what I can get in this world because no one is going to give me anything" (Brown, 1997).

In the classroom as well as in the workplace, Gen Xers often clash with baby boomer or traditionalist instructors. As an instructor, you may be the baby boomer trying to teach the Gen Xer. The common value that both generations share is the high priority placed on learning and developing new skills. This point poses a challenge to instructors, who must continually raise the bar in the classroom and perhaps adapt to cross-generational training. To accommodate the unique characteristics of Generation X learners, instructors need to step outside the box and promote learning that has applications in the school, work, and community settings that are part of the Gen Xers' experience.

Generation Y

Generation Y (also known as Millennials) is the name given to those people who were born immediately after Generation X. There is no consensus as to the exact range of birth years that are included in Generation Y or whether this term is geographically specific. The only consensus appears to be that Generation Y includes those individuals born after Generation X.

The "Y" in Generation Y refers to the namelessness of a generation that has only recently come into an awareness of its existence as a separate group. Members of this generation typically feel dwarfed and overshadowed by the baby boomers and Generation X. Nevertheless, members of Generation Y are known for holding wide-ranging opinions on political and social issues, religion, marriage, gay rights, and behavior (**FIGURE 2-5**). This lack of a common perspective may have occurred because these individuals have not encountered a personal situation where their actions or reactions have forced them to consciously choose sides on issues and maintain consistent positions. Most Gen Yers are more tolerant of alternative lifestyles and unconventional gender roles than are previous generations. At the same time, Generation Y tends to be more spiritual than the generation that includes their parents, and disagreement on social issues between the more liberal and more conservative members of this generation is commonplace.

The following statements tend to apply to Generation Y learners:

- They are confident, quick to learn, and sometimes brash.
- They want to work; they tend to have a strong drive to find a career that is fulfilling and satisfying.
- The members of Generation Y have been nurtured and programmed with a slew of activities since they were toddlers, meaning they are both high performers and sometimes considered high maintenance (Armour, 2005).
- They tend to do well at multi-tasking.
- Adapting to and applying ever-evolving technology comes easily to them.

Learning in virtual classrooms and virtual labs is ideally suited to Generation Y, as is teaching by technical professionals, because these adult learners can interact with others and get hands-on experience. Virtual labs suit the "click on it and see what happens" experiential approach to learning and have the added benefit that the learner remains in a safe environment.

The Gen Yers in the classroom present a challenge to the instructor who is either a baby boomer or a Gen Xer. The instructor needs to have a general understanding of members of this generation and support their learner characteristics and needs. Instructors need to understand that success in teaching this generation requires gaining their trust by being a good example and "walking the walk" in what instructors say and do.

Generation Z

Generation Z (also called Generation Next, Post-Millennials, or iGen) refers to those born in 1995 and later. They were born into a world where computer use was commonplace, and as a result, have a unique style of learning. Many Generation Z students have had constant computer access from the start of their education, and their learning is based on quick and very visually oriented interaction. Perhaps because of the availability of online videos and the prevalence of instructional videos, this generation prefers to see a skill demonstrated and then try it. This group has grown up teaching themselves how to do things by searching for it online, watching a video, and then trying it themselves (Seemiller and Grace, 2017). They tend to process information more rapidly than previous generations do, and seem to have a greater ability to perform multiple tasks at once; however, they may also have shorter attention spans.

Generation Z learners have grown up in an educational setting that has provided regular standardized testing and feedback. They are used to knowing where they stand, not only in their own classroom, but as compared to their peers across the country (Seemiller and Grace, 2016).

Generation Z prefers to have individual control over their learning with the option to collaborate. They expect active participatory learning environments and student-centered learning environments. (Seemiller and Grace, 2016).

FIGURE 2-5 Many members of Generation Y have been nurtured with a constant flow of activities since they were toddlers, resulting in high-performance and high-maintenance multi-taskers.
Courtesy of Forest Reeder.

The Laws and Principles of Learning

Numerous theorists, including Edward L. Thorndike and his contemporaries Edwin R. Gutherie, Clark L. Hull, and Neal E. Miller, have studied learning and developed ideas that have become recognized as the basic laws or principles of learning. According to Thorndike, there are six laws of learning: readiness, exercise, effect, primacy, recency, and intensity. They are described here:

1. **The law of readiness:** A person can learn when he or she is physically, mentally, and emotionally ready to respond to instruction.
2. **The law of exercise:** Learning is an active process that exercises both the mind and the body. Through this process, the learner develops an adequate response to instruction and is able to master the learning through repetition. For example, a fire fighter can master ropes by practicing tying knots every night for a month. The opposite is also true: A skill that is not exercised after it is developed will be lost. In some areas, this concept is referred to as the principle of disuse.
3. **The law of effect:** Learning is most effective when it is accompanied by or results in a feeling of satisfaction, pleasantness, or reward (internal or external) for the student—for example, when the student receives an A on a pop quiz.
4. **The law of primacy:** This principle states that the first method or way a concept is taught will be the way the student learns and remembers it. As the saying goes, "First impressions are lasting impressions"—and when it comes to learning, this holds true. The concern here is that if a student learns a new skill incorrectly, then the instructor has to help the student "unlearn" and correct the knowledge.
5. **The law of recency:** Practice makes perfect, and the more recent the practice, the more effective the performance of the new skill or behavior. Running drills on new skills will reinforce and perfect training in fire fighters. This law is why we retrain fire fighters and review information that is not used on a daily basis. If a skill is not used, it goes away, just as a muscle that is not exercised loses its strength over time.
6. **The law of intensity:** This principle states that the more of the senses that are stimulated, the more likely the student will be to change his or her behavior—which is what learning is all about. A lesson that moves from lecture to real-world videos to hands-on application of the skill will increase the student's ability to learn. The more the material is presented in a real-world manner, the stronger the learning effect (Thorndike, 1932).

Use of Senses

Adult learners use the five senses—hearing, seeing, touching, smelling, and tasting—to obtain information (**FIGURE 2-6**). The type and number of senses used to learn determines how learning occurs and what is remembered. Of the five senses, seeing (e.g., a visual demonstration of a skill) is generally the most effective means of learning, followed by hearing (e.g., listening to a lecture), smelling (e.g., smelling smoke), touching (e.g., handling fire equipment), and tasting (rarely used in training fire fighters, especially due to hazardous materials and other safety concerns). As an instructor, the more senses you can engage during the learning process, the greater the students' ability to retain the information that was presented. Effective lesson plans will address this issue by allowing time for questions, hands-on training, and other methods that bring all of the senses together.

The Behaviorist and Cognitive Perspectives

Two schools of thought have developed to explain how behavior evolves: behaviorist and cognitive. According to the **behaviorist perspective**, learning is a relatively permanent change in behavior that arises from experience. Not all changes in behavior reflect classroom learning. Many changes in behavior take place owing to a person's maturation and physical changes as he or she ages (Rathus, 1999).

According to the **cognitive perspective** (mental), learning is an intellectual process in which experience

FIGURE 2-6 Adult learners use their senses to obtain information.
© Jones & Bartlett Learning.

contributes to relatively permanent changes in the way individuals mentally represent their environment. Cognitive theories stress the use of mental capabilities to change behavior. Cognition is defined as the acquisition of knowledge through the use of perception, encounters with ideas, and obtaining of experiences (Lefrancois, 1996). Thus, cognitive learning is demonstrated by recall of knowledge and use of intellectual skills. It entails the comprehension of information, organization of ideas, analysis and synthesis of data, application of knowledge, choice of alternatives in problem solving, and evaluation of ideas or actions. According to this perspective, learning is an internal process that is demonstrated externally by behavioral changes.

Competency-Based Learning Principles

Learning that is intended to create or improve professional behavior is based on performance, rather than on content. Such **competency-based learning** is usually tied to skills or hands-on training. Not surprisingly, much of what is accomplished on the fire ground occurs through hands-on activities that require proficiency with skills—for example, the ability to force entry on a door. Competency-based learning is based on what the adult learner will do or perform.

Competency-based learning originates with the identification and verification of the needed competencies to perform particular skills. These competencies are actually job requirements, and ideally, they will have been identified through a job task analysis. For the fire service, this job task analysis has been done by the National Fire Protection Association (NFPA) via its professional qualification standards, which are all written in job performance requirement (JPR) format.

To ensure the success of a competency-based course, it is essential to follow a standardized process. This process ensures that each successive step has the proper foundation. Without the proper foundation, what may look like an excellent course might not address the real skills issues of the fire department. The design of a competency-based course includes the following requirements:

- Competencies exist.
- Job task analysis is performed.
- Standards are identified and set.
- Course goals are identified and written.
- Lesson objectives are identified and written.
- Competency-based instruction is delivered.

The most critical facet of competency-based learning is the notion that the skills must be mastered by the adult learners for each of the competencies. To ensure that this requirement is met, students must be given enough time and appropriate instruction to both learn and master the skill. Both the instructor and the students must understand that with mastery comes the responsibility of maintaining a particular performance level. Thus, learning objectives, instruction, and the evaluation of the skills must be tailored to performance in the field, not just to the acquisition of abstract knowledge that is out of context.

A competency-based learning process must be flexible enough to be adapted to the needs of all students. Each student inevitably learns at his or her own pace and in his or her own individual fashion. Some students may take longer than others to learn the skill, so you must take each individual's abilities into account when teaching skills. Also, appropriate and immediate feedback needs to be provided to students to allow them to achieve mastery of the skill, thereby applying the principles of primacy, recency, and intensity (**FIGURE 2-7**).

Forced Learning

There is an important fact of which you need to be aware: Adult learners cannot be forced to learn. Unfortunately, too many adult learners attend training programs or engage in educational situations only because they are required to do so or because they are motivated by the wrong reason. Understanding the dynamics of the fire service, there always seem to be new continuing education credits to be accumulated for recertification or some new required certification to be obtained. Mandating learning often results in a lack of motivation related to classroom activities and a poor attitude toward the learning process in general,

FIGURE 2-7 Immediate feedback needs to be given on the training ground to ensure that each student masters the skill.
© Jones & Bartlett Learning.

both of which present quite a challenge to the instructor who is tasked with teaching the course.

The presence of students who are mandated to attend a class or training session should alert you to the possibility that the lesson plans might need to be altered. For example, if you are assigned to teach a cardiopulmonary resuscitation (CPR) recertification course to paramedic students, then you might look for ways to involve students in both the teaching process and the learning process. The creation of such dual roles will help capture their attention and enhance the learning process.

Areas of Interest

All adults have certain areas of interest. Tapping into these areas may help reinforce the desire to learn. The source of this interest is less important than its mere presence, because all students need to have encouragement if learning is to take place. Such positive reinforcement can serve to increase the motivation of students who are forced into your classroom. You may need to acknowledge the areas of special interest of reluctant students to guide them through the learning process and to keep them engaged in your class. Knowing your students and their areas of interest allows you to capitalize on their knowledge base and become a more effective instructor.

When teaching adults, you are really more of a facilitator than a traditional teacher. Here are some core qualities that will help you become a good facilitator:

- **Be genuine in your relationships with adult learners, rather than consistently adhering to the traditional role as "teacher."** Show that you really care about the students and their success in the classroom.
- **Accept and trust in the adult learner as a person of worth.** Provide positive reinforcement and engage the learner with respect and value. Remember—people learn more from individuals whom they trust than from individuals whom they mistrust. Establishing a mutual climate of trust is important in fostering learning.
- **Provide nonjudgmental understanding of adult learners' perspectives.** Respond with empathy to both their intellectual and emotional perspectives. Remember that people are more open to learning when they are respected and feel supported than when they feel judged or threatened.
- **Establish a climate for learning.** Openness and authenticity are essential for maintaining this supportive atmosphere. People will be more likely to examine and embrace new ideas in an open and authentic environment.

Learning Skills for Adult Learners

As an instructor, you may find yourself teaching adult learners who have not been in a classroom setting since high school or college or rookie school. Study habits that these students developed while in primary school, secondary school, or college could have become rusty over time. Perhaps the most important study skill to develop as a student (and instructor) is to have the right attitude toward learning and studying. A positive attitude toward learning will help you learn new material and enjoy the classroom environment.

To assist your students in becoming effective learners, you can share study skills that enhance what they learn during classroom and personal study time. Study tips can be viewed as belonging to one of three categories: classroom, personal study, and test preparation.

Classroom Study Tips

In the classroom environment, students can learn to develop the following skills to help them in the learning process:

- **Ask questions.** If students do not understand the material being presented by the instructor, they should be encouraged to ask questions. Students should not wait for someone else to ask a question or feel that they are the only student who does not understand. In most situations, other students have the same question.
- **Take notes.** A good study technique is to develop an outline format for notes that follows the syllabus or outline given by the instructor.
- **Participate in class.** Participate in classroom activities and discussions and group activities.
- **Turn off your phone.** If students want to get the most out of classroom instruction, they should turn off their cell phones and other technological distractions and listen to the instructor.
- **Connect with peers.** As appropriate, students should get the names and phone numbers of other students in the class to contact for notes or information in case of a missed class.

Personal Study Time

Outside the classroom, students will need to study on their own. This is the time when most adult learners really learn the new material and develop effective learning skills and habits, including the following:

- **Time management.** Adult learners must learn to budget their time. This can be done by setting

- **Study location.** If studying in the fire house after hours, the student should find a location to study that is free of distractions and will allow him or her to focus on the material to be learned.
- **Material prioritization.** A student should review what needs to be studied and determine what is the highest priority and what needs to be covered first.
- **Class notes review.** To aid the student in focusing personal study time, he or she should review notes from class.
- **Reading skills.** One technique that will help students retain what they have read is to take brief notes of the material read. These notes can be used later to help in test preparation.
- **Study groups.** Depending on the class format and the material being learned, study groups can be an effective method for learning new material. These groups can be formal groups assigned by the instructor or informal groups that are organized by students who desire to study together.

Test Preparation Skills

Testing and course evaluation during instruction are commonly used methods for instructors to determine whether students have learned the material presented in class. Testing can take the form of quizzes or questioning by the instructor. Many courses require a comprehensive test to be administered at the end of the course to assess the students' learning. The following techniques can be used to help students prepare for a test:

- **Participate in study groups.** Study groups are effective methods for students to quiz one another and help one another in test preparation.
- **Do not cram.** Preparing for a course exam should begin on the first day of class, not hours before the test by cramming. Students should not wait until the last hours before a test to study; this method does not work.
- **Use mnemonic devices.** Fire service members and paramedics have long used mnemonic tools to help them remember information, such as RECEO-VS or SAMPLE. (RECEO-VS: Rescue, Exposure, Confinement, Extinguishment, Overhaul-Ventilation, Salvage; SAMPLE: Signs/symptoms, Allergies, Medications, Past illnesses, Last oral intake, Events leading up to present illness)
- **Take a break.** One skill that students need to learn is knowing when to take a 10-minute break from studying and refresh yourself—let your mind stop thinking, get some fresh air, and then go back to work.

To be a successful instructor, you must truly know yourself. Some of us are excellent speakers, whereas others are effective writers. What are your strengths? What do you need to work on? Taking a personal inventory assessment such as the Myers–Briggs Type Indicator can clue you in to behaviors that you may need to improve. Sometimes you will find that the fire fighters in your department are ready and willing to learn and experience training as a group. When it comes down to individual fire fighters, however, blocks to learning may spring up. For example, a fire fighter may be tired and unable to concentrate because he worked an incident all night. When basic needs are not met, learning becomes a second priority, which can in turn create a safety issue. A student may perform an unsafe act due to his or her less-than-optimal physical or mental state, or a student may miss information (because of inattention) that he or she must know to be able to perform at actual incidents.

COMPANY-LEVEL INSTRUCTOR TIP

Think back to a dynamic and unique instructor you encountered during school. Which characteristics or approaches did that instructor use to motivate you to learn the material? Then recall an instructor who did not motivate you, and identify the characteristics that may have played into that. Pinpointing the instructor traits or techniques that you found positive or negative can help you determine the style you want to incorporate in your own approach.

Maslow's Hierarchy of Needs

Abraham H. Maslow was a psychologist who researched mental health and human potential. He is most often associated with the concepts of a hierarchy of needs, self-actualization, and peak experiences. Maslow's initial work in 1943 was followed up in 1954 by his development of what is commonly referred to today as the "hierarchy of needs." Maslow took a unique approach in conducting his research, called "biographical analysis," wherein he reviewed the biographies and writings of 21 people he identified as being "self-actualized." His subjects included such major figures as Albert Einstein, Jane Addams, Eleanor Roosevelt, and Frederick Douglass. Based on his research, Maslow determined that certain needs should be satisfied to enable a person to move forward toward self-actualization. He perceived human needs as being arranged like a ladder or a pyramid (**FIGURE 2-8**). According

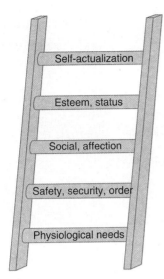

FIGURE 2-8 Maslow's hierarchy of needs.
© Jones & Bartlett Learning.

FIGURE 2-9 Fire fighters need periods of rest and rehydration to meet their most basic physiological needs.
© Jones & Bartlett Learning.

to Maslow, unfulfilled needs at the lower levels on the ladder will inhibit a person from climbing to the next step. As he pointed out, a person who is dying of thirst quickly forgets the thirst when deprived of oxygen.

Understanding the basic needs of people can facilitate an appreciation of which basic needs your students have. An instructor should keep these needs in mind when creating, adapting, or modifying lessons and the classroom dynamic.

Level One: Physiological Needs

The most basic human needs, at the bottom of the ladder, are physical—air, water, food, and shelter. It is hard to be creative or productive when you are exhausted and hungry. For the instructor, this means staying alert for conditions where the students are hungry, exhausted, dehydrated, too hot, or too cold. Although some of these conditions may arise in a long classroom session, they are more likely to occur in off-site training, during which reasonable breaks must be given for food, water, and rest. An example of a basic need that is too often overlooked is restroom facilities at a remote training site.

Sometimes fire fighters have to work at the extremes of discomfort and personal inconvenience during emergency activities, but these situations should be the exceptions and not a normal way of getting the job done. Making a fire company train or work without a rest and rehydration period will fail to meet their most basic needs (**FIGURE 2-9**).

Level Two: Safety, Security, and Order

Safety, security, and order are next on the ladder of needs. Safety is obviously a primary concern in the fire service, although a fire fighter's perceptions of safety are probably quite different from those of a typical office or factory worker. If students feel that their instructor is leading them in an unsafe manner or that a particular policy or practice in the fire department is exposing them to an avoidable risk, safety is likely to become a very significant issue.

The concept of security is more closely associated with maintaining employment or status within the organization. This need could be as subtle as a fear that a personality clash with a particular instructor will impede a fire fighter's professional development. An individual who feels threatened may do what is necessary to survive, but will not feel highly motivated to advance the fire department's objectives.

The need for security and order could become a factor when a fire department is undergoing a significant reorganization, such as a change in a standard procedure that the fire fighter was accustomed to over a long period of time. If a department adopts a new procedure or operational structure, the fire fighter may feel apprehension over new training responsibilities, evaluations, and scrutiny by upper levels of management. Any type of change in the established order is likely to arouse insecurities. The anticipation of significant changes in the organization can easily impede the fire fighter's creativity and motivation.

Level Three: Social Needs and Affection

Social requirements and affection, as identified by Maslow, are the psychological or social needs that are related to belonging to a group and feeling acceptance by the group (**FIGURE 2-10**). Group acceptance and belonging are particularly strong forces within the fire service. The majority of fire fighters have some

FIGURE 2-10 Group acceptance and belonging are particularly strong forces within the fire service.
© Jones & Bartlett Learning. Photographed by Glen E. Ellman.

FIGURE 2-11 An instructor can positively affect a fire fighter or company's esteem and status by making efforts to recognize new certifications or training milestones.
© Jones & Bartlett Learning.

type of identification on their personal vehicles that visibly declares their local fire department affiliation and membership in the fire service at large. Transferring a fire fighter away from a "good" assignment can be a strong demotivator, just as suspending a volunteer from responding to calls can have a significant impact on him or her at the social and affection levels. Fire fighters who are temporarily or permanently assigned to a company other than the one they have worked with and for over a long period of time will feel the same impact. For example, they must learn how they will fit into the formal and informal rank structures and work groups in the new station. In fact, many newly placed but experienced members feel like they do not fit into the company until they can begin to socialize with the new crew members.

Make your class its own community of social acceptance by acknowledging that all students present are to become the future of the fire service. This group acceptance leads the class to a greater cohesion even when many of its members come from differing backgrounds and departments.

Level Four: Esteem and Status

Promotions, certifications, and special awards are all symbols that relate to the student's esteem and status level (**FIGURE 2-11**). For example, membership in an elite fire department technical unit is often an indicator of advanced qualifications or achievements. Many fire fighters are willing to invest money and off-duty time to compete for a high-esteem or high-status position. For example, fire fighters may spend months preparing for a fire officer promotional exam or taking courses to qualify for a rescue company. Most elite fire units have more applicants than available positions. As an instructor, keep these motivational factors in mind. Every student is in your class for a reason, and it is your responsibility to guide them toward meeting the course objectives, and in doing so getting another step closer to their own goals.

As an instructor you need to be aware of how self-esteem plays out in the classroom environment and the importance of protecting your students from negative comments both from you as the instructor and from their fellow students. The fire service is known for joking in a "brotherly" fashion to foster camaraderie, but as the instructor you need to be careful not to allow this type of joking to cross the line where it affects a student's self-esteem.

Level Five: Self-Actualization

Peak experiences are profound moments of love, understanding, happiness, or rapture, when a person feels more whole, alive, self-sufficient, and yet a part of the world—more aware of truth, justice, harmony, and goodness. Self-actualizing people have many such peak experiences. These individuals tend to focus on problems outside of themselves and have a clear sense of what is true. Self-actualization in the learning environment is seen when students finally complete a skill or task they have been struggling with and can turn their attention to the next challenge. Graduation day at a recruit academy is an ultimate moment and point of self-actualization for those students, when they have completed a hard training process and receive their badges as fire fighters.

Learning Domains

In 1956, Benjamin Bloom and his research team identified three types of **learning domains**—that is, categories in which learning takes place (**FIGURE 2-12**).

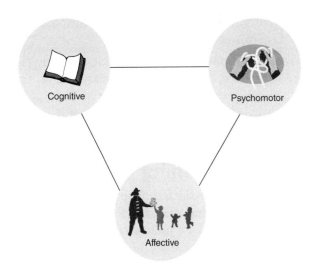

FIGURE 2-12 The three types of learning domains are cognitive, psychomotor, and affective.
© Jones & Bartlett Learning.

- **Cognitive domain**: knowledge
- **Psychomotor domain**: physical use of knowledge
- **Affective domain**: attitudes, emotions, or values

Today, this classification system is widely known as Bloom's Taxonomy of the Domains of Learning (Bloom, 1984 & Anderson et al., 2001). **Bloom's Taxonomy** has had a widespread influence in the field of learning research. Researchers have since developed new strategies of classifications and identified additional learning domains to build upon Bloom's original taxonomy.

Although Bloom identified three general learning domains, the learning process seldom takes place solely in one domain. Additionally, within each of the domains, the levels of learning build one upon the other. These levels, which have been expanded and refined by subsequent researchers, serve as the building blocks of the learning process. Within each level are found three basic sublevels: knowledge, application, and problem solving. Being aware of these levels and Bloom's three original learning domains will assist you in developing lesson objectives and course material to address a specific level of learning.

Cognitive Learning

The most commonly understood of the learning domains is cognitive learning, which results from instruction. Bloom's Taxonomy distinguishes six levels of cognitive learning:

- Remembering (Knowledge): remembering knowledge acquired in the past
- Understanding (Comprehension): understanding the meaning of the information
- Applying (Application): using the information
- Analyzing (Analysis): breaking the information into parts to help understand all of the information
- Evaluating (Evaluation): using standards and criteria to judge the value of the information
- Creating (Synthesis): integrating the information as a whole

Each of these levels builds on the previous one. The most basic is the knowledge level; the highest is the evaluation level. By understanding the cognitive domain of learning, you will be able to build more effective lesson plans and use students' thinking processes to encourage them to learn the information and apply that new knowledge. Adult learners should be able to fit the abstract thinking, facts, and information you present into the context of their jobs.

Consider these examples of cognitive learning and their application for fire fighters:

- Learning terms, facts, procedures, or principles: The fire fighter uses cognitive learning to learn the standard operating procedures (SOPs) of the fire department.
- The ability to translate and explain facts and principles: The fire fighter is taught the basic principles of fire and then asked to identify types of fuels.
- The ability to apply facts and principles to a new situation: The fire fighter is able to size up the scene, formulate a plan based on that analysis, and evaluate all possible outcomes.

Psychomotor Learning

For fire fighters, psychomotor learning is the most common learning domain. The term *psychomotor* refers to the use of the brain and senses (psycho) to tell the body what to do, and the use of the muscles (motor) to tell the body how to move. The psychomotor learning domain involves the ability to physically manipulate an object or move the body to accomplish a task or perform a skill. This kind of learning is also referred to as **kinesthetic learning**.

As in the cognitive domain, the learning phases in the psychomotor domain build progressively one upon the other. Building on Bloom's work, Ravindra Dave (1970) added the following six levels to the psychomotor domain model:

- Observation: watching the skill or activity being performed
- Imitation: copying the skill or activity in a step-by-step manner
- Manipulation: performing the skill based on instruction
- Precision: performing the skill or activity until it becomes habit
- Articulation: combining multiple skills together

- Naturalization: performing multiple skills correctly all the time

Each level of learning in Dave's psychomotor domain model builds upon the previous level: observation → imitation → manipulation → precision → articulation → naturalization. By understanding that the physical use of knowledge is the basis for students learning to perform the skill, you will be able to develop a lesson plan that involves demonstrating the skill to students, who then practice until they are able to perform that skill flawlessly. As students progress from one level to the next within the psychomotor domain, they may experiment with new ways to perform the skill. This trial-and-error process is completely natural as students learn how to adapt the skill and become more familiar with the activity.

Awareness of the following aspects of human behavior will assist you in teaching the psychomotor skill and developing an effective lesson plan:

- Gross body movement: the ability to move the arms, legs, and shoulders in a coordinated, controlled manner
- Fine motor control: the ability to use hands, fingers, hand–eye coordination, and hearing
- Verbal behaviors: the use of sound to communicate
- Nonverbal behaviors: the use of facial expressions and body gestures to communicate

In your delivery of lesson plans, ideally you will identify how a student should perform a skill or activity by carrying out a task analysis and identifying the steps involved in performing the activity. Here are some examples of psychomotor learning and its application:

- Comprehensive skill approach: The skill is demonstrated by the instructor and the fire fighter watches.
- Step-by-step approach: The fire fighter performs the skill in a precise step 1, step 2 manner until the entire skill activity is accomplished.
- Repetition of the skill or activity until it becomes habitual and is correct: The fire fighter performs the skill until it becomes second nature and he or she executes the skill flawlessly. The instructor then presents "what if" questions to help the student determine when to perform the skill.
- Combination of multiple skills: The fire fighter is able to adapt to changing situations and defend his or her choice of skills for the activity.
- Performing multiple skills correctly all the time: The fire fighter can perform the skill, and it is truly a natural activity regardless of the environment or circumstance in which the skill is performed.

For students to learn skills or activities properly, they need to practice, practice, practice (**FIGURE 2-13**). As the brain processes the attempts at performing the skill, it also learns to avoid what does not work or what was in error.

Affective Learning

Affective learning is the feeling or attitude domain. It focuses on those characteristics that make each person unique—that is, an individual's preferences, perceptions, and values. Many of these characteristics will have evolved in individual fire fighters over long periods of time, so attitudes may not change immediately after the introduction of a new concept. Instead, learning within this domain progresses from simple awareness, to acceptance, to internalization, and finally to acting out the attitude. Bloom's work with Masia and Krathwohl (1965) led to the development of the affective domain and identified the following five learning levels:

- **Receiving:** becoming aware of the skill or concept
- **Responding:** acknowledging the implications of the skill or concept and altering behavior accordingly
- **Valuing:** internalizing the skill or concept and having it become part of everyday life
- **Organizing:** comparing and contrasting skills or concepts
- **Characterizing:** adopting and personalizing skills or concepts

Each level of affective learning in Bloom's *Taxonomy of Educational Objectives: Book 2* builds on the previous level: receiving → responding → valuing →

FIGURE 2-13 For students to learn skills or activities properly, they need to practice.
© Jones & Bartlett Learning.

organizing → characterizing. Because much of this learning takes place inside the mind, it is important to look for subtle changes in students to confirm that affective learning is occurring. Here are some examples of affective learning:

- **Acquisition of new values:** The fire fighter is willing to learn the information and attends class regularly.
- **Acknowledgment of the concept:** The fire fighter studies for tests and participates in class activities.
- **Internalization of values:** The fire fighter volunteers to participate in an extra-credit activity.
- **Internalization of the organization of information:** The fire fighter decides to pursue continuing education because the additional training can keep him or her abreast of new developments and technologies.
- **Full adoption of the new values:** The fire fighter decides to pursue a degree in fire service management.

The affective domain is difficult to measure, but it should not be overlooked. The instructor must be observant to identify a student's reaction to the learning situation and changes in his or her value system. Another way of looking at the affective domain is as the marriage of cognitive knowledge with the psychomotor skills and their application in the real world by the student.

Learner Characteristics

How Does an Individual Learn?

When preparing for a class, you must ask, "What is learning?" as well as "How does a particular individual learn?" To answer these questions, you need to recognize that each student has different learner characteristics—that is, a way in which he or she prefers study and consider information and the ways he or she views the world around them. To use the terminology introduced by Bloom's Taxonomy, learning styles comprise characteristic cognitive, psychomotor, and affective behaviors.

Additional information on how students learn and retain new information and knowledge was provided by Jean Piaget, a biologist from Switzerland who developed his learning theory based on his observations of children and how they learn. Piaget's theory (Piaget and Inhelder, 1958) states that children learn "through motor actions, develop intelligence by using their natural intuition, develop cognitively through the use of logic and finally develop the ability to think abstractly." While Piaget's theories apply to adult learners in many ways, most of all they help us see how to help a student learn by going back to basics.

Although learning is an invisible mental and emotional process, the results of the learning process are frequently visible in that they may be traced back to certain experiences (e.g., a training session) and are evident in behaviors (e.g., performance of a skill).

Every individual has different characteristics and abilities as a learner. A successful instructor must take into account individual learner characteristics and abilities to ensure that all students' needs are met. This concept of considering learner characteristics and abilities is part of **student-centered learning**. Student-centered learning shifts the focus of teaching from the student as a passive observer watching a teacher to the student as an individual with an active role in the learning process (Student-Centered Learning Definition, 2014).

While learners have individual characteristics, groups of students may share common characteristics. Discovering these characteristics in the preparation phase of planning a lesson might involve contacting the organizer ahead of time and gathering specific information about the audience's demographics, their learning backgrounds, and their experience levels.

Of course, it is inappropriate to make assumptions about a group based on a single group identifier. For example, if you were assigned to teach a group of senior citizens in an assisted living center and assumed that all of them would be sitting in chairs all day, you would be very surprised to learn that most of them cook, play golf, and lead very active lives.

Groups of learners may have subsets within the larger group. To see how this works in practice, consider a group of fire fighters in a class on fire-ground support operations. Although the majority of the class wants to start placing ladders immediately and practicing ventilation, a minority prefer to watch a video and discuss the skills before actually practicing the operations (**FIGURE 2-14**). Both groups want to learn, but the preferred method is different for each.

Your students may have different ways they prefer to study or review information and different ways they think about concepts. In learning environments, using varying modes of instruction benefits all students. Using a variety of means and methods allows individual students to better grasp ideas by presenting the same ideas in different ways. Multimedia is the most common way of incorporating multiple modes of presentation. The availability of computer programs and educational websites has significantly expanded the ease of incorporating multimodal learning options.

FIGURE 2-14 Both fire fighters want to learn how to raise the ladder; they simply have different learner characteristics.
© Jones & Bartlett Learning.

Effective Teaching and Learning for All Students

Fortunately, despite the differences in learners, there are some common features to learning and effective learning experiences that inform teaching practice. Many fields of study contribute to the body of knowledge known as *learning science*. Our understanding of how people learn is incomplete and evolving, but more than a hundred years of research has led to a strong base of knowledge in the discipline. Being aware of some key concepts of learning science is useful in developing your skills as an instructor and in meeting the needs of all students. The following five learning science concepts are particularly useful for the fire and emergency services instructor:

1. **Starting points are important.** The process of learning involves connecting the new ideas and concepts to existing ones (National Research Council, 2000). Every student you will encounter brings with them preexisting ideas and perspectives. It is not uncommon for students to have an incomplete or even erroneous understanding of concepts that can interfere with their ability to add and integrate the knowledge you are presenting. As an example, students who have seen fire sprinklers go off on television may believe that a typical sprinkler activation includes multiple heads all over a building going off simultaneously. It may be harder for that student to comprehend a lesson on sprinklers because their existing knowledge will not fit with the new information. Likewise, students who have watched fire fighters work in a burning structure on television may have an unrealistic expectation of visibility in a fire setting that impedes their attention to the importance of specific communication details of a search and rescue lesson. An important step in making sure the lesson will be effective for all is assessing the base knowledge of your students and clearing up any misconceptions that may exist.

2. **Real learning is an active process.** Adults and children learn when they are involved in **active learning**. Active learning can be defined as "anything that involves students in doing things and thinking about the things they are doing" (Bonwell and Eison, 1991). "Doing things" does not mean just performing physical tasks such as raising a ladder. Doing things can include having a discussion, drawing a diagram, or working in a group. Students are doing things when they are using the information to construct their understanding of a concept. Listening and reading are passive and do not require the student to make any use of the information. To deliver effective instruction, you must do more than stand in front of the class, show a slide presentation, and talk. Learning environments that provide opportunities for students to use and integrate knowledge have more successful outcomes.

   The second aspect of this definition is *metacognition*, meaning "thinking about thinking." Students need to think about what they are doing and consider how they learn as individuals. Good instruction helps learners to be aware of their thought processes. Recall from earlier in this chapter the idea of critical reflection. This is a form of metacognition. Providing learners opportunities to develop awareness of how they learn, what helps them learn, and ways to meet their needs is part of a successful learning environment (Hacker, Dunlosky, and Graesser, 2009).

3. **Organization of knowledge is a critical part of learning.** To gain mastery in an area, students must not only know facts and concepts, but must be able to organize them into a conceptual

framework and retrieve it as usable knowledge (National Research Council, 2000).

Good teaching provides students with guidance in developing conceptual frameworks. It is important to make sure you are presenting the big picture and emphasizing the important overall concepts. Visually organizing the lesson can help students to see the connections between concepts. Pay attention to how students are thinking about the concepts being presented. Ask students to explain their thinking or "think out loud" and give feedback and guidance. Provide a model of what more advanced thinking about the topic looks like. For example, explain how an experienced fire fighter would think about conditions present on arrival at a fire. Scaffolding, which is discussed in more detail later in this chapter, is an effective way to provide frameworks for students.

4. **Practice matters.** Deliberate, goal-directed practice with feedback, is necessary for learning. This is intuitive to many fire instructors when thinking about practical skills. We have all experienced a situation in which we were shown a skill, practiced it with guidance to correct our mistakes, and finally mastered it. What is less intuitive is that this is true for not only physical skills but mental skills as well. Students need to practice using and applying knowledge (National Research Council, 2000). Again, this is easy to see when we think about specific instances. If you are teaching students a formula for determining friction loss, it is important to provide the opportunity for them to practice not only using the formula with guidance, but also determining when to apply the formula. It is less apparent, but equally as necessary, when presenting a complex idea. For example, in a lesson about combustion, students need to have opportunities to practice retrieving and applying the concept. In the classroom, this may involve asking students studying combustion to predict what would happen if they tried to light a match on the moon. After letting them apply their knowledge to make a prediction, discuss the correct answer and why it is correct. This allows them to practice applying the concepts and get feedback on how well they did.

Part of being able to provide students with the feedback they need to learn is assessment. Assessment can be informal, like asking a question, or more formal, like testing. Often, students and teachers see testing and assessment as a required behavior at the end of a session that is unrelated to the classroom. In reality, assessment is fundamental to helping students determine what they know and what they have not yet mastered. Assessment is not just performed by the instructor for the students. Encouraging students to do continual self-assessment is part of metacognition and helps them develop their learning skills (Ambrose, Bridges, DiPietro, Lovett, and Norman 2010).

5. **Learning is social.** The majority of our learning takes place with the input of others. Consider the exchange of information and skill that occurs when you are engaged in an activity with friends. Golfing, fishing, a school organization, a sports team, an office group: whether working or playing, we transfer knowledge and skills within groups. Humans use peers to clarify their understanding and learn new things on a regular basis. You have probably used this technique yourself multiple times. Maybe you read the directions to set up something on your phone but do not quite understand, so you ask a friend for help and you walk through it together. Working together helps people to refine concepts.

Explaining things to others is a prime example of "doing things" with new knowledge that active learning requires. It turns out that explaining a concept to others is a very effective way of developing understanding. Many people have experienced the "a-ha!" moment when something becomes clear as they are explaining it to another person. Given this, it should come as no surprise that research indicates that peer interaction and cooperative learning improve outcomes (Crouch and Mazur, 2001).

Student-Centered Learning

Think about teaching and learning. What comes to mind first? Probably a scene where students are seated in rows, focused on the instructor, occasionally looking at a whiteboard or overhead presentation and exhibiting boredom, fatigue, and apathy. Contrast this picture with a classroom where students and instructor are engaged equally, where the instructor is facilitating the work of the students who are working in groups and practicing a skill, creating energy, synergy, and excitement.

Teacher-centered passive learning requires little involvement on the part of the student. Learners simply receive information by listening, watching, reading, or observing. The instructor disseminates information and is at the center of activity. Passive involvement leads to limited retention of knowledge by students. Unfortunately, this is historically the type of teaching fire fighters have been subject to. Because most fire and emergency services instructors had limited, if any, teaching experience and little exposure to the field of learning science, the fire service developed a pattern of instructor-centered teaching that provided the least effective avenue for retention.

Over the past 20 years, there have been significant advances in the field of teaching and learning. As learning science has provided more insight into what works, educational practices have moved from this older instructor-centered learning model to a more effective student-centered learning model. Teaching in a student-centered learning setting is called *learner-centered* or *student-centered teaching*.

In a student-centered learning model, the focus is on the students and their learning rather than the instructor and teaching. Responsibility for learning is a shared responsibility of both the instructor and the student.

Student-centered learning is active learning. Active learning is anything students do more than simply receiving instruction and passively listening to a lecture or watching a movie. Students are required to participate by reading, writing, discussing, solving problems, and engaging in higher-level thinking. Activities may include group discussion, questioning, brainstorming, problem solving, participation in demonstrations, or a variety of other hands-on techniques for involvement. Frequent assessment opportunities are provided to assist the student in understanding how well they have mastered skills and concepts.

In a student-centered learning environment, the role of the instructor is that of a facilitator. The instructor's job is to work with the students to facilitate their learning by providing guidance and feedback to enable students to develop skills and knowledge. This should not be interpreted as a philosophy that removes all responsibility for the educational process or outcomes from the instructor or as a system that lets students "do whatever they want," because this is not the case at all. Learner-centeredness is a partnership model where the instructor and the students work together to achieve the course goals.

One of the keys to student-centered learning is providing frequent feedback so students know how they are doing. Assessment is one of the main ways instructors and students get feedback. Assessment does not mean testing; while testing is one method of assessment, there are a wide variety of other ways that learner progress can be measured. Assessment can be as simple as watching a student do a task and providing feedback on his or her performance. In addition to being useful to students, assessments are helpful to instructors in knowing where to focus their attention. Providing frequent low-stakes assessments can ensure that assessment is viewed as a positive experience by both the learner and the instructor.

Lessons involving active, student-centered learning take more planning and organization but are always worth the effort in the end. An Instructor I must be prepared to use a wide range of student-centered teaching methods and strategies that are called for in lesson plans.

Student-Centered Teaching Methods and Strategies

The best teaching includes a variety of methods, often referred to as strategies, used to meet the diverse needs of the learners. The decision to use one strategy over another will depend on the time available, number of students, objectives of the lesson, type of information to be presented, availability of materials and resources, domains and levels of learning, and personal preferences of the instructor. Most instructors tend to teach in the ways that they themselves learn best. It takes a knack to select an instructional method that best fits both the instructor's teaching style and the lesson situation. Often, one teaching method will naturally flow into another, all within the same lesson. Experienced instructors have the skills to make this process seamless and productive for students without overloading their ability to process information. There is no one "right" method for teaching, but certain criteria specific to each style help the instructor make the best choice possible for the learners. We will discuss some of the more common teaching strategies in this chapter in terms of their advantages and disadvantages as well as some tips for implementing each.

Discussion

A discussion format is different from a lecture-and-discussion format in that learners are as much a part of the presentation as the instructor. In fact, the instructor's role shifts to that of facilitator. Traditional discussion can tend to be instructor centered. A true discussion creates an exchange of ideas to provide two-way communication throughout the session, allowing everyone to participate. The educator speaks directly to the learners, asking questions or making comments and waiting for a response, or groups of learners can discuss views with each other. This method is very useful when trying to engage the affective domain and when working with concepts, ideas, and feelings. The participants can express themselves and learn from each other. This method deals less with facts, although the facts may be presented in the beginning as a reference for discussion. Discussion groups are effective in combination with or after a demonstration or film presentation.

This method, although very productive, is not recommended for groups larger than 20. Even in smaller groups, some participants may withdraw, not feeling comfortable enough to participate. Other group members may dominate discussions or become overbearing when voicing their opinions. It takes careful planning on the part of the facilitator to prevent these bumps in the road. When planning to use this method, be sure to design questions that evoke critical thinking and a sense of cooperation. Be clear that all ideas will be considered and that no idea or suggestion will be considered incorrect. The facilitator should allow time for a well-constructed summary that includes information gathered during the session. Another technique that can foster more engaged discussion is called think-pair-share, wherein the question is directed to the class, students formulate responses, share with another student, then share with the class.

> **COMPANY-LEVEL INSTRUCTOR TIP**
>
> An effective way to reinforce learning with a group that meets for more than one session is to use a sum-up. Sum-ups are short paragraphs or fill-in-the-blanks that capture the main ideas of a presentation. At the beginning of the next session, review the sum-ups from the last class. These can serve as an evaluation tool to discover whether the main points made their way to the learners.

Small Discussion Groups

Small discussion groups offer some of the same advantages as the larger group discussion, but they may even provide for more engagement and interaction. Some people feel more comfortable in smaller groups and will become more verbal. Each member will have more opportunities to participate, which will generate more viewpoints. To ensure that each person is engaged and contributing, assign roles such as reporter, scribe, timekeeper, and organizer. Sometimes these roles will be taken by participants and need not be assigned; other times, the instructor will need to be aware of the group dynamics and assign the roles. When handled correctly, the small-group discussion method facilitates active learning and can create energy conducive to learning in the room. Groups that develop their own synergy are capable of a heightened level of problem solving and creativity. Small-group discussions lend themselves well to the method of guided discovery, wherein participants are coached as they work to come up with solutions to problems on their own. Problem solving is an important element of learning; participants in the experience are more likely to remember the solutions when they were involved in the process of discovery. However, it takes a skilled facilitator to keep several small groups focused and on task. There is a tendency for the groups to get side tracked and run out of time before the objectives are met. Planning a session with small-group discussions requires preparing specific, well-crafted tasks or questions for the groups to address. The makeup of the groups must be considered. Will the instructor assign groups or allow members to form their own groups? It will also be important to build in time during and after the group time for reporting and summarizing. The facilitator should stroll around the room to make sure groups are focusing on the problems, while still allowing for an atmosphere of fun and creativity. The clock will be ticking, so occasional reminders to the groups will keep them cognizant of the amount of time they have left to finish their work.

Demonstrations

The purpose of a demonstration is to transmit the big picture to a group in a short period of time. This method is more instructor centered than student centered, but it is useful for teaching skills or situations that are difficult to fully explain through a verbal description alone. By seeing a task performed, learners become more aware of what materials are needed, better remember the steps in the process, and are able to observe the final outcome. Demonstrations generally take more preparation and usually need to be supported by audiovisual materials or other equipment, but the benefits will be well worth the effort for the

> **COMPANY-LEVEL INSTRUCTOR TIP**
>
> When teaching a group that meets more than once you can 'flip' the classroom. Instead of showing your students a video or giving a lecture during class time, assign those as pre-class work to the students. Have them watch the video or lecture or read the information before the class. Then when they come to class, use the class time to check for understanding and practice the skills. This is flipping the classroom. Think about people learning to play a musical instrument. They practice on their own and then see the instructor to get feedback on how well they are doing and correct mistakes. The instructor's time is not used to watch the student practice but rather to check progress, clarify questions, set goals, and evaluate mastery.
>
> A critical part of this is that the students must know that completion of the pre-class work is required and essential for their learning. Instructors can help bring that point home by holding students accountable for being prepared by checking for the preparation and for understanding in class and attaching value such as points or grades to the preparation.

participants. Field trips, tours, and simulations are all variations of the demonstration method.

Instructors using demonstrations should make sure all students are in a position where they can see and hear. Decide how many students and how much room and equipment will be needed. To hold the audience's attention, demonstrations should not exceed 20 minutes. It is important to practice and time the demonstration before presenting it to an audience. Without an actual run-through, presenters are often surprised that their planned demonstration takes far less or much longer than they anticipated. Either situation can be perceived as unprofessional, especially if the instructor does not have backup materials to fill the gap. Handouts with pictures or written instructions and information serve to support and clarify the demonstration for the audience.

Fire instruction has many opportunities to present demonstrations of interest to students. Personnel need to remember that although students are intrigued by what they see, educational objectives should be at the core of the presentations. The objectives will provide the reasons for giving the presentation in the first place and dictate the arrangement of the information from beginning to end.

> **COMPANY-LEVEL INSTRUCTOR TIP**
>
> Give students a short pre-lesson quiz. This can be informal through a show of hands at a station training session, it can be done on paper with partners correcting it, or software or web-based applications can be used. Doing this will not only improve the students' learning but it will also help you assess what level the students are starting at and what you might need to focus more attention on in class.

Questioning

One of the best ways to evoke a response from an audience is by asking good questions. Socrates knew that he was able to get students to think more critically by challenging them with questions rather than relying solely on a direct lecture approach. Generally, questioning is used in conjunction with other methods. The learner becomes more actively involved and begins to think at higher levels. Questions and their responses also allow an instructor to evaluate the learners' level of understanding and make adjustments immediately. Instructors should be aware that there is an art to using questions to produce the greatest learning. Poorly crafted questions limit the success of using this method. The give and take of a questioning session will also take more time than straight lecture (**TABLE 2-1**). You also can pose a question or topic for the class, and then allow them the opportunity to research and develop responses, instead of simply asking the question.

Following are several tips to keep in mind when using questioning as a method of instruction:

- Ask challenging, open questions. This means avoid phrasing questions that are closed or require straightforward factual answers unless you are only checking for retention. Questions that begin with who, when, or where and yes-or-no questions do not require learners to think as critically as when they are asked questions that begin with why, how, and what if. For example, you might ask a group of students if they should perform ventilation at a fire. The answer is simply "yes" and could be answered with almost no thought at all. It would be better to ask "What could happen if a fire fighter started ventilation before a water supply was established?" More than one answer exists, and the feedback will generate further discussion.
- Closed questions, or those requiring a specific answer, will work best in the classroom with students who cannot focus well.
- Ask uncluttered questions. Questions with many sub-questions or interspersed with background information can confuse students because it becomes unclear as to what is being asked of them.
- Make sure the questions are crafted at a level the students are prepared to understand. Ask if the question is clear and write questions down where they can be seen by everyone, if possible.

Role Playing

Role playing brings students into a lesson and encourages them to become part of the learning and teaching. The instructor and the student take on roles to act out situations, providing for more active learning than direct teaching methods and a more interesting way to present material. For a brief time, the classroom becomes a stage where an audience member is given an opportunity to practice newly introduced knowledge or skills. As the instructor, you are able to introduce problem situations more dramatically and observe students' reactions. The students can explore alternative solutions, and you can see how the material plays out in a fairly realistic scenario. This method is particularly effective with skills that involve an interaction between people, such as taking an emergency call or interviewing a witness. The experience can be rewarding because learners retain more by observing members of their

TABLE 2-1 Questioning Techniques

Type of Question	Purpose and Strength
Factual	Soliciting simple, straightforward answers based on facts, observation, or awareness. Answers are usually right or wrong. Example: How often should hose be tested?
Convergent	Looking for answers that fall within a very finite range of acceptable accuracy. The student can make some inferences based on personal awareness or other forms of information. Utilizes organizing, planning, sequencing, and structuring skills. Example: If you were sitting in front of the fire station and a person ran up and reported a fire, how would you handle the situation? What would you do first?
Divergent	Requiring students to explore different avenues and create many different variations, alternatives, and scenarios based on existing knowledge as a springboard. Students may arrive at answers through intuition, imagination, inference, projection, or conjecture. May not be a right or wrong answer. Higher-level cognitive thinking is necessary than for factual or convergent questions. Divergent questioning lends itself to better discussions and engagement. Hypothetical questions provide interesting discussions. May become essential questions for a lesson. Example: You are applying high volumes of water to a warehouse fire through an open loading dock door, but it does not appear to be having the expected effect. Why do you think the suppression efforts are not working?
Evaluative	Asking for more sophisticated levels of cognitive and/or emotional judgment. Students will combine multiple logical and/or affective thinking processes, or comparative perspectives, to synthesize information for conclusions. Example: What are reasons that fire fighters might fail to follow safety procedures like wearing seatbelts?
Combination	A blend of any of the types of questions in this table. Example: If you were asked to prepare a list of objectives for a community class about fire safety, what five messages would you consider, in order of importance, and why?

own groups; however, be aware of some of the pitfalls of using this approach. It is difficult to use in large groups. Some students may be too self-conscious to fully engage in the activity. Not everyone wants to be the center of attention. To prepare for successful role play experiences, define the problem situation and roles clearly, give clear instructions, watch the time, and plan for correcting incorrect responses as they occur.

Peer Tutoring

Peer tutoring occurs when a student with adequate or excellent mastery of an idea helps another student while a lesser degree of mastery. Peer tutoring works because the students are likely to more readily express opinions, ask questions, and risk untested solutions. They also may speak a more similar language with each other than in an instructor–student relationship. This approach provides many positive benefits for both the tutor and the tutee in all three domains of learning. Fire and emergency services instructors can easily make use of this teaching method when the student group has a mixed level of experience or when some of the group is mastering skills or knowledge faster than the rest. For example, if an instructor presents a lesson on knot tying to a group of fire fighters, members new to knot tying can be paired up with members with knot tying experience to practice with feedback.

Storytelling

Everyone loves a good story—especially fire fighters. Stories do have a place in the learning environment. Told correctly, they can be an effective hook, reeling in the audience to the world of the speaker. The storyteller must ensure the tale is relevant to the lesson content and interesting to the audience. Tell a story in which all audience members are in the story. Using imagery, walk them through to generate a mental picture. For example, begin with, "Imagine you are on the second floor of a home with a fire in the attic. You are about to open the door to the attic stairs. Picture the hallway and your crew. How is everyone positioned?" If you are animated and believable, the method is sure to set the stage for learning. But if the story is too long or irrelevant or if stories are overused during a presentation, this method will fail. Keep the lesson objectives in mind when choosing and preparing a story for presentation. Instructors should avoid telling too many stories during one presentation, especially about themselves; use these stories sparingly to maintain interest! Effective storytelling is an art and should be practiced before using it in a presentation.

Brainstorming

Brainstorming is a fast-paced activity in which group members arrive at answers or solutions by bringing ideas to the table for discussion and resolution. Learning flourishes when the audience members are active learners sharing in a spirit of cooperation. Full participation is encouraged because all ideas are welcomed and recorded. One idea often sparks a series of questions, new ideas, and dynamic creativity. The activity draws on the entire group's knowledge and experience. Again, the instructor becomes a facilitator or coach, and learning is student centered.

Instructors new to this technique may find that participants have difficulty getting "off the ground" or that the activity can easily become unfocused and chaotic; therefore, actual brainstorming sessions should be around 5 to 7 minutes long.

Sessions must be well facilitated to avoid criticism and discussions going off on tangents. Facilitators need to carefully select and focus the issues and prepare ideas for stimulating the groups that are slow to get the main idea. Most importantly, brainstorming sessions require the instructors to prepare guidelines and objectives and make them clear to the participants. Instructors must also be ready to intervene if the groups get bogged down. It will always help the "brainstormers" to have guidelines posted on an easel or a whiteboard as a reference to center their thinking. Flip charts for each group are also helpful for groups to record their ideas as they evolve. Following are some rules for brainstorming:

1. Define the problem and set directions.
2. Encourage participation from everyone. All ideas are valuable. All ideas become the property of the group rather than of a particular individual.
3. Build on the ideas put forth by others. Use the word "yes" instead of "but" when responding to another idea. Piggyback on ideas rather than discounting them.
4. Suspend judgment. Not everyone is comfortable with brainstorming. Aim to minimize barriers and inhibitions.
5. Encourage outside-the-box thinking. Be creative. No idea is too absurd. Wild and exaggerated ideas can lead to other possibilities.
6. Go for *quantity*, not quality. Reality checks and fine tuning will come later.
7. Record and display ideas. Be complete in writing down ideas; use more than one word so the intent is not forgotten. Sticky notes and flip charts work well.
8. When the session is finished, circle 20 percent of the most workable or interesting ideas. Think in terms of the 80/20 rule: look for 20 percent of the ideas that will accomplish 80 percent of the desired results.
9. Watch the clock. Start and finish on time. Encourage a brisk pace to maximize investment in the process.
10. Have fun! Keep the atmosphere light; it increases creativity.

> **COMPANY-LEVEL INSTRUCTOR TIP**
>
> The K-W-L (what I *know*, what I *want to know*, and what I *learned*) is a predictive approach to learning, but it can easily be incorporated into a strategy for discussion formats. Create a simple form with three columns, titling them "What I Know," "What I Want to Know," and "What I Learned" and provide the form to each member of the group. Lead students through brainstorming on the topic. You can participate as a contributor of relevant questions. Ask each participant to fill in the "K" and "W" columns with what they already know or believe about the topic and what more they would like to know about it. Show a video or give a brief presentation. After the presentation, have group members go back to check which questions were answered by the information provided; these can be written in the last column.

Motivation and Learning

Motivational factors make students feel enthusiastic and eager to learn. Motivation affects how students practice, what they observe, and what they do—all of which are critical to learning. As an instructor, you need to understand how learning and motivation work together to make a great fire fighter. Instructors will face various motivations themselves that will require them to become more vocal, animated, or intense to emphasize why information is important to the student.

Learning is the potential behavior, whereas motivation is the behavior activator. Learning and motivation combine to determine performance in training. Without a motivated student and an expert instructor, learning will not take place. If the student's motivation is poor, then his or her performance in training will be sloppy and inaccurate. If the student's performance in training is sloppy and inaccurate, then the fire fighter's response at an incident will be sloppy and incorrect. This sort of shortcoming is not acceptable on the fire ground.

Motivation of Adult Learners

Cyril Houle (1961) identified three types of adult learners: goal oriented, activity oriented, and learning oriented. Each type is associated with different motivational issues, which in turn affects how learning takes place (Draper, 1993).

Goal-oriented learners are the most common type of adult learner. Their learning decisions are guided by clear-cut goals, as the learner takes a "utilitarian" or "pure" stance toward education. An example of a goal-oriented learner is a fire instructor attending a workshop on using distance learning to enhance a department's training program.

The second type of adult learner is the activity-oriented learner. For these individuals, participation is based on reasons other than the stated purpose of the learning activity. Examples of activity-oriented learners include a full-time worker who attends evening noncredit courses with a friend or a fire fighter attending a conference with other department members just to attend social events after the conference.

The third type of adult learner, the learning-oriented learner, seeks knowledge for the purpose of learning something new. For this learner, education is a way of life and, as such, is an intrinsically valued activity. Because this type of learner focuses on the content of a learning activity, he or she can participate either within a group or individually (Houle, 1961).

Consider the example of two fire fighters who are attending a CPR training class. A fire fighter who is a goal-oriented learner might attend the class in an effort to improve her job skills, whereas the fire fighter who is an activity-oriented learner might take the class to spend time with his friends and possibly learn CPR at the same time. Although the learning agendas and goals of these fire fighters are quite different, both are in the same class. Learner-oriented adults focus on the act of being involved in learning.

Another factor that greatly influences success in the classroom is whether the adult learners grasp how the material relates to the real world and how this learning can be beneficial to job performance. This factor is well documented as affecting the learning outcome (Wlodkowski, 1999). Experienced instructors use real-world applications and can point directly to how information translates to use in the workplace. Theory into practice is a concept that can work to an instructor's advantage in this sense. For example, as part of their training, recruits learn that a gallon of water weighs a little more than 8 pounds. That statistical information is fairly simple to recall, but if you show recruits that a master stream flows 500 gallons per minute $\times$ 8 pounds per gallon, the students can quickly see that when they use a master stream in the real world, they are putting 4000 pounds of live load per minute into a structure, which may cause collapse.

Yet another key factor is the adult learner's psychological state. Mild anxiety can help the adult learner to focus on a particular task. Conversely, intense negative emotions (e.g., anxiety, panic, rage, insecurity) may detract from motivation, ultimately either interfering with learning or contributing to a low performance. To counteract this possibility, make your students feel welcome in the class session or at the drill ground by explaining the context and content of the training and allowing for a positive-outcome-based environment.

The current position of adult learners in the life cycle also influences the quality and quantity of learning. Research suggests that each phase of life is characterized by the need for learning certain behaviors and skills. Because they are adults, adult learners invariably want to be treated as autonomous, independent learners.

Geographical changes (moving from one location to another), physical changes (reaction times), visual and auditory acuity, and intellectual functioning influence the adult learner's performance as well. The adult learner's internal world of ideas, beliefs, goals, and expectations for success or failure can either enhance or interfere with the quality and quantity of thinking and information processing. In addition, the adult's attitudes about his or her ability to learn and judgment as to whether learning is an important life goal affect motivation. Moreover, motivational and emotional factors influence the quality of thinking and the processing of information.

No matter where the adult learner is and what he or she believes about his or her abilities, the adult learner continues to acquire knowledge both intentionally and unintentionally (Butler and McManus, 1998). The effectiveness of the learning process relates directly to the learning environment and the adult learner's interactions with that environment. In short, the adult learner's mental and emotional processes for learning can either help or hinder the fire fighter during training.

Motivation as a Factor in Class Design

Motivation is the primary learning component for adult learning. It generates the question "Why?" Why is the adult learner in the class? Why does the adult learner want to take (or not want to take) this class? When you can identify the why, it will allow you to tap into the how—the precise motivation and techniques that will help the adult learner achieve the goal of learning.

For example, knowing that attendance at a training class is mandatory allows you to design the lesson plan for the captive audience to maximize motivation. Unfortunately, most adult learners who are mandated to attend a class do poorly during such training, simply because they are unmotivated. Some adult learners may not know why they must learn the information or what is expected from the training or educational program, whereas other adult learners may have high expectations of the class. The unmotivated adult learners, the questioning adult learners, and the adult learners with high expectations all share the same classroom.

Find out from your students why they are in your class and what they expect. This information will enable you to design the lesson plans to meet every student's needs and sow the seeds of self-motivation in every student. You will inevitably need to employ a variety of motivational techniques to stimulate students. To learn, a student must be stimulated by the subject (McClincy, 2002).

If you are motivated, your enthusiasm may prove contagious in the classroom. When you are dynamic, outgoing, charismatic, and generally interested in the topic, your lessons will be taught with energy. Your personality will shine through in the classroom and will serve as a motivational factor for your students.

The use of audiovisual aids, teaching strategies, teaching methods, and even the classroom itself may serve as motivational factors. How you incorporate any or all of these motivational factors into the classroom will be reflected in the success of the learning process. For example, live-action video used to highlight a key point or term will motivate and engage the learner to observe the action much more than a still photo will, and even more than text on a slide will. Nevertheless, you must ensure that your video enhances your presentation but does not become the presentation. The next two chapters will expand on this topic.

Motivational factors are part of each level of instruction, from design to delivery to evaluation. The best courses you have attended motivated you to learn and provided you with the topic's relevance and application. Take time to consider ways to engage your learners throughout the presentation. Students are motivated to learn when they can understand how the material applies to them and how they will use it. An adult learner's motivation may change throughout a session. The number of factors that can affect motivation is limitless. You need to be able to recognize the changing motivational needs of adult learners. Frequently cited course improvement suggestions include more periodic input on student grades, ongoing monitoring of progress in a course, and the ability to provide more feedback between the instructor and the student.

Learning Disabilities

As an instructor, you may encounter students with a wide variety of learning disabilities. The Americans with Disabilities Act (ADA—federal legislation that was originally passed in 1990) classifies learning disabilities into these major categories:

- Reading disabilities range from the inability to understand the meaning of words, to the inability to read or comprehend, to **dyslexia**. This disability could become apparent in class if a student is asked to read a section of course material aloud or to summarize a paragraph from a textbook.
- Some students lack the ability to write, to spell, or to place words together to complete a sentence. This type of disability, known as **dysphasia**, might present itself when a student submits work for a course and the work is poorly written.
- Other students might suffer from **dyscalculia**, or difficulty with math and related subjects. Students with this disability might struggle on a hydraulics course or test.
- **Dyspraxia** is the inability to display physical coordination of motor skills. This condition could be observed on the training ground as a student's inability to complete a task such as climbing a ladder.
- A disorder that affects children but can be carried into adulthood is known as **attention-deficit/hyperactivity disorder (ADHD)**.

A person with ADHD has a chronic level of inattention and an impulsive hyperactivity that affects his or her ability to function on a daily basis. This disorder is a documented condition identified by the Centers for Disease Control and Prevention (CDC). Although it can be treated, there is no known cure for ADHD.

- Other types of disabilities that could affect the learning process include color blindness and poor vision, even if it is just difficulty in seeing the screen in a classroom.
- Some adult learners may have poor hearing and cannot properly hear material presented in the classroom. Hearing loss is a common problem that occurs with aging, so instructors may often teach older fire fighters who are dealing with being unable to hear.
- English as a second language (ESL) is a factor that some organizations are facing that can affect the learning environment and create a challenge to the instructor.

Knowing and understanding each of these learning disabilities will help you teach. Unfortunately, in most cases you will not know that students have these problems until a class has already begun. If you are a guest lecturer, you will not get the opportunity to know your students until after the first few hours of instruction. Once you have identified students with some type of learning disability, you need to adjust your teaching as necessary. There are many ways to help students in your class who are struggling to learn. For example, a few positive words in private or suggestions on a written assignment might help a struggling student understand a concept that is just beyond his or her reach. Students who have learning disabilities are not unintelligent or unable to learn; they simply learn in different ways. They usually have average to high intelligence, but may perform poorly on tests because of their disabilities if accommodations are not made for their special needs.

A student who is a hesitator may be suffering from a learning disability. Such a student is often shy, reluctant, at a loss for words, and quiet. Although he or she may know the material and have much to offer, the individual's shyness, fear, or lack of confidence may keep him or her from participating in class activities or answering questions. You can engage this student by asking nonthreatening questions and offering encouragement. You need to let a hesitator know that his or her contributions are worthwhile and important. Every student's participation in the classroom is important, so provide reassurance to this type of student as necessary.

Making accommodations for students with learning disabilities is largely a fact-based situation. Under the ADA, reasonable accommodations must be made for students with disabilities; however, accommodations are not necessary if they create an undue hardship to the employer or agency or pose a direct threat to the safety and security of others.

Instructing Students with Disabilities

Instructors can employ a variety of strategies to assist students with learning disabilities. For example, breaking the material into smaller steps or components will allow a student to comprehend the material more easily. Supplying regular, positive feedback in a manner that does not single out the individual is very effective in encouraging a student with a disability to learn. Use of diagrams, graphics, and pictures to support a lecture is effective in tying learning material together.

A new teaching technique that has been developed over the past few years is called scaffolding. Scaffolding is intended to assist students with learning disabilities, but could be used with any student. Think of metal scaffolding that surrounds a building as it is constructed, serving as a support system. As the structure takes shape and is able to stand on its own, the scaffolding is removed until we see the final building. When working with a student who has a learning disability, the instructor who is using scaffolding establishes a support system when presenting the new material. The instructor gives the student the content, motivation, and foundation of the new material. In building and establishing the scaffolding, the instructor can use some of the following strategies:

- Activate the student's prior knowledge with questions or ask for others' experiences.
- Offer the student a situation, involving the subject being taught, that gives the student a chance to think and give a response from his or her own experiences.
- Break complex course material into easier pieces that will allow a student learning "successes" in understanding the new material.
- Show a student an example of the desired outcome before he or she completes the task.
- Allow a student to "brainstorm" or think out loud while seeking a solution to questions or problems.
- Teach the student mnemonic techniques to help in memorizing material or procedures.
- Teach key vocabulary terms before reading to help improve what is learned while reading.

When using this technique, you offer students support to learn the new material and then slowly or strategically remove the scaffolding to allow the students to stand on their own. For learning to progress, scaffolds should be removed gradually as instruction continues, so that students will eventually be able to demonstrate comprehension independently.

Disruptive Students

Other types of students may pose challenges as well. For example, some students, for one reason or another, feel the need to establish their presence in the class by acting out in other ways (also discussed in Chapter 3, *Methods of Instruction*).

- The class clown feels the need to make comments or jokes about the course material or perhaps others in the classroom.
- The class know-it-all has been there and done that already, regardless of the topic.
- The gifted learner is very familiar with the course material or has the ability to read quickly. This individual becomes bored because he or she is so far ahead of the rest of the class.

Most often, these types of students have other underlying issues that cause them to seek out attention in a negative way. A variety of methods may be used to deal with these types of students, but in general the most effective method is to minimize their impact on the class by not giving into the same type of behavior. Most often a stern look is enough to get a class clown to stop cracking jokes (**FIGURE 2-15**). If this measure fails to quell the disruptive behavior, then engage the student with a question that is meant to bring him or her into the class discussion. If this technique does not work, then at the next break a one-on-one discussion might be appropriate. The last step in dealing with a student who continues to disrupt the learning environment would be to ask him or her to leave the class. In using this form of progressive discipline, the goal is to bring the student back into the classroom in a positive way that does not embarrass either the student or you as the instructor.

The student who is bored or quiet may be displaying that behavior because of a particular circumstance. Sources of this problem may include a lack of interest in the subject, an objection to being forced to take the class, unfamiliarity with the terminology in the class, boredom, and long and technical lectures. This student may just drift off mentally and refuse to participate in the classroom or become disruptive in other ways. As the instructor, you need to keep all students interested and focused on the material. Make your lectures interactive and concise. See Chapter 3, *Methods of Instruction*, for more on this topic.

Other students may not be interested in the topic. They may be in the class simply because they have to be. These students lack energy and attention, and you should try to determine whether this is a regular problem. If so, these students may have a disability or other problem that you may need to discover and address through counseling, tutoring, or other means.

A student who is a slow learner may be the most difficult for you to deal with. Such an individual has trouble keeping up and may not understand some or all of the material. One technique for handling this situation is to encourage input from other students—hearing the information differently may help. During a break, have a respectful one-on-one talk with the student to ensure understanding and comprehension.

Many students have some kind of impediment to learning, and most of those students have found unique ways to compensate for these obstacles. We all have times when our attention drifts or our energy is low, but if this failure to engage in the learning process is a recurring event it needs to be addressed because it will interfere with long-term learning. Some students may need help in the form of tutoring or individualized instruction, for example. These options should be discussed with students who have learning disabilities, and together you and the student should design a learning plan.

FIGURE 2-15 Occasionally, an instructor will have to deal with inappropriate behavior. How might you deal with this situation?
© Jones & Bartlett Learning.

Training BULLETIN

JONES & BARTLETT FIRE DISTRICT
TRAINING DIVISION
5 Wall Street, Burlington, MA, 01803
Phone 978-443-5000   Fax 978-443-8000

Instant Applications: Lesson Plans

Drill Assignment

Apply the chapter content to your department's operation, training division, and your personal experiences to complete the following questions and activities.

Objective

Upon completion of the instant applications, fire and emergency services instructor students will exhibit decision making and application of job performance requirements of the instructor using the text, class discussion, and their own personal experiences.

Suggested Drill Applications

1. During several upcoming company training sessions, evaluate your shift members and identify their learner characteristics.

2. Identify and demonstrate different methods of dealing with learning disabilities.

3. Prior to participating in a company-level training session, identify several of the influences that may affect the ability of your shift members to learn during the session.

4. Review the laws and principles of learning from the text, and identify which of those appear to have the greatest impact on your shift-level training.

Incident Report: NIOSH Report F2007-09

Drill Assignment

Review the information in this incident report and prepare a practice presentation for class delivery at the direction of your instructor. Your presentation should include a summary of the incident facts and a review of the NFPA 1403 compliance and noncompliance findings. Use an outline to organize your thoughts. You may be evaluated on your communication skills during your presentation.

Baltimore, Maryland—2007

The Baltimore City Fire Department was conducting a live fire training exercise in a row house following the department's Fire Fighter I program. The program had not been successfully completed by all participants, and some of the recruits did not successfully complete the physical performance requirements during training. Some of the recruits had also never participated in interior live fire training evolutions.

The involved building was a three-story row house, and was one of a series of very narrow row houses, only 11 ft 4 in. (3.5 m) wide. It was a 1200 ft^2 (111.5 m^2) unit, built as a single-family home. It had been vacant for almost 7 years, and condemned for 3 years. Each unit had separate exterior walls and the side walls touched on one or two sides of the adjacent unit(s). The units were trapezoidal in shape, and this was an end unit (**FIGURE 2-A**).

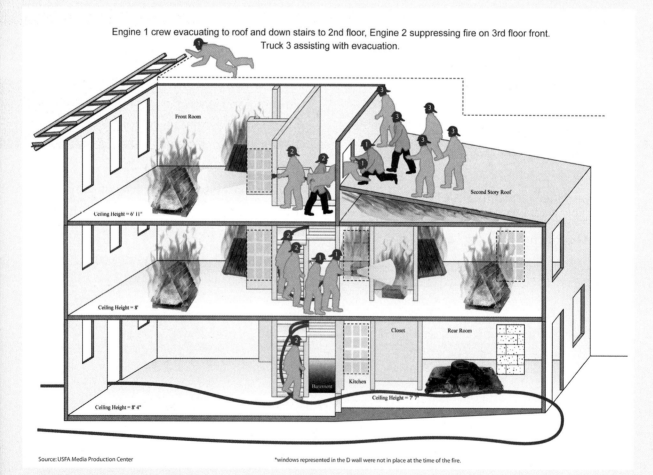

FIGURE 2-A Baltimore floor plan. Note fire locations.
Courtesy of the USFA Media Production Center.

The live fire exercise consisted of multiple fires on all floors. The investigation indicates that approximately 11 bales of excelsior and at least 10 wooden pallets were used for the fires. Pallets were propped against the walls in several locations, with excelsior underneath. Excelsior was also placed in openings in the ceilings and in walls. In the backroom on the first floor, an automotive tire, two full-size mattresses, one twin-size mattress, one foam rubber chair, tree branches, and other debris were piled up and burned. The exact number of fires set could not be definitely determined, but there were at least nine separate fires.

It appears that the ignition of the fires was not coordinated with the preparation of students entering. Several students were not ready when the fires were ignited, and that delay allowed the fires to burn unimpeded while crews made final preparations to enter.

Five separate crews were expected to operate simultaneously as engine and truck companies. A sixth crew was designated as the rapid intervention crew (RIC), but they were not briefed and did not have a hose line.

An adjunct instructor and four students of the first crew, designated as Engine 1, entered through the front door. All of the students had personal alert safety system (PASS) devices, but the adjunct instructor did not, nor did he have a portable radio. Per the direction of the instructor-in-charge, Engine 1 began making their way to the third floor and expected a second hose line to follow. They advanced an initially uncharged 1¾-in. (44-mm) hose line to the centrally located stairs on the first floor. There is disagreement as to when the line was actually charged.

Another crew of recruits was to advance a hose line to the second floor from the rear door. When they entered the rear of the house, they encountered the large pile of debris burning, which inhibited their ingress and egress. The fire was spreading across the ceiling, and they extinguished the fire before making a delayed advance to the second floor. As with the first crew, the four students had PASS devices, but the adjunct instructor did not, nor did he have a portable radio.

The Engine 1 crew advanced to the second floor and encountered heavy fire conditions. Although directed to go to the third floor, the instructor determined it was necessary to knock down the fire before proceeding upstairs to the third floor. Two of the crew's members stayed on the landing between the second and third floors to pull hose, while the remainder of the crew advanced upstairs. Conditions on the third floor became too hot, and the instructor did not have a radio to advise command of their situation or of the interior conditions. The instructor lifted himself up and climbed out of a high window onto the back roof of the second floor. The first student was able to lift her upper body out of the window. The instructor then grabbed her self-contained breathing apparatus (SCBA) straps to pull her out the rest of the way onto the second floor roof. She was then hoisted to the third floor roof by Truck 3, where she told them that her crew needed help.

The other student was now at the window as the instructor attempted to pull her through, using her SCBA harness as he had done with the other fire fighter. She was initially able to talk to the instructor to tell him that she could not help, and that she was burning up. It appears that she still had her SCBA face mask on at this time. The instructor lost his grip, and the student fell back into the room, landing on her feet. When he was able to get hold of her again, she was still conscious, but her mask was partially dislodged or removed. Her face had visibly started to blister. The instructor did not have a radio, but yelled for help. He lost his grip on her a second time, and the student fell back into the room. Shortly after regaining his grip on her for the third time, she became unresponsive. At this point, the crew jumped about 6 ft down from the third floor roof to help pull the student out, but they were unable to do so. With additional assistance, they were eventually able to help lift her onto the roof, where a student used a portable radio to advise of the fire fighter down.

Assistance was requested from the on-duty battalion, engine, and truck units that had come by to watch. The engine company engaged the fire, while the truck company entered the structure to ensure that the students were all out.

Two members of the on-duty truck company climbed the aerial ladder to the roof; they assisted in placing the victim in a Stokes basket and carried her down the ladder. An advanced life support ambulance initiated advanced life support and transported her to a shock trauma center, where her care continued until she was pronounced dead.

Incident Report continued

Postincident Analysis — Baltimore, Maryland

NFPA 1403 Noncompliant

NOTE: Near-total lack of compliance with NFPA 1403. 50 issues considered to be violations of NFPA 1403 by investigative team.

- Flashover and fire spread unexpected
- Walk-through was not performed
- Multiple fires on different floors
- Excessive fuel loading
- Fire was beyond the training and experience of the students to participate in live burn exercises in an acquired structure
- No incident safety officer
- Noncompliant gear on victim
- Instructors without PASS
- Instructors not equipped with portable radios
- Adjunct instructors had little to no prior instructional experience

NFPA 1403 Compliant

- 22 students, 11 instructors
- EMS (ALS transport) on standby on scene

Other Contributing Factors

- Students not informed of emergency plans/procedures
- Rapid intervention crew staffed by students and unequipped or unprepared

After-Action REVIEW

IN SUMMARY

- As an instructor, you are a leader and have a responsibility to teach information, to hone skills, and to promote positive values and motivation in your students.
- According to Edward L. Thorndike, there are six laws of learning: readiness, exercise, effect, primacy, recency, and intensity.
- Of the five senses, seeing (e.g., a visual demonstration of a skill) is generally the most effective means of learning, followed by hearing (e.g., listening to a lecture), smelling (e.g., smelling smoke), touching (e.g., handling fire equipment), and tasting (rarely used in training fire fighters).
- Two schools of thought have developed to explain how behavior evolves: behaviorist and cognitive.
  - According to the behaviorist perspective, learning is a relatively permanent change in behavior that arises from experience.
  - According to the cognitive perspective, learning is an intellectual process in which experience contributes to relatively permanent changes in the way individuals mentally represent their environment.
- Learning that is intended to create or improve professional behavior is based on performance, rather than on content. Such competency-based learning is usually tied to skills or hands-on training.
- Mandating learning often results in a lack of motivation in classroom activities and a poor attitude toward the learning process in general, which present quite a challenge to the instructor who is tasked with teaching the course.
- To assist your students in becoming effective learners, you can share study skills that enhance what they learn during classroom and personal study time. Study tips can be viewed in three categories: classroom, personal study, and test preparation.
- Abraham Maslow determined that certain needs should be satisfied to enable a person to move forward toward self-actualization:
  - Physiological needs
  - Safety, security, and order
  - Social needs and affection
  - Esteem and status
  - Self-actualization
- Benjamin Bloom and his research team identified three types of learning domains:
  - Cognitive domain (knowledge)
  - Psychomotor domain (physical use of knowledge)
  - Affective domain (attitudes, emotions, or values)
- Adult learners can be divided into three types: learner oriented, goal oriented, and activity oriented. Each type is associated with different motivational issues, which in turn affects how learning takes place.
- Researchers focused on learner characteristics are divided into two distinct camps: those emphasizing the cognitive learning domain and those highlighting other learning-related factors, such as motivation and personal preferences.
- Once you have identified students with some type of learning disability, then you need to adjust your teaching as necessary to ensure their learning. There are many ways to help students in your class who are struggling to learn.
- Most often, disruptive students have other underlying issues that cause them to seek out attention in a negative way.

KEY TERMS

Access Navigate for flashcards to test your key term knowledge.

Active learning Anything students do beyond simply receiving instruction and passively listening to a lecture or watching a video.

Affective domain The domain of learning that affects attitudes, emotions, or values. It may be associated with a student's perspective or belief being changed as a result of training in this domain.

Andragogy The art and science of teaching adults.

Attention-deficit/hyperactivity disorder (ADHD) A disorder in which a person has a chronic level of inattention and an impulsive hyperactivity that affects daily functions.

Audience analysis The determination of characteristics common to a group of people; it can be used to choose the best instructional approach.

Baby boomers The generation born after World War II (i.e., 1946–1964).

Behaviorist perspective The theory that learning is a relatively permanent change in behavior that arises from experience.

Bloom's Taxonomy A classification of the different objectives and skills that educators set for students (learning objectives).

Cognitive domain The domain of learning that effects a change in knowledge. It is most often associated with learning new information.

Cognitive perspective An intellectual process by which experience contributes to relatively permanent changes (learning). It may be associated by learning by experience.

Competency-based learning Learning that is intended to create or improve professional competencies.

Dyscalculia A learning disability in which students have difficulty with math and related subjects.

Dyslexia A learning disability in which students have difficulty reading due to an inability to interpret spatial relationships and integrate visual information.

Dysphasia A learning disability in which students lack the ability to write, spell, or place words together to complete a sentence.

Dyspraxia A lack of physical coordination with motor skills.

Generation X People born after the baby boomers.

Generation Y People born immediately after Generation X.

Generation Z People born immediately after Generation Y, starting in the late 1990s.

Kinesthetic learning Learning that is based on doing or experiencing the information that is being taught.

Learner characteristics Designating a target group of learners and defining those aspects of their personal, academic, social, or cognitive self that may influence how and what they learn. (NFPA 1041)

Learning A relatively permanent change in behavior potential that is traceable to experience and practice.

Learning domains Categories that describe how learning takes place—specifically, the cognitive, psychomotor, and affective domains.

Motivation The activator or energizer for an activity or behavior.

Motivational factors States of the person that are relatively temporary and reversible and that tend to energize or activate the behavior of the individual.

Pedagogy The art and science of teaching children.

Psychomotor domain The domain of learning that requires the physical use of knowledge. It represents the ability to physically manipulate an object or move the body to accomplish a task or use a skill.

Student-centered learning Educational methodologies that focus on student engagement and require students to be active, responsible participants in the learning experience. (NFPA 1041)

REFERENCES

Ambrose, S. A., M. W. Bridges, M. DiPietro, M. C. Lovett, and M. K. Norman (Eds.). 2010. *How Learning Works: Seven Research-Based Principles for Smart Teaching.* San Francisco: Jossey-Bass.

Anderson, Lorin W., Krathwohl, David R., Airasian, Peter W., Cruikshank, Kathleen A., and Mayer, Richard E. (Eds.). 2001. *A Taxonomy for Learning, Teaching, and Assessing: A Revision of Blooms Taxonomy of Educational Objectives.* Boston: Allyn & Bacon.

Armour, Stephanie. 2005. "Generation Y: They've Arrived at Work with a New Attitude." USA Today, November 6, 2005. http://www.usatoday.com/money/workplace/2005-11-06-gen-y_x.htm.

Bender, D. 1999. *Unique characteristics of adult learning.* www2.bw.wdu/~dbender/csu/adultlearn.html. Accessed August 10, 2018.

Bloom, Benjamin S. 1984. *Taxonomy of Educational Objectives.* Boston: Allyn and Bacon/Pearson Education.

Bloom, Benjamin S., David R. Krathwohl, and Bertram B. Masia. 1965. *Taxonomy of Educational Objectives: Book 2.* London: Longman.

Bonwell, C. C., and J. A. Eison. 1991. Active learning: Creating excitement in the classroom (ASHE–ERIC Higher Education Rep. No. 1). Washington, DC: The George Washington University, School of Education and Human Development.

Brown, B. L. 1997. *New learning strategies for Generation X.* ERIC Clearinghouse. http://www.ed.gov/databases/ERIC.

Butler, Gillian, and Frida McManus. 1998. *Psychology.* Oxford: Oxford University Press.

Connick, G. 1997. "Beyond a Place Called School." *NLII Viewpoint* (Fall/Winter).

Crouch, C. H., and E. Mazur. 2001. "Peer Instruction: Ten years of experience and results." *American Journal of Physics,* 69 (9): 970. doi:10.1119/1.1374249

Dave, R. H. 1970. *Developing and Writing Behavioral Objectives.* Tucson: Educational Innovators Press.

Draper, J. 1993. *The Craft of Teaching Adults.* Toronto: Culture Concepts.

Elias, J. L., and S. Merriam. 1980. *Philosophical Foundations of Adult Education.* Malabar, FL: Krieger.

Erikson, E.H., 1959. Identify and the life cycle: Selected papers. *Psychological Issues,* 1: 1–171.

Hacker, Douglas J., John Dunlosky, and Arthur C. Graesser (Eds.). 2009. *Handbook of Metacognition in Education.* London: Routledge.

Houle, C. 1961. *The Inquiring Mind.* Madison, WI: University of Wisconsin Press.

Knowles, M. 1984. *The Adult Learner: A Neglected Species.* 3rd ed. Houston, TX: Gulf Publishing.

Knowles, M. 1990. *The Adult Learner: A Neglected Species.* Revised ed. Houston, TX: Gulf Publishing.

Laskowski, Lenny. A.U.D.I.E.N.C.E. Analysis—It's Your Key to Success. Presentation-Pointers website, copyright 2000-2010. http://www.presentation-pointers.com/showarticle/articleid/248/. Accessed August 17, 1018.

Lefrancois, Guy R. 1996. *The Lifespan.* 5th ed. Belmont, CA: Wadsworth.

Levinson, Daniel J. 1977. The Mid-LIfe Transition: A Period in Adult Psychosocial Development. *Psychiatry,* 40: 99–112.

McClincy, W. D. 1995. *Instructional Methods in Emergency Services.* Englewood Cliffs, NJ: Brady Prentice Hall.

McClincy, W. D. 2002. *Instructional Methods in Emergency Services.* 2nd ed. Englewood Cliffs, NJ: Brady Prentice Hall.

Merriam, Sharan B., and Rosemary S. Caffarella. 1999. *Learning in Adulthood.* San Francisco, CA: Jossey-Bass.

National Fire Protection Association. 2019. *NFPA 1041: Standard for Fire Service Instructor Professional Qualifications.* Quincy, MA: National Fire Protection Association.

National Fire Protection Association. 2018. *NFPA 1403: Standard on Live Fire Training Evolutions.* Quincy, MA: National Fire Protection Association.

National Research Council. 2000. *How People Learn: Brain, Mind, Experience, and School: Expanded Edition.* Washington, DC: The National Academies Press.

Piaget, Jean, and Barbel Inhelder. 1958. *The Growth of Logical Thinking from Childhood to Adolescence.* London: Routledge.

Rathus, Spencer A. 1999. *Psychology in the New Millennium.* 7th ed. Fort Worth, TX: Harcourt Brace College.

Seemiller, Corey, and Meghan Grace. 2017. "Generation Z: Educating and Engaging the Next Generation of Students." *About Campus* 22 (3), 21–26.

Seemiller, Corey, and Meghan Grace. 2016. *Generation Z Goes to College.* New York: San Francisco: Jossey-Bass.

Stewart, D. 2003. "Computer and Technology Skills." http://www.ncwiseowl.org/Kscope/techknowpark/Kiosk/index.html.

"Student-Centered Learning Definition." *The Glossary of Education Reform,* Great Schools, 7 May 2014, www.edglossary.org/student-centered-learning/.

Thorndike, Edward L. 1932. *The Fundamentals of Learning.* New York: Teachers College Press.

Wlodkowski, Raymond J. 1999. *Enhancing Adult Motivation to Learn: A Comprehensive Guide for Teaching All Adults.* Revised ed. San Francisco, CA: Jossey-Bass/Pfeiffer.

CHAPTER PRESENTATION EXERCISE

Instructions: Your course instructor will assign individual or group discussions on the key points and teaching tips of this chapter. You or your group will review the chapter teachings and identify the major learning points from the chapter. You should be able to discuss the points, why they are important, and how/where they apply to your responsibilities at your level of instructor training.

REVIEW QUESTIONS

1. Two schools of thought have developed to explain how behavior evolves: behaviorist and cognitive. Describe each of these.
2. Learning takes place in three domains. List and explain them.
3. Provide examples of *convergent* and *divergent* questioning techniques.

CHAPTER 3

Methods of Instruction

NFPA 1041 JOB PERFORMANCE REQUIREMENTS

Note: An asterisk denotes that the 1041 standard contains further information in its annex section.

4.3.2 Review instructional materials, given the materials for a specific topic, target audience, learner characteristics, and learning environment, so that elements of the lesson plan, learning environment, and resources that need adaptation are identified.

(A) Requisite Knowledge.
Recognition of student learner characteristics and diversity, methods of instruction, types of resource materials, organization of the learning environment, and policies and procedures.

(B) Requisite Skills.
Analysis of resources, facilities, and materials.

4.3.3* Adapt a prepared lesson plan, given course materials and an assignment, so that the needs of the student and the objectives of the lesson plan are achieved.

(A)* Requisite Knowledge.
Elements of a lesson plan, selection of instructional aids and methods, and organization of the learning environment.

(B) Requisite Skills.
Instructor preparation and organization techniques.

4.4.2 Organize the learning environment, given a facility and an assignment, so that lighting, distractions, climate control or weather, noise control, seating, audiovisual equipment, teaching aids, and safety are addressed.

(A) Requisite Knowledge.
Learning environment management and safety, advantages and limitations of audiovisual equipment and teaching aids, classroom arrangement, and methods and techniques of instruction.

(B) Requisite Skills.
Use of instructional media and teaching aids.

4.4.3 Present and adjust prepared lessons, given a prepared lesson plan that specifies the presentation method(s), so that the method(s) indicated in the plan are used and the stated objectives or learning outcomes are achieved, applicable safety standards and practices are followed, and risks are addressed.

(A)* Requisite Knowledge.
The laws and principles of learning, methods and techniques of instruction, lesson plan components and elements of the communication process, and lesson plan terminology and definitions; learner characteristics; student-centered learning principles; instructional technology tools; the impact of cultural differences on instructional delivery; safety rules, regulations, and practices; identification of training hazards; elements and limitations of distance learning; distance learning delivery methods; and the instructor's role in distance learning.

(B) Requisite Skills.
Oral communication techniques, methods and techniques of instruction, ability to adapt to changing circumstances, and utilization of lesson plans in an instructional setting.

4.4.4* Adjust to differences in learner characteristics, abilities, cultures, and behaviors, given the instructional environment, so that lesson objectives are accomplished, disruptive behavior is addressed, and a safe and positive learning environment is maintained.

(A)* Requisite Knowledge.
Motivation techniques, learner characteristics, types of learning disabilities and methods for dealing with them, and methods of dealing with disruptive and unsafe behavior.

(B) Requisite Skills.
Basic coaching and motivational techniques, correction of disruptive behaviors, and adaptation of lesson plans or materials to specific instructional situations.

KNOWLEDGE OBJECTIVES
After studying this chapter, participating in a structured learning environment, and completing assigned assessments, you will be able to:

- Describe methods of instruction typically used in adult and fire service education. (**NFPA 1041: 4.4.3**, pp 75–77)
- Describe the four-step method of instruction. (pp 77–78)
- Explain how distance learning can enhance instruction. (**NFPA 1041: 4.4.3**, pp 78–79)
- Describe communication techniques that will improve your presentation. (**NFPA 1041: 4.4.3**, pp 79–82)
- Explain methods for controlling the physical learning environment. (**NFPA 1041: 4.3.2**, pp 82–84)
- Describe how to create a learning environment that facilitates the learning process. (**NFPA 1041: 4.4.2**, pp 85–92)
- Identify issues unique to outdoor learning environments and explain how to address them. (**NFPA 1041: 4.3.2**, pp 89–92)
- Identify considerations when arranging learning environments for testing functions. (**NFPA 1041: 4.3.2**, p 89)
- Describe the effect of demographics on the learning environment. (**NFPA 1041: 4.4.3**, pp 93–94)

SKILLS OBJECTIVES
After studying this chapter, participating in a structured learning environment, and completing assigned assessments, you will be able to:

- Deliver an instructional session using prepared course material. (**NFPA 1041: 4.4.1; 4.4.3**, pp 75–82)
- Manage disruptive and unsafe behaviors in the classroom. (**NFPA 1041: 4.4.4**, pp 82–84)
- Organize the learning environment to meet the requirements of the students and the material to be presented. (**NFPA 1041: 4.3.2; 4.4.2**, pp 85–92)
- Present material to a diverse audience following acceptable presentation methods and techniques. (**NFPA 1041: 4.3.3**, pp 85; 93–94)

You Are the Fire and Emergency Services Instructor

As the newest member assigned to your shift, your shift officer has asked you to conduct a training session about the department's fire fighter safety and survival program. Prior to the session, however, he warns you that several of the more veteran fire fighters on your shift have no interest or buy-in to the topic as your department has never experienced a fire fighter line-of-duty death or even a close call. Knowing that you are going to be evaluated on how well you present this topic, address the following:

1. Identify several motivational techniques or methods you can use to promote interest and buy-in among your audience, especially the more veteran fire fighters.
2. List the steps of the four-step method of instruction and how you can use these steps to deliver a successful training session.
3. Consider different evaluation and questioning techniques you can use to assess the learning process during your training session.

Access Navigate for more practice activities.

Introduction

As a fire and emergency services instructor, you must constantly ask, "How can I best help my students to learn?" It is your responsibility to present information and have a clear understanding of the course objectives. Instruction is the single most important factor in students' success: Teaching has a greater impact on achievement than all other factors combined.

Whether the class is a large introductory course for recruits or an advanced course for fire officers, it makes good sense to start off well. Students decide very early—maybe even within the first class—whether they will like the course, its contents, the instructor, and their fellow students.

Fire fighters are adults. This point may seem obvious, but sometimes instructors can lose sight of that fact. Some of the strategies that work with second graders also work well with adults, but there are important differences in the way adults and children learn. Never treat your fire service students like children—you are teaching adults. Remember, too, that although you are teaching to a class, the individual may require a particular learning strategy.

Methods of Instruction

The **method of instruction** is the process of teaching material to students. You can also identify it as how your lesson plan will be given to your students. Many different and diverse methods of instruction exist, and each applies to different types of materials as well as different types of students. Each of the various student demographics and categories discussed earlier in this chapter will respond differently to the material according to how they are motivated and the method(s) of instruction used. If you are an Instructor I, the method of instruction will be identified on your lesson plan by the lesson plan developer. Chapter 5, *Using Lesson Plans*, discusses formatting of and placeholders in the lesson plan that developers can use to identify the methods of instruction they have determined to be most appropriate for the material. The same chapter also discusses the parts of a lesson plan and provides guidance on how lessons are constructed, considering the method of instruction to be used.

Determining which method of instruction should be used to deliver information is accomplished based on several factors:

- The experience and educational level of the students
- The classroom setting or training environment
- The objectives of the material to be taught
- The number of students to be taught

As an instructor at the Instructor I level, you should understand the different methods of instruction and master those methods. Based on this fundamental knowledge, when a lesson plan states that Instructor I personnel are to deliver a "lecture," they understand the method of instruction to be used. For Instructor II personnel, as they develop a lesson plan, they will select the most appropriate and effective method to deliver the material in the lesson plan they have created. Any level of instructor should be able to adapt a lesson plan to the learning environment as well as to the audience who will receive the instruction so as to make it most effective. All instructors will perform an audience analysis by looking at the demographics of the audience, including age, learning level, previous experience, and other factors, and then adapt the lesson plan accordingly.

Ask yourself three basic questions when reviewing and planning for the delivery of a lesson plan:

- What are the goals of the lesson?
- Who is the intended audience?
- What are the resources and learning environment characteristics?

Methods of instruction vary among disciplines, but for the fire service the following techniques are known to be the most effective in teaching new and in-service personnel.

Lecture

The lecture method is used widely in the fire service. It consists of the instructor being the primary source of information and delivering the course material while using various multimedia tools (**FIGURE 3-1**). As part of your Instructor I job duties, you will be required to present an illustrative lecture in a practice teaching environment to a group of fellow students. An illustrative lecture allows the instructor to use a visual aid such as a presentation slide to illustrate key points, which is important to effective communication and is probably the most common and familiar method of fire service instruction.

Demonstration or Skill Drill

This method is used when teaching skills or hands-on learning. This type of learning is used when the instructor demonstrates the use of a new tool or demonstrates how to perform a specific task-oriented skill (**FIGURE 3-2**). Several variations in the sequence of how these types of drills should progress are possible, especially when learning a new skill. You might

FIGURE 3-1 The lecture method of instruction consists of the instructor being the primary source of information and delivering the course material while using various multimedia tools.
© Jones & Bartlett Learning.

FIGURE 3-2 Skill drills allow the student to learn by actively participating and physically using the new tool or practicing the new skill.
© Jones & Bartlett Learning.

the new skill. While the student engages in hands-on application during this method of learning, the instructor needs to monitor student safety. This method of instruction references a skill drill and requires the instructor to ensure that personal protective equipment (PPE) is in use and proper student-to-instructor ratios are maintained to ensure student and instructor safety. Chapter 7, *Training Safety*, discusses appropriate levels of safety considerations that must be applied to this method of instruction.

Discussion

The discussion method is used to facilitate the learning process by having a two-way open forum, from instructor to student and from student to student (**FIGURE 3-3**). This method fosters interaction among the instructor and the students; the instructor's role in this method is more of a facilitator or monitor to ensure the discussion stays on track. Time keeping will be an essential element when this method of instruction is employed. Discussion is typically used as the method of instruction when consensus is needed or when new ideas are necessary to explore higher levels of learning and application. In most cases, this method requires students to have some prerequisite level of topic knowledge.

Based on the material to be covered and the make-up of the class, most instructors use a combination of these methods while delivering course material. For example, instructors might deliver course material via lecture and then engage in a discussion to determine the level of understanding from the students. When teaching a new skill or use of a tool, the instructor might begin with a lecture and then demonstrate

view the training as a crawl–walk–run sequence or engage in an instructor-led demonstration—that is, you might demonstrate the skill at full speed for the students, then slowly explain each step and stop at each step point, then perform it one more time at full speed while explaining the step. This approach is also known as a psychomotor lesson or hands-on method of instruction. As part of the Instructor 1 job duty area, you may be required to present a demonstrative presentation where a student will perform a task from a prepared skill sheet as part of the student instructor's practice presentation.

Most fire fighters and other adult learners respond very well to hands-on training. This method gives each student the opportunity to learn by actively participating and physically using the new tool or practicing

FIGURE 3-3 Discussion is used to facilitate the learning process by having a two-way open forum between instructor and student.
© Jones & Bartlett Learning.

how to use the tool; the class might then move to the training ground to practice using the new tool.

The Four-Step Method of Instruction

While reviewing and preparing for class with your lesson plan, the **four-step method of instruction** is the primary process used to relate the material contained in the lesson plan to the students (**TABLE 3-1**). These four steps are found within the lesson plan. The four-step method of instruction is the most basic method of instruction used in the fire service.

TABLE 3-1 The Four-Step Method of Instruction

Step in the Instructional Process	Instructor Action
1. Preparation	The instructor prepares the students to learn by identifying the importance of the topic, stating the intended outcomes, and noting the relevance of the topic to the student. This should not be confused with the time used by the instructor to prepare course material and review lesson plan content.
2. Presentation	The presentation of content is usually organized in an outline form that supports an understanding of the learning objectives. Varied presentation techniques should be used to keep the students' interest and to maximize learning.
3. Application	The instructor provides opportunities for the students to *apply* what is being presented through questions, exercises, and discussion. Typically, this is not a graded activity, since it is the opportunity to practice under the supervision and support of the instructor.
4. Evaluation	The students' understanding is evaluated through written exams or in a practical skill session.

Step 1: Preparation

The preparation step is the first phase in the four-step method of instruction. The preparation step—also called the motivation step—prepares or motivates students to learn. When beginning instruction, you should provide information to students that explains why they will benefit from the class. Adult learners need to understand what they will get out of attending the class, because very few adults have time to waste in sitting through a presentation that will not directly benefit them.

The benefit of a class can be explained in a few ways:
- The class may count toward required hours of training.
- The class may provide a desired certification.
- The class may increase students' knowledge of a subject.

Whatever the benefit may be, you should explain it thoroughly during the preparation step. In a lesson plan, the preparation section usually consists of a paragraph or a bulleted list describing the rationale for the class. During the preparation step, the Instructor I needs to grab students' attention and prepare them to learn when he or she begins presenting the prepared lesson plan. Adult learners like to learn quickly how the class material will affect them: Will it make them safer or more knowledgeable about their job? Will it improve their efficiency on the fire ground or make them better leaders? This step in the process is critical as a motivational tool to encourage the students to participate in the learning process, not just sit in on the class as a way to pass the time.

Step 2: Presentation

The presentation step is the second step in the four-step method of instruction; it comprises the actual presentation of the lesson plan. During this step, you lecture, lead discussions, use audiovisual aids, answer student questions, and perform other techniques to present the lesson plan. In a lesson plan, the presentation section normally contains an outline of the information to be presented. It may also contain notes indicating when to use teaching aids, when to take breaks, or where to obtain more information.

Step 3: Application

The application step, which is the third step in the four-step method of instruction, is the most important step because it is during this phase that students apply the knowledge presented in class. The root word *apply* is very important to illustrate what the instructor needs to let the students do: apply what they are learning to examples or actual hands-on practice. Normally, this is where learning occurs, as students practice skills, perhaps make mistakes, and retry skills as necessary. You should provide direction and support as each student performs this step. You must also ensure that all safety rules are followed as students engage in new behaviors.

In a lesson plan, the application section usually lists the activities or assignments that the student will perform. In the fire service, the application section of a psychomotor objective lesson often requires the use of skill sheets for evaluation purposes. The experienced instructor uses the application step to make sure that each student is progressing along with the lesson plan. This step also allows students to participate actively and to remain engaged in the learning process.

Step 4: Evaluation

The evaluation step is the final step of the four-step method of instruction. It ensures that students have correctly acquired the knowledge and skills presented in the lesson plan. The evaluation may, for example, take the form of a written test or a skill performance test. No matter which method of evaluation is used, the student must demonstrate competency without assistance. In a lesson plan, the evaluation section indicates the type of evaluation method and the procedures for performing the evaluation.

Pre-Course Survey and Prerequisites

Some instructors use programs that require students to take a pre-course survey. This survey might try to determine student motivation for taking the class or the students' educational level, work experience, or previous training. Its results can also be used to adapt the lesson plan based on the students' needs.

A tool used in some classes is a pre-course test or quiz, which is used to gauge the students' knowledge of the material. After reviewing the results, you can adjust the course material accordingly to meet the needs of the students better. These tests or quizzes do not count toward the final grade, but they do provide an understanding of how the instructor can best help each student.

Some programs require that students take prerequisite classes before attending certain advanced classes. The fact that students have completed such preliminaries ensures that the students taking your course have a certain base level of training and education. It also ensures that students do not get into a course that is beyond their experience level. Success in a class will be impaired if a student is already behind the knowledge level for the course. You must ensure that your students have the correct competencies before moving to the next level.

Enhanced Instructional Methods

Instructors have other tools at their disposal to enhance the learning process and increase the retention of the material being taught. These tools include the use of role play, labs/simulations, case studies, and assignments to be done after class, such as handouts or assignments in a course workbook:

- **Role play:** Students become part of the lesson by assuming roles and "acting out" a situation or scenario meant to reinforce material from the lesson. Many leadership and management classes feature a senior instructor role-playing a problem employee or angry citizen to allow students to practice handling confrontations.
- **Labs/simulations:** Students are placed in a "real-life" situation, albeit in a controlled environment. A table-top fire-ground lab where the student can run a fire scene is an example of such a tool. Placing students under pressure in a controlled scenario strengthens information learned in the classroom. Many simulation programs that create a very realistic learning environment now exist to simulate actual fires and incidents (**FIGURE 3-4**). A simple online

FIGURE 3-4 Many new simulation programs that create a very realistic learning environment now exist to simulate actual fires and incidents. Pictured here is a Command Training Center using the Blue Card Certification program.
Courtesy of Orland Fire Protection District.

search of fire scene simulations will reveal multiple sources of simulation-based programs.
- **Case studies:** Instructors often take advantage of examples from the real world to teach material. Using an incident report that was generated after a line-of-duty death and allowing students to read the report, discuss the material, and then determine ways to apply the knowledge is an effective way to work with a case study. The National Institute for Occupational Safety and Health (NIOSH) fire fighter fatality reports provide tremendous learning opportunities in many types of training sessions. The Charleston, South Carolina, furniture store fire investigation, for example, has become a staple in case study presentations discussing fire behavior, risk assessment, and tactical options. Most fire fighters can recall a presentation that included case studies when such scenarios relate to the lesson being presented, especially if the scenarios are from incidents with which they are familiar. All fire fighters should be aware of the significant fire fighter fatality investigations that have shaped our profession (such as the Hackensack, New Jersey, car dealership fire; the Worcester, Massachusetts, cold storage warehouse fire; and the Phoenix, Arizona, supermarket fire), as they have greatly influenced modern firefighting knowledge.
- **Out-of-class assignments:** Assignments given to students to be completed outside of class are another effective method of helping students learn. Many courses require pre-course work to be completed prior to class; the instructor might then give assignments in a workbook to be completed during the course. If the instructor does use this tool as part of the coursework, the instructor needs to make sure that these assignments are evaluated and that feedback is provided to the students. Failure to follow through with these measures will give the impression that the material was not important to learn.

Three other methods of instruction are becoming prevalent in today's fire service: online learning, independent study, and blended learning. These methods of learning are gaining popularity for many reasons (such as budget constraints or distance to a training location) and are being used in many disciplines for new or continuing education training.

Online learning (or distance learning) is conducted on the internet and takes the form of students entering a "classroom" that is hosted by an organization or campus. Delivery methods for the classroom consist of either a "synchronous" or "asynchronous" delivery method. In **synchronous learning**, the entire class meets at the same time, and the instructor delivers the course material in real time by voice and/or video. Students interact with the instructor and class verbally or with instant messaging. In **asynchronous learning**, in contrast, the course material is provided in the online classroom and students enter and exit it on their own schedule. Classroom discussion in an asynchronous class is facilitated by use of discussion boards where the instructor will post discussion questions or assignments and students are required to post responses to the instructor or to classmates and complete assignments that students then upload to the instructor.

Independent study has been used for many years (primarily in the military) to help students complete course material at distant locations. It entails registering for a course, receiving course material in the mail, completing assignments, and submitting them back to an instructor for evaluation. This method is one example of distance learning for those who might not have access to the internet.

A third method of instruction combines online/independent study with a face-to-face meeting with the instructor. This **blended learning** format allows for students to learn material on their own at remote locations. Then, at certain benchmarks within the curriculum, the class gathers to complete the hands-on portion of the training. Chapter 6, *Technology in Training*, further discusses technological resources, distance delivery, and creative ways for instructors to enhance their presentations.

All of these methods of instruction are dependent on the instructor knowing how to use each method and applying them while delivering the material as outlined on the lesson plan.

Lesson Presentation Skills and Techniques

No matter which methods of instruction you use or which generations you teach, the following communications techniques will help you improve the delivery of your information to your class:

- **Dress professionally.** Students should see you as a professional and recognize you instantly as a subject-matter expert. Even though we like to believe that we don't "judge a book by its cover," you will be sized up by your students. A professional appearance lays the groundwork for making a good first impression.
- **Welcome each student.** Your first impression begins the moment that students enter the

FIGURE 3-5 A good way to make a positive impression is to greet your students as they enter the classroom.
© Jones & Bartlett Learning.

classroom. Shake students' hands as they arrive, introduce yourself, and have a warm smile on your face—this will immediately convey that you are an approachable person who cares about your students (**FIGURE 3-5**).

- **Call students by name.** Making every student feel important is essential in creating motivation and maintaining attention. Calling students by their names is one of the most effective methods of demonstrating to them that they are valued. Although it may not be possible in large, lecture-style courses that meet only once, most fire service courses are small enough to allow you to learn the students' names. Many of us struggle to learn names, but a variety of methods have been developed that can help you with this task. One of the most basic strategies is to use name tents or cards with the students' names. Another technique is to keep a seating chart in front of you. Name association may also work well. Each method has its own advantages and disadvantages, but determining a method that works best for you will go a long way toward helping you manage your class.
- **Use eye contact.** In our culture, it is important to make eye contact with people to whom you are speaking. Making eye contact with each fire fighter for a few moments in the course of your presentation will help students feel that you are communicating directly with them. Resist the temptation to look at the wall, at the clock, or over students' heads. Lack of eye contact makes people think you are not willing to be straight with them or that you are not confident in what you are saying. In fact, a lack of eye contact is probably one of the most disturbing mannerisms you can display.
- **Use your normal speaking voice but instill enthusiasm in your message.** As an instructor, you have a strong motivation to build a positive relationship with your class. If you are committed to the importance of what you are teaching, then that commitment will be revealed in your voice and by the enthusiasm that you display for the topic. Remember that enthusiasm is contagious—let it show!
- **Speak with respect.** You are teaching adults, and it is important that they perceive that they are respected as adults. When you are respectful to them, they will be respectful to you.
- **Use pauses when you speak.** A short pause can be used to signal an important point or a transition in your lecture. A pause can also indicate the conclusion of one topic and the beginning of another. It allows you the opportunity to view your students and visually check for comments, interest levels, or questions.
- **Use information appropriately.** It is important to gauge your students' level of understanding correctly. For example, Fire Fighter I students will have a different understanding of information than seasoned fire officers. Make sure you speak at a knowledge and vocabulary level that students will understand and explain concepts or information on the same level. You may have to spend time ensuring that the level of training or education is appropriate for all participants in the course.
- **Use gestures sparingly, for emphasis.** Avoid meaningless gestures—for example, jingling keys or change in your pocket, playing with a pencil, or tapping mindlessly on a podium or desk. Try to avoid playing with things, chewing on the ends of pencils, or repeatedly tapping your foot. Hide those tempting items if playing with them has become an unconscious habit for you. Eventually, these habits will become intrusive—and distracting to your students—and students will be placing bets on whether you will swallow the pencil or how many times you will tap your foot in a minute.
- **Practice positive mannerisms.** Well-placed mannerisms can increase your effectiveness in presentation of your lectures, and appropriate gestures can add to the verbal messages. Vary your position and movement, because your ability to change the pace of the lesson will increase involvement. Your body language and facial expression can give a fire fighter support and encouragement even when you are not saying anything aloud.

- **Choose your teaching aids appropriately.** The technology that you have at your fingertips in the classroom is simply a way to communicate with students and reinforce information. Eraser boards and paper pads have some unique advantages in communicating, as they can be used to convey a variety of information during a single lesson. If you use these tools, make sure that your writing is legible. These teaching aids also provide a visual reinforcement of your verbal message. Slide presentations, charts, models, and videos are other means of communicating information to students and can reinforce information being presented verbally. Your wise use of technology can enrich your classroom and assist you in teaching. Even so, you should not rely on technology alone to teach: It can malfunction, so always have a backup plan.
- **Know how to use your chosen teaching aids in the classroom.** Your communication with your students is interrupted when you do not handle technology smoothly. Each time you are guilty of mishandling your teaching aids, you will weaken the impact of the lesson by sending an unintended message ("I don't know what I'm doing").
- **Transition techniques during presentations.** While presenting your lesson plan, make sure that you make smooth transitions between your media and verbal communication techniques. Distributing handouts before you want to have the students read them is an example of using media (nonprojected materials/handouts) at the wrong time. If you want to hold students' attention, then use media in the correct manner and use transition techniques that enhance the learning process rather than distract from it.

Transition techniques also are used during a presentation to move from one topic to the next. Many students get lost during a lecture when the instructor moves on and the students do not realize or understand the instructor has changed topics. To help students follow a presentation, instructors need to learn transition techniques. Five effective techniques are delineation, words/phrases, pictures, voice, and body language:

- **Delineation.** This is the simplest way to make a transition. As an instructor, you tell your class that you will be covering four points and then begin by stating "point number one," and then move to point two, point three, and point four, and then summarize the material.
- **Words/phrases.** Use phrases such as "the next point we need to discuss," or summarize the topic and then use a phrase like "the next step in the process . . ."
- **Pictures.** Because so many of today's lessons use PowerPoint presentations, an effective tool to use within such a presentation is the insertion of historic or unique pictures between points or topics. This visual signal will allow you to pause and switch topics, and the students will quickly pick up on the fact that the topic has changed.
- **Voice.** Use your voice by changing your tone or pitch, by using different volumes to emphasize change from one idea to another, or by inserting an effective pause. Pausing will draw students' attention to you and will develop anticipation about what is coming next.
- **Body language.** Effective use of body language will assist you in moving from one topic to the next. Moving from the podium to one side of the classroom with a change in voice will clearly inform students that the topic has changed or is very important. Make your movements conscious and meaningful, and try to avoid nervous pacing.

All of these techniques or methods will allow you to make transitions during a presentation, but they need to be practiced to be used effectively. Some instructors will develop all of these transition techniques; others will not. The goal is to develop transition techniques that work best for you.

Effective Communication

Effective communication is fundamental to ensuring that the students learn—that they receive and remember the information you intended to transmit to them. As a general guideline consider that what student read is the basic level of learning but as you increase other senses such as hearing, seeing, combined with what is spoken words and hands on application increases the learning and communication process.

It is important to choose the right kind(s) of communication when you are preparing for class. It is also important to use the selected communication methods effectively. Understanding how the communication process works will help you develop lesson plans or adapt lesson plans based on your students' learning characteristics. For example, if your students are visual learners, then you may need to include more visual presentations. See Chapter 4, *Communication Skills*, for a discussion of the elements and role of the communication process.

Questioning Techniques

An important part of the learning process is to assess the learning process during class. One simple method used over and over is to ask students questions from the material that is being taught from the lesson plan. Many lesson plans will build in questions for the instructor to ask the students to determine whether the students are picking up the new information and understanding what is being taught.

When using questions during the teaching process, it is important to ask the correct question at the correct time. Again, many lesson plans will identify the appropriate time to ask questions to support what has just been taught. Another appropriate place to seek student feedback is at the end of a lesson during a summary activity. There are times during a presentation at which an instructor can use a question to stimulate discussion, to bring class focus to a new topic, or to break up side discussions in the classroom. When used correctly, a question can also be used to challenge a disruptive student positively and bring him or her back into the classroom.

Numerous types of questions can be used during the learning process, including the following:

- **Direct questions.** Direct questions are those beginning with *who*, *what*, *where*, *when*, *why*, or *how*. They are meant to elicit specific information and limit discussion. This type of question is used when the instructor is seeking an answer from a specific student and is often effective in dealing with a student who is being disruptive or whose attention seems to be somewhere besides the classroom. It is important to exercise caution with this type of questioning, as it could be perceived as singling out a student.
- **Rhetorical questions.** This type of question is usually asked to stimulate discussion or to prepare the learners' minds for a new topic. Often a rhetorical question does not have an answer, or it may have multiple correct answers.
- **Open-ended questions.** This type of question could be answered with varying correct answers based on the question and the situation presented.
- **Closed-ended questions.** This type of question seeks a specific answer based on the question asked.

Other types of question formats could be added to this list, but these are the most common forms of questions used within a classroom during a presentation. Knowing which question to ask and when to ask the question is very important, but dealing with a student's response to a question is just as important. Instructors are there to teach students and to foster a positive learning environment, so responding to a student's answer is critical to maintaining a good learning environment. If the student gives the correct response, the instructor can build on the response and move on with the lesson. But what happens when a student's response or answer to a question is incorrect? Effective instructors should use the answer that was given and highlight the positive and downplay any negatives. Instructors who make light of a student's incorrect response can destroy a student's desire to learn and dampen the learning environment. Students are in a classroom to learn; if they respond incorrectly, then the instructor has an opportunity to correct the information, reinforce the positive, and help the student learn new material.

> **COMPANY-LEVEL INSTRUCTOR TIP**
>
> One of the instructor's most important responsibilities is to maintain a positive learning environment where students can feel comfortable with the instructor and their peers and feel free of judgment.

Managing Disruptive Behavior

There are many types of disruptive behavior, and your strategy for handling the problematic situation will vary depending on the type of behavior. The most important thing to remember is that you are in charge of the classroom. If you do not control your classroom, you are setting the stage for problems. You may be perceived by your students, peers, and administration as being a less competent instructor. Disruption in the classroom can significantly reduce or limit the amount of learning time for students (**FIGURE 3-6**). It can also lead to unsafe and negative learning that could have serious implications when students apply what they have learned in the field. You are responsible for providing a safe learning environment and proper, safe instruction to all students.

The type of disruptive behavior a student exhibits might fall into one of the general categories described here:

- The *monopolizer* has a tendency to take control of classroom discussions and dominate class time. To deal with this type of student, make certain to call on other students during classroom discussions. You can also call a break or, when a class break occurs, pull the monopolizer aside and discuss your concerns about his or her "over-participation"

FIGURE 3-6 You are responsible for providing an appropriate learning environment to all of your students, which includes minimizing disruptive behavior.
© Jones & Bartlett Learning.

during the class. Instructors have been successful positioning themselves between the monopolizer and other students as a physical barrier in an effort to engage other students. More direct approaches may be to ask an overhead question: "Anyone other than 'John' have an idea how this term applies to fire department operations?"

- The *historian* has some experience and wants to make sure that everyone is aware of it. Real historians have the benefit of years of experience and share their experiences to help illustrate an instructor's lesson; they focus on helping the instructor and other students. By contrast, artificial historians focus on helping themselves and interject their opinions to show other students how much they know. Real historians can be disruptive if their stories become distracting for the other students, in which case you should carefully guide the discussion back to the main topic. Another way to redirect real historians is by pairing them with struggling students. Artificial historians are almost always distracting. Assertively redirecting the discussion back to the lesson plan generally corrects this issue. If not, take the student aside and discuss the need for the class to stay focused on the lesson plan. Treat your adult students as adults—with respect.
- The *daydreamer* is preoccupied with anything other than the lesson. He or she may have been forced to attend the training class or may be participating as part of a job assignment. Often, this student is focusing on life outside of the classroom. Daydreamers may not directly disrupt other students, but are more of a distraction to you, the instructor. Try to draw such students into the class by using direct questioning techniques and making maximum eye contact.
- The *expert*, or the *know-it-all*, is a student who may become confrontational or monopolize the class in an attempt to show off his or her brain power and perceived expertise, often as a measure of competitiveness. The expert will provide up-to-date information, factoids, applications, and case information about the topic and can easily subvert your position as the classroom authority. You may have to discuss his or her contributions to the class during a private break.
- The *class clown* is the student who seeks attention by acting out or making jokes about course content, the instructor, or other students in the class. Most often the reason for this type of behavior is to draw attention to the individual. Two effective methods for dealing with this type of behavior are to ignore the class clown or to engage him or her with a direct question. Sometimes a stern look will deflate this type of student. Another option is, during a break, to take a moment to engage the student in conversation and to get to know him or her better and seek the individual's help in setting an example for others in the class.
- Another type of student who can be a challenge in the classroom is the *gifted learner*. This type of student is usually not disruptive, but becomes a challenge to the instructor because he or she seems to be disengaged in the learning process. These students might appear to be shy or daydreaming, but in reality, they are usually intelligent and are "bored" with the material being presented because they have read the material already or have good working knowledge of it. An effective method to manage this type of student behavior is to have such students assist you as an instructor during the course.

Some behaviors are clearly inappropriate in the classroom because they are unsafe or illegal. Other worrisome behaviors may include actions or words that are threatening or offensive, or that create risks to students, instructors, or property. These behaviors must be dealt with immediately and without hesitation.

Behavioral problems in the classroom arise for many reasons, ranging from complex to simple. For example, the student who falls asleep in class may have been on a working fire during all or part of the previous night and is just plain tired. This issue is relatively easy to identify and is not a persistent problem. By contrast, the student who falls asleep in class every day may have a medical disorder that causes him or her to be unusually drowsy. Learn more about human behavior, behavioral problems, and their causes to become a more effective instructor.

Behavioral problems displayed by the disruptive student may be rooted in other problems as well. As the instructor, you may be put in the position where you must deal with behaviors that are the result of situations outside the instructional environment. These problems may be as simple as hunger or thirst, illness, or medical needs. On other occasions, disruptive students may be experiencing personal problems. Your first priority is to create a positive learning environment for the whole class so all behavioral expressions can be addressed. Your second priority is to ensure the existence of a positive learning environment for the individual student. If you are able to determine the reason for the behavioral issue, and it is one that you can address, this will allow the individual student to learn as well.

Other behaviors that may prove disruptive to the class include calling out, asking irrelevant questions, and giving excessive examples. In these cases, students may be seeking attention from the class or from you. Some students may seek to gain power by being argumentative, lying, displaying a temper, or not following directions. These students need to be dealt with quickly and directly before their behavior gets out of hand.

Some of these problem behaviors can be easily addressed. The hungry student may need to take a break so he or she can eat. The bored student may need a more stimulating lesson to motivate him or her to learn. No matter what the problematic student behavior is, understanding the sources of some basic behaviors will help you control the classroom environment.

As the instructor, you have the opportunity to make positive behavioral changes possible for your students. Here are some simple ways to do so:

- **Lead by example.** You set the tone for the class through your own behavior; in doing so, you set a standard for the class. Act as a professional.
- **Take steps to prevent undesired behaviors.** By preventing the behavior from occurring, you will have changed it before it happens and disrupts the class. For example, in your opening discussion with the class, you can outline the rules that govern the classroom and tell students which kind of behavior is (and is not) expected of them (Bear, 1990).
- **Know what to expect.** By putting the rules in writing, you will have anticipated what the behaviors may be. When formulating these policies, it is important to include the consequences for their violation and to let the students know what rights they have in the classroom situation.
- **Stick to the rules.** Do not contradict the facility's or administration's rules.
- **Communicate.** Talk to problem-causing students and explain that it is unfair for their behavior to continue to disrupt the class.
- **Reward good and appropriate behavior.** This can be done through special assignments or may be as simple as a pat on the back—in front of the class, of course (Walker and Shea, 1991).
- **Act like a professional.** This includes being prepared, approachable, and good natured.
- **Do not either overreact or underreact.** Some issues may bother some people and not bother others. Conversely, problems left unaddressed seldom resolve themselves.

When all else fails, know when it is "time to clean house." If a student continues to be disruptive after you have attempted to speak with him or her one-on-one and used the chain of command to resolve disruptive behavior, the time might come when the instructor requests that the student leave the class. This is an extreme step and one that should be used only when all other efforts have failed to correct disruptive or inappropriate behavior.

Transitioning Between Methods of Instruction

The Instructor I should acquire skills to effectively apply problem-solving techniques, to facilitate and lead conferences, and to employ discussion methods during presentations. These techniques are frequently used to conduct small-group sessions where participants have advanced knowledge and experience in the subject matter and the goal is to reach a group solution to a problem or issue. When students become engaged in the learning process and actively participate in a class, the instructor can shift from a lecture format to a guided discussion. The same effect can occur when a student has adapted to a basic skill level in a hands-on training session, allowing the student to perform simultaneous actions and progressive skills.

Some disconnects in learning occur when a student who has learned in high-tech environments is suddenly dropped into a fire service classroom that is a traditional lecture-based format. The Generation Z student may easily become bored and not engaged unless the instructor is able to motivate them using many application questions and more hands-on methods. Instructors may use more directed study applications and preload the lesson plan content by providing resources and information ahead of time to the students.

Whereas a chalk board and overhead projector were common to a Generation X student, an interactive board and flat panel would be more recognizable to the Generation Z student. This underlines the importance of knowing your audience characteristics and adapting the learning environment to them to increase retention and understanding of class material. As more and more Generation Z students enter the work force and the fire and emergency medical services field, instructors will have to adapt the learning environment to varied generations of students who will all bring different learning backgrounds and different ways that learning is retained and put to use in the future. Chapter 2, *The Learning Process*, discusses the different generational traits that are important considerations when determining your teaching approach.

Audience and Department Culture

To be effective in teaching, you must know your audience, the message you are trying to convey, your best delivery method, and the ideal atmosphere in which to deliver your message. This information is not always easy to come by. You may not know exactly with whom you are going to speak or even the context in which you will be presenting. You must be able to think on your feet and be prepared to take the presentation in the direction it needs to go so that the learning objectives are met. Remember—a good instructor always has a backup plan.

Audience

Depending on whether you are presenting at a conference with attendees from multiple jurisdictions or presenting locally to your own department, the manner in which you share the material with your audience will change. When presenting to a diverse audience, your presentation may need to be more generic. Avoid using specifics, such as the exact procedures of any one department. Consider speaking in generalities or giving examples instead of describing mandates or specific procedures. Taking this kind of broader perspective will allow each audience member to adapt your message for his or her own department. For example, when speaking about a procedure, give examples of several procedures used in fire departments across the country. Even when you are presenting to one department, be careful not to mistake instruction for department policy.

You may also need to adapt your presentation style to better match the learning environment. As your audience and your space increase in size, your presentation loses its intimacy. If you typically encourage a lot of student interaction, you may have to adapt your style to deliver a more traditional lecture because the larger room may not allow for personal student interaction. Avoid the phrase, "If I were you, I would do it this way." The best instructors help lead students into making the best decisions for their own departments based on the knowledge that the instructors provide.

Department Culture

When making a presentation or teaching a class, you must also be aware of culture. Here we speak not of customs associated with a particular national origin, but rather of departmental culture. Every department has its own culture. It is what makes fire fighters different, yet exactly the same. It is based on the mission of each and every fire department in the world. Departmental culture specifies what we will do, what we can do, and what we will not do. When you instruct students, you must put aside your personal culture. If you are brought in to teach, you must be able to separate facts from opinions.

Your misgivings about a technique do not automatically make that approach a bad idea. For example, some fire fighters might insist that positive-pressure ventilation causes a building to burn down; they may even show examples and argue that this technique introduces air into a hostile environment. These people may have had unpleasant experiences with positive-pressure ventilation, but that does not mean the technique is at fault. The misapplication or misuse of any technique will end poorly. This is the information a student needs—not personal opinions.

You must listen to the culture in which you are instructing and adapt your training accordingly. Every fire department says it is in existence to save lives. Likewise, every department says it is in existence to protect property. But how will the members of that culture save lives and property? To what extent are they willing to risk their own lives? These are the questions a departmental culture must answer.

As fire fighters, a major portion of our job is risk management. Risk management must be viewed within the context of the fire department's culture so that you can place the correct emphasis on the decision-making process. For example, although you may not agree with the decision to go offensive, you must respect and acknowledge the department's decision.

The Learning Environment

There is an old saying: When two people are together, one is always the leader. In training, whenever two fire fighters are together, one is always the instructor. The setting is not important, the time is not important, and the subject is not important—what *is* important is that you use the environment to ensure that learning takes place.

The **learning environment** is not always a classroom, because learning takes place in many different locations. In many company officer training classes, the concept of tailboard chats is encouraged. Many fire officers report that a great deal is learned sitting around the kitchen table in a firehouse or standing around a charred table in a kitchen that has just been overhauled (**FIGURE 3-7**). The place is not as important as recognizing the opportunity to share a message or thought and to deliver a piece of knowledge with others. Training might start in the classroom lecture or a demonstration by the instructor. From the classroom, the demonstration moves to the training ground, where the students get to apply hands-on training to further develop the skill set. Skill sets become building blocks that develop into scenarios, and scenarios move to larger-scale evolutions that are based on training as close to reality as possible. This process encompasses a variety of learning environments that require the instructor to know how to teach in each venue.

Many agencies are now using online training for their personnel for either initial or continuing education needs. As an instructor, the time may come when you will present a lesson in an online environment, which is different than the traditional classroom but is now considered standard for many agencies. The learning environment for an online course is different because you do not have a classroom to prepare or set up. However, depending on the type of online learning platform, you can have a big impact on how your virtual classroom is set up and how the students feel in that environment. Chapter 6, *Technology in Training*, has more information on online courses.

FIGURE 3-7 Learning can take place almost anywhere, and in some cases it can be even more effective in informal settings than in a classroom or on a drill ground.
© Jones & Bartlett Learning.

To be an effective instructor, you need a solid understanding of the benefits and weaknesses of potential learning environments, and you must be able to select the best environment for presenting your message. First and foremost, you need to ensure the safety of your students, yourself, and others involved in the training. Eliminating all safety hazards and concerns should be priority number one. After evaluating the learning environment for safety, you next need to address the environment to ensure that learning can take place.

The Physical Environment

The perfect environment for students is one in which all of their needs are met. In Abraham Maslow's hierarchy of needs, the first need relates to creature comforts and physiological needs. Warmth and shelter make you feel safe in your environment. In the teaching sense, this need would be met when the classroom has enough light, the temperature is comfortable, and the seating and work space are adequate. In the case of an outdoor classroom, it may mean a place of shelter is available where students can warm up in cold weather and cool off in hot weather.

Learning takes place in two general settings: formal and informal. The formal environment consists of a classroom plus learning objectives, lesson plans, lesson outlines, and a final evaluation. By comparison, informal learning takes place every day in every location. Controlling the learning environment is a critical task for instructors. Consider the following factors to ensure you have a positive learning environment that will facilitate the learning process:

- Classroom environment
- Classroom setup
- Safety
- Lighting
- Teamwork and self-actualization
- Minimizing distractions

If you are instructing locally, the venue could be anything from the back of a fire engine to a formal classroom in a training center. Knowing where you will present the information is important so that you can tailor your presentation style to the venue. The size of the room is one factor; the seating in the room is another. The receptivity of students to your message will vary depending on whether the instruction takes place in a formal classroom with tables and chairs or in an apparatus room where students are just standing around. A formal training center with tables and chairs automatically puts students in the learning mind frame; that is, students are more focused in this venue. Because this venue represents a more formal situation, student behavior tends to be more formal. By comparison, the more casual atmosphere of the apparatus room puts students at ease and minimizes the formality of the student–instructor relationship. This opens up opportunities for the discussion to go more in the direction of the students' interests.

The concept of tailboard chats is not a new one. These discussions have been used for many years and can be an effective method of educating fire fighters. The **tailboard chat** is an informal gathering of fire fighters at the back of an engine where the company talks about various subjects. The tailboard is the equivalent to the water cooler in a business setting. It puts all members of the company at ease and on the same level when you instruct them there. For example, you might use the previous call as an opportunity to reinforce a learning objective. These tailboard chats can happen either back at the station or on the scene after the call is over.

Some instructors use the structure that the company was just at during a call as an example and "simulate" a fire to reinforce learning objectives. The tailboard chat may begin with you saying, "If there were a fire on the third floor of this building, in the B-C corner, where would we place the engine and the truck? What would the first engine company be responsible for doing? What would the first truck company do? What if this fire was at 2:00 AM? At 4:00 PM?" This is tactics and strategy training at its best.

> **COMPANY-LEVEL INSTRUCTOR TIP**
>
> To maximize the learning experience, add realism to a lesson. It helps if the lesson is based on a local area or one that is otherwise known to the students and relevant to their field. This strategy improves both leaning and retention. Real-life examples and experiences are more compelling than far-fetched scenarios.

The Indoor Classroom

Perhaps the most common setting for conducting a class is in a traditional classroom. Thus, when an instructor thinks of a "classroom environment," he or she probably thinks of a classroom in a building. Most often, classes start in an indoor setting and either remain there for the entire class or, once introductory material is covered, move outside to a training ground.

An important consideration in many classrooms today is the addition of new technology that affects the learning process. Most classrooms today have projectors and computers that are used for presenting lesson material. Many agencies have provided wireless internet service to assist students with the learning process or to complete assignments online. Of course, with this technology comes the possibility that the student might be distracted by "surfing" the internet or go to websites that are not class related. Issues related to classroom internet use can be addressed with agency policies or by turning the wireless access on or off as appropriate.

Temperature is very important to the learning process. If a room is too hot or too cold, students will focus more intently on the temperature than on what is being taught. You cannot make every student happy, so you must go with the majority opinion. If possible, lean toward keeping the room cooler versus warmer. In a colder environment, students are less likely to become drowsy. Also, if the temperature is too cool, students can always put on a sweater or light jacket.

You must also provide time for breaks so that students can attend to their personal needs. An old instructor's rule states, "The brain can learn only as long as the bottom will allow it." Some instructors (and students) may consider these breaks to be a waste; nevertheless, they are necessary both for the students and for you. In addition to enhancing students' personal comfort, the informal conversations that take place during a break can actually enhance the learning process.

Classroom Setup

Setting up the classroom is as important to the learning process as copying handouts or creating a slide presentation. Arranging the classroom sets the tone for the learning process. The first question you need to ask is, "Which type of learning do I want to accomplish?" If you plan to lecture to the students, then the room can be set up in rows facing a monitor or dry erase board (**FIGURE 3-8**). **TABLE 3-2** lists various classroom setups, along with their advantages and disadvantages (**FIGURE 3-9**).

FIGURE 3-8 Your style of presentation will determine the best classroom setup.
© Jones & Bartlett Learning. Photographed by Glen E. Ellman.

TABLE 3-2 Classroom Setups

Method of Instruction	Learning Environment	Advantage	Disadvantage
Lecture	▪ Auditorium, traditional, or theater	▪ Good choice when students are able to see the front of the room ▪ Instructor has more control in this setting ▪ Can be effective for demonstrating skills	▪ Difficult for students to see one another ▪ Depending on how close chairs and tables are, the instructor may be very limited in moving around the classroom ▪ Depending on the size of the room, students may have a difficult time hearing the instructor or other students
Discussion	▪ Small U-shaped arrangement ▪ Hollow square ▪ Conference table	▪ Students have the opportunity to see one another ▪ Good environment for small-group interaction ▪ Good environment for instructors to work in smaller groups with students	▪ Students can easily be distracted by one another ▪ Difficult for all students to see the front of the room ▪ Difficult for instructors to control the teaching environment
Demonstration	▪ Large U-shaped/horse-shoe arrangement	▪ Allows students to see the demonstration of skills ▪ Allows the instructor to move about the room and interact with students	▪ Students can easily be distracted by one another ▪ Difficult for all students to see the front of the room

A

B

C

D

FIGURE 3-9 Sample classroom configurations. **A.** Traditional. **B.** U-shaped (horse shoe). **C.** Conference table. **D.** Tabletop.

A & B: © Jones & Bartlett Learning; C: Courtesy of L. Charles Smeby, Jr./University of Florida; D: © Jones & Bartlett Learning. Photographed by Glen E. Ellman.

Safety

After the basic needs are met, the next-highest-priority needs are those relating to safety. In a classroom environment, the term *safety* is used frequently.

Safety in the classroom is just as important as safety on the fire ground, but the hazards are much different. Ensuring that a classroom is safe involves removing both physical hazards (e.g., the trip hazard of a cord, book-bags around students' feet, wet floors from rain or spilled coffee). Physical hazards need to be controlled to ensure a safe classroom environment.

The second type of safety provided in the learning environment is emotional safety of the students. Students need to have an environment free of judgment or harassment, which will allow them to learn and express their thoughts and opinions. For example, students in a learning environment need to feel that if they make a mistake, there is an opportunity to learn from the mistake. In training, whether in the confines of a classroom or on the expanse of the drill ground, students must feel confident that if they make a mistake, you as the instructor will not belittle or otherwise demean them. Your task is to support students and to help them learn from their mistakes. Football coach Vince Lombardi once said that a person is not measured by how many times he is knocked down, but rather by how many times he gets back up. Your students must believe that you will help them get back up, and this sense of comfort can be provided in the right learning environment.

Lighting

In addition to the arrangement of tables and chairs, having the proper lighting for the room is important. Depending on the type of presentation that you are giving, you may need to adjust the lighting. In some cases, lights may be left on during a PowerPoint presentation or during other projected presentations. In the ideal classroom, the lights will be split, allowing lights in the front of the room to be dimmed while lights in the back of the room remain on to allow students to take notes. Your podium should have a light source so that you can read your notes. If the classroom does not have split lighting or adjustable lighting over the students, then you need to decide which type of lighting is best for your students.

Minimizing Distractions

Proper selection of the learning environment will determine the success or failure of the class in meeting the learning objectives. One challenge that all instructors face is the variety of distractions offered by today's communications technology. Phones, text messaging, pagers, and station alarms—all of these ubiquitous devices can cause students to be distracted from learning. Many of these distractions can be eliminated at the beginning of class by asking students to turn off or silence all electronic devices. If you have on-duty crew members in class, then make arrangements with them to turn down their pagers or radios.

Other distractions, such as people entering and leaving the classroom during a lecture, can be addressed before the class begins by posting a sign on the door stating that class is in session. Request that latecomers wait until a break to enter the classroom. If training is to occur with on-duty crews, consider having one student monitor a portable radio with the volume turned low.

If training is taking place outdoors, then controlling the learning environment is more difficult. Noise and distractions can disrupt the learning environment, so try to conduct your training in an isolated place.

Classroom Arrangement for Testing

The setup of the indoor classroom typically changes when students are required to take a test. Specifically, the classroom should be arranged to provide each student with enough space to put both the testing booklet and the answer key on a writing surface. It should also be designed to minimize the chance of wandering eyes, while simultaneously allowing the instructor to observe all students. Given that most testing is individualized, the classroom should remain quiet. If the testing is group based, you must be able to distinguish between testing noise and distractions from a group. The personal needs of students should be recognized. If a student needs to leave the room to attend to personal needs, the security of the testing process should be maintained. Because testing and evaluating the learning process are important to the fire service, having the right testing environment will ensure that students can focus on the test.

The Outdoor Classroom

Although it might appear that the work conducted in the outdoor classroom is more important than the instruction that occurs in the indoor classroom, both are needed to ensure that students achieve the learning objectives. Your challenge is to match the environment to the learning objectives to make sure that the student is always in the correct environment in which to learn.

Outdoor classrooms differ in more respects than just their location (**FIGURE 3-10**). When moving the learning process to an outdoor venue, the instructor needs to address issues such as weather, training

FIGURE 3-10 Outdoor locations can present a set of challenges for the instructor.
© Jones & Bartlett Learning.

FIGURE 3-11 Technical rescue training environments are often found in the areas where the skills will be used.
© Jones & Bartlett Learning.

and ambient noise, and logistical issues such as equipment needs, parking, or rehabilitation facilities. The key to training outdoors is to plan ahead and to conduct the proper pretraining assessment to ensure a safe and effective learning experience for your students.

When curriculum or training needs require an outdoor training environment, the instructor needs to be prepared to teach in this environment. These outdoor classrooms can be organized on a developed and organized training ground, which could contain a fixed burn building to teach interior fire attack, a mobile prop used to teach fire behavior, or a pond to teach drafting skills. Another type of outdoor classroom could be an off-site location such as a mountain or a slope used to teach high-angle rope rescue skills (**FIGURE 3-11**).

Outdoor locations, whether on a training ground or an off-site location, pose similar challenges and considerations for the instructor to address in order to facilitate the learning process. These challenges include the weather, outside noise, logistics/resources, and the ability to teach so that students can learn in the outdoor classroom. As mentioned previously, distractions are a common factor in the outdoor setting. To help with this, keep your groups small, so that every student is more closely engaged in the learning process with the oversight of an instructor.

A classic example of an outdoor learning environment that would present several challenges would be training at an acquired structure. These buildings present a great learning opportunity for students, but also create many difficulties for instructors. Following NFPA 1403: *Standard on Live Fire Training Evolutions* will assist the instructor in correctly preparing for such a training event.

Communicating Outdoors

Communicating in the outdoor classroom can be a challenge to any instructor. Learning to establish an environment that will allow you to teach outdoors is just as important as it is when in a traditional classroom.

For example, when teaching in a live fire training scenario, communication can be difficult. In this environment, you first need to be cognizant of your own safety. Taking off your self-contained breathing apparatus (SCBA) mask to communicate in a smoke-filled environment not only puts you at risk but also sends a message to the students that the use of a SCBA mask is not important. Communication in the live fire environment is essential, however, and can be accomplished by first gaining the students' attention and then by using a calm and controlled voice to communicate what you are attempting to teach.

A live fire training scenario is also an environment in which teaching is a constant process. You need to be close enough to the students to observe their behaviors and far enough back to observe the scene and allow the students to perform and learn.

Learning how to communicate in the outdoor classroom is a skill that you will need to develop and always be mindful of when outdoors. Observing your students will be critical to determining whether your voice and the content of your message are reaching everyone. Observing students' body language will be a good indicator of whether your message is getting to everyone.

Safety

One point cannot be emphasized enough: Safety should be your greatest concern in the outdoor classroom. During outdoor training, the stakes become higher and the attention to detail becomes more intense. When performing live fire training, for example, the environment needs to be as safe as possible while allowing learning to take place. You hold students' lives in your hands, and students need reassurance that you take this responsibility seriously. It is important to follow standards such as NFPA 1403, among others. Having the right number of personnel, in the right positions, doing the right things, with an emphasis on safety, makes the learning environment the safest place for fire fighters in which to work and learn.

The purpose of many outdoor training sessions is to teach or evaluate students' psychomotor skills; these skills are the ones that fire fighters use to protect the lives and property of citizens. For that reason, when training and evaluating students, you need to be in the most realistic situation possible while maintaining a safe environment for students. In rope rescue, this concept is called being "redundantly redundant"—that is, there is a safety for everything.

Make sure that students' psychomotor skills are practiced and completed perfectly. A skill is either acceptable or unacceptable—there is no middle ground. If a fire fighter is asked to don a full protective ensemble and forgets to put the hood up, he or she has failed in that skill. Vince Lombardi said, "Practice doesn't make perfect; perfect practice makes perfect." Another relevant saying is, "The more we sweat in training, the less we bleed in battle."

Weather

A major consideration outdoors is the weather. Obviously, outdoor classrooms are subject to the weather conditions. The physical location of your outdoor classroom can determine how you prepare for and handle the class. If you are in the South or Southwest, you may need to deal with high heat levels. If you are in the Northeast or Midwest, you may be more concerned about cold temperatures and snow. No matter where you are, the threat of rain is a consideration when an outdoor classroom will be used for instruction.

The safety of your students should always be your highest concern. Many fire departments use either a temperature/humidity (heat) index or a wind chill index as a guide when determining whether outdoor training is feasible or acceptable within department policy. NFPA 1584: *Standard on the Rehabilitation Process for Members During Emergency Operations and Training Exercises* offers guidelines on temperature and physical exertion levels relating to how much training, PPE, and rehabilitation is required during extreme temperature and weather.

Always have a "plan B" in case inclement weather forces the class to use an alternative location that is unaffected by the weather, such as an apparatus bay floor or a large, unused warehouse. Plans to use these areas should be put into place before the training occurs, rather than decisions being made spontaneously during the training as an emergency measure. Rain is a fact of life. A small amount of rain may not be an issue during outdoor training, but torrential downpours and lightning are definitely reasons to seek an alternative environment. Warming or cooling areas and shelter from wind, rain, or other weather phenomena are necessities for your students' safety.

Noise

Noise in the outdoor environment is also a concern during training. In most cases, you will need to project your voice so that all of the students can hear you without the use of an electronic device such as a microphone. Environmental factors such as wind can be a deterrent in allowing your voice to carry. Other distractions—such as vehicle traffic, ambient noise from the field, or other training activities going on around you—can add to the difficulty of communicating in outdoor environments. During your preplanning of the outdoor classroom, look for locations in which you can present verbal information effectively. Many training grounds have outdoor pavilions that are designed for teaching and for sharing information. Shielding yourself near a fire apparatus can help minimize noise and project your voice. It might not be possible to eliminate all outside noise, so it falls on you as an instructor to use your voice and other teaching skills to help communicate the information and knowledge you are attempting to teach.

Logistics and Resources

When the classroom moves to the outdoors, there are several other issues to address. As part of your classroom setup, ensure that all logistical arrangements have been secured prior to moving outdoors. Reviewing your lesson plan and ensuring that all required equipment and materials are in place before class are steps that are expected of you as the instructor. Knowing the location of restroom facilities and having a rehabilitation area established are just as important as having the right equipment ready for your class. At an established training ground, these resources are usually in place and have been designed into the facility. The real challenge comes when training moves to an off-site location, because you need to convey all of your resources to that site. Making arrangements before class to address the physical needs of your students for restroom facilities and rehabilitation becomes an important challenge when moving to an off-site location. The need to take required and extra equipment with you is also important so downtime is minimized if something breaks or additional equipment is needed.

Contingency Plans

You must have a plan B regardless of the environment. What do you do if the projector fails to work? What do you do if lightning begins to flash in the sky on a live fire training day? What happens if an assistant instructor does not arrive on a practical skill day? All of these contingencies need to be considered and planned for. Many times the solutions are simple—for example, keeping a spare projector bulb or a spare projector, or having a rain date for outdoor training. In any event, you as the instructor need to be able to adjust your training plan so that your training plan can continue. If an evolution or training session cannot continue due to some type of issue, many rescheduling or make-up logistics will be involved in finishing the training. Without question, safety and proper delivery will always take precedence over any of these considerations. A best practice may be to have a set of "ready-to-go" backup drills in the case that training must be stopped. You do not want to lose momentum or the opportunity to train when your department meets only a few times per month.

Another way to look at this situation is much like you would approach an emergency scene. Do you have the resources in place if something should go wrong? For example, suppose you are teaching a psychomotor drill and a trainee gets hurt. Is there an ambulance available? Who is notified? Which forms must be completed? How will the training continue if you are missing one student? On an emergency scene, fire fighters constantly plan for contingencies.

Always have a contingency plan. Things can and will go wrong, and having at least considered these issues will allow you to adapt to the changing circumstances quickly.

Teamwork and Self-Actualization

In Maslow's hierarchy of needs, Maslow speaks of the issue of belonging. A main focus of fire training is teamwork, and the learning environment you create must be one that supports and recognizes the crucial nature of teamwork. Many of the best friends you have in the fire service are likely to be people whom you met as you trained to become a fire fighter. During your training, you built a strong sense of camaraderie, which ultimately helped you in completing a very difficult task. An important part of the instruction should be providing time either before, during, or after the formal instruction to allow students to network.

The next level in Maslow's hierarchy of needs is the need for self-esteem or accomplishment in the student. You should create an environment that celebrates students' victories as they are accomplished. Status and recognition can be very powerful motivators in the learning environment. In his book *The One-Minute Manager*, Ken Blanchard talks about catching people doing things right (Blanchard and Johnson, 1982). In a good learning environment, this idea is expanded to not only catching students doing things right but also recognizing and celebrating that accomplishment. People want to be in an environment where they can succeed.

The highest-priority need in Maslow's hierarchy is the need for self-actualization. At this level, the student is able to synthesize material by taking multiple learning situations and putting them together to solve a problem. The need for self-actualization is satisfied in the environment where you perform live fire simulations or even live fire training. Look to challenge your students to take your message and use it to solve a problem.

COMPANY-LEVEL INSTRUCTOR TIP

As an instructor, you should be aware of how demographics affect the classroom. Always exercise caution and common sense in the classroom. Never make a comment that refers to someone's background, such as where they attended school, their marital status, or their ethnicity.

Demographics in the Learning Environment

Demographics include the age, gender, marital status, family size, and educational background of a fire fighter. Demographic considerations extend beyond issues of race and national origin; in the fire service, they require looking in a more holistic manner at everything a fire fighter brings to the department.

Assess a fire fighter's demographics and ask yourself a few questions: Which type of skills does this fire fighter bring? Could these skills be a benefit to the others in the learning environment? Are you dealing with a highly educated individual or a fire fighter who has met only the basic educational requirements? Does the same class include a mix of experienced and inexperienced fire fighters? Determining the answers to these types of questions will help you grasp the basic demographics of your students, shaping your teaching approach.

Another demographic consideration unique to the fire service is the different types of staffing encountered among fire departments. The fire service includes career, volunteer, combination, and paid-on-call fire fighters. The learning objectives for paid-on-call or volunteer fire fighters do not differ, but the presentation to these groups does. Because volunteer and paid-on-call fire fighters generally have full-time jobs, training accommodations must be made to fit their schedules. By gearing instruction toward the demands of the students' schedules, you allow each student to focus fully on the subject at hand and not be distracted. For example, if a class is made up of all career fire fighters, the class may be 6 to 8 weeks long, 8 hours a day. By contrast, if the same class is made up of all volunteers, the class may meet Tuesday and Thursday nights for 3 hours and all day on Saturdays for 6 months. Both classes should cover the same learning objectives to the same levels of proficiency, but the differing demographics of the participants require different schedules.

> **COMPANY-LEVEL INSTRUCTOR TIP**
>
> Today's public safety learning environments are made up of a highly diverse group of students. A major area of focus for the instructor is to identify the characteristics of each learner and build a lesson that will meet the needs of each of these students.

Gender Considerations

Gender is a hot-button issue today. One change that has affected the fire service in the last few years is the abandonment of the term *fireman*. Many qualified women have joined the ranks of this once male-dominated profession, so it is more appropriate to refer to students as *fire fighters*. Firefighting does not, according to the fire, have a gender. Although learning objectives do not change with the gender of the student, you do need to be cognizant of students' gender. If you use the term *guys* to address your students, your students need to understand that "guys" includes both men and women. Remember to be gender neutral when developing learning objectives. The key is to be respectful of all your students at all times.

Offensive Language, Gestures, and Dress

The use of offensive language should be avoided—period. Some people may contend that to make a point, sometimes you must swear or use slang terms; others contend that this is part of the fire service culture. This is not the case. Jokes and language of a sexual or explicit nature are unacceptable in the learning environment. As an instructor, you should have a mastery of the English language. You should be able to make your point without the use of profanity. You will find that students gain a deeper respect for you when you demonstrate eloquence in the classroom.

Do not participate in improper situations. A wise instructor once shared a key piece of wisdom when he said, "If you hear an off-color joke and grin, you're in." If you hear an offensive joke, show that you do not agree with what was said. It is your responsibility to stop any unsafe or inappropriate behavior immediately. It is important to take pride in your department, and one source of that pride involves showing respect to everyone.

Sometimes, regardless of our awareness and best intentions, we inadvertently offend one another. If it happens that you offend a student, apologize to the student in private and then in front of the class. This can go a long way toward making amends.

An equally important step is to establish a policy that covers acceptable behavior and language. Your organization should have a harassment policy that addresses these issues, including remedies for offenses, reporting of incidents, and conduct. The best policy is to understand the audience to whom you are speaking. Ensure that your presentation occurs in a context that is appropriate for your audience. A wise instructor

FIGURE 3-12 How you present yourself is a reflection on the fire department.
© Jones & Bartlett Learning.

used to say, "Make sure that your language would be acceptable to your grandmother."

Like words, gestures and dress can sometimes be offensive. How you dress and how you move are representations of your position (**FIGURE 3-12**). Always maintain a neutral position, which allows the student to feel free to express his or her opinion and concerns in a safe environment.

Adapting the Lesson Plan Based on Demographics

The information that you learn as a result of students' verbal introductions may require slight modifications to the lesson plan. Perhaps students will need to spend more time in one area of the class and less time in another. The information gained in this way can also help you decide the depths you need to go to in the various areas of the presentation, perhaps allowing you to lengthen parts of the program and shorten others to meet the needs of the class. For example, if you are presenting a disaster preparedness course in Arizona, you may not need to spend an entire hour on blizzards. Even in a small company training setting, you should be aware of the skill, experience, and ability levels of the members of the class to gain the best results.

If a student requests that you cover a certain topic, you should not rewrite the entire course; however, you should keep the student's request in mind. If time and circumstances permit, you could briefly present the topic and then ask the student, "Did that meet your need?" By doing so, you ensure that students know you are available to instruct them on whatever they need to know to meet the learning objectives of the course. Bear in mind that as an instructor, you must be prepared either to adapt your lesson plan or to tell the student that the topic is outside the scope of the class.

If you indicate that you will cover an area or an objective, you need to ensure that you fully address any concerns or questions about that topic. Meeting this goal requires knowing and understanding your audience. Many fire departments have target hazards or situations that are unique to their locations. For example, many communities in the Northeast include row houses, whereas these structures are rarely found in the Midwest. When asked, "How would you handle this situation?" you should be prepared to render an opinion. By anticipating what the questions will be, you will be prepared to offer an opinion because you have had time to research the question and present an intelligent response. If you do not know the answer, do not—under any circumstance—make up the answer. It is completely acceptable to say, "I do not know, but I will try to find out for you." Fire fighters respect honesty in their instructors.

Training BULLETIN

JONES & BARTLETT FIRE DISTRICT
TRAINING DIVISION
5 Wall Street, Burlington, MA, 01803
Phone 978-443-5000   Fax 978-443-8000

Instant Applications: Methods of Instruction

Drill Assignment
Apply the chapter content to your department's operation, training division, and your personal experiences to complete the following questions and activities.

Objective
Upon completion of the instant applications, fire and emergency services instructor students will exhibit decision making and application of job performance requirements of the instructor using the text, class discussion, and their own personal experiences.

Suggested Drill Applications
1. Identify several NIOSH fire fighter fatality reports that utilized rapid intervention team activations or identified survivors that used skills learned in a fire fighter safety and survival program and discuss them with your shift.
2. Review your department's Rapid Intervention/Safety and Survival SOG and discuss it with your shift, identifying strengths and weaknesses and ways that it can be incorporated into your company-level training.
3. Evaluate the culture of your department's training and identify ways that it can be improved.
4. Discuss the challenges of conducting a fire fighter safety and survival course outdoors and what can be done to make the training more realistic, relevant, and results oriented.

Incident Report: USFA Technical Report, USFA-TR-015/October 1987

Drill Assignment

Review the information in this incident report and prepare a practice presentation for class delivery at the direction of your instructor. Your presentation should include a summary of the incident facts and a review of the NFPA 1403 compliance and noncompliance findings. Use an outline to organize your thoughts. You may be evaluated on your communication skills during your presentation.

Hollandale, Minnesota—1987

Search and rescue drills had been conducted for some time in a small two-bedroom house without problems. Leaves burning in a 35-gallon (132.5-liter) drum were used to create smoke. The drum had a metal cover to reduce the danger of fire spread, and the drum was situated above the floor. Caution was exercised to prevent fire from spreading from the container. Two fire apparatus were on scene with 1400 gallons (5299 liters) of water on one of the fire units and 1000 gallons (3785 liters) in a portable tank. The participants carried a charged 1½" (38.1-mm) hose line.

After the interior search evolutions, the intent was to destroy the house by burning it down. Four crew members wearing full protective clothing, equipped with breathing apparatus, and protected by a charged 1½" (38.1-mm) hose line entered the structure using the east entry that had been used during the drill (**FIGURE 3-A**). They attempted to light a fire in the back bedroom but were unable to get it to ignite, most likely due to the dampness from the water. They moved to the living room and splashed #2 fuel oil on the wall and carpet. Two of the fire fighters moved to the kitchen while another used a propane torch to ignite the fire in the living room. The entire room flashed, and a third fire fighter crawled out, thinking his partner was with him. Three of the fire fighters crawled to the outside. The fourth made it to the kitchen and stopped, where he apparently removed his SCBA mask to call for help and then reentered the fire area and was killed. His Nomex hood, propane torch, and fuel can are later found in the kitchen. The three fire fighters that had exited the house reported the emergency and entry was immediately forced into the front entrance. The fire was quickly extinguished, and the victim was found dead.

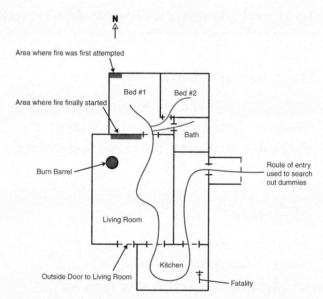

FIGURE 3-A Floor plan of the Hollendale structure.
Reproduced from the Minnesota Department of Labor and Industry, Minnesota OSHA Compliance.

It was uncovered in the investigation that the living room paneling was made of a lightweight combustible material found in mobile homes at the time. The smoke barrel with live fire may have preheated the combustibles in the living room (**FIGURE 3-B**). The report noted that the NFPA 1403 standard should have been followed. Investigators believed that if the deceased had remained in the kitchen, he most likely would have survived.

The fire department was cited by the state OSHA for failure to "examine the building more closely, failure to brief the participants on the building's layout, failure to have a building evacuation plan, and failure to have a qualified instructor on the scene." This was the first fire fighter fatality after the release of NFPA 1403.

FIGURE 3-B Photo of rear of house in the Hollandale fire. The kitchen is below the area with the flat roof. Note the minor amount of smoke and heat damage above door.
Courtesy of the Minnesota Department of Labor and Industry, Minnesota OSHA Compliance.

Postincident Analysis Hollandale, Minnesota

NFPA 1403 Noncompliant

- Combustible wall paneling and carpeting contributed to rapid fire spread
- Flammable and combustible liquids used
- Flashover and fire spread unexpectedly
- No direct egress from living room
- No formal evacuation plan including evacuation signal
- No incident safety officer
- No interior hose line
- No EMS on scene
- No written emergency plan
- Participants not informed of emergency plans and procedures
- No preburn plan
- No walkthrough performed with students and instructors

After-Action REVIEW

IN SUMMARY

- Adult students need realistic, practical, and knowledgeable lessons that deliver information in a variety of ways to accommodate their differing learning characteristics.
- When you can communicate to students that you know the information, and that you understand their unique needs and motivations, you will find that it is much easier to achieve the goal of teaching the information and having the students learn.
- Determining which method of instruction should be used to deliver information is accomplished using several factors:
  - The experience and educational level of the students
  - The classroom setting or training environment
  - The objectives of the material to be taught
  - The number of students to be taught
- Ask yourself three basic questions when reviewing and planning for the delivery of a lesson plan:
  - What are the goals of the lesson?
  - Who is the intended audience?
  - What are the resources and learning environment characteristics?
- Methods of instruction vary among disciplines, but for the fire service the following are most common and effective in teaching new and in-service personnel:
  - Lecture
  - Demonstration/skill drill
  - Discussion
  - Role play
  - Labs/simulations
  - Case studies
  - Out-of-class assignments
  - Online learning, independent study, and blended learning
- While reviewing and preparing for class with a lesson plan, the four-step method of instruction is the primary process used to relate the material contained in the lesson plan to the students:
  - Preparation
  - Presentation
  - Application
  - Evaluation
- Five effective transition techniques are delineation, words/phrases, pictures, voice, and body language.
- An important part of the learning process is to assess the knowledge gained during class. This is most commonly done by asking the students questions about the material you have presented. Several types of questions may be used:
  - Direct questions
  - Rhetorical questions
  - Open-ended questions
  - Closed-ended questions

- Disruption in the classroom can significantly reduce or limit the amount of learning time for students. It can also lead to unsafe and negative learning that could have serious implications when students apply what they have learned in the field. As the instructor, you are responsible for providing a safe learning environment and proper, safe instruction to all students.
- When you develop your lesson plan, it should be detailed and complete enough to achieve the identified goals and desired outcomes, yet flexible enough that it can be adjusted to meet students' needs on the fly.
- Because learning can take place in many locations—such as a traditional classroom, fire-ground training facilities, or another location such as an acquired structure—you as the instructor need to be keenly aware of your environment.
- When making a presentation or teaching a class, you must be aware of both the audience and their departmental culture.
- Some programs require that students take prerequisite classes before attending certain advanced classes. The fact that students have completed these preliminaries ensures that the students taking your course have a certain base level of training and education.
- Establishing a learning environment is just as important as having a lesson plan. Without the proper environment, the learning process will be limited and can be ineffective.
- Matching the correct classroom arrangement to the needs of the lesson plan will allow students to learn.
- When presenting a lesson in an indoor classroom, consider the classroom environment and setup, safety, lighting, and ways to minimize distractions. Adjustments may need to be made for testing.
- When presenting a lesson in an outdoor classroom, considerations include communication, safety, weather, background noise, logistics, and resources.
- You must have a plan B when using any environment. All potential contingencies need to be considered and planned for.
- Understanding demographics extends beyond issues of race and national origin. In the fire service, taking demographics into account during instruction requires looking in a more holistic manner at everything a fire fighter brings to the department.

KEY TERMS

Access Navigate for flashcards to test your key term knowledge.

Asynchronous learning An online course format in which the instructor provides material, lectures, tests, and assignments that can be accessed at any time. Students are given a time frame during which they need to connect to the course and complete assignments.

Blended learning An instruction method that combines online/independent study with face-to-face meetings with the instructor.

Demographics Characteristics of a given population, possibly including such information as age, race, gender, education, marital status, family structure, and location of agency.

Four-step method of instruction The primary process used to relate the material contained in a lesson plan to the students; the steps are preparation, presentation, application, evaluation.

Independent study A method of distance learning in which students order the course materials, complete them, and then return them to the instructor.

Learning environment A combination of a physical location (classroom or training ground) and the proper emotional elements of both an instructor and a student.

Method of instruction Various ways in which information is delivered to student, both in a classroom and on the training ground. (NFPA 1041)

Synchronous learning An online class format that requires students and instructors to be online at the same time. Lectures, discussions, and presentations occur at a specific hour and students must be online at that time to participate and receive credit for attendance.

Tailboard chat An informal gathering where fire fighters discuss various issues.

REFERENCES

Bear, George G. 1990. "Models and Techniques That Focus on Prevention." In *Best Practices in School Psychology*, edited by Alex Thomas and Jeff Grimes, 652. Silver Spring, MD: National Association of School Psychologists.

Blanchard, Kenneth, and Spencer Johnson. 1982. *The One-Minute Manager*. New York: William Morrow and Company, Inc.

Dale, Edgar. 1969. *Audiovisual Methods in Teaching*. 3rd ed. New York: Dryden Press.

National Fire Protection Association. 2019. *NFPA 1041: Standard for Fire and Emergency Services Instructor Professional Qualifications*. Quincy, MA: National Fire Protection Association.

National Fire Protection Association. 2018. *NFPA 1403: Standard on Live Fire Training Evolutions*. Quincy, MA: National Fire Protection Association.

National Fire Protection Association. 2015. *NFPA 1584: Standard on the Rehabilitation Process for Members During Emergency Operations and Training Exercises*. Quincy, MA: National Fire Protection Association.

Walker, J. E., and Shea, T. M. 1991. *Behavior management: A practical approach for educators*. 5th ed. New York: Macmillan.

CHAPTER PRESENTATION EXERCISE

Instructions: Your course instructor will assign individual or group discussions on the key points and teaching tips of this chapter. You or your group will review the chapter teachings and identify the major learning points from the chapter. You should be able to discuss the points, why they are important, and how/where they apply to your responsibilities at your level of instructor training.

REVIEW QUESTIONS

1. Describe some of the communications techniques that will help you improve the delivery of information to your class.
2. Describe the steps that make up the four-step method of instruction.
3. Describe asynchronous learning.

CHAPTER 4

Communication Skills

NFPA 1041 JOB PERFORMANCE REQUIREMENTS

Note: An asterisk denotes that the 1041 standard contains further information in its annex section.

4.4.3 Present and adjust prepared lessons, given a prepared lesson plan that specifies the presentation method(s), so that the method(s) indicated in the plan are used and the stated objectives or learning outcomes are achieved, applicable safety standards and practices are followed, and risks are addressed.

(A)* Requisite Knowledge.
The laws and principles of learning, methods and techniques of instruction, lesson plan components and elements of the communication process, and lesson plan terminology and definitions; learner characteristics; student-centered learning principles; instructional technology tools; the impact of cultural differences on instructional delivery; safety rules, regulations, and practices; identification of training hazards; elements and limitations of distance learning; distance learning delivery methods; and the instructor's role in distance learning.

(B) Requisite Skills.
Oral communication techniques, methods and techniques of instruction, ability to adapt to changing circumstances, and utilization of lesson plans in an instructional setting.

KNOWLEDGE OBJECTIVES

After studying this chapter, participating in a structured learning environment, and completing assigned assessments, you will be able to:

- Identify and describe the elements of the communication process. (**NFPA 1041: 4.4.3**, pp 103–104)
- Compare and describe the different types and styles of communication. (pp 104–107)
- Describe communication techniques that will improve your presentation. (**NFPA 1041: 4.4.3**, pp 104–108)

SKILLS OBJECTIVES

After studying this chapter, participating in a structured learning environment, and completing assigned assessments, you will be able to:

- Demonstrate the ability to use various communication styles in the classroom. (**NFPA 1041: 4.4.3**, pp 104–107)
- Demonstrate effective oral communication techniques. (**NFPA 1041: 4.4.3**, pp 105–107)
- Demonstrate effective written communication techniques. (**NFPA 1041: 4.4.3**, pp 108–112)

You Are the Fire and Emergency Services Instructor

While studying several National Institute for Occupational Safety and Health (NIOSH) fire fighter fatality reports, you notice that a common theme among both the contributing factors and key recommendations is the continued mention of communication issues and recommended training. Communications plays a very important role in most everything we do in the fire service, from training to occupancy inspections to emergency incidents. Understanding this, you have decided to conduct a training session for your department on emergency scene communications.

1. Identify and describe the elements of the communication process.

2. Describe the role of communications in both the learning (training) process and emergency incident response.

3. Compare and describe the different types and styles of communications.

4. Identify barriers to effective communications, both in the classroom and on the emergency incident scene.

 *Access Navigate for more practice activities.*

Introduction

Communication is the most important element of the learning process. Communication in and of itself is a process that you must master to become an effective fire and emergency services instructor. When you have mastered communication skills, the environment in which you are teaching becomes less of a factor in the learning process. Even if the environment is less than perfect, strong communication skills will enable you to be an effective and dynamic instructor. Your goal is to make sure that your students leave the classroom with a greater knowledge base than they had when they came in.

According to professional educator Leah Davies:

> Being able to communicate is vital to being an effective educator. Communication not only conveys information, but it encourages effort, modifies attitudes, and stimulates thinking. Without it, stereotypes develop, messages become distorted and learning is stifled.

An instructor must be an effective communicator. One of the keys to the learning process is the conveyance of information. When you demonstrate effective communication skills, you can take the learning process to even greater heights by encouraging effort, modifying attitudes, and stimulating thinking.

This chapter addresses both the spoken and written communication processes as well as the environment's effect on communication. Students have different learning styles and use different cues when learning. Knowing how to access these diverse cues and take advantage of them to enhance the learning experience will make you a more effective instructor. See Chapter 2, *The Learning Process*, for more information on different learner characteristics.

The fire service classroom encompasses so much more than four walls, tables and chairs, and a dry erase board. It can also include the back of an engine after the call, a burn building, and even a kitchen table at the station surrounded by members who have just completed an emergency response. No matter what the environment, the end objective is the same: that the student learns and that the student's safety and ability to do his or her job improve. The instructor must have excellent communication skills and must be able to read the students' comprehension of the training through verbal and nonverbal cues. Although the fire service is certainly steeped in tradition, the means of communication used by fire fighters have changed dramatically over the years. The first fire officers used speaking trumpets to communicate to the members of their department at a fire scene. Even with today's many avenues for the rapid transmission of messages, the message itself remains the critical piece of the communication puzzle.

The Basic Communication Process

The basic **communication process** consists of five elements: the sender, the message, the medium, the receiver, and feedback (**FIGURE 4-1**). Surrounding these elements is the environment. The five elements of the communication process act like the links of a chain. For the chain to function properly, all the links must be attached to one another. If any one of the links breaks, then the chain falls apart.

The Sender

In the communications chain of the fire service classroom, the **sender** is you, the instructor. As the sender, you must know which style of communication to use. The choice of the style of communication is based on many factors, including your comfort level. If you are more comfortable using oral (verbal) communication, then you are likely to use that style most often. Many other methods of instruction that center on the sender and how he or she delivers training material are discussed in Chapter 3, *Methods of Instruction*.

The Message

The next link in the communication chain is the **message**. What are you trying to communicate to your students? You must encode the point of the lesson and then transmit it. This may sound simple, but it is actually the most complex part of the communication process. For students to understand your message, they must hear it, decode it or put it into their own terms, and then ensure that their terms match your terms. Consider the command, "Put out that fire now." A student might hear this message as "Get a hose and spray some water right now," when the message really was "Get a fire extinguisher and suppress those sparks."

The Medium

How the message is transmitted is as important as the message itself. The means of conveying a message constitutes the **medium**. For example, your tone of voice, volume, speed of voice, and other qualities all convey information. If you give a lecture about a new piece of personal protective equipment (PPE) while sitting behind a desk, fiddling with a paper clip, and speaking in a very relaxed manner, you will convey to the students the message, "This new piece of PPE really isn't that important," even though the message you are stating verbally might be "This piece of equipment will save your life."

The Receiver

The next link in the communications chain is the **receiver**. In the learning environment, this link is the student. The receiver plays an important, active role in the communication chain; that is, the student needs to listen actively and maintain focus to decode your message. At the start of each training session, remind students of their role in the communications chain and continue to keep them engaged throughout the learning process. Fulfilling this role means leaving all distractions at the door and focusing solely on the information you present. Ensure that all students understand their active role in the communications process.

A student's previous experience and knowledge base play a major role in determining whether the student is able to receive your message. Adapt your style and presentation for your students. If you "talk down" to students or if the presentation goes over their heads, the message is not in a form that they can receive.

Feedback

Feedback is the link that completes the communication process. It is often considered the most important step in the communication process because it allows the sender to determine whether the receiver understood the message. Feedback can take the form of either verbal or nonverbal response. An example of the verbal feedback process is when orders are given over the radio. Consider this example:

Incident Commander: Engine 2, stretch a line to Division C and provide exposure protection.
Engine 2: Engine 2 copies—we are stretching a line to Division C for exposures.

An example of nonverbal feedback would be blank stares from students, indicating that they do not understand the message. Collectively asking a classroom of students if they understand is not the best method of determining whether the message was properly received. The natural response is "yes," because no student wants to admit that he or she does not understand. Asking a student to describe your message in his or her own words may be an effective

FIGURE 4-1 The five elements of the communication process act like the links of a chain.
© Jones & Bartlett Learning.

way to make sure that the student understands your message. However, this can depend significantly on the dynamics of the class. If the student's description is incorrect, gently explain that it's not quite right and ask for other students to offer the information.

Face-to-face conversation is the most effective means of conveying many types of information. With this approach, the sender and the receiver can engage in a two-way exchange of information that incorporates body language, facial expressions, tone of voice, and inflection. In contrast, when verbal communication must be conducted over radio or telephone, supplementary expressions are sacrificed. When discussing communication in the fire service, particularly within the incident command system, face-to-face communication is considered the preferred method. Usually, a face-to-face exchange is required if there is a transfer of command so that the sender and the receiver can pick up on any nonverbal indicators of confusion.

Another type of feedback may be apparent when you are giving feedback on a research paper. Writing notes in the margins of the paper allows the student to review and process your comments carefully, as opposed to just saying, "Nice job," and handing the student the paper.

The Environment

Surrounding all of the links of the communication chain is the environment. The environment—which includes physical, social, and environmental factors—can dramatically affect the communication process.

The physical environment is the room or area in which communication is taking place. It could be a classroom, the training ground, or even cyberspace. The physical environment can either inhibit or enhance the learning process. Teaching in a classroom with radios blaring and phones ringing directly outside, for example, will inhibit the communication process. When students cannot hear or see you, your message has a limited chance of being understood. Whether you are working in a classroom or on a busy training ground, make sure that all your students can see and hear you.

The social environment is the context in which the communication takes place. If the class is not enthusiastic about the message, the message will have a limited chance of being received. Mandatory training usually has this effect. You must make sure that your students want or need to receive your message—this is the art of motivation. Show your students why they need your message. Instructors must also expect the students to ask why and must not react by getting defensive. This challenges the instructor to find additional methods of explaining the material. Although students may ask why for different reasons, keep in mind that they have been engaged in what you were saying and participating in the preparation step of the learning process.

The final component is the environmental factor. The receiver must first feel that his or her needs are being met. An environment that is physically difficult for its inhabitants to tolerate is distracting for students. For example, if the room is too cold or too hot, then your message will not be received properly. Select an appropriate location for instruction. Having an impromptu lesson in a truck bay with heaters running may distract students and cause them to lose interest in your message quickly. Lessons that continue too long may also cause discomfort for students and negatively impact the learning process.

Nonverbal Communication

According to Charles R. Swindoll (2003), an American writer and clergyman, "Life is 10 percent what happens to you and 90 percent how you react to it." The same can be said about nonverbal communication: Communication is 10 percent what you say and 90 percent how you say it.

Nonverbal communication is a difficult communication skill to control because you generally do not watch yourself in the mirror all day. This type of communication includes the tone of your voice, your eye movement, your posture, your hand gestures, and your facial expressions. For example, eye contact can show sincerity. If you can look a student in the eye, you are demonstrating trustworthiness and caring. If you cannot maintain eye contact, you lessen the feeling of trust. You can also use your eyes to elicit a reaction: If you continue to look at a student, then that student is compelled to focus on you.

Standing straight and tall is a sign of confidence. You should work on your posture for this reason alone (**FIGURE 4-2**). In addition to showing confidence, such a stance helps maintain your back in good health during long days of instruction.

Hand gestures can express either aggression or enthusiasm. Rapid hand movements or repeated movements of the hands to the face can be signs of deception or uncertainty.

Active and Passive Listening

As an instructor, you must be well versed in both active and passive listening. **Active listening** is the process of hearing and understanding the communication sent and demonstrating that you are listening and have understood the message. It requires you to

FIGURE 4-2 A. Good posture is critical for good communication. **B.** Poor posture is unprofessional.
© Jones & Bartlett Learning.

FIGURE 4-3 Pay attention to what your students are telling you with their body language.
© Jones & Bartlett Learning.

keep your mouth closed and your ears wide open. As part of active listening, it is also very important that you hear the entire message. All too often, instructors formulate an answer before the student finishes asking the question. If you find yourself jumping in with an answer, try restating the student's question before answering. "Why do we use foam on a chemical fire? We use foam on a chemical fire because . . ."

Passive listening is listening with your eyes and your senses without reacting to the message. Observe the student's body language and facial expressions **FIGURE 4-3**. Nervous movements can be indicative of misunderstanding or confusion, whereas a relaxed brow and alert eyes can indicate comprehension and interest.

Carefully monitor what you are communicating nonverbally. Unknowingly, you may communicate through your body language or your attire that students should take unnecessary risks on the training ground. One of the best ways to develop your communication skills is straightforward: Practice them. One method is to have another instructor sit through your presentation and evaluate you. This evaluation should include how you communicated your message, which nonverbal cues you gave, and how your tone of voice affected the delivery of your message. Through such a peer evaluation, you can see the student's perspective of you.

Another technique is to make a video of your presentation. Watch the recording and look for any nonverbal cues that do not match your message. Many of the nonverbal cues you project become more obvious when you play the recording back at a faster-than-normal speed. Finally, close your eyes and listen to your voice at normal speed. Is your voice really sending your intended message?

> **COMPANY-LEVEL INSTRUCTOR TIP**
>
> Always strive to improve your presence in the learning environment by avoiding habitual and annoying distractions such as the use of repetitive words or physical gestures like playing with keys in your pocket.

Verbal Communication

Verbal, or oral, communication is more than just talking. Many people talk, yet never actually communicate. Great communicators, by contrast, use many tools to convey their message. Watch and listen to the speeches of John F. Kennedy and Martin Luther King, Jr.—both were highly effective communicators, inspiring their followers to take action. At his inauguration in 1961, Kennedy said, "And so, my fellow Americans, ask not what your country can do for you; ask what you can do for your country" (Kennedy, 1961). The inflection in his voice added emphasis to his words and reinforced his message. In his "I Have a Dream" speech at the Lincoln Memorial in 1963, King used personalization so that the entire audience would put themselves in his shoes for a moment (King, 1963).

Many factors, including language and tone, affect verbal communication. The volume of your speech is important. By lowering or raising your voice, you can emphasize a point. You can also use volume to gain the attention of your students. A loud, forceful tone will let the class know that you are very serious. Thus, when a student faces imminent danger, your voice must be loud and forceful.

Likewise, the importance of speaking softly cannot be overstated. When speaking to a student privately, maintain both a level of decorum and a softer tone to your voice. In addition, a softer voice is less threatening and does not put the student on the defensive. If the student is defensive, he or she is less likely to receive your message.

Bringing your voice to a loud crescendo and then making it softer can serve to emphasize your message. This technique is especially effective when you are using compare-and-contrast strategies. For example, during a discussion of tone of voice with fire officers, you might want to compare a dominant tone versus a supportive tone. You would speak loud and forcibly to show dominance, and then speak calmly and softly to show support. The same technique can also be used when describing an escalating event. For example, when describing a flashover, you would speak at normal volume to describe the fire building. As the fire builds, your volume would increase until the flashover occurs, with your voice booming.

COMPANY-LEVEL INSTRUCTOR TIP

When posing questions, an effective practice is to call on a student for the answer before asking the question. This allows the student the opportunity to focus on the question and formulate a response. In most instances, early engagement in the active listening process produces effective results.

The learning environment is also a factor when it comes to voice volume. If you are instructing outside or in a noisy environment, the volume of your voice may have to increase to project it over any **ambient noise**. Of course, sometimes ambient noise is chattering students. In those cases, you should use volume to gain your students' attention and silence.

Projecting your voice is important. If you are stationary or do not have the ability to walk around, be aware of the need to make your voice carry to the back

FIGURE 4-4 Project your voice and maintain good eye contact and posture while working in large classrooms.
© Jones & Bartlett Learning.

of the room to each student (**FIGURE 4-4**). This feat can be accomplished by using a microphone. If a microphone is not available, use your voice as a tool to convey your message. If you are more mobile, you may be behind or among the students, so be sure to project your voice so that everyone in the class can hear you.

The difference between instructors can sometimes be as simple as the level of enthusiasm projected through the voice. Enthusiasm breeds enthusiasm. It creates interest both in the topic and in you. Focus on teaching the courses that fascinate you—this will add passion to your presentation. For those topics that may not be as exciting, research why the topic is important. Understanding the underlying value of the topic will help you make it more interesting for the students. The dynamic use of voice is one of the most effective methods of holding students' attention. Knowing when and how to change your voice is a skill learned through practice. Practice making a conscious effort to really use your voice to convey your message until it becomes automatic.

It is important to be aware of any students who have any level of hearing impairment. For a student with hearing difficulty, you may need to consider your body positioning so that the student is able to see your face while you are speaking, or use an auxiliary device to assist the student in hearing the lesson.

Language

The fire service has its own unique language. We use terms that mean something completely different to the general public. For example, when we say "company," we mean a group of personnel assigned to perform a task; by contrast, the general public defines *company* as a business. Within the fire service, terminology can change from department to department and from region to region. A rescue in one department is called an ambulance in another. It is imperative that you use the terminology of your own department when discussing equipment, apparatus, or procedures. Use words that are familiar to your audience and explain new terms thoroughly. Mentioning the problems of fighting a fire in a "group home," for instance, may not be effective if the students do not understand what a group home is.

The use of appropriate language is critical. Avoid language that does not put you or your department in the best light. Ask yourself if you would use the same language around the family dinner table. Once spoken, words cannot be taken back. Exercise restraint and caution when speaking. Always be on guard with your speech: As an instructor, you are held to a higher standard. Many students have expectations of what an instructor should say and do, so never let your guard down.

The tone of your voice can express more than your words. The purpose of tone is to express emotion. Your tone should always be positive, while expressing the passion of conviction. If your words say that you believe in something but your tone expresses apathy, the students will hear apathy, no matter what your words say. To see how this effect works, watch an actor on stage. He may be having a bad day, but when he steps on stage, he portrays good humor and joy. As the instructor, you must put your personal feelings aside and use your tone of voice to express the message properly. Ethically speaking, fire fighters routinely use language that they would never use in front of their grandmothers. Is it ethical to use that same language when instructing? While your first instinct may be to communicate directly in the students' everyday language, it is inappropriate to do so in the classroom. You are judged by your actions. Always take the high road.

The fire service classroom can be as simple as a room with chairs, tables, and any number of audiovisual devices to enhance the learning experience, or it can be as complex as a state-of-the-art "techno-heaven" with multimedia consoles that respond to your requests with only a touch of a button. Sometimes the classroom may be a bunkroom that serves as a multipurpose room with an overhead projector and slide projector, but it can also be the back of a fire engine or the base of the training tower. Many facilities have "dirty" classrooms—that is, designated areas allowing fire fighters in full turnout gear to discuss the training session before or after it occurs.

The purpose of the classroom is to facilitate the learning process. Consequently, the physical location is less important than the communication process. The back of an engine can be a great learning environment if the proper communication skills are used.

Many company officers take time after a call to discuss how the call went. This mini-critique can be of great value to the company, even without the traditional classroom or the audiovisual devices; only good communication is necessary for this kind of education.

Audience Analysis

As a good instructor, you need to determine how your message needs to be communicated before deciding on the appropriate format for your presentation. You also need to evaluate your audience to determine how they will best receive your message.

At the most basic level of firefighting, the lecture is usually the most appropriate presentation method. Because students at the entry level have very little experience to draw upon, it is best to use an informative lecture to communicate concepts and objectives. You need to pass the information to the student in a very direct manner. It will be repetitive in nature—some would say almost rote memorization.

In speaking to more experienced fire fighters, the lecture is more persuasive in nature. With these students, you may be taking a skill learned in basic training and discussing it in finer detail at a higher level. This type of class requires students to be convinced about the importance of the message; otherwise, they might tune out the lecture because they mistakenly believe that they already know everything possible about the skill.

Another type of audience encountered by fire instructors is made up of experienced fire fighters and fire officers. These students tend to require discussions rather than informative lectures. Discussions engage these experienced students and encourage them to incorporate their knowledge and experiences into the learning process. This concept, which is sometimes referred to as a "tailboard chat," is used with more experienced fire fighters because it takes their existing knowledge and experience and melds it to a new concept (**FIGURE 4-5**). By examining and reexamining the way we do things, and by discussing experiences and options, new ideas and new ways of doing things are developed. This sort of exchange can be very exciting and rewarding because you can witness an idea developing in front of you. In fact, many of the tools used in the fire service today evolved out of discussions that arose when fire fighters saw a problem or a situation that needed to be improved and made adaptations to rectify the need. This style of instruction is sometimes referred to as facilitation.

Alternative communication methods should also be considered. For example, role playing can often reinforce learning objectives. In role playing, you can

FIGURE 4-5 The tailboard chat can provide an opportunity for a quick training exercise or to analyze the recent response.
© Jones & Bartlett Learning.

take the position of someone with whom the student interacts. This method requires you to have an idea of how a student will respond in this situation and to be prepared for that response. You should be prepared for several potential responses from the student and be able to lead the student toward the learning objective. At the end of the role playing, you should explain the learning objective of the exercise.

Another possible communication method consists of group exercises. With this method, you use the collaborative efforts of the group to benefit all members of the group. Such a technique exploits the natural tendency of fire fighters to work in groups. With group work, you can evaluate whether students can work collaboratively to solve a problem. In most cases, this is the way a fire company solves problems in the field. It is imperative that you listen to the conversations in the groups. Walk around to each group and stand near or actually sit down with the group and listen. You can see if the group is moving in the right direction or needs to be redirected. If necessary, help the group along by saying, "Have you considered . . ." Steering the discussion is not equivalent to giving students the answer, but rather assists them in remaining focused on the message. Be sure to solicit outcomes from each group involved in the exercise; adult learners will want to have their work discussed and will not appreciate "busy work" during a class.

Your communication style will be situational, such that you may have to use several different styles during the same presentation. When teaching a class on fire extinguishers, for example, you might begin with a lecture style and then move into a psychomotor session. When giving the lecture, you are simply presenting facts: There is no pressure and no sense of danger. When you move into the psychomotor session, however, it may be necessary to speak authoritatively to ensure that the students understand the issues of safety.

Issues of safety will inevitably affect your communication style. When you explain to new recruits that they must stay down in a fire situation, they generally nod their heads as you explain the science of thermal layering. When you get on the fire ground, be more forceful in your tone and action to keep your students focused on safety.

When an important objective needs to be reinforced, you can highlight it by changing your communication style. For example, by adding a group exercise in the middle of a lecture, you highlight the learning objective.

Time is another factor when considering communication styles. If students spend 2 hours straight sitting and listening to a lecture, retention of the material can diminish. Nothing is worse for a student than to have a monotone instructor lulling him or her to sleep with an endless lecture. Retention of material is increased by adding activities. Engaging the student in the learning will pay big dividends. For example, if a student has to speak and do something, the information will be retained longer and more completely. This consideration is especially important after a meal. The more activities you can provide to students after a lunch break, the more likely they are to retain the information. The cone of learning illustrates the importance of communication skills and how students are able to remember information (**FIGURE 4-6**). The more active the student is in the learning, the more effective and real the learning is to them. Lesson plans are discussed in Chapter 5, *Using Lesson Plans*; activities or chances for student practice or participation are incorporated in the application step of a lesson plan.

Written Communications

Written communication is needed to document both routine and extraordinary fire department activities. It establishes institutional history and is the foundation of any activity that the fire department wishes to accomplish. As a senior chief explained, "If it is not written down, it did not happen." The instructor is responsible for many types of written communications, including narratives, which are included in training records, and more comprehensive training reports, such as annual reports of training activities. Although e-mail has replaced many formal written letters and other communications, a written communication

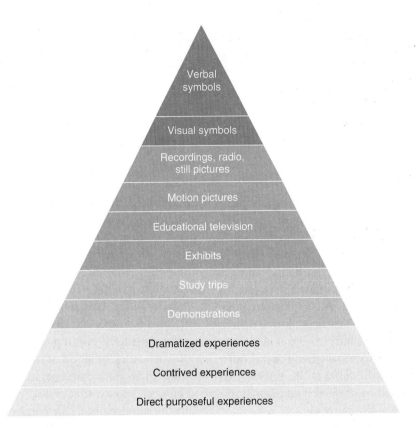

FIGURE 4-6 Cone of learning.
Modified from: Dale, Edgar. *Audio-Visual Methods in Teaching*, 1E. © 1969 Wadsworth, a part of Cengage Learning, Inc.

serves as a permanent record of an event or happening. Each agency should specify guidelines that outline which types of written documentation are necessary for specific occurrences, such as reports of injuries and documentation of training issues.

Reading Level

Most newspapers in the United States are written at approximately a sixth-grade reading level to allow for understanding by the general public. The newspaper does not intend to talk down to its readers who have a higher level of education; instead, it is simply ensuring that the majority of readers can easily understand the information presented.

Do you know the reading levels of each of your students? Most fire fighters are required to have a high school education or a GED, so the expectation is that all students can read and write at the high school level. Most educational textbooks are written at a reading level that the majority of fire fighters should understand. Technical editors and select fire service personnel review textbooks to establish their readability before they are printed and bound.

Most word processing programs can perform an evaluation of the document's reading level based on the Flesch–Kincaid readability index. With this index, a higher number indicates that the document is easier to read. Therefore, the higher the Flesch–Kincaid number, the lower the reading level. The index is based on the average number of words per sentence and the number of syllables per 100 words. Be aware of the length of your sentences and the complexity of the words you use. Always keep your students' reading levels in mind when writing.

Writing Format

It is also important to know which format you should use when writing. In general, when writing to individuals within your organization, you can use a memo (short for *memorandum*). Such a document is designed for in-house communications only. For example, a memo from the chief of your training division to all instructors might inform them of an upcoming training drill. Written communications intended for readers outside the organization should appear in letter format, using departmental letterhead, and should be signed by the sender. A letter would be the appropriate form of communication to Mr. Jones confirming a preincident planning appointment at his coffee shop, for example.

The Five Ws of Writing

The basic rule of thumb for writing is to include the "five Ws":

1. Who
2. What
3. Where
4. When
5. Why

Add an "H"—How—to expand the discussion further.

By including these points in your writing, you will ensure that you address the information necessary for readers to understand your message properly. For example, suppose you need to write a memo about a live burn that will take place at a donated structure on the outskirts of town (**FIGURE 4-7**). You want your students and additional instructors to know where and when to meet on the live burn day. You also want the entire fire department to know when the live burn is taking place, who is scheduled to be onsite, and whether permission must be granted if additional fire fighters want to be on the live burn site.

The memo in Figure 4-7 contains all five Ws (plus H) of writing:

- Who: Lavalle Fire Department
- What: Live burn training exercise
- Where: 1015 Willow Way
- When: July 10 at 10:00 AM
- Why: Permission must be granted to be on the live burn site
- How: Fire fighters must see Officer Archie Reed for a permission request form

The Rules of Writing

Good, clear, concise writing is critical to getting your message across. Fortunately, it is easy to improve your writing: If you follow the rules of good writing in step-by-step fashion and practice for 15 minutes each day, you can become an effective written communicator. As a fire fighter, you train on a regular basis to maintain your firefighting skills. Similarly, as an instructor, you need to practice your writing skills on a regular basis. This step involves writing and allowing fellow instructors to read your writing to ensure that your message is coming through loud and clear.

The rules of good writing begin with the paragraph. Each paragraph begins with an introduction—a sentence or two that states what you plan to discuss in the paragraph. Then comes the main body of the paragraph—the frame. Like the frame of a structure, the frame of the paragraph is the support that holds everything together. The conclusion reviews what was discussed in the paragraph and reinforces the frame, as explained here:

> The fire officer must first consider the intended audience for the report before writing. For example, a fire officer may be assigned to prepare a study that proposes closing three companies and using quints that can respond as either engine or ladder companies. If the report is intended for internal use, the technical information would use normal fire department terminology. Conversely, if the report is intended for the fire chief to deliver to the city council, with copies going to the news media, many of the terms and concepts would need to be explained in simple terms for the general public. Always consider who your audience is before writing a report.

To practice constructing a solid paragraph, consider writing a two- to three-paragraph summary after every class. What was your message for the class? Did your students receive it?

Once you have mastered the building of a paragraph, you are on your way to constructing a complete written structure—anything from a two-paragraph memo, to a full-page letter, to a 10-page report for the city council. The paragraph is your basic building

---

To: Lavalle Fire Department
Cc: Chief Michael Deforge
From: Officer Archie Reed
Date: May 23
Subject: Live Burn on July 10

On July 10, the Lavalle Fire Department will conduct a live burn at 1015 Willow Way. Fourteen students from the Fire Suppression Course will be present. Students will participate in supervised training using medium-diameter hose lines during defensive operations.

Students and Fire Service Instructors Thomas, Kelly, Hinkler, and Andrews will assemble at the South Street Fire Station at 10:00 a.m. to be transported to the live burn structure. Safety procedures will be reviewed at the site prior to the live burn. In addition to the students and the fire service instructors, Engine Company 3 will be on site.

If you are not a student in the Fire Suppression Course, an assigned fire service instructor, or a member of Engine Company 3, you may not be on the live burn site without written permission. Please see Officer Archie Reed for a permission request form.

FIGURE 4-7 A sample memo using the five Ws of writing.
© Jones & Bartlett Learning.

block of writing. All good writing has an introduction, a frame, and a conclusion. In an eight-paragraph report that is intended for the city council and explains why the fire department's training budget should be increased by 10 percent, for example, the report would be broken down in the following way:

- Paragraphs 1 and 2: Introduction. These paragraphs give an overview of why the fire department needs to increase the training budget.
- Paragraphs 3 through 7: Frame. These paragraphs discuss in depth the specific reasons why the training budget should be increased. For example, the city might have grown by 20 percent in 3 years and needs more fire fighters to provide good service.
- Paragraph 8: Conclusion. This paragraph summarizes the arguments for increasing the fire department's training budget.

Style

Most of the writing you do as a fire fighter and as an instructor will be technical in nature. In such a case, your sole objective is to convey information—not to express emotion or to inspire passionate feelings. Writing in the fire service should not be boring or uninspiring, but it should remain true to the premise that you write to impart facts.

Once words are printed, they are difficult to retract. Read an editorial column in a newspaper to see how writing can affect people's thoughts and emotions. Words can be construed in different ways by different people, so it is extremely important to choose your words carefully. The meaning of words can change based on geographical location. A person's individual experiences also can change the meaning of a word. Know your audience before you begin to write. Your goal is to make certain that your message is clearly and properly conveyed to every reader. Choose your words carefully when writing. Make every effort to ensure that alternative interpretations cannot occur. Failure to do so can mislead students and ultimately lead to undesired performances.

Informal Communications

Informal communications include internal memos, e-mails, instant messages, and messages transmitted via mobile data terminals. Informal reports have a short life and are not archived as permanent records. Instead, they are used primarily to record or transmit information that will not be needed for reference in the future. Be aware that some agencies, as public agencies, are required to keep archived records of even informal records, and such communications may be subject to freedom of information requests.

Some informal reports are retained for a period of time and may become a part of a formal report or investigation. For example, a written memo could document an informal conversation between a fire instructor and a fire fighter regarding a training drill attendance issue. The memo might note that the student or crew member was warned that corrective action is required by a particular date. If the issue is corrected, the warning memo is discarded after 12 months. If the fire fighter continues to demonstrate poor attendance, the memo becomes part of the formal documentation of progressive discipline.

In recent years, the storage and recall of e-mails has been a topic of much discussion. Policies will vary from agency to agency, so know your department's policy about e-mails, instant messages, and related informal communication. Ignorance will not usually work as a defense in a court of law, so know your agency's policy regarding the use of this type of informal communication.

Formal Communications

A **formal communication** is an official fire department document printed on business stationery with the fire department letterhead. If it is a letter or report intended for someone outside the fire department, the document is usually signed by the fire chief or a designated staff officer to establish that it is an official communication. Subordinates often prepare these documents and submit them for an administrative review to check the grammar and clarity. A fire chief or the staff officer who is designated to sign the document performs a final review before it is transmitted.

The fire department maintains a permanent copy of all formal reports and official correspondence from the fire chief and senior staff officers. Formal reports are usually archived.

Standard Operating Procedures

Standard operating procedures (SOPs) are written organizational directives that establish or prescribe specific operational or administrative methods to be followed routinely for the performance of designated operations or actions. SOPs are intended to provide a standard and consistent response to emergency incidents as well as personnel supervisory actions and administrative tasks. They serve as a prime reference source for promotional exams and departmental training.

SOPs are formal, permanent documents that are published in a standard format, signed by the fire chief, and widely distributed. They remain in effect permanently or until they are rescinded or amended. Many fire departments conduct a periodic review of all SOPs so they can revise, update, or eliminate any that are outdated or are no longer applicable. Any changes in the SOPs must also be approved by the fire chief. SOPs should always be referenced as part of a training session, especially when demonstrating skills and ability.

General Orders

General orders are formal documents that address a specific subject, policy, condition, or situation. They are usually signed by the fire chief and can be in effect for various periods, from a few days to permanently. Many departments use general orders to announce promotions and personnel transfers. Copies of the general orders should be made available for reference at all fire stations.

Announcements or Communications

The formal organization may use announcements, information bulletins, newsletters, websites, or other methods to share additional information with fire department members. These methods are generally used to distribute short-term and nonessential information that is of interest.

Legal Correspondence

Fire departments are often asked to produce copies of documents or reports for legal purposes. Instructors must always remember that training records are also legal documents and, as such, are available for scrutiny through the legal system. Sometimes the fire department is directly involved in the legal action, whereas in other cases the action involves other parties but relates to a situation in which the fire department responded or had some involvement. It is not unusual for a fire department to receive a subpoena—a legal order demanding all of the documentation that is on file pertaining to a particular incident or person. Gathering such documents is a task that can require extensive time and effort. In any situation for which there is a request for reports or documentation, consult with legal counsel.

The fire fighter or officer who prepared a report may be called on, sometimes years later, to sign an affidavit or appear in court to testify that the information provided in the report is complete and accurate. He or she might also have to respond to an **interrogatory**, which is a series of written questions asked by someone from an opposing party. The fire department must provide written answers to the interrogatories, under oath, and produce any associated documentation. If the initial incident report or other documents provide an accurate, factual, objective, complete, and clear presentation of the facts, the response to the interrogatory is often a simple affirmation of the written records. The task of dealing with the interrogatory can become much more complex and embarrassing if the original documents are vague, incomplete, false, or missing.

Reports

Reports should be accurate and present the necessary information in an understandable format. All reports should be proofread before being submitted. The purpose of the report might be to brief the reader, or it might provide a systematic analysis of an issue. If the fire chief wants only to be briefed on a new piece of equipment and the fire officer delivers a fully researched presentation, both parties would be frustrated.

Some reports are prepared on a regular schedule, such as daily, weekly, monthly, or yearly. Other types of reports are prepared only in response to specific occurrences or when requested. To create a useful report, you must understand the specific information that is needed and provide it in a manner that is easily interpreted.

Some reports are presented orally; others are prepared electronically and entered into computer systems. Some reports are formal, whereas others are informal. The most common form of reporting is verbal communication from one individual to another, either face-to-face or via a telephone or radio. To be effective, the transfer of information must be clear and concise, using terminology that is appropriate for the receiver.

Monthly Activity and Training Report

The monthly activity and training report documents the company's activity during the preceding month. These reports typically include the number of emergency responses, training activities, inspections, public education events, and station visits that were conducted during the previous month. Some monthly reports include details such as a list of the number of feet of hose used and the number of ladders deployed during the month. The number of training hours completed, members who receive certifications or attend classes, and any number of training-related projects or meetings should be documented in a regular format to highlight the role and importance that training plays in an organization.

In many cases, the officer delegates the preparation of routine reports that do not involve personnel actions or supervisory responsibilities to subordinates. In such a case, the officer is still responsible for checking and signing the report before it is submitted.

Some fire companies or municipal agencies post a version of their monthly reports on their public website. These reports often include digital pictures of the incidents and the people involved.

Special events or unusual situations that occurred during the month might also be included in the report. For example, the monthly report might note that the company provided standby coverage for a presidential visit or participated in a local parade.

COMPANY-LEVEL INSTRUCTOR TIP

Both instructors and administrative staff share the responsibilities for completing written documentation. Instructors should be aware of any written documentation that is required for their program and the timeline in which these forms need to be submitted.

Infrequent and Special Reports

Infrequent reports usually require a fire instructor's personal attention to ensure that the report's information is complete and concise. Such reports include the following types of documents:

- Fire fighter injury report
- Student complaints
- Property damage or liability-event report
- Vehicle accident or equipment damage report
- New equipment or procedure evaluation
- Suggestions to improve fire department operation
- Response to a grievance or complaint
- Fire fighter work improvement plans
- Research report or recommendation

A chronological statement of events is a detailed account of activities, such as a narrative report of the actions taken at an incident or accident. A **recommendation report** is a document that suggests a particular action or decision.

Some reports incorporate both a chronological section and a recommendation section. For example, the fire fighter line-of-duty death investigations produced by NIOSH include a chronological report of the event, followed by a series of recommendations that could prevent the occurrence of a similar situation.

Writing a Decision Document

The goal of a decision document is to provide enough information and persuasion, so the intended individual or body accepts your recommendation. This type of report includes recommendations for employee recognition or formal discipline, for a new or improved procedure, or for adoption of a new device. A decision document usually includes the following elements:

- **Statement of problem or issue:** One or two sentences.
- **Background:** Brief description of how this became a problem or an issue.
- **Restrictions:** Outline of the restrictions affecting the decision. Factors such as federal laws, state regulations, local ordinances, budget or staff restrictions, and union contracts are all restrictions that could affect a decision.
- **Options:** Where appropriate, provide more than one option and the rationale behind each option. In most cases, one option is to do nothing.
- **Recommendation:** Explain why the recommended option is the best decision. The recommendation should be based on considerations that would make sense to the decision maker. If the recommendation is going to a political body and involves a budget decision, it should be expressed in terms of lower cost, higher level of service, or reduced liability. The impact of the decision should be quantified as accurately as possible, with an explanation of how much money it will save, which new levels of service will be provided, or how much the liability will be reduced.
- **Next action:** The report should clearly state the action that should be taken to implement the recommendation. Some recommendations may require changes in departmental policy or budget. Others could involve an application for grant funds or a request for a change in state or federal legislation.

Training BULLETIN

JONES & BARTLETT FIRE DISTRICT
TRAINING DIVISION
5 Wall Street, Burlington, MA, 01803
Phone 978-443-5000 Fax 978-443-8000

Instant Applications: Communication Skills

Drill Assignment

Apply the chapter content to your department's operation, training division, and your personal experiences to complete the following questions and activities.

Objective

Upon completion of the instant applications, fire and emergency services instructor students will exhibit decision making and application of job performance requirements of the instructor using the text, class discussion, and their own personal experiences.

Suggested Drill Applications

1. Review a recent NIOSH fire fighter fatality report that involves communications in either the Contributing Factors or Key Recommendations areas and identify what your department could do to address the issue.
2. Evaluate a recent training session you conducted or participated in during which you experienced communication difficulties. Identify solutions for avoiding the communications pitfalls in future sessions.
3. Identify the different means of written communications employed by your organization and discuss ways to improve the communications process.
4. After conducting a training session with your shift, discuss your own personal communications traits and ask for feedback on how you can improve your delivery during training.

Incident Report: NIOSH Report: F2002-34

Drill Assignment

Review the information in this incident report and prepare a practice presentation for class delivery at the direction of your instructor. Your presentation should include a summary of the incident facts and a review of the NFPA 1403 compliance and noncompliance findings. Use an outline to organize your thoughts. You may be evaluated on your communication skills during your presentation.

Poinciana, Florida—2002

Just 10 months after the incident in Lairdsville, and less than 3 weeks after the sentencing of the Lairdsville chief officer, a live fire training session trapped two fire fighters in Poinciana, Florida, near Kissimmee.

The 1600-ft^2 (148.6-m^2) cement block house had three bedrooms, one of which was converted from a one-car garage with the large door removed, the wall blocked up, and a window installed (**FIGURE 4-A**).

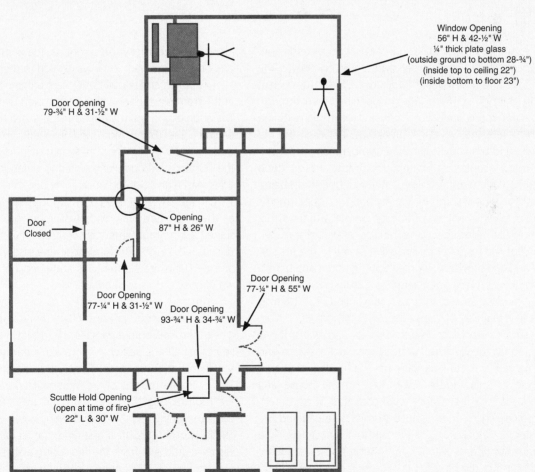

FIGURE 4-A Poinciana floor plan.
Courtesy of the Florida State Fire Marshal.

Incident Report continued

FIGURE 4-B Two consecutive offset turns and this narrow hallway made it difficult for fire fighters to maneuver.
Courtesy of the Florida State Fire Marshal

FIGURE 4-C Interior of fire room in Poinciana.
From the NIOSH Fire Fighter Fatality Investigation and Prevention Program, Investigation Report F2002-34.

The room had block walls, one door, and a fixed ¼-in. (6.4-mm) thick, commercial-grade glass window. The exit from the room was through two consecutive offset turns, through a 26-in. (65-cm) opening, then through two small rooms, and out another door into the dining room (**FIGURE 4-B**). A breezeway between the old garage and the main house had also been enclosed. Unlike many acquired structures, this house was in good condition and was part of an entire neighborhood being razed for a new campus. Several area fire departments were involved in a series of training events, and the structures available included other houses, as well as a motel.

All of the instructors were experienced in the fire service and had previously worked together on training fires. Safety crews were not briefed at the beginning of this exercise because of their past involvement. The training officer walked all of the participants through the structure and explained the safety aspects and goals of the training evolution, which were to conduct a search and rescue with an actual fire burning and find a mannequin in bunker gear hidden somewhere in the house and remove it. The first search team consisted of an experienced lieutenant and a trained recruit (state-certified Fire Fighter II), who were to search without a hose line. Two suppression teams with 1¾-in. (44-mm) hose lines were in position in the house with four interior safety officers broken into two teams.

The fire was started in the converted bedroom near the only doorway to the fire room (**FIGURE 4-C**). Two piles, practically vertical with pallets, wood scraps, and hay, were almost adjacent to each other, one inside and one outside the open closet. After the fire started, and with the instructor-in-charge's agreement, a foam mattress from one of the other bedrooms was added to the pile.

The search and rescue (SAR) team entered the structure at the front (east) door with a suppression team following. The suppression team stopped in the small room located between the dining room and the bedroom, where the training fire was located. The search team continued into the burning bedroom, encountering deteriorating conditions with high heat and no visibility due to heavy smoke.

On the exterior, a second suppression team waited at the front doorway, with two more fire fighters assigned as the rapid intervention crew (RIC), and a third, uncharged 1¾-in. (44-mm) hose line also was available outside. One fire fighter was stationed on the exterior waiting for orders to ventilate.

With near zero visibility and increasing heat conditions, two of the interior safety officers monitoring the activities of the fire room area later stated that they heard the lieutenant of the search team ask his partner if he had searched the entire room. They heard the answer "yes." Shortly thereafter, one of the safety officers yelled into the fire room asking if the search team was out. Although someone answered "yes," no one knows who replied. The interior safety officer assumed that he had missed the search team's exit from the fire room, as there were several fire fighters present by the dining room. He began to search the rest of the structure in an attempt to find them.

The instructor-in-charge ordered the front window of the fire room to be broken. After the window was broken out, heavy black smoke followed, which ignited very quickly with flames forcibly venting from the window. The suppression crew closest to the fire room applied water in short bursts, but increasing heat and steam forced the safety officers and the suppression crew to back out. Both safety officers were forced to exit the structure, after receiving burn injuries.

The second suppression team was ordered to replace the first team and engage the fire. The second team and an interior safety officer entered the fire room and extinguished the fire.

During this time the instructor-in-charge called the missing fire fighters several times and received no answer.

While the second suppression team was overhauling the fire area, they found a body in fire fighter bunker gear facedown on the floor. Both of the suppression team fire fighters initially thought it was the rescue mannequin, not realizing it was the lieutenant of the SAR team.

After no radio response from the SAR team, the instructor-in-charge ordered a personnel accountability report (PAR) and ordered the RIC to enter the structure and find the SAR team.

The suppression team that found the lieutenant dragged him to the front window of the fire room and removed him to the outside. The missing fire fighter was also found inside that window, and he too was removed to the outside. The entire event, from the time the SAR crew entered to when the first of the two fire fighters was located, was under 14 minutes. The two fire fighters died despite working in teams with experienced instructors, two interior staffed hose lines, two interior safety teams, an RIC with its own hose line, a participant walk-though, and other safeguards in place.

NIOSH investigated the fire. Two separate investigations were conducted by the state fire marshal: one for criminal violations, and an administrative investigation of state training codes. No criminal charges were filed; however, the findings included the following:

- "All of the participants stated . . . they did not have any concerns regarding the conditions of the fire inside the structure and it appeared to them as normal fire behavior."
- Although NFPA 1403 was already required in state code, it was under the environmental laws and so the law enforcement department of the state fire marshal's office did not have the authority to enforce it.
- The fire was started in a room with too much fuel for the size of the room, including a foam mattress that was added after ignition. Flashover was precipitated by too high a fuel load and inadequate ventilation.
- National Institute of Standards and Technology (NIST) determined that the mattress was contributory to considerable smoke production; however, in testing, the room flashed over in roughly the same time without the mattress present.

Postincident Analysis Poinciana, Florida

NFPA 1403 Noncompliant

- No written preburn plan prepared
- Accountability was not maintained at the point of entry to the fire room
- Safety crews did not have specific assignments or emergency procedures
- Fire started near the only doorway in and out of the fire room
- Excessive fuel loading
- No communications plan in place

NFPA 1403 Compliant

- Experienced instructors had worked together on training fires before
- Primary and secondary suppression crews inside with hose lines
- A walk-through of the structure was conducted
- Crews were alerted of the possibility of a victim
- Structure in good condition

Other Issues

- Radio communications less than optimal
- Egress limited by offset turns, a 26" (66-cm) opening, and multiple rooms and turns

After-Action REVIEW

IN SUMMARY

- Communication is the most important element of the learning process.
- The basic communication process consists of five elements: the sender, the message, the medium, the receiver, and feedback.
- Be aware of the message that you are sending to students nonverbally through your body language.
- Your communication style must be adapted to match the situation. You may have to use several different styles during a presentation.
- As an instructor, you must be well versed in both active and passive listening.
- Many factors, including language and tone, affect verbal communication.
- Written communication is needed to document both routine and extraordinary fire department activities. It establishes institutional history and serves as the foundation of any activity that the fire department wishes to accomplish.
- The most common form of reporting is verbal communication from one individual to another, either face-to-face or via a telephone or radio. To be effective, the transfer of information must be clear and concise, using terminology that is appropriate for the receiver.
- Reports should be accurate and present the necessary information in an understandable format. The purpose of the report might be to brief the reader, or it might provide a systematic analysis of an issue.
- Most presentations for an instructor involve informative speeches, or lectures. As a good instructor, you must determine how your message needs to be communicated before selecting the appropriate format for your presentation.

KEY TERMS

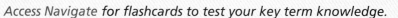

Access Navigate *for flashcards to test your key term knowledge.*

Active listening The process of hearing and understanding the communication sent; demonstrating that you are listening and have understood the message.

Ambient noise The general level of background sound.

Communication process The process of conveying an intended message from the sender to the receiver and getting feedback to ensure accuracy.

Feedback The fifth and final link of the communication chain. It allows the sender (the instructor) to determine whether the receiver (the student) understood the message.

Formal communication An official fire department communication. The letter or report is presented on stationery with the fire department letterhead and generally is signed by a chief officer or headquarters staff member.

General orders Short-term documents signed by the fire chief that remain in force for a period of days to 1 year or more.

Informal communications Internal memos, e-mails, instant messages, and computer-aided dispatch/mobile data terminal messages. Informal reports have a short life and are not archived as permanent records.

Interrogatory A series of formal written questions sent to the opposing side of a legal argument. The opposition must provide written answers under oath.

Medium The third link of the communication chain, which describes how you convey the message.

Message The second and most complex link of the communication chain; it describes what you are trying to convey to your students.

Passive listening Listening with your eyes and other senses without reacting to the message verbally.

Receiver The fourth link of the communication chain. In the fire service classroom, the receiver is the student.

Recommendation report A decision document prepared by a fire officer for the senior staff. The goal is support for a decision or an action.

Sender The first link of the communication chain. In the fire service classroom, the sender is the instructor.

REFERENCES

Davies, Leah. "Effective Communication." Kelly Bear. http://www.kellybear.com/TeacherArticles/TeacherTip15.html. Accessed August 17, 2018.

Kennedy, John F. 1961. "Inaugural Address." John F. Kennedy Presidential Library and Museum. http://www.jfklibrary.org/Asset-Viewer/BqXIEM9F4024ntFl7SVAjA.aspx

King, Martin Luther Jr. 1963. Martin Luther King's speech: "I have a dream"—the full text. http://abcnews.go.com/Politics/ martin-luther-kings-speech-dream-full-text/story?id= 14358231&page=2

National Fire Protection Association. 2019. *NFPA 1041: Standard for Fire and Emergency Services Instructor Professional Qualifications.* Quincy, MA: National Fire Protection Association.

National Fire Protection Association. 2018. *NFPA 1403: Standard on Live Fire Training Evolutions.* Quincy, MA: National Fire Protection Association.

Swindoll, Charles R. 2003. *Strengthening Your Grip.* Cengage.

CHAPTER PRESENTATION EXERCISE

Instructions: Your course instructor will assign individual or group discussions on the key points and teaching tips of this chapter. You or your group will review the chapter teachings and identify the major learning points from the chapter. You should be able to discuss the points, why they are important, and how/where they apply to your responsibilities at your level of instructor training.

REVIEW QUESTIONS

1. The communication process consists of five elements. Explain what each of these is.

2. Explain passive and active listening.

3. Describe some of the elements that are in a report or decision document.

CHAPTER 5

Using Lesson Plans

NFPA 1041 JOB PERFORMANCE REQUIREMENTS

Note: An asterisk denotes that the 1041 standard contains further information in its annex section.

4.2.2 Assemble course materials, given a specific topic, so that the lesson plan and all materials, resources, and equipment needed to deliver the lesson are obtained.

(A) Requisite Knowledge.
Components of a lesson plan, policies and procedures for the procurement of materials and equipment, and resource availability.

(B) Requisite Skills.
None required.

4.2.3 Prepare requests for resources, given training goals and current resources, so that the resources required to meet training goals are identified and documented.

(A) Requisite Knowledge.
Resource management, sources of instructional resources and equipment.

(B) Requisite Skills.
Oral and written communication, forms completion.

4.2.4* Schedule single instructional sessions, given a training assignment, AHJ scheduling procedures, instructional resources, facilities and timeline for delivery, so that the specified sessions are delivered according to AHJ procedure.

(A) Requisite Knowledge.
AHJ scheduling procedures and resource management.

(B) Requisite Skills.
Training schedule completion.

4.3.2* Review instructional materials, given the materials for a specific topic, target audience, learner characteristics, and learning environment, so that elements of the lesson plan, learning environment, and resources that need adaptation are identified.

(A) Requisite Knowledge.
Recognition of student learner characteristics and diversity, methods of instruction, types of resource materials, organization of the learning environment, and policies and procedures.

(B) Requisite Skills.
Analysis of resources, facilities, and materials.

4.3.3* Adapt a prepared lesson plan, given course materials and an assignment, so that the needs of the student and the objectives of the lesson plan are achieved.

(A)* Requisite Knowledge.
Elements of a lesson plan, selection of instructional aids and methods, and organization of the learning environment.

(B) Requisite Skills.
Instructor preparation and organization techniques.

4.4.3 Present and adjust prepared lessons, given a prepared lesson plan that specifies the presentation method(s), so that the method(s) indicated in the plan are used and the stated objectives or learning outcomes are achieved, applicable safety standards and practices are followed, and risks are addressed.

(A)* Requisite Knowledge.
The laws and principles of learning, methods and techniques of instruction, lesson plan components and elements of the communication process, and lesson plan terminology and definitions; learner characteristics; student-centered learning principles; instructional technology tools; the impact of cultural differences on instructional delivery; safety rules, regulations, and practices; identification of training hazards; elements and limitations of distance learning; distance learning delivery methods; and the instructor's role in distance learning.

(B) Requisite Skills.
Oral communication techniques, methods and techniques of instruction, ability to adapt to changing circumstances, and utilization of lesson plans in an instructional setting.

KNOWLEDGE OBJECTIVES

After studying this chapter, participating in a structured learning environment, and completing assigned assessments, you will be able to:

- Identify and describe the components of learning objectives. (**NFPA 1041: 4.3.3**, pp 124–125)
- Describe the parts of a lesson plan. (**NFPA 1041: 4.3.2, 4.4.3**, pp 126–130)
- Describe the instructional preparation process. (**NFPA 1041: 4.2.2, 4.2.3, 4.3.2, 4.3.3**, pp 130–131)
- Describe the lesson plan adaptation process and limits for the Fire and Emergency Services Instructor I. (**NFPA 1041: 4.3.1, 4.3.3**, pp 131–134)
- Identify the factors to consider when scheduling a training session. (**NFPA 1041: 4.2.4**, pp 130–131)

SKILLS OBJECTIVES

After studying this chapter, participating in a structured learning environment, and completing assigned assessments, you will be able to:

- Prepare to teach a lesson. (**NFPA 1041: 4.2.2, 4.2.3, 4.3.2**, pp 126–130)
- Review a lesson plan and identify the adaptations needed. (**NFPA 1041: 4.3.3**, pp 131–134)
- Adapt a lesson plan so that it meets the needs of the students and ensures that learning objectives are met. (**NFPA 1041: 4.3.3**, pp 131–134)

You Are the Fire and Emergency Services Instructor

Recently, your community has annexed a neighboring area of the county that is more rural in nature than your primary response area. As a result, your shift captain has asked for you to deliver a training session to your shift on water shuttle operations, a subject you haven't trained on before primarily because your jurisdiction is fully covered with a municipal water distribution system. While researching the internet for training materials, you have encountered a lesson plan published by an established training academy in another state that appears to suit your needs.

1. Identify and describe the parts of a lesson plan.
2. Describe the lesson plan adaptation process for the Instructor I.
3. Identify the learning objectives necessary for the training session that you found.
4. Describe how you, as an Instructor I, can adapt the lesson plan you found to better suit the needs of your department and ensure that learning objectives are met.

Access Navigate for more practice activities.

Introduction

When most people think about the job of a fire and emergency services instructor, they picture the actual delivery of a presentation in front of the classroom. Although lectures are an important aspect of instruction, they are not the only part of the job. Most instructors spend many hours planning and preparing for a class before students ever arrive in the classroom. There are several details to address when planning a class:

- What are the expected outcomes (objectives) of the training session?
- How much time will the class take?
- How many students will attend the class?
- Are there student prerequisites required to understand the objectives?
- Which training aids and equipment will be needed?
- In what order will the instructional material be presented?
- Will the delivery schedule be affected by availability of specific resources?

All of these questions and more are answered during the planning and preparation for the class. This information is compiled into a document called a **lesson plan**. A lesson plan is a detailed guide used by the instructor for preparing and delivering instruction to students. An instructor who uses a well-prepared and thorough lesson plan to organize and prepare for class greatly increases the odds of ensuring quality student learning. An Instructor I uses a lesson plan that is already developed, usually by an instructor who is certified as an Instructor II or higher. The Instructor II has received training in how to develop his or her own lesson plan and may be responsible for developing all parts of the lesson plan, including objectives, lesson outline, suggested student activities, methods of evaluation, and many other components of a properly crafted class session.

As an Instructor I, you are responsible for delivering training from a prepared lesson plan. An Instructor I is able to **adapt** a lesson plan to meet specific student-centered needs but will not modify the *content* of the lesson plan—that remains the responsibility of an Instructor II or above. Instructing from a prepared lesson plan is the core outcome of your instructor training. You already understand the learning process, various methods of instruction, and communication skills. That knowledge will allow you to use the lesson plan as a tool to influence a change in behavior. This chapter sets the stage for you to be successful in delivering effective training by using a standard form and content layout to ensure you meet the lesson goals.

Why Use a Lesson Plan?

Many fire fighters who are assigned to instruct a class have a lot of emergency scene experience and may have even participated in group discussions on department operations, personnel evaluations, or the budget process. However, working from prepared lesson plan materials is a process with which many fire fighters have no experience, as they may have participated as students and not in delivering information *to* students.

Many people without experience in the field of education may not understand the importance of a lesson plan. Attempting to deliver instruction without a lesson plan is like driving in a foreign country without a map (**FIGURE 5-1**). The goal in both situations is to reach your intended destination. In a lesson plan, the learning objectives are the intended destination. Without a map (the lesson plan), you most likely will not reach the destination. If you attempt to shoot from the hip without a prepared lesson plan that details the expected outcomes, content may be skipped, safety points may be omitted, and inconsistency between deliveries will occur. Also, without a lesson plan that contains learning objectives, you may not even know what the destination for the class is. In other words, if you do not have clearly written learning objectives for your class and a plan for how to achieve them, there is a high probability that you will not be successful.

Occasionally, an instructor may use a lesson plan from an established or ongoing training program. Such lesson plans may have been adapted for previous deliveries based on a specific audience and factors related to that class. If an instructor becomes complacent and fails to check the existing lesson plans fully against his or her own class requirements, the result may be an inferior lesson delivery for the students.

FIGURE 5-1 Using a lesson plan keeps the instructor organized and provides a clear direction for the class.
© Jones & Bartlett Learning.

Written lesson plans also ensure consistency of training throughout the various companies of a large fire department, or when a class is taught multiple times, especially by different instructors. In such cases, a common lesson plan ensures that all students receive the same information. Lesson plans are also used to document what was taught in a class. When the class needs to be taught again in the future, the new instructor will be able to refer to the existing lesson plans and achieve the same learning objectives.

> **COMPANY-LEVEL TRAINING TIP**
>
> In the case of legal action against the instructor or educational institution, the lesson plan may become a critical piece of documentation. When writing a lesson plan, you must ensure the information is technically correct, not just for educational purposes, but potentially for legal reasons.

Understanding Learning Objectives

All instructional planning begins by identifying the desired outcomes. What do you want the students to know or be able to do by the end of class? These desired outcomes are called objectives. A **learning objective** is defined as a goal that is achieved through the attainment of a skill, knowledge, or both, and that can be observed or measured. Sometimes these learning objectives are referred to as performance outcomes or behavioral outcomes, for a simple reason: If students are able to achieve the learning objectives of a lesson, they will achieve the desired outcome of the class. Effective instructors always start their presentations by discussing and reviewing the objectives of the presentation with the students.

A **terminal objective** is a broader outcome that requires the learner to have a specific set of skills or knowledge after a learning process. An **enabling objective** is an intermediate objective and is usually part of a series of objectives that direct instructors on what they need to instruct and what the learners will learn to accomplish the terminal objective. Consider the enabling objectives to be the steps that allow you to reach the top floor—that is, the terminal objective. Terminal and enabling objectives are developed from job performance requirements (JPRs), which is covered at the Instructor II level.

The Components of Learning Objectives

Many different methods may be used for writing learning objectives. One method commonly employed in the fire service is the **ABCD method**, where the acronym stands for **A**udience (Who?), **B**ehavior (What?), **C**ondition (How or using what?), and **D**egree (How well?). (Learning objectives do not always need to be written in that order, however.) The ABCD method was introduced in the book *Instructional Media and the New Technologies of Education* written by Robert Heinich, Michael Molenda, and James D. Russell (1998).

Audience

The **audience** of the learning objective describes who the students are. Are your students experienced fire fighters or new recruits? Fire service learning objectives often use terms such as *fire fighter trainees*, *cadets*, *fire officers*, or *students* to describe the audience. In the case of this class, you will see objectives with an audience of *fire and emergency services instructor*. Some lesson plans will contain multiple objectives to meet the course goal. In this case, it is acceptable to reference the audience once at the beginning of the objective listing, as long as all of the objectives relate to the same audience members. If the objective is written correctly, the audience demographics will be evident in the structure of the objective.

Behavior

Once the students have been identified, typically the **behavior** is listed next. The behavior must be an observable and measurable action. A common error in writing learning objectives is using words such as *know* or *understand* for the behavior. Is there really a method for determining whether someone understands something? It is preferable to use words such as *state*, *describe*, or *identify* as part of learning objectives—these are actions that you can see and measure. It is much easier to evaluate the ability of a student to identify the parts of a portable fire extinguisher than to evaluate how well the student understands the parts of a portable fire extinguisher. The importance of the behavior portion of the objective will be discussed later in the chapter.

The behavior may identify the type of presentation or class that will be conducted. Words such as *describe* or *state* in the objective imply that the student will know something, whereas words such as *demonstrate* or *perform* indicate that the student will be able to do something. This is where the terms **cognitive objective** or **psychomotor objective** are applied in a properly formatted objective (**TABLE 5-1**). Instructors should blend presentation styles to enhance the learning environment whenever possible. Appealing to multiple senses and allowing for many application opportunities enhances learning.

TABLE 5-1 Examples of Learning Objectives and Their Parts

	Complete Objective	Parts
Cognitive Objectives	The fire fighter will identify hose loads used by the department from photos of finished hose beds with 100 percent accuracy.	■ Audience 　■ Fire fighter ■ Behavior 　■ Identify hose loads ■ Condition 　■ Examples of finished hose beds ■ Degree 　■ 100 percent accuracy
Psychomotor Objectives	The fire fighter will rebed a flatbed using a minimum of 300 ft (91 m) of 3-in. (76-mm) supply hose according to department hose manual without errors.	■ Audience 　■ Fire fighter ■ Behavior 　■ Rebed flat load ■ Condition 　■ 300 ft (91 m) of 3-in. (76-mm) supply hose ■ Degree 　■ 100 percent accuracy

Condition

The **condition** describes the situation in which the student will perform the behavior. Items that are often listed as conditions include specific equipment or resources given to the student, personal protective clothing or safety items that must be used when performing the behavior, and the physical location or circumstances for performing the behavior. For example, the following phrases in an objective specify the condition:

> ". . . in full protective equipment, including self-contained breathing apparatus (SCBA)."
> ". . . using the water from a static source, such as a pond or pool."

Degree

The **degree** is the last part of the learning objective; it indicates how well the student is expected to perform the behavior in the listed conditions. With what percentage of completion is the student expected to perform the behavior? Total mastery of a skill would require 100 percent completion—this means perfection, following every step on the skill sheet. For example, a student performing a ladder raise needs to complete all of the steps on the skill checklist to raise the ladder safely. Skipping any step (failing the skill) may result in a serious injury, damage to the ladder, or death. In contrast, many times knowledge-based learning objectives are expected to be learned to the degree stated in the passing rate for written exams, such as 70 percent or 80 percent. Another degree that is frequently used is a time limit, which can be included in learning objectives dealing with both knowledge and skills.

Using the ABCD Method

Strictly speaking, well-written learning objectives should contain all four elements of the ABCD method. Nevertheless, learning objectives are often shortened because one or more of the elements are assumed to be known. If a lesson plan is identified as being used for teaching potential fire and emergency services instructors, for example, every single objective may not need to start with "the fire and emergency services instructor trainee." The audience component of the ABCD method may be listed once, at the top of all the objectives, or not listed at all.

The same principle applies to the condition component. If it is understood that a class requires all skills to be performed in full personal protective gear, it may not be necessary to list this condition in each individual learning objective. It is also common to omit the degree component, as many learning objectives are written with the understanding that the degree will be determined by the testing method. If the required passing grade for class written exams is 80 percent, it is assumed that knowledge learning objectives will be performed to that degree. Similarly, if the skill learning objectives for a class are required to be performed perfectly, a 100 percent degree for those learning objectives can be assumed.

Learning objectives should be shortened in this way only when the assumptions for the missing components are clearly stated elsewhere in the lesson plan. Of course, a learning objective is unlikely to omit the behavior component because it is the backbone of the learning objective. Some curricula do require the ABCD to be stated each time, in sequence, and in complete form. For enabling objectives, a shortened form may sometimes be permitted; otherwise, many of the components in the series of objectives would simply be restated over and over again, which may confuse or frustrate a learner trying to identify course expectations.

ABCD learning objectives do not need to contain all of the parts in the ABCD order. Consider the following example:

> In full protective equipment including SCBA, two fire fighter trainees will carry a 24-foot (7-m) extension ladder 100 feet (30 m) and then perform a flat raise to a second-floor window in less than 1 minute and 30 seconds.

Here the audience is "the fire fighter trainees." The behaviors are "carry a 24-foot (7-m) extension ladder" and "perform a flat raise." Both carrying and raising are observable and measurable actions. The conditions are "full protective equipment including SCBA," "100 feet (30 m)," and "to a second-floor window"; they describe the circumstances for carrying and raising the ladder. The degree is "less than 1 minute and 30 seconds." The fire fighter trainees must demonstrate the ability to perform these behaviors to the proper degree to meet this learning objective successfully.

The ABCD method is a clear and appropriate way to address objectives. Inclusion of all four of its elements is essential in the construction of the terminal objective, which is the main idea for the lesson. This objective should contain all the components to inform the students what will be taught, the method of evaluation, and the resources consulted for the information presented. Subsequent (enabling) objectives may assume certain points previously stated in the main objective, such as audience, degree, and references, as long as these points are clarified in the main objective. Each enabling objective allows the student to meet the intent or goal of the terminal objective.

Parts of a Lesson Plan

Many different styles and formats for lesson plans exist. No matter which lesson plan format is used, however, certain components should always be included. Each of these components is necessary for you to understand and follow a lesson plan (**FIGURE 5-2**).

Lesson Title

The **lesson title** or lesson topic describes what the lesson plan is about. For example, a lesson title may be "Portable Fire Extinguishers" or "Fire Personnel Management." Just by the lesson title, you should be able to determine whether a particular lesson plan contains information about the topic you are planning to teach. A title page may be used to highlight or preview the content of the lesson plan package (**FIGURE 5-3**). It may serve as a summary of all of the contents and help prepare you for the class.

Level of Instruction

It is important for a lesson plan to identify the **level of instruction**, because your students must be able to understand the instructional material. Just as an elementary school teacher would not use a lesson plan developed for high school students, so you as an instructor must ensure that the lesson plan is written at an appropriate level for your students. Often the level of instruction in the fire service corresponds with National Fire Protection Association (NFPA) standards for professional qualifications. If you are teaching new recruits or cadets, you would use lesson plans that are designated as having a Fire Fighter I or II level of instruction. If you are teaching fire service professional development classes, you may use lesson plans that are specified as having a Fire Officer I or II level of instruction. Another method of indicating the level of instruction is by labeling the lesson plan with terms such as *beginner*, *intermediate*, or *advanced*. No matter which method is used to indicate the level of instruction, you should ensure that the material contained in a lesson plan is at the appropriate level for your students.

Another component of the level of instruction is the identification of any prerequisites. A **prerequisite** is a condition that must be met before the student is permitted to receive further instruction. Often, a prerequisite is another class. Ensure that the proper prerequisites are met by each of your students. Failure to do so may mean that a student performs tasks that he or she is not qualified or prepared to perform.

For example, a Fire Service Administration class would be a prerequisite for taking an Advanced Fire Service Administration class. A certification or rank may also be a prerequisite. Before being allowed to receive training on driving an aerial apparatus, for example, the department may require a student to hold the rank of a Driver and possess Driver/Operator—Pumper certification.

CHAPTER 5 Using Lesson Plans

Instructor Guide
Lesson Plan

Lesson Title: Use of Fire Extinguishers
Level of Instruction: Fire Fighter I

Method of Instruction: Demonstration

Learning Objective: The student shall demonstrate the ability to extinguish a Class A fire with a stored-pressure water-type fire extinguisher. (NFPA 1001, 4.3.16)

References: Fundamentals of Fire Fighter Skills, 4th Edition, Chapter 7

Time: 50 Minutes

Materials Needed: Portable water extinguishers, Class A combustible burn materials, skills checklist, suitable area for hands-on demonstration, assigned PPE for skill

Step #1 Lesson Preparation:
- Fire extinguishers are first line of defense on incipient fires
- Civilians use for containment until FD arrives
- Must match extinguisher class with fire class
- FD personnel can use in certain situations, may limit water damage
- Review of fire behavior and fuel classifications
- Discuss types of extinguishers on apparatus
- Demonstrate methods for operation

Step #2 Presentation	Step # 3 Application
A. Fire extinguishers should be simple to operate. 　1. An individual with only basic training should be able to use most fire extinguishers safely and effectively. 　2. Every portable extinguisher should be labeled with printed operating instructions. 　3. There are six basic steps in extinguishing a fire with a portable fire extinguisher. They are: 　　a. Locate the fire extinguisher. 　　b. Select the proper classification of extinguisher. 　　c. Transport the extinguisher to the location of the fire. 　　d. Activate the extinguisher to release the extinguishing agent. 　　e. Apply the extinguishing agent to the fire for maximum effect. 　　f. Ensure your personal safety by having an exit route. 　4. Although these steps are not complicated, practice and training are essential for effective fire suppression. 　5. Tests have shown that the effective use of Class B portable fire extinguishers depends heavily on user training and expertise. 　　a. A trained expert can extinguish a fire up to twice as large as a non-expert can, using the same extinguisher. 　6. As a fire fighter, you should be able to operate any fire extinguisher that you might be required to use, whether it is carried on your fire apparatus, hanging on the wall of your firehouse, or placed in some other location. B. Knowing the exact locations of extinguishers can save valuable time in an emergency. 　1. Fire fighters should know what types of fire extinguishers are carried on department apparatus and where each type of extinguisher is located. 　2. You should also know where fire extinguishers are located in and around the fire station and other work places. 　3. You should have at least one fire extinguisher in your home and another in your personal vehicle, and you should know exactly where they are located. C. It is important to be able to select the proper extinguisher. 　1. This requires an understanding of the classification and rating system for fire extinguishers. 　2. Knowing the different types of agents, how they work, the ratings of the fire extinguishers carried on your fire apparatus, and which extinguisher is appropriate for a particular fire situation is also important. 　3. Fire fighters should be able to assess a fire quickly, determine if the fire can be controlled by an extinguisher, and identify the appropriate extinguisher. 　　a. Using an extinguisher with an insufficient rating may not completely extinguish the fire, which can place the operator in danger of being burned or otherwise injured. 　　b. If the fire is too large for the extinguisher, you will have to consider other options such as obtaining additional extinguishers or making sure that a charged hose line is ready to provide back-up. 　4. Fire fighters should also be able to determine the most appropriate type of fire extinguisher to place in a given area, based on the types of fires that could occur and the hazards that are present. 　　a. In some cases, one type of extinguisher might be preferred over another.	Slides 7–10 Ask students to locate closest extinguisher to training area Review rating systems handout—Have students complete work book activity page #389 What happens if wrong type or size extinguisher is used?

FIGURE 5-2 The components of a lesson plan (indicated in bold print).
© Jones & Bartlett Learning.

D. The best method of transporting a hand-held portable fire extinguisher depends on the size, weight, and design of the extinguisher.
  1. Hand-held portable fire extinguishers can weigh as little as 1 lb to as much as 50 lb.
  2. Extinguishers with a fixed nozzle should be carried in the favored or stronger hand.
    a. This enables the operator to depress the trigger and direct the discharge easily.
  3. Extinguishers that have a hose between the trigger and the nozzle should be carried in the weaker or less-favored hand so that the favored hand can grip and aim the nozzle.
  4. Heavier extinguishers may have to be carried as close as possible to the fire and placed upright on the ground.
    a. The operator can depress the trigger with one hand, while holding the nozzle and directing the stream with the other hand.
  5. Transporting a fire extinguisher will be practiced in Skill Drill 8-1.
E. Activating a fire extinguisher to apply the extinguishing agent is a single operation in four steps.
  1. The P-A-S-S acronym is a helpful way to remember these steps:
    a. Pull the safety pin.
    b. Aim the nozzle at the base of the flames.
    c. Squeeze the trigger to discharge the agent.
    d. Sweep the nozzle across the base of the flames.
  2. Most fire extinguishers have very simple operation systems.
  3. Practice discharging different types of extinguishers in training situations to build confidence in your ability to use them properly and effectively.
  4. When using a fire extinguisher, always approach the fire with an exit behind you.
    a. If the fire suddenly expands or the extinguisher fails to control it, you must have a planned escape route.
    b. Never let the fire get between you and a safe exit. After suppressing a fire, do not turn your back on it.
  5. Always watch and be prepared for a rekindle until the fire has been fully overhauled.
  6. As a fire fighter, you should wear your personal protective clothing and use appropriate personal protective equipment (PPE).
  7. If you must enter an enclosed area where an extinguisher has been discharged, wear full PPE and use SCBA.
    a. The atmosphere within the enclosed area will probably contain a mixture of combustion products and extinguishing agents.
F. The oxygen content within the space may be dangerously depleted.
 	
Step #4 Evaluation:
1. Each student will properly extinguish a Class A combustible fire using a stored-pressure type water extinguisher. (Skill Sheet x-1)
2. Each student will return extinguisher to service. (Skill Sheet x-2)
Lesson Summary:
• Classifications of fire extinguishers
• Ratings of fire extinguishers
• Types of extinguishers and agents
• Operation of each type of fire extinguishers
• Demonstration of Class A fire extinguishment using a stored pressure water extinguisher
Assignments:
1. Read Chapter 7 prior to next class.
2. Complete "You are the Firefighter" activity for Chapter 7 and be prepared to discuss your answers.
	Display available types of extinguishers
	
	Have students demonstrate steps using empty extinguisher
	
	Complete skills sheet #7–9 for each student
	
	Review PPE required for extinguisher use
	
	Discuss hazards of extinguishing agents

FIGURE 5-2 *Continued*

Method of Delivery

As addressed in Chapter 3, *Methods of Instruction*, the method in which the lesson plan is intended to be delivered is an important element in the lesson plan. At a quick glance, you can see whether the lesson plan will be a classroom lecture or a hands-on training class. Some agencies who review training delivery will actually measure the number of hands-on sessions (psychomotor, manipulative) compared to classroom (cognitive, knowledge) to ensure that there is balance in the types of training delivered by the department. Some examples of methods of instruction include lecture, demonstration, and small group discussion.

Learning Objectives

As mentioned earlier in this chapter, learning objectives are essential to the lesson plan. These may also be called behavioral objectives, performance objectives, or learning outcomes. All lesson plans must have learning objectives. Many methods for determining and listing learning objectives are available. The specific method used to write the learning objectives is not as important as ensuring that you understand the learning objectives for the lesson plan that you must present to your students. The Instructor II will use JPRs to develop the learning objectives in the ABCD format, and the Instructor III will write course objectives.

JBL Fire Department
Division of Training

Lesson Title

Training Date:

Firehouse Code:

Location of Training:

Safety Plan Required? Y ☐ N ☐

Topic:	Instructor(s):
Teaching method(s):	Time allotted:
Handouts:	
AV needs:	Teaching resources:
Level of instruction:	Evaluation method:
NFPA JPR's:	Equipment needed:

LEARNING OBJECTIVES: *Upon completion of the class and study questions, each participant will independently do the following with a degree of accuracy that meets or exceeds the standards established for their scope of practice:*

General class activities for student application

| **Safety Briefing** | **SAFETY RED FLAGS** |
| 1. | **ALL STOPS:** |

FIGURE 5-3 A sample cover sheet for a lesson plan. This gives the user an "at-a-glance" view of the presentation they will deliver.
© Jones & Bartlett Learning.

References

Lesson plans often contain an outline of the information that must be understood to deliver the learning objectives. Instructors who are not experts in a subject may need to refer to additional references or resources to obtain further information on these topics. The references/resources section may contain names of books, websites, or even names of experts who may be contacted for further information. By citing references in the lesson plan, the validity of the lesson plan can be verified.

Materials Needed

Most lesson plans require some type of instructional materials to be used in the delivery of the lesson plan. Instructional materials are tools designed to help you present the lesson plan to your students. For instance, audiovisual aids are the type of instructional material most frequently listed in a lesson plan—that is, a lesson plan may require the use of a computer. Other commonly listed instructional materials include handouts, pictures, diagrams, and models. Also, instructional materials may be used to indicate whether additional supplies are necessary to deliver the lesson plan. For example, a preincident planning lesson plan may list paper, pencils, and rulers as the instructional materials needed. The actual equipment students will be using or operating could be shown and passed around during the classroom portion of the lesson (e.g., nozzles, hand tools, personal protective equipment [PPE]).

Some lesson plans will indicate the approximate time it will take to deliver the lesson. This is a subjective estimate in most cases as there are many factors that influence delivery time, including the amount of questions or discussions that take place during the training. Additionally, knowing the number of slides used in a cognitive lesson may be helpful to the instructor.

> **COMPANY-LEVEL TRAINING TIP**
>
> Instructors are the "building contractors" for their courses. The construction materials or tools are the items the instructor needs to compile to make a presentation effective. As the instructor, you must design and make lesson plans work for you.

Lesson Outline

The **lesson outline** is the main body of the lesson plan. This element is discussed in detail later in this chapter. The lesson outline comprises four main elements: *preparation*, *presentation*, *application*, and *evaluation*, which are the components of the four-step method of instruction. Each area fulfills a specific purpose in the delivery of instruction. Instructors who engage their students effectively use each part to ensure that the objectives of the lesson plan are properly met to the level intended. See Chapter 3, *Methods of Instruction*, for more information on the four-step method of instruction.

Lesson Summary

The **lesson summary** simply summarizes the lesson plan. It reviews and reinforces the main points of the lesson plan. The lesson summary plays an important role in the overall lesson, allowing you to enhance the application step by asking summary questions on key objective and lesson points. You may view this as the instructor asking the student, "What did I just teach you or show you?"

Assignment

Lesson plans often contain an **assignment**, such as a homework-type exercise that will allow the student to further explore or apply the material presented in the lesson plan. Be prepared to explain the assignment, its due date, the method for submitting the assignment, and the grading criteria to be used.

Instructional Preparation

Once you have a lesson plan, the instructional preparation begins. Which materials are needed for the class? Which instructional technology tools will be used? Where will the class be conducted? How much time will be needed? These and many other questions must be answered during instructional preparation. The information contained in the lesson plan should be used as a guide for instructional preparation.

Student Preparation

Students should prepare for instruction by coming to a class prepared and ready to learn. Certain classroom or drill-ground rules may exist that prohibit bringing cell phones, newspapers, or other reading materials into the learning environment. The instructor should monitor the preparedness of the students as they come to the classroom or drill ground, and may enhance their readiness to learn by providing class information and objectives ahead of time. The expectations and outcomes may be better if the students take the time to prepare before the class. Bringing textbooks, notebooks, and writing supplies is another important part of the student's preparation.

Organizational Techniques

Identify the time available for you to plan and prepare for the class. The time available for preparation is usually the amount of time from the point when the lesson plan is identified until the day when the class is scheduled to be taught. Identify the milestones that must be accomplished as part of this timeline. Depending on the lesson plan, milestones may include obtaining audiovisual equipment, purchasing materials, reserving a classroom, or previewing audiovisual aids.

Procuring Instructional Materials and Equipment

Most classes take advantage of instructional materials or equipment. The method of obtaining these instructional materials and equipment differs from fire department to fire department. A common method of procuring materials is for the instructor to contact the person in the fire department who is responsible for purchasing training materials, such as a training officer or someone assigned to the training division. You may be required to provide a list of needed materials to the training officer. Often, this list of materials must be submitted well in advance of when the class is scheduled to begin, due to the purchasing requirements of your agency. The key is to know your agency's time frame and to work within it to allow you to follow policy. The training officer, in turn, compiles the materials either by purchasing new materials or by securing materials already available at the training division. The training officer contacts the instructor when all class-related materials are available.

Another common method for procuring class equipment is the equipment checkout process, which is typically managed by the fire department's training division. For example, if you need a multimedia projector for a class, you would submit a request for the projector in which you indicate the date and time when it is needed. The training division would then reserve the projector for you. Depending on the organizational procedures, you might be required to pick up the projector and set it up, or the training division might set up the projector at the class location for you. Regardless of how you obtain your projector, you should make sure you understand how to use it and how to troubleshoot any problems before class begins.

Preparing to Instruct

The most important part of the instruction's preparation is preparing for actual delivery of the lesson plan in the classroom or on the drill ground. If you obtain the necessary materials, equipment, apparatus, drill tower, or classroom, but you do not prepare to deliver the lesson plan, the class will not be successful. You should be thoroughly familiar with the information contained in the lesson plan, which may require you to consult the references listed in the lesson plan and research the topic further.

No matter which method of instructional delivery is used, you should always rehearse your presentation before delivering it to a training session full of students. A class is destined for failure if you are seeing the presentation material for the first time in front of the class. Successful instructors have a sound understanding of the information that they are delivering and can adapt to the particular needs of their class because they always know what is coming next.

Scheduling

An Instructor I may be called upon to schedule a single instructional session. This type of instruction may focus on a subject that is covered only one time based on a needs assessment, or it may be part of an overall training program. Address the following factors when scheduling a single instructional session:

- Define and understand the goal of the session.
- Determine the time necessary to deliver the session and the department scheduling process.
- Identify and locate all training materials for the session, including lesson plans, skill sheets, and equipment needed to meet the objectives. Prepare requests for resources through the proper channels and ensure that expendable materials are replaced and accounted for in the agency.
- Gain approval from supervisors as necessary for members to attend the session. Make sure that members can attend. Determine response procedures with supervisors and officers.
- Schedule and deliver sessions according to the authority having jurisdiction. Make sure notice is given to any instructors who will assist.
- Complete training record reports as required by department policy. Create a through training record report, with all required data, objectives, and narratives of the session. Once complete, forward all reports through the proper channels for recordkeeping purposes.

Adapting Versus Creating a Lesson Plan

One of the most important distinctions between an Instructor I and an Instructor II is the Instructor II's ability to create and modify a lesson plan. A lesson plan is a guide or a roadmap for delivering instruction, but rarely is it implemented exactly as written. To understand what can and cannot be changed by each level of instructor, let us review what the JPRs for NFPA 1041: *Standard for Fire and Emergency Services Instructor Professional Qualifications*, say about creating and adapting a lesson plan:

Fire and Emergency Services Instructor I

4.3.2 Review instructional materials, given the materials for a specific topic, target audience, learner characteristics, and learning environment, so that elements of the lesson plan, learning environment, and resources that need adaptation are identified.

4.3.3 Adapt a prepared lesson plan, given course materials and an assignment, so that the needs of the student and the objectives of the lesson plan are achieved.

The Instructor I should not alter the *content* or the lesson objectives. Prior to the beginning of the class, the instructor should be able to evaluate local conditions, assess facilities for appropriateness, meet local standard operating procedures (SOPs), and recognize students' limitations. As the Instructor I, you should be able to change the *method of instruction* and the *course materials* to meet the needs of the students and accommodate their individual learner characteristics, including making adaptations as necessary due to the learning environment, audience, capability of facilities, and types of equipment available. NFPA 1041 charges the Instructor II with creating a lesson plan, which includes making modifications to the plan as needed.

Fire and Emergency Services Instructor II

5.3.2 Create a lesson plan, given a topic, learner characteristics, and a lesson plan format, so that learning objectives, a lesson outline, course materials, instructional technology tools, an evaluation plan, and learning objectives for the topic are addressed.

Put simply, an Instructor II can create and make basic or fundamental changes to the lesson plan, whereas an Instructor I can adapt only to local conditions without altering objectives. Fundamental changes include changing the performance outcomes, rewriting the learning objectives, modifying the content of the lesson, and so on.

As the Instructor I, you can make the lesson plan fit the situation and conditions. These conditions include the facility, the local SOPs, the environment, limitations of the students, and other local factors.

NFPA 1041 specifically states that an Instructor I may adapt the method of instruction and course materials to meet the needs of the students and accommodate the individual instructor's style. Here are a few real-life examples:

- An Instructor I *may change a lesson plan's method of instruction* from lecture to discussion if he or she determines that the latter method would be a better presentation format because of the students' level of knowledge.
- An Instructor I *may adapt the classroom setting* if the facility cannot meet the seating arrangement listed in the lesson plan.
- An Instructor I *may adapt the number of fire fighters* performing an evolution in a lesson plan from three to four to meet local staffing SOP requirements.
- An Instructor I *cannot* modify a lesson plan learning objective that states a fire fighter must raise a 24-ft (7.3-m) extension ladder because he or she feels the task is too difficult for one fire fighter.
- An Instructor I *cannot* change the JPR of developing a budget in a Fire Officer lesson plan because he or she does not feel comfortable teaching that subject.

As with all other positions within the fire service, it is important that instructors perform only those actions within their level of training. As an Instructor I, you must recognize what you can and cannot do. Acting outside your scope of training may lead to legal liability. If you are ever unsure whether you have sufficient knowledge or skills to teach a topic, discuss this issue with your superior.

Lesson plans must remain dynamic in both the short term and the long term. In the short term, you should understand when it is appropriate to adjust a lesson plan during its delivery based on students' learning styles, changing conditions, timing considerations, and students' progress. In the long term, you should provide input to your supervisor regarding the success of the delivery. If problems occurred or improvements are needed, report this feedback as well.

One critical component of lesson plan adaptability is the break times. Break times are built into course schedules and may need to be adjusted to reflect the training environment or to accommodate other classes being conducted at the same time.

If you make adjustments to the delivery of a lesson plan, it is critical that you ensure that all learning objectives are still covered. For example, many times activities must be scheduled around the activities of other courses. Scheduling use of resources in the field with other instructors can help reduce conflicts. The program coordinator should be advised if an instructor intends to move portions of the program around so that the coordinator can ensure the change does not affect other programs and shared resources.

Reviewing Instructional Materials for Adaptation

There are many ways for an Instructor I to obtain a lesson plan: fire service websites, commercially published curriculum packages, the National Fire Academy, the instructor's fire department's training library, or other fire departments. No matter which method is used, the lesson plan must be reviewed, and any areas that need adaptation must be identified. This is true even for lesson plans that were originally developed within your own fire department. Over time, standards and procedures change; a lesson plan that was completely correct for your department when it was created may be out-of-date in just a few months.

After obtaining a lesson plan, you must review the entire lesson plan and determine whether adaptations are needed to make it usable for your class. As part of the class planning and preparation process, lesson plan adaptations must be scheduled and completed before you deliver the presentation to the class. A lesson plan might need adaptations for many reasons, such as differences related to the learning environment, the audience, the capability of facilities, and the types of equipment available. Always be prepared to adapt the lesson plan to accommodate last-minute classroom or equipment changes. While these occurrences should be rare, they do happen.

Instructors may find they need to adapt a lesson plan in situations where all members of the audience do not come from one department, as in a regional or academy delivery. If you are teaching from a prepared lesson plan within your department only, the audience factors, prerequisite knowledge, and abilities are likely already known, so you may have to adapt the lesson plan only if it is from an outside source or to tailor it to the available delivery time.

Evaluating Local Conditions

The main focus when adapting a lesson plan is to make minor adjustments so it fits your local conditions and your students' needs. To accomplish this, you must first be familiar with your audience. The Instructor I should contemplate the following questions when reviewing a lesson plan for any adaptations needed to accommodate the intended audience:

- Which organizational policies and procedures apply to the lesson plan?
- What is the current level of knowledge and ability of your students?
- Which types of tools and equipment will your students use in performing the skills within the lesson plan?

The second area pertaining to the local conditions that must be considered is you, the Instructor I:

- What is your experience level and ability? Can you identify resources and materials to improve your knowledge and background? If not, be prepared to ask for assistance from another instructor.
- How familiar are you with the topic that will be taught? Teaching material you are not familiar with can pose severe safety and credibility hazards. If you are not familiar with the topic, do not attempt to fake it, as wrong or incomplete information could be dangerous.
- What is your teaching style?

The answers to these types of questions will allow you to adapt the lesson plan so that you deliver the lesson in the most effective way given your own abilities.

Evaluating Facilities for Appropriateness

You should also review and adapt the lesson plan based on the facilities that will be used when delivering the class. Several factors—for example, the equipment available, student seating, classroom size, lighting, and environmental noise—must be considered as part of this evaluation. For example, a lesson plan may call for students to sit at tables that have been moved into a U-shaped arrangement. However, if the local classroom has desks that are fixed to the floor, you will not be able to arrange the seating as indicated in the lesson plan. The lesson plan would then need to be adapted to meet the conditions of the facility and the seating arrangement changed accordingly. You should make this adaptation, keeping in mind the reason for the indicated seating arrangement.

Meeting Local SOPs

A lesson plan must be reviewed to ensure that it meets and follows local SOPs. This is one of the most important considerations when adapting a lesson plan. You should never teach information that contradicts a SOP. Not only would this lesson be confusing for the students, but it would also create a liability for you. If a student were to be injured or killed while performing a skill in violation of a SOP, you would be held responsible. At a minimum, you would be disciplined within the organization—but it is also possible that you might be held legally responsible in either criminal or civil court.

When reviewing a lesson plan, make note of the SOPs that may cover the topics in the lesson plan. After completely reviewing the lesson plan, research the SOPs and ensure that no conflicts exist. If your research turns up conflicting information, you should adapt the lesson plan to meet the local SOPs as long as it does not change the lesson objectives. If you are not familiar with your local SOPs, contact someone within the department who can assist you with ensuring that the lesson plan is consistent with local SOPs.

Evaluating Limitations of Students

The lesson plan should also be reviewed in light of student limitations and adapted to accommodate those limitations if possible. The lesson plan should be at the appropriate educational level for the students, and the prerequisite knowledge and skills should be verified. For example, if you were reviewing a lesson plan to teach an advanced hazardous materials monitoring class, students should have already undergone basic hazardous materials training. If you were training new fire fighters and reviewing this lesson plan, you may not be able to adapt it. Instead, you would most likely have to require students to undergo additional training before the lesson plan would be appropriate to present to them.

Adapting the Method of Instruction

The method of instruction is the one area that an Instructor I may readily modify. Such a modification may be needed to allow you to deliver the lesson plan effectively, but it should not change the learning objectives. For example, you may not be comfortable using the discussion method to deliver a class on fire service sexual harassment as indicated by the lesson plan. Instead, you might adapt the lesson plan and change the method of instruction to lecture. This would allow the same information to be taught, just in a different format, and the same learning objectives would still be achieved.

Instructors regularly adapt material to meet their departmental needs and to improve the curriculum. Is it ethical to alter material to reflect your personal opinions when those opinions run counter to the traditional way of thinking? The solution to this dilemma is not as simple as it may seem. Imagine where the fire service would be today if only a few years ago bold instructors did not step up to the plate and refuse to present material that was not based on safe practices. "Doing the right thing" is what ethical decisions are all about—but there is a fuzzy line between "the right thing" and "my way is the only right way."

For example, changing a course by eliminating the use of fog nozzles because you believe that these nozzles lead to hand burns is as dangerous as another instructor eliminating the use of smooth-bore nozzles from a lesson plan. The reality is that students must understand the appropriate use, benefits, and dangers of each type of nozzle. What might seem like a simple adaptation could have serious consequences for a student who is not trained thoroughly and properly to department standards.

Accommodating Instructor Style

In addition to ensuring that the method of instruction best suits your abilities, lesson plans may be adapted to accommodate your personal style. A lesson plan often reflects the style of the instructor who wrote it. When reviewing and adapting a lesson plan, consider whether the lesson plan—and especially the presentation section—fits your own style. For example, a lesson plan may call for a humorous activity designed to establish a relationship between the instructor and the students. If you are teaching a military-style academy class, this may not be the best style, so you may need to adapt the presentation accordingly.

Meeting the Needs of the Students

All adaptations should be done with one purpose in mind—namely, meeting the needs of the students. As with all lesson plans, the main goal is to provide instruction that allows students to obtain knowledge or skills. This goal should be verified after you review and adapt a lesson plan.

Training BULLETIN

JONES & BARTLETT FIRE DISTRICT
TRAINING DIVISION
5 Wall Street, Burlington, MA, 01803
Phone 978-443-5000   Fax 978-443-8000

Instant Applications: Lesson Plans

Drill Assignment

Apply the chapter content to your department's operation, its training division, and your personal experiences to complete the following questions and activities.

Objective

Upon completion of the instant applications, fire and emergency services instructor students will exhibit decision making and application of job performance requirements of the instructor using the text, class discussion, and their own personal experiences.

Suggested Drill Applications

1. Using a sample lesson plan included in supplemental course material, identify the components of the lesson plan.

2. Using the same sample lesson plan, adjust the lesson plan based on the needs of different audiences.

3. Analyze an existing lesson plan from your department. Are the components complete and accurate?

4. Review the Incident Report in this chapter and be prepared to discuss your analysis of the incident from a training perspective and as an instructor who wishes to use the report as a training tool.

Incident Report: USFA-TR-015/October 1987

Drill Assignment

Review the information in this incident report and prepare a practice presentation for class delivery at the direction of your instructor. Your presentation should include a summary of the incident facts and a review of the NFPA 1403 compliance and noncompliance findings. Use an outline to organize your thoughts. You may be evaluated on your communication skills during your presentation.

Milford, Michigan—1987

Four days after the fatal training fire in Hollandale, Minnesota, a training session that was supposed to teach fire fighters to recognize physical evidence of arson fires trapped six fire fighters on the second floor of a 120-year-old house (**FIGURE 5-A**). The wood-framed house had low-density ceiling tiles with lightweight wood

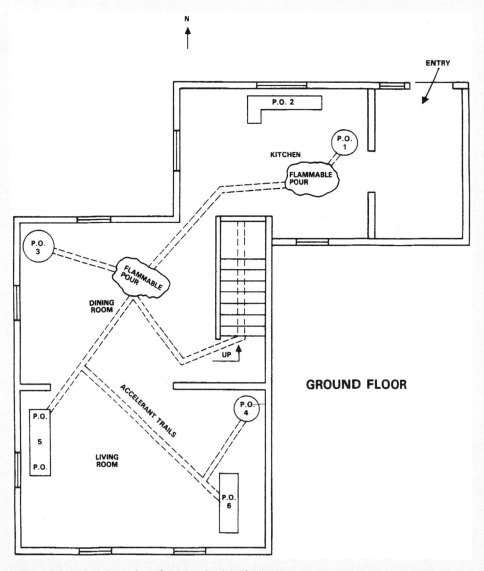

FIGURE 5-A Floor plan of the involved Milford structure.
Courtesy of the U.S. Fire Administration.

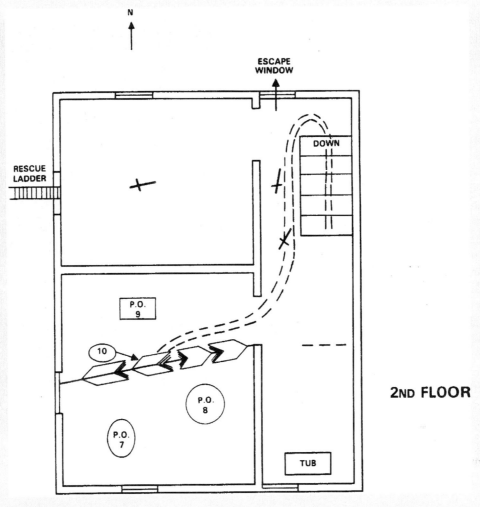

FIGURE 5-A *Continued*

paneling on the first floor. There were numerous holes in the walls and ceilings.

Various arson scenarios were set up in each of the rooms on both floors using furniture, clothing, and other items. Both flammable and combustible fluids were used. Multiple hose lines were charged and ready on the outside of the structure. Portable water tanks were set up and filled with tenders, prepared for a water shuttle operation.

Fire fighters toured the building to see the scenarios before ignition. Initial ignition efforts failed, and several windows on the second floor were broken to improve ventilation. A fire set on a couch in the southwest corner of the living room using flammable or combustible liquids was openly burning and had breached the exterior wall and entered the attic space.

One fire fighter was already inside trying to ignite the fires on the second floor. Four fire fighters were directed to enter the house, along with the assistant chief. Without a hose line, these fire fighters passed multiple burning fire sets that were producing little heat or smoke. Most likely, they did not see the fire in the living room or realize it had spread to the attic. They met with the assistant chief and another fire fighter on the second floor

Incident Report continued

to ignite a fire in one of the bedrooms. At this time, the other upstairs bedroom was already burning. The assistant chief directed them to exit the house as fire conditions rapidly intensified. The escape route down the stairs to the first floor was cut off and, under very adverse conditions, the fire fighters were able to locate a window on the second floor for egress (**FIGURE 5-B**). Three of them were able to exit through the window and onto the first-floor roof below. Reportedly, the SCBA face piece was melting off the last fire fighter able to exit.

Outside, fire fighters observed the change in conditions and the rescue of the assistant chief and the two other fire fighters. They initiated suppression and rescue operations. Ladders were raised to the second-floor windows in the now almost fully involved house.

The first trapped fire fighter was located and removed in approximately 10 minutes. The others were all located on the second floor shortly thereafter. Per the NFPA report, "One of the fire fighters who was able to escape did not know that he would be part of an interior training until he was instructed to 'suit up,' and even then he was unsure of his specific assignment." Three fire fighters died in the fire, which was the first multiple-death training incident in the United States since the release of *NFPA 1403: Standard on Live Fire Training Evolutions* in 1986.

FIGURE 5-B Stairs to the second floor of the Milford house show significant fire involvement that blocked escape from second floor.
Reproduced with permission from NFPA, © 1987, National Fire Protection Association.

Postincident Analysis — Milford, Michigan

NFPA 1403 Noncompliant

- Combustible wall paneling and ceiling tiles contributed to rapid fire spread
- Flammable and combustible liquids used
- Multiple simultaneous fires on two floors
- No interior hose line
- Numerous holes in walls and ceilings allowed fire spread
- Interior stairs only exit other than windows (Note: a violation due to the limited normal means of egress, which may have precluded the use of the upper floor without additional provisions put in place)
- Flashover and fire spread unexpected
- Fire chief claimed no knowledge of NFPA 1403

After-Action REVIEW

IN SUMMARY

- An instructor who uses a well-prepared and thorough lesson plan to organize and prepare for class greatly increases the odds of ensuring quality student learning.
- All instructional planning begins by identifying the desired outcomes, called objectives.
- In the ABCD method of writing learning objectives, ABCD stands for **A**udience (Who?), **B**ehavior (What?), **C**ondition (How or using what?), and **D**egree (How well?).
- A lesson plan includes the following parts:
  - Lesson title, or lesson topic
  - Level of instruction
  - Behavioral objectives, performance objectives, and learning outcomes
  - Instructional materials needed
  - Lesson outline
  - References/resources
  - Lesson summary
  - Assignment
- While reviewing and preparing for class with a lesson plan, the four-step method of instruction is the primary process used to relate the material contained in the lesson plan to the students:
  - Preparation
  - Presentation
  - Application
  - Evaluation
- Preparing for instruction is very important. You may need to spend several hours preparing to teach a class, including reviewing the lesson plan, reserving classrooms and instructional aids, and purchasing materials.
- An Instructor I can use a lesson plan to teach a class and may adapt the lesson plan to the local needs of the class.
- The method of instruction is the one area that an Instructor I may readily alter.
- The main focus when adapting a lesson plan is to make minor adjustments so it fits your local conditions and your students' needs. To accomplish this goal, you must be familiar with your audience.
- You should review and adapt the lesson plan based on the facilities that will be used when delivering the class.
- A lesson plan must be reviewed to ensure that it meets and follows local SOPs. After completely reviewing the lesson plan, research the SOPs and ensure that no conflicts exist.
- The lesson plan should be reviewed based on student limitations and adapted to accommodate those limitations if possible.
- Reviewing and adapting a lesson plan should be a formal process. You should document in writing which adaptations have been made.
- All adaptations should be done with one purpose in mind—namely, meeting the needs of the students.

KEY TERMS

Access Navigate for flashcards to test your key term knowledge.

ABCD method A process for writing lesson plan objectives; it includes four components: audience, behavior, condition, and degree.

Adapt To make fit (as for a specific use or situation).

Assignment The part of the lesson plan that provides the student with opportunities for additional application or exploration of the lesson topic, often in the form of homework that is completed outside of the classroom.

Audience Who the students are.

Behavior An observable and measurable action for the student to complete.

Cognitive objective A statement that defines a learning outcome that is based on knowledge and understanding.

Condition The situation in which the student will perform the behavior.

Degree The last part of a learning objective, which indicates how well the student is expected to perform the behavior in the listed conditions.

Enabling objective An intermediate learning objective, usually part of a series that directs the instructor on what he or she needs to instruct and what the learner will learn to accomplish the terminal objective.

Learning objective A goal that is achieved through the attainment of a skill, knowledge, or both, and that can be measured or observed.

Lesson outline The main body of the lesson plan; a chronological listing of the information presented in the lesson plan.

Lesson plan A detailed guide used by an instructor for preparing and delivering instruction.

Lesson summary The part of the lesson plan that briefly reviews the information from the presentation and application sections.

Lesson title The part of the lesson plan that indicates the name or main subject of the lesson plan. Also called the lesson topic.

Level of instruction The part of the lesson plan that indicates the difficulty or appropriateness of the lesson for students.

Prerequisite A condition that must be met before a student is allowed to receive the instruction contained within a lesson plan—often a certification, rank, or attendance of another class.

Psychomotor objective A statement that defines a learning outcome that is based on increasing the learner's ability to perform a specified motor skill.

Terminal objective A broader outcome that requires the learner to have a specific set of skills or knowledge after a learning process.

REFERENCES

Heinich, Robert, Michael Molenda, and James D. Russell. 1998. *Instructional Media and the New Technologies of Education*. 6th ed. Upper Saddle River, NJ: Prentice Hall College Division.

National Fire Protection Association. 2018. *NFPA 1403: Standard on Live Fire Training Evolutions*. Quincy, MA: National Fire Protection Association.

National Fire Protection Association. 2019. *NFPA 1041: Standard for Fire and Emergency Services Instructor Professional Qualifications*. Quincy, MA: National Fire Protection Association.

CHAPTER PRESENTATION EXERCISE

Instructions: Your course instructor will assign individual or group discussions on the key points and teaching tips of this chapter. You or your group will review the chapter teachings and identify the major learning points from the chapter. You should be able to discuss the points, why they are important, and how and where they apply to your responsibilities at your level of instructor training.

REVIEW QUESTIONS

1. Effective instructors always start their presentations by discussing and reviewing what with their students?

2. Explain the components of the ABCD method of a learning objective.

3. Describe the main goal of a lesson plan.

CHAPTER 6

Technology in Training

NFPA 1041 JOB PERFORMANCE REQUIREMENTS

Note: An asterisk denotes that the 1041 standard contains further information in its annex section.

4.3.3* Adapt a prepared lesson plan, given course materials and an assignment, so that the needs of the student and the objectives of the lesson plan are achieved.

(A)* Requisite Knowledge.

Elements of a lesson plan, selection of instructional aids and methods, and organization of the learning environment.

(B) Requisite Skills.

Instructor preparation and organization techniques.

4.4.2 Organize the learning environment, given a facility and an assignment, so that lighting, distractions, climate control or weather, noise control, seating, audiovisual equipment, teaching aids, and safety are addressed.

(A) Requisite Knowledge.

Learning environment management and safety, advantages and limitations of audiovisual equipment and teaching aids, classroom arrangement, and methods and techniques of instruction.

(B) Requisite Skills.

Use of instructional media and teaching aids.

4.4.5 Operate instructional technology tools and demonstration devices, given a learning environment and equipment, so that the equipment functions, the intended objectives are presented, and transitions between media and other parts of the presentation are accomplished.

(A) Requisite Knowledge.

Instructional technology tools, demonstration devices, and selection criteria.

(B) Requisite Skills.

Use of instructional technology tools, demonstration devices, transition techniques, cleaning, and field level maintenance.

KNOWLEDGE OBJECTIVES

After studying this chapter, participating in a structured learning environment, and completing assigned assessments, you will be able to:

- Describe the types of multimedia tools available for the fire and emergency services instructor. (**NFPA 1041: 4.3.3, 4.4.2**, pp 145–151)
- Describe the advantages and limitations of audiovisual equipment and other teaching aids. (**NFPA 1041: 4.4.2**, pp 142–151)
- Describe how to use multimedia tools. (**NFPA 1041: 4.4.5**, pp 145–151)
- Describe when to use multimedia tools in a presentation. (**NFPA 1041: 4.4.2**, pp 145–151)
- Describe how to maintain instructional technology tools. (**NFPA 1041: 4.4.5**, p 151)

SKILLS OBJECTIVES

After studying this chapter, participating in a structured learning environment, and completing assigned assessments, you will be able to:

- Demonstrate how to use multimedia tools during a classroom presentation. (**NFPA 1041: 4.4.5**, pp 145–151)
- Demonstrate how to maintain instructional technology tools. (**NFPA 1041: 4.4.5**, p 151)

You Are the Fire and Emergency Services Instructor

You are a relatively new instructor but have already learned that the methods of instruction used can influence students' retention of material and participation in class. In reviewing the training calendar for the next training period, you notice a variety of instructional opportunities in which you can participate. One class will include primarily young fire fighters who are in the early stages of the fire academy; another class will consist of a mix of new and experienced members. Your experience with technology-based instruction is sound, and you contemplate the methods that might be used to deliver these classes. You can foresee the use of blended learning where the students complete some of the coursework online and other parts in traditional environments.

1. Which type of technology-based instruction is the younger generation of fire fighters accustomed to in their education so far?

2. To which traditional methods of instruction have the experienced members become accustomed?

3. How does a blended learning approach maximize your ability to share information with your students? What are the downfalls of this approach?

Access Navigate for more practice activities.

Introduction

The use of training aids has been part of the training process, regardless of the discipline, since the beginning of humankind. Ancient warriors learned how to fight with practice swords and shields before they used real ones; commanders would draw scenarios in the sand while trainees would gather around to observe. Table-top simulators were used to run mock battles while commanders looked on and learned what was to happen before ever stepping foot on the battlefield. All of these activities are forms of training aids to assist the instructor in training and the students in learning.

The training aids we use in the fire service have evolved from basic chalkboards and erasers to a presentation platform such as the most commonly used ones like PowerPoint (PPT), Prezi, Apple Keynote, and computer simulations run in a classroom or over the internet. You would be hard pressed to participate in a lecture today without some type of presentation management or presentation delivery system.

Although training aids are fundamental in assisting the instructor in the learning process, they are merely "aids" to help the instructor impart the knowledge the fire fighter needs to know and understand. As an instructor, it is your responsibility to learn how and when to use these tools during your presentations and, when necessary, to be able to troubleshoot or repair them in the field. The key is to remember that these tools are there to "assist" you during your presentation—but ultimately, you are the instructor, and the students learn from you.

Technology-Based Instruction

One of the most dynamic and fast-changing concepts in education and training is **technology-based instruction (TBI)**. TBI may also be called other names, such as *e-instruction*, *internet-based instruction*, or *distance learning*. In general, TBI is defined as presenting training and education using the internet or a cloud-based file management system from a student would download files for their coursework. As is true for traditional methods of instruction, there are both advantages and disadvantages to TBI. TBI tends to be a student-centered instructional method because in most cases students can control the pace, the rate of delivery, and even the time at which they view lessons and information. This flexibility is highly desirable to many busy adults, fire fighters and emergency medical services (EMS) providers included. Learning should be student-centered, and when TBI is used, each student can adapt the material being presented to their individual style. Research also supports the student-centered nature of TBI as being more effective, in that students can repeat lesson areas they are struggling with, while instructors enjoy the ability to monitor each student's participation and progress more closely.

Consider a traditional slide presentation given by even the most dynamic instructor. Some students may disengage from the learning process, while other, more engaged and participatory students carry the load in answering questions and giving examples. Unless the instructor carefully engages and involves everyone, some students could sit through an entire presentation and receive a passing grade simply by attending the session, without truly participating in it. Consider this example: The instructor displays a slide of a house fire. The instructor asks for a student to perform a size-up and initial radio report. In a class of 25 students in a typical classroom-based delivery, chances are that only a few students will be able to participate in answering the question. Using TBI, the instructor can set up a system that shows the same slide, but then requires all students to perform the size-up and initial radio report in a narrative example. The TBI method will certainly involve all students, whereas the more traditional approach may not.

Distance Learning

Distance learning simply consists of instruction "a distance away." A student who may be located at home or in a remote classroom receives instruction and assignments from a training center or college/university. Distance learning can even take place without an instructor being present. Distance learning programs can be as simple as a package of printed materials sent through the mail, or as complex as Flash animation presented during a video conference. Today, the majority of distance learning is accomplished through a combination of reading assignments, online activities and assessments, video conferencing, video presentation, testing with proctors, and written assignments. Student evaluations are processed through an instructor.

Computer-Based Training and Learning Management Systems

Computer-based training (CBT) software provides a virtual course environment with tools for course preparation, delivery, and management. With this software, you have the tools to prepare course materials and efficiently manage day-to-day teaching tasks. Many CBT programs offer features that allow you to facilitate collaborative learning, personalize content based on students' unique needs, and positively affect learning outcomes. CBT is not limited to students in colleges and universities; it has also changed the way in which private industry and public safety agencies train their staff members.

CBT can be a powerful way of teaching. It can inform, illustrate, test, and demonstrate complex concepts simply. For example, students can view Flash animations with a voiceover that explains how a fire pump works internally. They can also click on parts of the fire pump image to see labels or learn more details. If real-world experience is desired, some systems have built-in simulators to enable students to experiment with key concepts while using different parameters. This allows students to benefit from real-world learning while ensuring that they can make mistakes without endangering personnel or equipment.

E-learning (short for "electronic learning") is a form of computer-based training—one that has enjoyed increased popularity and prevalence in recent years. Web-based learning management systems are at the heart of the e-learning experience. The following terms will help you understand the systems that support CBT:

Learning management system (LMS) A learning management system is a web-based software application that allows online courseware and content to be delivered to learners, that also includes classroom administration tools such as activity tracking, grades, communications, and calendars. An LMS is sometimes also referred to as a virtual learning environment (VLE) or learning course management system (LCMS).

Synchronous learning Synchronous learning happens when learners are all engaged in or receiving educational content simultaneously. An example might be a live webcast, webinar, or online chat session.

Asynchronous learning Asynchronous learning simply means that all learners are not consuming content at the same time. In a traditional face-to-face classroom, all learners would hear a lecture simultaneously. In an asynchronous learning environment, however, the same learners might consume educational content at different times, usually at their own pace, provided they adhere to established submission deadlines set by the instructor.

Hybrid learning Sometimes referred to as "blended" or "distributed" learning, hybrid learning occurs when some content is available for completion online and some learning requires face-to-face instruction or demonstration.

Courseware Courseware is any educational content that is delivered via a computer.

Flipped classroom An instructional approach that reverses or "flips" the standard learning environment by delivering online content for self-study and reserving the physical classroom for mentored practice that traditionally would have been assigned as homework. The in-person practice sessions allow for more direct application of class content by the student with instructor involvement versus a traditional homework assignment the student would do alone.

Virtual classroom A virtual classroom is a digital environment where content can be posted and shared, and where instructors can create quizzes and assignments to be completed and tracked online via an internet-connected computer.

Webinar An online meeting, the word *webinar* is the portmanteau of "web" (denoting the online component) and "seminar." These meetings occur in real time and usually allow for participants to share their desktop screens, web browsers, and documents as the meeting is occurring.

Social learning Social learning is an informal method of learning using technologies that enable collaborative content creation. Examples include blogs, social media (e.g., Twitter, Facebook), collaborative discussions, and wikis.

Instructor-led training (ILT) In the world of e-learning, ILT usually refers to a scenario in which a human instructor facilitates both the in-person sections and the online component of the course. Instructors usually set deadlines for submission, create quizzes and assignments, and track student progress.

Self-directed study Prepackaged learning modules can be designed to allow learners to engage in self-directed study—to stop and start as they desire, progress to completion of the module, and submit a final score to determine whether they passed or failed the module, all without an instructor's involvement. An example would be the online Incident Command System (ICS) courses freely available from the Federal Emergency Management Agency's (FEMA's) Emergency Management Institute.

The popularity of e-learning has exploded in the past few years. In fact, many colleges and universities have doubled or tripled their courses offered via the web. All degree levels can now be completed over the internet, from Associate to Doctoral, in many fire and emergency service disciplines. Recognizing this fact, many students now expect courses to be offered in this manner. Colleges, universities, fire departments, and other training organizations are all considering how best to integrate e-learning into the instructional experience.

Of course, no technology can ever replace the wisdom and experience of a seasoned fire and emergency services instructor. E-learning is not a threat to instructors, but rather a tool that can be used to address challenges posed by live training. For example, in the future, virtual reality simulators could be used instead of live burns in the early stages of training. The skill will eventually need to be performed to be truly learned, but work with simulators can increase students' comfort level with the scenarios before encountering them with all dangers present.

An LMS is a web-based software application that acts as both a digital content warehouse and a classroom management tool. Learning objects (content) reside within the LMS and are accessible to any member of a course who has internet access. LMS platforms allow instructors to upload (and thereby save and store) documents into their online classroom and then link those resources in a meaningful way for learners to consume. In this regard, the LMS acts like a repository for your content.

Uploading documents to a website, however, hardly provides the entire solution with regard to managing an online classroom. Just as you would do in a traditional classroom, as the instructor, you need to track *how well* your e-learning students are performing, *what* they are consuming, and *whether* relevant objectives have been met. That is where the classroom management piece of an LMS comes into play. **FIGURE 6-1** shows possible course management tools.

Although an LMS can be used to facilitate fully online learning, in the context of fire service instruction it is critical that students demonstrate comprehension of topics as well as a capability to perform critical tasks. Given this requirement, the hybrid learning approach is preferable.

When you consider the difficulties facing today's instructor—shrinking budgets, departments located at a geographical distance from the nearest training facility, increased hours and curriculum requirements, and the composition of the fire service (70 percent volunteer, meaning they have other primary responsibilities)—a compelling argument for the inclusion of technology begins to emerge.

CHAPTER 6 Technology in Training

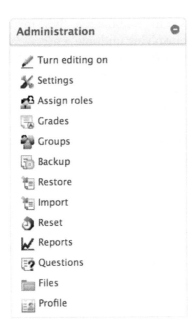

FIGURE 6-1 Course management tools.
© Jones & Bartlett Learning.

FIGURE 6-2 When used appropriately, multimedia tools enhance the learning process.
© Jones & Bartlett Learning. Photographed by Glen E. Ellman.

An LMS is *not* a replacement for the human instructor, hands-on evolutions, or face-to-face interactions. In the world of public safety education, the LMS should be seen as an extension of the instructor—a tool to augment what you are already doing in the classroom, which is why the concept of hybrid learning is both viable and appealing. It represents an alternative to totally internet/computer-based training. It takes the best parts of the learning environment and partners them with computer-based elements so that the student has greater flexibility in completing assignments. Texts, assignments, and exercises can be completed online, and discussion and hands-on learning can be completed in the classroom. This creative alternative is useful in attempting to meet the needs of both the student and the fire service instructor. It can be difficult for any fire fighter—not just volunteers—to be in a classroom at the same time, 5 days per week.

Multimedia Applications in Instruction

A key element in most presentations is the use of a multimedia tool in some form, ranging from handouts to slide presentations. Using multimedia tools enhances the learning process for students and can make your job easier (**FIGURE 6-2**). At the same time, multimedia tools can be a distraction if they are poorly developed or used improperly. Knowing how to use multimedia tools effectively is just as important as knowing the lesson plan. When preparing to give a presentation, part of your efforts should focus on reviewing the instructional technology tools of the lesson plan.

Understanding Multimedia Tools

Multimedia tools and applications come in a variety of types and are intended for a variety of uses. There are PowerPoint and Keynote software programs; videos; CBT programs; internet-based training; distance learning; National Fire Academy training programs; tablet computers; digital audio players; and smart phones. It might seem that it would be impossible to choose from such an abundance of multimedia tools, but each fills a specific niche based on the type of learning that is to take place.

Learn how to use multimedia tools before you apply them in the learning environment. It may take a fair amount of preparation to master the tools you choose to make your presentation. Sometimes use is as simple as pressing the "play" button; at other times it is as complicated as learning a whole new "language" or software program or transitioning among various types of media. As an instructor, it is up to you to learn how to use these tools correctly and incorporate them smoothly into your presentation skills.

COMPANY-LEVEL INSTRUCTOR TIP

Don't fall into the trap of blaming a poor presentation on your inability to use the equipment, and don't depend on multimedia tools to instruct for you. Practice with these tools to make them an effective medium that enhances your existing instructional skills.

When to Use Multimedia Presentations

Multimedia presentations enhance student learning by allowing students to see, hear, and (in some cases) touch the educational content. Multimedia applications should be used to simplify complex theories or hypotheses. They can be used to reinforce materials through exercises, pictures, or words that relate to the information you are presenting. They can summarize previously integrated material for use in the current program so that you do not waste a lot of time covering prerequisites. In combination with the learner characteristics of today's students, multimedia presentations help maintain students' interest.

Despite all of these advantages, there are occasional drawbacks to using multimedia in training. Sometimes, the failure of the media to perform as designed can put the user at a loss. As an instructor, you must be versatile enough to be able to overcome these glitches and not let failure impede what you are trying to present. If a technology failure does happen, then you must have a plan B.

One issue facing instructors is determining what type of multimedia to use and when it is appropriate to use. Multimedia resources are not *always* needed. The delivery format must be appropriate for the audience and appropriate to the nature of the content being presented.

The learning environment has long been known as a place where networking and information combine to create the learning experience. This milieu allows the students to learn from one another as well as from the instructor and the materials presented. At its best, this environment tends to encourage students to learn for themselves. Think of the saying, "You can lead a horse to water, but you cannot make it drink." The same is true for students: You can present all the material in the world, but if students do not want to learn, they won't. Multimedia tools try to break down the resistance to learning by engaging the student.

Multimedia should be used whenever its application makes it easier for students to grasp the material. It does not matter if the application means bringing the emergency incident into the classroom, demonstrating a nonthreatening environment, exemplifying safety concerns, or using simulation or distance education. The only reason to use multimedia tools is to help students learn.

Selecting the right media application is critical. When used as an instructional aid, the right tool allows you to clarify the information you present to students. Essentially, the various multimedia features organize the way in which students receive the material. Diagrams, flowcharts, and graphics impart information that can be retained by students who are visual learners. The lecture or audio cues assist auditory learners. Used in combination, these methods engage students by involving multiple senses.

Emphasis alone will not guarantee that every student retains all of the information presented in a course, but it certainly makes it possible to clarify difficult concepts. Through visual representations, students can travel to places that are unreachable in real life. For example, imagine the advantage of being able to follow a drop of blood through the body to understand the circulatory system. In this case, words are not an adequate medium to get the point across, yet the desired effect is easily achieved through simulation. Likewise, students could recognize the different ways that heat is transferred by allowing them to feel the heat from a light bulb to clarify the concept of radiant heat transfer.

Multimedia Presentations: Best Practices

During the course of your career, you will find that the use of multimedia tools can greatly enhance your ability to convey your message. But use of technology can also come with some drawbacks. To minimize the potential pitfalls, consider these issues when using multimedia tools:

- Do not read from the slides. All instructors make this mistake at one time or another. Most slides are outlines of the lesson plan and serve to emphasize critical points. The only time it is acceptable to read from the slide is if you are reading a quote, legal definition, standard operating procedure, or legal statement.
- Select the appropriate level of lighting for the type of multimedia equipment you are using. If you want students to take notes during the presentation, the room must be dark enough to see the slides, yet light enough to write.
- There is an art to transitioning from the lecture to the multimedia technology—and practice makes perfect.
- When lecturing with slides being projected behind you, do not stand in front of the projection screen. It is difficult for students to see what you are talking about if you are in the way.
- Always have your multimedia equipment set up before class and the media queued. Video or audio recordings should be ready to go at the touch of a button.
- If an internet connection is required, test it for access, and confirm that it can adequately handle the media in your presentation.

- Always have a backup plan. What if you have problems with the computer or projector? Always have a plan B.

Technology and Software
Presentation Programs

Presentation programs include a variety of software programs developed by software vendors. Products such as PowerPoint, Keynote, Prezi, SlideDog, and Google Slides are software programs that are used in creating portable presentations for business, education, and government. Each presentation program has its own unique features and benefits that you can use to enhance your instruction. Some are easier to use than others; some allow for greater flexibility. For the purposes of this chapter, we refer to presentation programs with the acronym PPT, reflecting the fact that, at this time, most fire departments use the PowerPoint program (**FIGURE 6-3**). If you use another software program, the basic style rules are still relevant.

The ability to embed video and audio clips into your presentation is one advantage of using PPT, and it adds a new layer of sophistication and professionalism to your presentation. A disadvantage is the potential for over-reliance on technology as part of your instruction. If your computer doesn't work or problems crop up between the computer and projector, then you cannot use the presentation. If your instruction relies solely on your slides and the technology fails, then you may have to reschedule your presentation if you do not have backup technology available.

Another disadvantage associated with these kinds of presentations is the overload factor. Some slide presentation creators tend to include too much information, too many sounds, and too many visual effects on their slides. Keep your presentation simple and concise for greatest effect.

Suggestions for Uses of Slides

If you plan to use slides to illustrate or otherwise support a lecture, remember that lecture notes displayed on your slides play a different role in the lecture than handwritten lecture notes that only you can see. Don't try to make them play the same role, or you may find students reading your slides instead of listening to you.

You can use slides in the lecture to list the major points of your presentation. Several major points might stay on the screen as you develop each of them in turn, providing a way for students to place each point within the larger context. Again, one slide with several terms may remain on the screen for some time, allowing you to refer to each term as you introduce it.

In addition, slides can be used to create prompts for group work, being projected for all groups to see as the students do their work. This approach helps students move from one stage to the next when you manually change the slide or through the use of a timer built into the slide presentation. Slides can also contain breakouts that indicate how each group will handle different aspects of a scenario.

Always give credit where credit is due. If materials came from a journal or other publication, remember to credit the authors. Regardless of how you found the material, how long ago the information was developed, and whether you had to obtain permission, give credit to its author. This will assist you in presenting credible and defendable information.

Presentation-Ready Programs

Prepackaged software programs are a valuable resource when you are planning your instruction. Educational publishers provide instructors with presentation-ready programs for use on computers at the local level. This material includes slide presentations, testbanks, handouts, and other course-specific material that will assist the instructor. Thanks to these programs, you do not need to spend valuable preparation time developing your own presentation; instead, you can simply use the professionally developed materials. Of course, you must still familiarize yourself with the presentation before class, rather than seeing it for the first time in front of the class. Professionally developed presentations are intended to be dynamic instructional tools, used to enhance students' learning experience, but will never replace the experienced instructor. No presentation-ready programs should truly be considered ready-to-use. Instead, instructors must review

FIGURE 6-3 A sample PowerPoint slide.
© Jones & Bartlett Learning.

materials ahead of time and adapt them to the students who will use them. Some of the most detailed presentations will customize stock backgrounds with department logos and department scenes to make the material more relevant to the students.

Data Projectors

Liquid crystal display (LCD) projectors and digital light processor (DLP) projectors (also known more simply as data projectors) are used with video players, computers, and visual projectors to project documents, images, and motion pictures so that a group of students can easily view the material simultaneously. Data projectors can present material on screens ranging in size from 2 ft (0.6 m) wide to 30 ft (9.1 m) wide (**FIGURE 6-4**). Important features to consider when selecting a data projector are the ability to show videos directly, audio capabilities, screen resolution, and the brightness of light emitted by the projector (measured in lumens). You should also consider the price of the equipment and the price and lifespan of the bulb.

The size of a data projector is compact, so unless mounted to a wall, it can be moved from among facilities with little effort. The various pros and cons should be examined before purchasing a data projector. As with an overhead projector, having an extra bulb available is a good idea.

You should also consider the issues that can arise in the classroom when you use a data projector. First, make sure that the cables match the output of the computer and the input of the projector, such as VGA, HDMI, or DVI. Also, make sure that the cables are long enough, because short cables will hinder your setup and use of the projector. It is still common to find projectors whose inputs do not work with more current technology, so make sure you have an adapter that works with your equipment.

You must learn the proper start-up procedures for your data projector. Generally, the data projector must be connected to the computer before the computer is turned on. Learn how to get the computer to send its output to the data projector. Some computers will automatically recognize the data projector device, but often you must press a control-function key to activate the connection. Given these caveats, always test your presentation before class.

Digital Audio Players and Portable Media Devices

Digital audio players, such as the iPod, are creating a new horizon for digital recording. In the past, the digital audio player was used to download music and music videos. Through the creative work of many instructors, digital audio players are becoming a mainstream technology for students at most universities. These devices are so compact that students can carry them easily and download lectures or pictures that can be reviewed while outside of class (**FIGURE 6-5**).

Some digital audio players allow students to use an audio record feature. Several media outlets are actively engaged in offering downloads, or podcasts, to these

FIGURE 6-4 A data projector can enhance your presentation.
© Goygel-Sokol Dmitry/Shutterstock.

FIGURE 6-5 Digital audio players allow students to listen to lectures and view images outside the classroom.
© Jones & Bartlett Learning.

devices featuring such notable fire service figures as Chief Billy Goldfeder.

Students may also find that digital audio players allow them to complete class assignments. In some courses, students may be able to use desktop video editing software to create and edit projects for course assignments. The results can be uploaded to the training center's computer system for presentation, or content playback can occur immediately via classroom instructional technology equipment. Digital audio player devices can store not only images and audio, but also entire presentations created in PowerPoint, word processing documents, or files created with other software.

As an instructor, you can benefit greatly from these devices. The possibilities for their use are endless: classroom performance and practice, group discussions, field interviews, interviews with content experts, classroom lectures, and discussions or presentations. The possibility of a podcast replacing a webcast allows you to make the materials for a class more widely available, too.

As with other technologies, we must wait to see which benefits this technology will ultimately yield. As more digital systems become available and their costs are driven lower by greater demand, fire service instruction may be transformed from a system of learning confined to a specific locale to one where information can be obtained "on the fly." As technology adapts, improves, and advances, new opportunities will be found to augment educational media.

Tablet Computers

A tablet is a wireless portable computer with a touch screen that generally is smaller than a notebook computer but larger than a smart phone (**FIGURE 6-6**). These devices either include the keyboard on the screen or allow an external keyboard to be attached to the device. Such computers offer a wide range of capabilities and features such as cameras, phones, e-books, videoconferencing, and GPS satellite navigation. Many fire and EMS agencies have adopted these devices for fire inspections, preplanning activities, and gathering information during a medical emergency. Students can use these devices to take notes, complete assignments and e-mail them to the instructor, view course material, and conduct research during class. In some instances, instructors may send the students assignments by e-mail, which are then completed and returned via the tablet. An emerging trend in many college and academic programs is to issue tablet computers to their students that come preloaded with textbooks, online course materials, and other interactivities ready to use. Many instructors use tablets during lectures that connect to their cloud storage to access their developed media presentations.

Interactive Boards

An interactive board, typically mounted on the wall, uses touch detection for user input, much in the same way a personal computer does with a mouse (**FIGURE 6-7**). A projector is used to display a computer's video output on the interactive board, which then acts as a large touch screen. Interactive boards come with pens that allow the instructor to use "digital ink" to write on the board during a presentation. While using the system, a computer records the material, which can be saved and reviewed at a later time or emailed to students.

Dry-Erase Boards and Easel Pads

Low-tech, simple, and usually readily accessible, dry-erase boards (whiteboards) and easel pads remain

FIGURE 6-6 Students can use tablet devices to take notes, complete assignments and e-mail them to the instructor, view course material, and conduct research during class.
© Jones & Bartlett Learning. Photographed by Glen E. Ellman.

FIGURE 6-7 SMART Board® is brand of an interactive board that uses touch detection for user input.
Courtesy of Candy Brignac, Gonzales Fire Department.

a common and practical way to illustrate critical points in the learning environment.

Dry-erase boards use special markers. When using dry-erase markers, do not confuse them with permanent markers. Using a permanent marker on a dry-erase board will damage the whiteboard, so make sure you use the correct marker.

The easel pad can be developed before, during, or after a presentation, thereby becoming a flexible learning device. The greatest advantage to the easel pad is its flexible use in small-group settings. Members of each group can write their thoughts and ideas in response to the activity and post them for others to see, review, and discuss. Each group can write its responses to activities and post them for later discussion. The pads can also be used for later discussion to review information or to refer back to while clarifying a point or concept. Some brands provide a sticky side so that sheets in the pad can be torn off and stuck to a wall or other object to prevent the instructor from having to flip back and forth through the pages in the pad.

Virtual Reality and Simulation

Virtual reality devices allow students to interact with a computer-simulated environment. By displaying images either on a computer screen or via a headset, the goal is to immerse the student in an environment that can be controlled by the student and the programmer. This sensory experience is designed to force feedback from the student as he or she interacts with the environment. Although this technology is, in essence, a simulation, the visual, auditory, and tactile sensations create an artificial world wherein the student can make mistakes without causing injury.

The virtual environment can be designed to mimic a real-world scenario such as a burning grocery store, or it can be set up to resemble a video game. This technology is developing daily by quantum leaps, and it may become the standard means of training in the near future. Unfortunately, the currently available technology does not achieve the realistic look and feel instructors want for their students. In the future, as faster processors and more cost-effective systems emerge, instructors may be able to apply the best possible technology for the benefit of their students.

Like virtual reality systems, simulators allow students to interact safely with hazards in a controlled environment. This practical application was developed by the military to train soldiers for tactical operations during peacetime. Students today have the benefit of growing up with simplified simulators. With the advent of popular video games, many students are accustomed to being immersed in an imaginary world. Today, gaming has taken on new dimensions, as newer systems mimic the movements of the operator to accomplish a mission in the game. Simulations—no matter how simple or complex—allow students to be challenged and to learn from their mistakes.

Simulations have been developed both commercially and noncommercially. Commercially available simulators provide emergency responders with the opportunity to develop skills in command, control, coordination, and communications. Incidents faced in simulations may include structure and vehicle fire responses, aircraft accidents, terrorist acts, hazardous materials spills, fires, and natural disasters, with the programs being designed to test and validate students' understanding of emergency operations.

While this educational activity is not the same as live fire training, after a simulation the student walks away with concepts that can be applied and internalized long before he or she faces a live fire event. As with an actual live fire, replication of exact circumstances is difficult at best, yet with the simulator the same conditions can be repeated until the outcomes are acceptable. Simulators are no substitute for live fire training, of course.

Online certification courses such as Blue Card Hazard Zone Management use simulations for both instruction and evaluation purposes. Directed and distance learning opportunities can capture training opportunities with this technology and bring realism to a learning environment.

Audio Systems

Depending on the class size and the arrangement of the classroom, a microphone and audio system may be needed. Although this technology is not new, it is often misunderstood. Remember who your audience is. Can everyone see and hear your presentation? Do all of your students have the same hearing abilities? Just as some people don't like to wear bifocals when their vision begins to deteriorate with age, some do not want to admit that their hearing just isn't what it used to be.

When deciding which kind of audio system you will use, keep your presentation style in mind. Do you move around a lot, or do you stay at the podium with your notes? Wireless microphones offer you the most freedom to move around the classroom, but they have some problems, too. Wireless systems may have an issue with frequency usage, such that you might hear voices broadcast that are not supposed to be part of your program. Wireless microphones may also have limitations in terms of how far the microphone can be from the remote receiver and still operate properly.

The greatest limitation for wired systems relates to the cable. Is it long enough for you to move around effectively? Some wired microphones are permanently attached to the podium, in which case you lose the freedom to move around the classroom. Another disadvantage of wired systems is the potential for mishap if your feet become entangled in the wire.

Maintaining Technology

With any technology, maintaining its quality is an important consideration. Some vendors charge an additional fee to repair or replace damaged materials. When purchasing multimedia tools, get the maintenance contract in writing and understand what the contract covers in terms of repair and maintenance. Although this contract may seem to be just an additional step or an extra cost, it can pay dividends when you need it. Remember—like everyone else, fire fighters can and do break things.

Some multimedia products require less maintenance than others. A dry-erase board will not get a computer virus—but it also cannot transition from one slide to the next at the touch of a button. Be aware of the trade-offs between high technology and good traditional methods. You need to decide based on several factors (e.g., cost, ease in transportation, audience) what is appropriate for your event or presentation.

Instructional technology tools, like any other equipment, needs to be kept clean and in good order so that it will work when you need it. The simplest and safest way to make sure you are cleaning equipment properly is to follow the manufacturer's recommended maintenance practices, for any device. Most up-to-date equipment will simply need an occasional lens cleaning and dusting near the fan area to ensure reliable operation. Inspect cable connections regularly since they can become bent or frayed quite easily.

Another important maintenance item is to make sure your device has the latest software, as this factor can affect its ability to "talk" with other devices.

If your organization is considering building a training room or training center, consult with a reputable company with technical expertise in sound, lighting, and visual effects to develop an optimal and comfortable training environment. Many departments will build a new training room or space and have to go back later, at an increased cost, and retrofit sound systems, lighting, or audiovisual products because they were unaware of the deficiencies they would later experience.

Troubleshooting Common Multimedia Problems

Many things can go wrong when using multimedia applications. As a consequence, instructors need to know and practice a few alternatives in case problems crop up with this technology. **TABLE 6-1** can assist with some of the more common problems. The simplest things that can go wrong can often be easily fixed by taking a deep breath and thinking about the problem and potential solutions. Take the time to work with your equipment and know its limitations. Be prepared for the possibility that something may go wrong. Remember, always have a plan B.

Sometimes solving a problem is not as easy as referring to a simple chart. That is, sometimes you need to be able to diagnose the problem and verbalize it. This step is necessary to use the *help* files contained within most software programs. To find these files, look at the menu bar at the top of your screen. It typically lists options such as File, Edit, and View; at the very end of that list is Help. You can either double-click on the word Help or press the Alt and H keys simultaneously to open this script. Pressing the F1 key also brings up the help menu.

TABLE 6-1 Common Troubleshooting Procedures

Try each "solution" in order. There is no need to continue down the list if the first step solves your problem. If your problem remains unsolved, you may need more expert assistance.

Symptom	Solution
Presentation projector does not turn on/off	1. Point the remote directly at the projector and try again. 2. The remote may need new batteries.
No audio from podium	1. Make sure the audio system is turned on. 2. Make sure the microphone is turned on.
No audio from wireless microphone	1. Make sure the audio system is turned on. 2. Make sure the body pack is turned on. 3. The microphone may need new batteries.
No audio from DVD or computer	1. Check the volume controls to make sure they are not turned down. 2. Make sure the audio system is turned on. 3. Make sure connections are made to the audio output on the device and the input on the audio system.
No image from computer/laptop	1. Check that the correct input source has been selected. 2. Make sure the computer has not been turned off. 3. Change the video output from the computer (press the Alt and F5 keys simultaneously or other Alt Function keys, depending on the computer).
Projector light will not come on	1. If you hear a fan, make sure the lens cover is off. 2. If you hear a fan and the lens cover is off, you may need to replace the bulb.
Computer will not read media	1. Make sure the computer has the proper software needed for the media to work properly.

Training BULLETIN

JONES & BARTLETT FIRE DISTRICT
TRAINING DIVISION
5 Wall Street, Burlington, MA, 01803
Phone 978-443-5000   Fax 978-443-8000

Instant Applications: Technology in Training

Drill Assignment
Apply the chapter content to your department's operation, training division, and your personal experiences to complete the following questions and activities.

Objective
Upon completion of the instant applications, fire and emergency services instructor students will exhibit decision making and application of job performance requirements of the instructor using the text, class discussion, and their own personal experiences.

Suggested Drill Applications

1. Using the internet as your primary resource, research the best practices and recommendations for developing slide presentations. Using this information, develop a student handout outlining this information. In addition, take the information shared on the handout and develop a slide presentation for use in a classroom setting designed to introduce and familiarize students with PPT.

2. Select a current slide presentation and add additional graphics (photos, WordArt, illustrations) to better clarify or explain the topic of the original presentation.

3. Explore online fire incident simulator programs and determine the best of use of such technology, the cost of investing in the programs, and the ease of use of the respective programs.

4. Review the Incident Report for this chapter and be prepared to discuss your analysis of the incident from a training perspective and as an instructor who wishes to use the report as a training tool. Also discuss the technology options available to incorporate this Incident Report into a training session.

Incident Report: NIOSH Report F2001-38

Drill Assignment

Review the information in this incident report and prepare a practice presentation for class delivery at the direction of your instructor. Your presentation should include a summary of the incident facts and a review of the NFPA 1403 compliance and noncompliance findings. Use an outline to organize your thoughts. You may be evaluated on your communication skills during your presentation.

Lairdsville, New York—2001

The training evolution in Lairdsville, New York, involved a duplex that was to be razed. A search and rescue scenario with a rapid intervention crew (RIC) deployment was planned to simulate the entrapment of fire fighters on the second floor of one unit. For the purpose of the drill, the stairs to that unit would be considered impassable, so the RIC would have to enter from the other unit of the duplex and breach a second floor wall to reach the trapped fire fighters who were under light debris. Smoke from a burn barrel was used to obscure vision. The burn barrel was located in the back bedroom of the same unit as the trapped fire fighters, also on the second floor.

Neither of the two new fire fighters posing as victims had worn SCBA in fire conditions, and one (the deceased) was a brand new recruit fire fighter with no training. Both fire fighters wore full protective clothing and SCBA for this drill.

All of the crews assembled initially at the duplex. One engine was connected to a large water tanker on-site but no hose lines had been set up. A 1¾" (44.5 mm) hose line was supposed to be placed at the rear, with another at the front doors. An engine and heavy rescue were located off-site and would simulate a response to the scenario.

Safety officers were assigned and placed throughout both sides of the duplex. One officer was assigned to light the fire and place the new fire fighters in the front bedroom on the second floor. He was also instructed to guide the RIC if necessary. Another officer was assigned to a room on the second floor of the duplex unit, which the RIC was to enter first. He was assigned to guide the RIC if necessary, making sure they did not go through a wall opening that went to the outside. He was equipped with a 20-lb (9.1-kg) fire extinguisher. The 1st assistant chief was located on the first floor of the unit. The fire chief checked the scenario to make sure that accelerants were not used. He took command out front and the training was ready to begin.

The smoke barrel upstairs was ignited, but it was not creating the desired smoke conditions. Downstairs, the 1st assistant chief struck a road flare and ignited the foam mattress on a sleeper sofa that was located next to the open stairs (**FIGURE 6A**). The fire spread very rapidly across the ceiling, producing heavy smoke and flames, and flames quickly extended up the stairs and out the front windows.

The fire spread was so rapid that the officer in the adjoining unit could not enter to reach the new fire fighters, and he had to exit using a ground ladder. The 1st assistant chief exited the duplex and went to the rear exterior but he could not find a hose line. He then went to the pumper and pulled a 200' (61 m) preconnected hose line to the rear of the structure.

The fire fighter on the same floor as the trapped fire fighters had left his position to investigate the sound of another road flare being lit. Upon seeing the fire, he went back and reached the two new fire fighters. The flames were now entering the second floor windows. He led the

FIGURE 6-A A couch at the base of the stairs in the Lairdsville duplex.

From the NIOSH Fire Fighter Fatality Investigation and Prevention Program, Investigation Report F2001-38 NY.

new recruits to the stairwell, which was now fully engulfed in flames from the first floor. He lost his gloves and immediately burned his hands. At this point, he lost contact with the two recruit fire fighters. He was able to reach the back bedroom with heavy smoke and tenable heat. He searched for the boarded-up window and was able to break it open with his hands and jump from the second floor.

The 1st assistant chief advanced the hose line to the second floor, knocking down the fire. Arriving units were advised it was no longer a training and two RIC units were deployed to the second floor. One RIC forced entry through the involved unit's front door. One fire fighter was found and brought to the outside. The second RIC found the other fire fighter unresponsive. They removed him and transported him with advanced life support care to a local hospital, where he was later declared dead. The rescued fire fighter and the fire fighter who jumped from the window were both flown by EMS helicopters to a burn unit, due to the severity of their burn injuries.

During the ensuing investigation, the Lairdsville fire chief claimed he didn't know that a structure fire would be set as part of the training exercise, saying, "It was only supposed to be smoke." The 1st assistant chief was indicted and found guilty of criminally negligent homicide. He was sentenced to seventy-five days in jail and instructed to avoid contact with any fire department under a five-year term of probation. His attorney argued that the NFPA standard was not known by the department, and that it was the state's responsibility to distribute the guidelines to every volunteer fire fighter.

Postincident Analysis Lairdsville, New York

NFPA 1403 Noncompliant

- Students not sufficiently trained to meet job performance requirements (JPRs) for Fire Fighter I in NFPA 1001
- Debris from the structure used as fuel material
- Students acting as victims within the structure
- Deviation from the preburn plan
- Fire ignited without charged hose line present
- Exterior hose lines not in place
- Rapid Intervention Crew (RIC) not in place (NOTE: This is not a requirement of 1403, but is an OSHA requirement and is referenced in NFPA 1500, NFPA 1710, and NFPA 1720.)

After-Action REVIEW

IN SUMMARY

- Although training aids are fundamental in assisting the instructor in the learning process, remember that they are still simply aids, designed to help you impart the knowledge your students need to know and understand.
- One of the most dynamic and fast-changing concepts in education and training is technology-based instruction (TBI), also known as e-instruction, internet-based instruction, or distance learning.
- TBI is a good example of student-centered learning because course materials can be delivered in a variety of methods, providing each student with the resources that appeal to them.
- Computer-based training (CBT) software enables instructors to prepare course materials and efficiently manage day-to-day teaching tasks. Many CBT programs offer features that allow you to facilitate collaborative learning, personalize content based on students' unique needs, and positively affect learning outcomes.
- A learning management system (LMS) is a web-based software application that allows online courseware and content to be delivered to learners, that also includes classroom administration tools such as activity tracking, grades, communications, and calendars.
- Using multimedia tools enhances the learning process for students and can make the instructor's job easier. At the same time, multimedia tools can be a distraction if they are poorly developed or used improperly.
- Professionally developed presentations are intended to enhance students' learning experience, but do not replace the experienced instructor. Instructors must review materials ahead of time and adapt them to the students who will use them.
- Digital audio players have become a mainstream technology for students at most universities. Students can carry these devices easily and download lectures or pictures that can be reviewed outside of class.
- Virtual reality devices allow students to interact with a computer-simulated environment. The goal is to immerse the student in an environment that can be controlled by the student and the programmer.
- Like virtual reality systems, simulators allow students to interact safely with hazards in a controlled environment.
- As with any technology, maintaining the quality of multimedia tools is important.
- Multimedia tools should be used to simplify complex theories or to assist instructors in reinforcing their message through exercises or images.
- Selecting the right media application is critical. When used as an instructional aid, the right tool allows you to clarify the information you present to students.

KEY TERMS

Access Navigate for flashcards to test your key term knowledge.

Asynchronous learning An online course format in which the instructor provides material, lectures, tests, and assignments that can be accessed at any time. Students are given a time frame during which they need to connect to the course and complete assignments.

Courseware Any educational content that is delivered via a computer.

Flipped classroom An instructional approach that reverses or "flips" the standard learning environment by delivering online content for self-study and reserving the physical classroom for mentored practice that traditionally would have been assigned as homework.

Hybrid learning Learning environment in which some content is available for completion online and other learning requires face-to-face instruction or demonstration; sometimes referred to as "blended" or "distributed" learning.

Instructor-led training (ILT) A learning environment in which a human instructor facilitates both the in-person sections and the online components of coursework. Instructors usually set deadlines for submission, create quizzes and assignments, and track student progress.

Learning management system (LMS) A web-based software application that allows online courseware and content to be delivered to learners, that also includes classroom administration tools such as activity tracking, grades, communications, and calendars.

Self-directed study A learning environment in which learners can stop and start as they desire, progress to completion of a module, and submit a final score to determine whether they passed or failed the module, all without an instructor's involvement.

Social learning An informal method of learning using technologies that enable collaborative content creation. Examples include blogs, social media, collaborative discussions, and wikis.

Synchronous learning An online class format that requires students and instructors to be online at the same time. Lectures, discussions, and presentations occur at a specific hour and students must be online at that time to participate and receive credit for attendance.

Technology-based instruction (TBI) Training and education that uses the internet or a multimedia tool.

Virtual classroom A digital environment where content can be posted and shared, and where instructors can create quizzes and assignments to be completed and tracked online via an internet-connected computer.

Webinar An online meeting, occurring in real time, which usually allows for participants to share their desktop screens, web browsers, and documents live as the meeting is occurring.

REFERENCES

Capterra. 2017. "The Top 20 Most Popular LMS Software." https://www.capterra.com/learning-management-system-software/#infographic. Accessed August 17, 2018.

National Fire Protection Association. 2018. *NFPA 1403: Standard on Live Fire Training Evolutions*. Quincy, MA: National Fire Protection Association.

National Fire Protection Association. 2019. *NFPA 1041: Standard for Fire and Emergency Services Instructor Professional Qualifications*. Quincy, MA: National Fire Protection Association.

U.S. Department of Education. 2010. "Evaluation of Evidence-Based Practices in Online Learning: A Meta-Analysis and Review of Online Learning Studies." https://www2.ed.gov/rschstat/eval/tech/evidence-based-practices/finalreport.pdf. Accessed August 17, 2018.

CHAPTER PRESENTATION EXERCISE

Instructions: Your course instructor will assign individual or group discussions on the key points and teaching tips of this chapter. You or your group will review the chapter teachings and identify the major learning points from the chapter. You should be able to discuss the points, why they are important, and how and where they apply to your responsibilities at your level of instructor training.

REVIEW QUESTIONS

1. Describe the main benefits of using a learning management system.

2. How can the instructor minimize the amount of content that a student reads from slide, instead focusing the students on the instructor's message?

3. What elements make a strong multimedia or slide presentation? What elements detract from the quality or effectiveness of the presentation?

CHAPTER 7

Training Safety

NFPA 1041 JOB PERFORMANCE REQUIREMENTS

Note: An asterisk denotes that the 1041 standard contains further information in its annex section.

4.4.2 Organize the learning environment, given a facility and an assignment, so that lighting, distractions, climate control or weather, noise control, seating, audiovisual equipment, teaching aids, and safety are addressed.

(A) Requisite Knowledge.

Learning environment management and safety, advantages and limitations of audiovisual equipment and teaching aids, classroom arrangement, and methods and techniques of instruction.

(B) Requisite Skills.

Use of instructional media and teaching aids.

4.4.3 Present and adjust prepared lessons, given a prepared lesson plan that specifies the presentation method(s), so that the method(s) indicated in the plan are used and the stated objectives or learning outcomes are achieved, applicable safety standards and practices are followed, and risks are addressed.

(A)* Requisite Knowledge.

The laws and principles of learning, methods and techniques of instruction, lesson plan components and elements of the communication process, and lesson plan terminology and definitions; learner characteristics; student-centered learning principles; instructional technology tools; the impact of cultural differences on instructional delivery; safety rules, regulations, and practices; identification of training hazards; elements and limitations of distance learning; distance learning delivery methods; and the instructor's role in distance learning.

(B) Requisite Skills.

Oral communication techniques, methods and techniques of instruction, ability to adapt to changing circumstances, and utilization of lesson plans in an instructional setting.

4.4.5 Operate instructional technology tools and demonstration devices, given a learning environment and equipment, so that the equipment functions, the intended objectives are presented, and transitions between media and other parts of the presentation are accomplished.

(A) Requisite Knowledge.

Instructional technology tools, demonstration devices, and selection criteria.

(B) Requisite Skills.

Use of instructional technology tools, demonstration devices, transition techniques, cleaning, and field level maintenance.

KNOWLEDGE OBJECTIVES

After studying this chapter, participating in a structured learning environment, and completing assigned assessments, you will be able to:

- Discuss the relationship between training and fire fighter safety. (pp 160–161)
- Identify the life safety initiatives that relate to training activities. (pp 161–162)

- Describe how to ensure safety in the learning environment. (**NFPA 1041: 4.4.2, 4.4.5**, pp 162–163)
- Describe how to promote and teach safety by example. (p 162)
- Describe the laws and standards pertaining to safety during live fire training. (pp 170–171)
- Discuss how to develop safety as part of your department's culture. (pp 171–172)
- Identify the legal considerations surrounding training safety. (**NFPA 1041: 4.4.2**, pp 172–175)

SKILLS OBJECTIVES

After studying this chapter, participating in a structured learning environment, and completing assigned assessments, you will be able to:

- Assess the level and use of personal protective equipment (PPE) necessary for hazards present during training. (p 163)
- Control safety hazards in the physical learning environment. (**NFPA 1041: 4.4.2**, pp 163, 169–170)
- Model safety conscious behaviors during lesson presentation. (**NFPA 1041: 4.4.3**, pp 162–164)
- Follow safety policies, practices, and procedures during training events. (**NFPA 1041: 4.4.3**, pp 164–169)

You Are the Fire and Emergency Services Instructor

Recently an instructor at a neighboring training facility was injured during a training drill that left the instructor unable to return to work. During the investigation, it was determined that the instructor had failed to wear all the correct personal protective equipment (PPE) and that this failure had caused his injuries to be more severe than if he had been wearing his PPE. This and other related incidents on a national level have prompted your training division staff to review your agency's policies and procedures—not only those dealing with PPE, but also those pertaining to instructor conduct during training.

1. As an instructor, how do you ensure the safety of fellow instructors working with you during training?

2. How do you ensure the safety of your students during training evolutions?

3. Which policies and guidelines do you have in place to ensure the safety of your fellow instructors and the students for whom you are responsible?

Access Navigate for more practice activities.

Introduction

Each year in the United States, countless fire fighters are injured and, on average, 10 fire fighters are killed during training-related activities. This trend has accelerated over the past few years and is a growing concern among fire service leaders. In addition to the higher number of injuries to students during training, the number of injuries to instructors who teach fire fighters has increased.

As an instructor, you must take measures to stop the preventable injuries and deaths that occur during training. Of course, for training to be effective, it has to be realistic and, therefore, a certain level of danger will exist

FIGURE 7-1 It is your responsibility to provide meaningful drills that allow fire fighters to apply their skills, knowledge, and abilities while keeping your students safe.
© Jones & Bartlett Learning. Photographed by Glen E. Ellman.

in all training events. Live fire training tops the list of high-risk training events, but this same high risk can be found in many other training activities (NIOSH, 2016).

You must always place students' safety ahead of any performance expectations for the training session. This mandate includes everything from assigning safety officers, to implementing rehabilitation protocols, to decreasing instructor–student ratios, to ensuring that enough trained eyes are watching out for the safety of all students. The standards, technology, and information needed to protect your personnel from training tragedies are readily available. You also need to apply relevant checklists, standards, and control measures to your training sessions correctly to keep your students safe. Fire fighters want meaningful drills that allow them to apply their skills, knowledge, and abilities. It is your responsibility to provide this training while keeping your students safe (**FIGURE 7-1**).

The 16 Fire Fighter Life Safety Initiatives

In 2004, the National Fallen Firefighters Foundation (NFFF) brought together fire service experts and leaders to discuss ways to slow the number of line-of-duty deaths (LODDs) experienced by the fire service. Their work culminated in 16 initiatives that were given to the fire service with the goal of reducing LODDs by 50 percent over a 10-year period (**TABLE 7-1**). These initiatives should provide both guidance and motivation to you while you are developing and conducting any training. They highlight the importance of training as a strategy that will help reduce the number of

TABLE 7-1 Sixteen Fire Fighter Life Safety Initiatives

1. Define and advocate the need for a cultural change within the fire service relating to safety, incorporating leadership, management, supervision, accountability, and personal responsibility.

2. Enhance the personal and organizational accountability for health and safety throughout the fire service.

3. Focus greater attention on the integration of risk management with incident management at all levels, including strategic, tactical, and planning responsibilities.

4. All fire fighters must be empowered to stop unsafe practices.

5. Develop and implement national standards for training, qualifications, and certification (including regular recertification) that are equally applicable to all fire fighters based on the duties they are expected to perform.

6. Develop and implement national medical and physical fitness standards that are equally applicable to all fire fighters, based on the duties they are expected to perform.

7. Create a national research agenda and data collection system that relates to the initiatives.

8. Utilize available technology wherever it can produce higher levels of health and safety.

9. Thoroughly investigate all fire fighter fatalities, injuries, and near misses.

10. Grant programs should support the implementation of safe practices and/or mandate safe practices as an eligibility requirement.

11. National standards for emergency response policies and procedures should be developed and championed.

12. National protocols for response to violent incidents should be developed and championed.

13. Fire fighters and their families must have access to counseling and psychological support.

14. Public education must receive more resources and be championed as a critical fire and life safety program.

15. Advocacy must be strengthened for the enforcement of codes and the installation of home fire sprinklers.

16. Safety must be a primary consideration in the design of apparatus and equipment.

Courtesy of National Fallen Firefighters Foundations Everyone Goes Home® Program.

fire fighter LODDs. As an instructor, you will be specifically charged with carrying out initiative 5, which states, "Develop and implement national standards for training, qualifications, and certification . . . that are equally applicable to all fire fighters based on the duties they are expected to perform." This initiative should be applied every time you instruct a course and in every drill-ground evolution you carry out.

Leading by Example

As an instructor, you will use many methods of delivery to teach. One of the most powerful methods is leading by example. Fire fighters, regardless of their rank, years on the job, or past experiences, will look to you and make mental notes on how the job *should* be done (**FIGURE 7-2**). As an instructor and mentor, you play an extremely influential role in determining the performance of your students. Given this responsibility, you must ensure that your own practices and principles clearly demonstrate the safety values employed in the fire service. If you do not correct an unsafe practice during training, your students will mistakenly believe that you condone the error. You must set a positive tone and embrace safety as the number-one priority of every lesson plan delivered. Both in the classroom and on the drill ground, use the controlled environment of the training arena to point out where things can go wrong in real-world applications. Prior to class, survey the classroom or teaching environment and correct any safety concerns that may exist.

Safety in the Learning Environment

Safety during training begins with safety in the learning environment. First, consider the location of the classroom. While you may not have complete control over where the class is held, you can still affect some things. Survey the general learning environment. Are there inadequate fire exits or other unsafe conditions, such as remodeling in the area or icy steps leading to the classroom? If the classroom is in an engine bay, is the area subject to exhaust fumes, or is there oil on the floor? Evaluate the lighting conditions in the classroom. Not all students have good vision, and either low or high lighting may create eye strain. Excessive noise can be a problem as well—not only because it interferes with learning but also because it can cause unnecessary damage to students' hearing.

Next, review the physical arrangements of the classroom. Unsecured extension cords running throughout the classroom may trip either you or the students. Prior to each class, reroute or tape down extension cords. Also, be aware that fire stations may be the recipients of hand-me-downs from other city agencies. Often, chairs and tables that have outlived their usefulness in other areas are pressed into service in a fire department classroom and could be prone to collapsing. Make sure that each item is in good condition to prevent injuries to students.

At the beginning of class, be sure to raise safety awareness with the students. Go over the locations of the nearest fire exit and fire extinguishers. If you are teaching a class to an on-duty crew or a group of volunteer fire fighters, you should address what to do if an alarm sounds. For example, if some students in the class are on-call fire fighters, have them sit near an exit so that if they need to respond during class they can leave without interrupting the class. Explain the emergency plan to follow should a fire alarm activate

FIGURE 7-2 As the instructor, you will be demonstrating how the job should be done.
© Jones & Bartlett Learning. Photographed by Glen E. Ellman.

or a severe weather event or natural disaster occur. It is important to emulate the behavior you teach to the general public and to remind your students that safety is paramount.

Hands-on Training Safety

Hands-on training includes all training activities conducted outside the classroom environment. Whether it consists of a simple tailboard chat on nozzles or a full-blown live fire training exercise, hands-on training is often the most powerful method of instruction for a student. Unfortunately, hands-on training also produces hazards that must be considered and mitigated. Slippery surfaces, sharp or jagged metal, and products of combustion need to be accounted for. Risk management practices need to be applied to limit, reduce, or eliminate risk exposure to students. Environmental factors—such as heat, cold, wind, rain, and storms—must be considered as well. Mother Nature can pose a significant risk to your students.

Personal Protective Equipment

Determine the appropriate level of PPE after reviewing the lesson plan. Lesson plans often specify the level of PPE required. However, local standard operating procedures (SOPs) may also exist that specify the level of PPE for training. A general statement of the PPE requirements for the training session needs to be part of your presentation. As an instructor, you must always wear the PPE required for the drill being conducted. The appropriate level of PPE depends on both the potential hazards and your department's SOPs.

The best practice is always to use the same PPE in training as you would in real-world incidents. In some departments, it may be acceptable to conduct ladder drills using helmets, gloves, and boots as opposed to full fire gear and self-contained breathing apparatus (SCBA), due to the skill being taught or reinforced. This modification to the PPE policy may make the drill safer by reducing the potential for heat stress. Another option may be to begin the training session in full levels of PPE and then to reduce the PPE required after signs of heat stress appear or when the skill has been performed properly. When these adaptations are implemented, you must make it very clear to your students that the modification is only allowable in the training session and should never be employed when responding to actual emergencies. If any level of hazard exists in the session, then the highest level of PPE available should be used.

Rehabilitation Practices and Hands-on Training

NFPA 1584: *Standard on the Rehabilitation Process for Members During Emergency Operations and Training Exercises* requires that rehabilitation protocols be established and practiced at training events that involve strenuous activities or prolonged operations. As an instructor, you must be aware of the content of this standard and apply it to these types of training sessions. These practices further reinforce the importance of following appropriate rehabilitation practices on the fire ground (**FIGURE 7-3**). Training should mirror as closely as possible the actual fire-ground operation. Medical care and monitoring, hydration, and cooling/warming materials are just a few of the rehabilitation considerations that NFPA 1584 requires to be available during training events.

> **COMPANY-LEVEL TRAINING TIP**
>
> Instructors should encourage training participants to begin hydrating for the training 24 to 48 hours prior to the training event. It is imperative for instructors who conduct live fire training to follow this guideline.

Not to be overlooked is the importance of instructors participating in the rehabilitation process. Too often, instructors will rotate students through the rehab station but neglect to rehab themselves. In reality, participation in rehabilitation is a safety factor as well as an important way to set the proper example for your students. All instructors must be monitored for dehydration, heat stress, and general fatigue. A rehab rotation for instructors should also be included in the training plan.

FIGURE 7-3 Rehabilitation is an important part of training.
© Jones & Bartlett Learning. Photographed by Glen E. Ellman.

Safety Policies and Procedures for Training

Your role as the instructor mandates that you have a firm grasp of the safety policies mandated by your fire department. All safety policies required on the fire ground must be reinforced and practiced during training. Indeed, one of the more common causes of injuries during training is failure to follow established safety practices. The instructor has a responsibility to ensure that all policies of the department are followed during training. Policies such as establishment of an incident command system, fire-ground accountability, rehabilitation, and full use of PPE are followed on the emergency scene, and adherence to those policies must not be set aside in training. Enforce those policies in the same manner a safety officer or incident commander would at an emergency incident.

On occasion, personnel from another agency might train with you at your facility if you are conducting mutual aid training or a special course. It is important in these types of situations that all fire fighters know which policies will be followed during the training. If you are hosting the training, all visitors must follow your policies, and your visitors need to be aware of what they are expected to do and which policies they will follow. When conducting live fire training, fire fighters from other organizations participating in interior training evolutions should have a signed letter from their fire chief or employer stating that they have met the training requirements set forth by NFPA 1403: *Standard on Live Fire Training Evolutions* and health requirements set forth by NFPA 1582: *Standard on Comprehensive Occupational Medical Program for Fire Departments*. This documentation can reduce the liability on the lead instructor in the event of injury.

You are responsible for drafting and forwarding policies relating to training through the chain of command. At the minimum, a training policy relating to training safety will include procedures for instructor qualifications, PPE for both students and instructors, safe student-to-instructor ratios, and effective rehabilitation practices during high-risk/high-exertion exercises. All levels of instructors must develop and deliver training that adheres to applicable safety points and must enforce a strict, zero-tolerance policy for deviations from accepted safety standards or practices (**FIGURE 7-4**).

As an instructor, you must also be informed of and follow the manufacturer's recommendations for using equipment. Read the accompanying owner's manuals for equipment such as chainsaws and heavy hydraulic equipment. Often the limitations of these tools are assumed, and an injury occurs when the tool is used in a manner counter to its intended use. When new apparatus is delivered to your department, you should be present for the initial training by the apparatus builder. Essential information presented by the manufacturer needs to be recorded, practiced, and reinforced whenever that piece of equipment is used.

Influencing Safety Through Training

As an instructor, you play a vital role in setting a good example in terms of safe fire-ground practices. Constantly remind yourself that you serve as the role model of a safe and effective fire fighter. Everything from your fitness level to the way you wear your uniform should exemplify your values related to safety and adherence to rules. The example that you set sends a far more powerful message to your students than any theories and concepts you present in a lecture or drill.

You should be acutely aware of how influential a role the fire and emergency services instructor plays, especially with new fire fighters. New fire fighters are uncertain about what is expected of them—that is, what is considered right versus what is considered wrong. With these students, whatever you do or say will be construed as the right thing. If you use improper techniques or fail to follow safety protocols while teaching, students will perceive that as the right way to do things. This faulty knowledge becomes dangerous when students leave your protection and perform on the emergency fire ground.

Your primary responsibility during a drill or training session is to ensure the safety of all students attending the drill. The actions of a fire fighter on the fire ground are the result of training. If you allow students to use unsafe practices in training, they will repeat those same actions on the fire ground, because decisions and actions on the fire ground are based on previous training.

You have many responsibilities related to the safety of your students during training, but one that stands above all others is the responsibility to have students practice safe principles and practices during training so that they remain safe on the fire ground. As part of your duties as an instructor, you must correct poor skills or bad decisions as they occur. If you witness a skill being performed incorrectly or a safety procedure not being followed but you do not correct the student's error, then you are reinforcing the incorrect behavior. This constant monitoring and correction is important with new recruits but can be more difficult with

Training Safety Plan

Drill Date:　　　　　　　Time:　　　　　　　Shift:

Drill Location:

Type of Training: (check all that apply)

- ☐ Fire Suppression
- ☐ Live Fire Training
- ☐ Driver Training
- ☐ Apparatus Operation
- ☐ Physical Fitness Activity
- ☐ Other:

- ☐ EMS
- ☐ Vehicle/Machinery Extrication
- ☐ Other Acquired Structure Training
- ☐ Preplan Survey or Simulation

- ☐ Technical Rescue
- ☐ Hazardous Materials/WMD
- ☐ Water/Dive Rescue

Drill Risk Assessment:　☐ High　　☐ Medium　　☐ Low

Maximum Student/Instructor Ratio:　　to　　Safety Officer Needed: (Required on High-Risk Drills)

Instructor PPE Requirements:

- ☐ SCBA　　☐ Full PPE　　☐ Helmet　　☐ Eye Protection　　☐ Filter Mask　　☐ Hearing
- ☐ Gloves　　☐ Radio　　☐ Lights　　☐ High-Visibility Vest

Drill Objective(s): (brief explanation of objectives)

Description of Training: (e.g., extricate victim from vehicle, victim search in limited visibility)

PPE/Equipment Required for Each Participant: (check all that apply)

- ☐ Helmet
- ☐ Eye Protection
- ☐ Hearing Protection
- ☐ Gloves (Type):
- ☐ Bunker Coat
- ☐ Hood
- ☐ Bunker Pants
- ☐ Safety Boots
- ☐ Other (Specify):

- ☐ Personal Flotation Device
- ☐ Buoyancy Compensator
- ☐ Mask/Snorkel/Fins
- ☐ SCBA
- ☐ SCUBA
- ☐ Other Respiratory Protection (Type):
- ☐ Hazardous Materials CPC (Type):
- ☐ Radio

Department Related SOPs or Technical References: (list number and name)

FIGURE 7-4 A sample drill safety plan.
Courtesy of Forest Reeder.

Hazards and Control Measures: (check all Hazards AND write in control measure)

- ☐ Atmospheric (e.g., smoke, dust, low oxygen):
- ☐ Combustible/Flammable Environment:
- ☐ Confined Space:
- ☐ Electrical:
- ☐ Elevation:
- ☐ Hazardous Substances (e.g., asbestos, chemicals):
- ☐ Nighttime Conditions:
- ☐ Other:
- ☐ Sewage/Septic:
- ☐ Sharp Edges/Objects:
- ☐ Structural:
- ☐ Terrain:
- ☐ Traffic:
- ☐ Water:
- ☐ Weather:

Accountability: (check all that apply)

- ☐ Buddy System
- ☐ Visual
- ☐ Passport
- ☐ Dive Master Control Sheet
- ☐ Other:

Communications:

- ☐ Radio—Primary Frequency:
- ☐ Radio—Secondary Frequency:
- ☐ Hand Signals
- ☐ Rope Line
- ☐ Lights
- ☐ Other:

In Case of Emergency: (check all that apply)

- ☐ Code or Signal Used:
- ☐ RIT Assigned:
- ☐ ALS Standby:

Resources Assigned: (check all that apply AND fill in designated unit)

- ☐ Battalion Chief(s):
- ☐ Rehab Officer/Area:
- ☐ Rescue Unit(s):
- ☐ Safety Officer:
- ☐ Specialty Unit(s):
- ☐ Suppression Unit(s):
- ☐ Other Resources/Equipment:

Rehabilitation Plan: (describe rehabilitation plan and guidelines)

FIGURE 7-4 *Continued*

Training Event Safety Analysis

1. Identify level of required PPE for each participant.

2. List basic steps required to safely complete evolution.

3. Identify potential accidents or hazards that may occur.

4. Determine recommended safe procedures.

Safety Planning Notes: (e.g., site plan, drawings)

Lead Instructor:
Signature: Date:
Reviewed by (print):
Signature: Date:

FIGURE 7-4 *Continued*

experienced personnel. Correcting unsafe or poor behavior demonstrated by a veteran fire fighter can be very challenging—but it is a challenge you cannot ignore. Correcting an unsafe skill should be the same whether the fire fighter is a raw recruit or a seasoned veteran.

Planning Safe Training

You should begin to address safety at the earliest stages of course development and preparation, while formulating or modifying lesson plans. Analyze and critique your training to ensure that safety measures are correctly addressed and emphasized. It is too late to address safety issues after an injury has occurred. Preventing injury becomes much easier when the hazards are identified and addressed early in the planning process.

Many of the NFPA standards provide additional information concerning safety and training and should be consulted during the planning process. NFPA 1500: *Standard on Fire Department Occupational Safety, Health, and Wellness Program* would be an important document to review when planning a program of instruction. In general, almost all NFPA standards in the 1400 series relate to a training recommendation and focus on providing safe training sessions. NFPA 1584: *Standard on the Rehabilitation Process for Members During Emergency Operations and Training Exercises* is also a valuable resource that can assist in planning safe training evolutions.

Creating a safety-conscious student must be a primary goal of nearly all training in the fire service. While there are inherent dangers in their occupation, fire fighters should constantly strive to reduce those risks through good awareness and safety practices. When determining the appropriate level of protective gear to use during training, for example, the instructor should consider the exercise's safety implications for students:

- First, is PPE necessary to ensure students' safety during the training activity? If so, full PPE must be used and any inconveniences mitigated. For example, because full PPE is needed for live fire training, be sure to provide rehabilitation in such a scenario.
- Second, consider whether *not* using full PPE will make the training safer. For example, perhaps hoods need not be worn during ladder practice because of the potential for heat stress in July.

- Third, consider whether *not* wearing full PPE will create a false perception that real scenes can be safely mitigated without its use.
- Fourth, ask yourself if you are considering reduced PPE owing to peer pressure or if this decision is *truly* in the best interest of the student.

Each of these considerations should be evaluated prior to making a determination of the level of PPE to be used. Safety cannot be an afterthought or an unwanted formality: Instead, it must be the first and foremost consideration to ensure the lives and safety of all (**FIGURE 7-5**).

An important part of the planning process is determining how many instructors are needed to conduct the desired training. Two factors should be considered in making this decision: the type or risk level of the training and the skill level of the students.

Training can be classified based on risk as either high risk, medium risk, or low risk. An example of high-risk training would be live fire evolutions in an acquired structure or swiftwater rescue training. Medium-risk training would include fire extinguisher training or ladder evolutions. Low-risk training would include a classroom lecture or a knot-tying practice session (**FIGURE 7-6**).

The skill level of those being taught should also be considered. High-risk training of new recruits should generate different concerns than high-risk training involving veteran fire fighters.

Other areas of planning include assembling the equipment that may be needed. Sometimes outdated equipment or equipment that has been replaced by upgrades within the department is used when conducting drills. Using old or outdated equipment has long been an accepted practice in the fire service, but it has often led to injuries that could have been

A

B

C

FIGURE 7-5 While there are inherent dangers in their occupation, fire fighters should constantly strive to reduce those risks through good awareness and safety practices.
© Jones & Bartlett Learning. Photographed by Glen E. Ellman.

FIGURE 7-6 Examples of high-risk **(A)**, medium-risk **(B)**, and low-risk **(C)** training evolutions.
A: Courtesy of Captain David Jackson, Saginaw Township Fire Department; **B, C:** © Jones & Bartlett Learning; Photographed by Glen E. Ellman.

prevented. If the tool or apparatus does not meet the accepted performance standards, then it should not be used in training. You should also question whether using an outdated power saw, for example, is beneficial to students given that this type of saw is not used on current apparatus.

The instructor is also responsible for inspecting all equipment being used in the training to ensure that it meets the required standards prior to the drill. Do not assume that the equipment is in safe and operable condition. If your department does not have the equipment available to make the training safe, then it needs to find alternative equipment—perhaps by borrowing items from a neighboring fire department or the state fire training academy. Acquisition of the appropriate equipment should be arranged prior to the drill.

As the instructor, you are responsible for protecting students from physical and emotional harm in the classroom and on the drill ground. Most of this chapter focuses on means to protect students from physical injury. Nevertheless, instructors cannot overlook the psychological dangers students encounter during training both in and out of the classroom.

Hidden Hazards During Training

During training, students are asked to perform various tasks, recall policies, or demonstrate the ability to perform. When placed in an environment of peers, most students realize that their performance is being evaluated—not only by the instructor but also by their fellow classmates. Failure to accomplish the task or answer the question correctly can label the student as not competent or someone who cannot be trusted on the fire ground. This situation is particularly dangerous because the typical personality of fire fighters is driven by the need to succeed, regardless of the conditions or situation.

This scenario sets the stage for disaster if a fire fighter is placed in an overwhelming situation, such as being asked to conduct a simulated rescue of another fire fighter or a rescue mannequin that may be twice the candidate's size or weight. If the fire fighter is asked to perform this task while fellow classmates are watching, it forces the student to attempt the impossible. Often, the student makes a valiant attempt, only to suffer an injury in the process. You have a responsibility to construct the learning environment so that situations such as these do not arise.

Do not minimize your levels of performance or drop your benchmarks so that every student can achieve success; this is not realistic. At the same time, it is unacceptable for you to stand idly by as a student is placed into a scenario where he or she cannot perform successfully without being exposed to physical harm.

> **COMPANY-LEVEL TRAINING TIP**
>
> Using theatrical smoke during an evolution instead of live smoke during search-and-rescue training might allow for a candidate to have more confidence in their early training sessions. Theatrical smoke provides a non-IDLH atmosphere where, in the event of an emergency, the student is never in a dangerous environment.

Overcoming Obstacles

Money may be an obstacle in the training process, because instructors often face demands to train students while using limited resources. In the past, the fire service has been creative in developing methods and practices to accomplish drills or tasks with minimal cost. For example, when an acquired structure is not available for training on overhaul, it is easy to construct a wooden simulator from scrap 2-in. by 4-in. (2 × 4) lumber and cover it with discarded gypsum board that can be obtained from a lumber supply shop, often at very little expense (**FIGURE 7-7**). When foam is needed for a fire stream training session and expired or surplus foam is not available, you can make a homemade solution using dish soap liquid and water. Of course, if you are using something other than equipment or products that would normally be used in the evolution, make sure that they will not produce a safety hazard.

One item that cannot be minimized is personnel: An adequate number of personnel must be present to conduct the training safely. This is particularly true

FIGURE 7-7 If an acquired structure is not available for training on overhaul, it is easy to construct a wooden simulator of scrap 2 × 4 lumber and cover it with discarded gypsum board.
Courtesy of Forest Reeder.

when conducting live fire training. Live fire training is the pinnacle of all firefighting training, but it is also one of the most hazardous drills. The hazard becomes even more pronounced when the drill occurs with fewer personnel than are acceptable to ensure a safe environment.

Analysis of your audience will help you select the best method of instruction for your training session. Some training sessions will require you to demonstrate the proper techniques in a step-by-step fashion, pausing to allow for questions and to provide clarification. This method may be appropriate for entry-level or new-skills training. Other sessions will require you to review the objectives or desired outcomes of the training session along with key safety behaviors. This approach may be applied to training sessions for experienced fire fighters or when doing evaluations of skill levels.

Live Fire Training

Perhaps one of the most hazardous but exciting portions of training involves live fire training evolutions. Safety considerations during live fire training should be the highest priority for all instructors involved with this type of high-risk training. NFPA 1041 now contains language on live fire training instructor requirements. This is an example of the emphasis on training safety in high-risk training activities, such as live fire evolutions.

As a result of past fire fighter fatalities involved with live fire training, the NFPA developed a standard that is to be followed during all live fire training evolutions. NFPA 1403: *Standard on Live Fire Training Evolutions* describes all aspects of the training process involving live fire training. It identifies the positions that must be staffed when conducting live fire training and the prerequisite training that students must complete before they can participate in any live fire exercises addressed by this standard. This standard applies to all types of live fire training, including acquired structures, gas-fired and non-gas-fired live fire props, and exterior live fire props. It includes the permit process, pre/post training, and documentation of the training process. NFPA 1403 must be consulted before the live burn exercise and followed to the letter when conducting live fire drills. As an instructor, you have a responsibility to ensure that all positions required by this standard are staffed at your training sessions. Do not attempt the drill with anything less than a full complement of trained and qualified staff.

Positions that must be filled during live burn activities include a designated **safety officer**, who has received training and background in the responsibilities of this position during live fire training. This safety officer is responsible for intervention in and control of any aspect of the operation that he or she determines to present a potential or actual danger at all live fire training scenarios. An **instructor-in-charge** is required to plan and coordinate all training activities and is responsible for following NFPA 1403. The instructor-in-charge must meet the requirements of NFPA 1041, Instructor II, while all other instructors must meet the Instructor I requirements. An additional position identified in this standard is the **ignition officer**, who is responsible for igniting and controlling the material being burned. The ignition officer coordinates all fire-lighting activities with the instructor-in-charge, the safety officer, and the fire control team. Note that an instructor cannot serve as an ignition officer for more than one evolution in a row (NFPA 1403, 4.7.6.1). Specific detail should be applied to the acquired structure to prepare it for the live fire training evolution. NFPA 1403 maintains a check-off sheet to assist the instructor with all aspects of preparation and execution of the live fire training exercise.

The proper student-to-instructor ratio when conducting training involving live fire is recommended to be 5:1. In some training sessions involving multiple fire departments, you will need a large pool of instructors to assist you and maintain the 5:1 ratio of crew members to instructors. This 5:1 ratio must be maintained throughout the evolutions and is set so that the typical span of control is not exceeded.

NFPA 1403 states that the agency conducting the live fire training is responsible for identifying the qualifications for instructors who direct live fire training, but experience has demonstrated over time that senior instructors and fire fighters should staff each of the positions as identified in the standard. 1403 also requires the instructor-in-charge to meet Instructor II, while all other instructors meet Instructor I. In addition to being qualified, each person who carries out these functions should be properly trained.

NFPA 1403 identifies specific positions to be staffed during live fire training due to the nature of the risks involved while working around fire. This same "safety" mindset should apply during any type of high-risk hazard training, such as high-angle, swiftwater, or confined-space training. Having an experienced instructor-in-charge with appropriate safety officer(s) and a low student-to-instructor ratio will enhance a safe training environment and promote a safety culture.

Other Training Considerations

In many fire-ground incidents, if a particular position is not filled, then the incident commander is responsible for completing all of the functions of that position. Although normal ICS procedures allow tasks not delegated to be performed by the IC, NFPA 1403 requires that "The safety officer shall not be assigned other duties that interfere with safety responsibilities." (NFPA 1403, 4.5.6). If company officers are available to assist with training, the responsibilities associated with these positions must be covered in depth before training.

You must pay attention to details when planning and conducting drills where students may face a risk of injury. With practical drills, many variables come into play and many things can go wrong, resulting in an injury to a student. One way in which you can minimize that possibility is to pay special attention to every detail of the drill. Mentally walking through the evolution during the planning session and drawing upon your own experiences or case reports of actual incidents can help develop your awareness of what might potentially go wrong. This review will allow you to create backup plans or, in the best case, to eliminate the hazard altogether.

When the drill includes large numbers of people whom you may not know (e.g., in disaster drills where many outside agencies are invited to participate), your preplanning skills will be put to the test. Asking for assistance from other instructors or agencies that have conducted similar exercises will help you in covering all the bases during the planning phase. As the instructor-in-charge, run through each aspect of the session, always asking, "What might go wrong here?" Aspects of the training to be considered include the weather, topography, equipment being used, and steps of the task.

Additional instructors should be employed at training sessions where the following conditions are present:
- Extreme temperatures are expected.
- The qualifications or skill level of the students are unknown, which could require more one-on-one time.
- Large groups of people will participate.
- Training will take place over a long duration.
- Training involves complex evolutions and procedures where additional instruction and safety oversight may be valuable.

Developing a Safety Culture

Fire and emergency services instructors are in a position to select personnel to assist with their training programs. Often company officers are used to help deliver the session or watch over a skill being evaluated. As an instructor, you have the responsibility to know about the people you are asking to help conduct this training. What are their qualifications? What are their values concerning fire service safety? Despite your best efforts to keep your training safe, you cannot be everywhere at all times. Thus, you must rely on other instructors to help supervise the drill.

Instructors who are not competent in or confident with the subject being presented should not be placed in a situation where they attempt to teach that material to others. The motivation of your subordinate instructors must be assessed as well. An instructor who has a propensity to show off or try to prove that he or she is superior to students is an accident waiting to happen. Often the focus of such an instructor is not on the students or their safety. Know your staff, their attitudes toward safety, the way they operate on the fire ground, their motivations for being instructors, and their teaching methods. All support personnel for the drill, including those who simply assist in prop construction, should be included in this assessment, because this information is essential to conduct training safely.

Anticipating Problems

As part of the preparation for instruction, you need to assess potential problems and rectify them before they result in an injury. You need to train your eye to pick out these hazards on first glance. It may be easy for a company to throw up a ladder and leave it unheeled; you should be able to foresee the potential consequences of that action and take steps to make sure the hazard is addressed. Most experienced instructors have this quality already. The more fire-ground experiences you have, the better you will become at recognizing the small details that might escalate into a full-scale disaster.

Your fire-ground experiences translate to the drill ground, and you should apply your firefighting experiences to every evolution you instruct. Tell your fire fighter students what can go wrong, what to watch out for, and how to be prepared for changing conditions throughout the training evolution. If a smoke generator will be used for the training session, let students know how it compares to smoke at a real incident, if the situation may be significantly different from the training experience.

Often training tragedies are not the result of a single catastrophic event, but rather represent the culmination of several small safety violations that combine to create a fire training tragedy. The instructor-in-charge may identify that an attack line of a smaller size is appropriate for the amount of fuel being used if it is advanced without delay. But hose lines can get kinked, water may be slow to reach the nozzle from the pump, and conditions may deteriorate and lead to a delayed attack that can bring the fuel load above the capacity of the slowly advancing attack line. Collectively, these elements are a prescription for disaster.

Changes in wind or weather, crew fatigue, or student motivation may also affect the evolution. Your guard must stay up until the evolution is complete. You might imagine that if just one thing goes wrong you can quickly correct it, but the escalation of events often occurs so rapidly that you cannot gain control of the one small mistake before an injury takes place. If you see a potential hazard, correct it immediately.

Accident and Injury Investigation

Despite all of your careful preparation, sometimes an accident might happen. In such circumstances, it is important to conduct a comprehensive investigation of the incident. A key factor in any investigation is the attitude held by those conducting the investigation. The purpose of such an investigation should be to find the reasons for and contributing factors to the accident or injury—not to assign blame. If the main focus is on finding the cause of the accident, then individuals will be more honest and open to this important process.

The investigation should include the date, location, events prior to and during the incident, equipment involved, PPE in use at the time of the accident, and personnel who were present when the incident occurred. Anyone who witnessed the incident should provide a written statement. This information will help the investigator determine the cause of the injury and enable the agency and training personnel to put control measures in place to prevent future occurrences. Keep all injury investigation notes, doctors' reports, and any information generated during the investigation on file. Periodically review these files for any emerging trends that might cue you to change a policy or procedure that is contributing to injuries. When types of injuries are compared over a period of time, trends in their incidence often become apparent. You should see that modifications are made to the training program based on this information.

> **COMPANY-LEVEL INSTRUCTOR TIP**
>
> Review the causes of accidents for the type of training you will be conducting to ensure you have taken all appropriate preventive measures.

Student Responsibilities for Safety During Training

Every student participating in training has a stake in ensuring his or her own personal safety. Each student is responsible for his or her actions as well as the actions of other crew members, training team members, and fellow students when it comes to addressing hazardous conditions. Students should be knowledgeable about the training to take place and should meet all of the prerequisites specified before participating in the drill. During high-risk training, this background becomes even more important because of how an unprepared student's performance can affect other students involved in the training. Several case studies have dealt with improperly prepared students contributing to the injury of fellow students.

Students should also adhere to all PPE requirements specified for the evolution and wear all other assigned safety devices or equipment. Observance of all safety instructions and safety rules throughout the training session will assist in maintaining the overall safety profile of the training session.

Each student should also be honest regarding his or her strengths and weaknesses. If a student is not prepared for the skill or evolution, he or she should notify you so that other students or instructors are not put at risk.

Legal Considerations

Instructors are held accountable for planning, lack of planning, adherence to standards and codes, and decision making during the training process. There are many types of liability that arise during training. As an instructor, you have a responsibility to stay informed about all laws and standards governing fire fighter training and safety. These laws and standards change over time, so you must keep up-to-date with the most current editions. Instructor networks, websites, and electronic bulletin boards are all popular methods of staying abreast of these ever-changing rules.

In many cases, the standard of care or best practices relating to training safety may be tied to your department's SOPs. As mentioned in earlier chapters, those SOPs must be referenced within the scope of training along with the application to the individual job description.

Another legal test that may be applied to training is the "reasonable person" concept. This standard tests your decisions against the decisions made by others with similar training and background, asking what your peers would do when faced with the same situation.

Negligence, Misfeasance, and Malfeasance

Negligence, **misfeasance**, and **malfeasance** are terms you are likely to encounter in cases where a civil wrong is alleged. Generally, two types of wrongs can be alleged in civil cases when an injury occurs:

- Unintentional conduct, which is best equated to the term *accident* (negligence, misfeasance)
- Conduct that includes an element of intent or knowledge that a wrong is being committed (malfeasance)

Which conduct constitutes negligence/misfeasance and which conduct constitutes malfeasance depend on the facts of each situation. This determination is based on the conduct involved, the fire department's rules and regulations, prior training, and national guidelines. In several LODD cases involving live fire training, there have been civil and criminal cases accusing instructors of these principles.

Negligence/Misfeasance

Negligence and misfeasance are kindred spirits: They are void of the element of intent and are based on a duty to prevent harm or injury to others by exercising a degree of care commensurate with the task being performed. Negligence exists when there is a breach of the duty owed to another, there is an injury, and the breach is the proximate cause of the injury.

Once these three elements are established, you can become legally liable for your conduct absent some type of immunity. **Liability** in its simplest form means responsibility. If a person or entity is found liable, that person or entity is responsible for paying for the damages caused by its actions or inactions.

An instructor's conduct is judged according to the reasonable person standard, which is an objective (not subjective) standard. In other words, the question is asked: What would a reasonable person have done in the same or similar situation? In the case of instructors, what constitutes "reasonableness" is gleaned from the department's policies and procedures, national standards, and federal and state law. For this reason, it is important that you remain apprised of changes in the laws that affect the fire service. It is also important that any changes in the law, technology, and the national guidelines be considered and incorporated into the policies and procedures of the fire department when appropriate.

Misfeasance is the most commonly encountered type of wrong in the performance of otherwise lawful acts. Simply defined, misfeasance consists of improper or wrongful performance of a lawful act without intent through mistake or carelessness. An example would be the hiring of a family member—a practice that, unbeknownst to you, is a violation of the department's nepotism policy.

Most states offer some level of protection to public employees against lawsuits premised on claims of negligence or misfeasance. These protections often take the form of immunity laws that bar certain causes of action outright or require a complainant to allege and substantiate a higher level of proof. The higher standard of proof usually requires that an element similar to intent be proved to prevail on a claim. Quite often, the highest standard requires proof of willful and wanton conduct or establishment of gross negligence. Gross negligence could be found, for example, if an instructor fails to follow a written protocol or a standing order. For example, a standing order might require personnel to wear protective equipment when dealing with a live fire; failure to do so could constitute gross negligence. The purpose of the higher standard of proof is to allow public employees to perform their jobs without fear that every little mistake will expose them to potential liability.

Willful and wanton conduct is defined by some courts as an act that, if not intentional, shows an utter indifference or conscious disregard for the safety of others. For example, it might include the failure to follow a written protocol or a standing order such that this omission creates a substantial risk to another person. That standard can be established if it is shown that, after having knowledge of an impending danger, a person failed to exercise ordinary care to prevent the injury. In addition, if, through recklessness or carelessness, a danger is not discovered by the use of ordinary care, the willful and wanton conduct standard may be satisfied.

Similarly, **gross negligence** is defined as an act (or a failure to act) that is so reckless it shows a conscious, voluntary disregard for the safety of others. Gross negligence involves extreme conduct but not intentional conduct. It falls somewhere between a mere inadvertent act that causes harm and acting with an intent to harm.

Whether a particular act constitutes willful and wanton conduct or gross negligence depends on the facts of the situation. The most common evidence considered when analyzing whether an act is willful and wanton are your departmental policies and procedures, surrounding departments' policies and procedures, training records, and national guidelines or standards. It is vital that your policies and procedures are reviewed annually to ensure that they are current and up-to-date with national standards set for the fire service. The same is true for the training evaluations used. Prior to revising an evaluation, it is important to verify that it still satisfies the standards of the fire service.

Legal exposures exist in many training sessions that you will teach or in which you will participate. In the event of almost any accident or injury, the training records and training history of the participants may be examined by attorneys and investigators in an attempt to identify any potential errors or omissions in the training and educational process. Many instructors fail to realize the importance of the documentation part of the training process. When shoddy records are reviewed during investigations, the professional reputation of the instructor may suffer—not to mention that such lackluster recordkeeping presents the potential for litigation under the heading of "failure to train" or "inadequate training." Training development from established reference sources based on up-to-date SOPs is fundamental to ensuring legally defensible training development.

COMPANY-LEVEL INSTRUCTOR TIP

There is no substitute for a consultation with an attorney who is knowledgeable in the law or for the applicable industry standards on training safety such as the NFPA.

Malfeasance

Malfeasance, unlike negligence and misfeasance, is more than a simple mistake or accident. Most often malfeasance is associated with public officials who partake in unlawful conduct even though they know that the conduct is illegal and contrary to their duties as public employees. Such conduct usually results in criminal charges but could be made part of a civil suit. An example of malfeasance would be the acceptance of a bribe by an instructor in return for a passing grade.

Many of the NFPA standards in the 1400 series provide guidance in the application of different types of training exercises. For example, NFPA 1451: *Standard for a Fire and Emergency Service Vehicle Operations Training Program* provides information about the development and delivery of a vehicle training program. Within that standard and others such as NFPA 1500: *Standard on Fire Department Occupational Safety, Health, and Wellness Program*, it is required that training on vehicle driving take place before the fire fighter is allowed to respond to emergencies as a driver. If the department does not adopt that policy as a standard practice, the failure to do so might be held against the department in a lawsuit.

Most of the safety protocols that govern fire fighter training are embedded in laws that describe broader topics, such as Occupational Safety and Health Administration (OSHA) rules on respiratory protection (or SCBA) and hazardous materials. The NFPA standard specific to conducting live fire training safely is NFPA 1403: *Standard on Live Fire Training Evolutions*, which addresses conducting live fire training in acquired structures, gas-fired live fire training structures, non-gas-fired live fire training structures, exterior prop fires, and Class B fuel fires. Many of the requirements for these exercises must also be applied to skills training.

NFPA 1403 was developed as a result of fire fighter fatalities and injuries in live fire training accidents. This standard is almost entirely devoted to practices that must be followed to conduct live fire training safely. Although NFPA standards are not laws, you can be held accountable for adhering to them just the same. For example, a training officer was recently held criminally liable for a fire fighter fatality that occurred during live fire training because he did not follow the guidelines in NFPA 1403 when conducting the drill. Strictly adhering to all recognized standards and laws for the type of training you are conducting will ensure that such training is conducted in a manner that will be safe for all participants.

If the training will involve participants from other fire departments, additional documentation should be obtained for those fire fighters. Moreover, your fire department should enter into a **hold harmless** and **indemnification agreement** with any other department or agency involved in the program. Such an agreement provides that your fire department will not be held monetarily or legally liable for injuries or conduct of the other department or agency's employees. This document is necessary to protect your fire department from individuals over whom you have no control. As a representative of the hosting fire department, you should also verify that all outside participants are medically cleared to participate in the training, have met the necessary prerequisites, and possess the proper credentials.

Off-site training is filled with additional risks not encountered with on-site training. You should perform a site inspection well in advance of scheduling the training and prior to development of any handouts to identify any unusual or potentially unsafe conditions. In addition, contact should be made with all other entities (town, city, utility companies) to verify that the proposed training does not violate any law or create any unreasonable risks. For example, you should confirm that no ordinances or local laws prohibit the proposed activity. Document each conversation that you have with other entities as well as the steps you take after each conversation.

Laws and standards are intended to keep fire fighter training within a framework of safe practices. Some critics might suggest that the requirements place so many restrictions on the training that it is no longer realistic or relevant. This is simply not true. First and foremost, instructors have a responsibility to keep the training safe and to follow procedures (**TABLE 7-2**). The notion that you have to discard safety requirements to make the training realistic is contrary to the entire idea behind conducting training sessions. The standards, rules, and laws established for safe live fire activities must be adhered to during skills training, as well.

TABLE 7-2 General Training Safety Procedures

Training Location	Procedures
Classroom training	Make sure all applicable department standard operating guidelines (SOGs) are referenced within your lesson plan.
	Ensure all participants are aware of facility safety procedures (e.g., fire alarm, smoking, restricted access areas).
	Identify trip, slip/fall, and other facility hazards.
Hands-on training	Specify all learning objectives and performances. Use incident action plans as appropriate for high-risk training such as live fire evolutions.
	State all required PPE for instructors and students.
	Identify all department procedures for skills being performed (use the same incident command system as required during actual operations).
	Identify and demonstrate any hazard control activities that have been taken or are in place.
	Review all applicable safety procedures (e.g., emergency procedures, mayday, backup assignments, communications, rehab, emergency medical services [EMS] standby or aid, accountability, and rapid intervention teams).
	Check all communications equipment.
	Follow all applicable standards for the training session (NFPA 1403 for live fire training).
Keys to safe and successful training	Identify companies/members who are in or out of service.
	Identify and document the lead instructor.
	Notify supervising officers of all training events.
	Brief all participants on all training expectations.
	Monitor the previously mentioned practical training safety factors throughout (act as instructor/safety officer/incident commander).
	Immediately stop all unsafe acts or conditions; do not try to train through them.
	Debrief all participants on how training expectations were met and identify opportunities to improve individual and company performance.
	Return all equipment and personnel to service refueled, recharged, rehabbed, and ready for the next response.

Training BULLETIN

JONES & BARTLETT FIRE DISTRICT
TRAINING DIVISION
5 Wall Street, Burlington, MA, 01803
Phone 978-443-5000 Fax 978-443-8000

Instant Applications: Safety During the Learning Process

Drill Assignment

Apply the chapter content to your department's operation, its training division, and your personal experiences to complete the following questions and activities.

Objective

Upon completion of the instant applications, fire and emergency services instructor students will exhibit decision making and application of job performance requirements of the instructor using the text, class discussion, and their own personal experiences.

Suggested Drill Applications

1. Review local policies relating to safety in training.
2. Read NFPA 1403: *Standard on Live Fire Training Evolutions*, and review the sample checklists used for conducting live fire training exercises.
3. Research fire fighter line-of-duty deaths in training accidents and review the recommendations of the investigators regarding ways to prevent similar occurrences.
4. Review the Incident Report in this chapter and be prepared to discuss your analysis of the incident from a training perspective and as an instructor who wishes to use the report as a training tool.

Incident Report: NIOSH Report F2005-31

Drill Assignment

Review the information in this incident report and prepare a practice presentation for class delivery at the direction of your instructor. Your presentation should include a summary of the incident facts and a review of the NFPA 1403 compliance and noncompliance findings. Use an outline to organize your thoughts. You may be evaluated on your communication skills during your presentation.

Pennsylvania State Fire Academy—2006

The Pennsylvania State Fire Academy was conducting a Suppression Instructor Development Program (ZFID), an instructor development program for live fire training. The course covers instructional components including academy policies, NFPA 1403, building fires in the permanent live fire training facility, the instructor's role in student safety, emergency and rescue operations, and more. Instructors taking the course are already Instructor I and Fire Fighter II certified, and have completed the incident safety officer course. Candidates act as students, ignition officers, incident safety officers, incident commanders, and instructors, with an academy instructor evaluating each participant's performance.

The 2½-story residential non-gas-fired structure used for live fire training was built in 1993 and is used for structural firefighting training (**FIGURE 7-A**). The building has a basement, ground floor, second floor, and an attic space. Burn rooms are located on the second floor and in the basement, where this incident occurred (**FIGURE 7-B**). The burn room in the basement has only one entrance, which can be accessed by either the interior or exterior stairwell. Only wooden pallets and excelsior are used for fuel.

As part of the ZFID program, six scripted (planned) evolutions were organized in the burn building. Certain problems or emergencies were planned, and while the candidates had a basic understanding of the types of events, they did not know which event would occur during each evolution. The candidates were evaluated on their ability to handle the events.

The fire fighter who lost his life in this incident had already served as the evaluator of an instructor candidate in previous evolutions that day. On this last evolution, he was assigned to be the ignition officer for the burn room in the basement. He had more than 30 years in the fire service and had been an adjunct instructor at the facility for 7 years. He was wearing full protective clothing, including coat, pants, helmet, hood, gloves, boots, and

FIGURE 7-A The Pennsylvania State Fire Academy's non-gas-fired live fire training structure.
From the NIOSH Fire Fighter Fatality Investigation and Prevention Program, Investigation Report F2005-31.

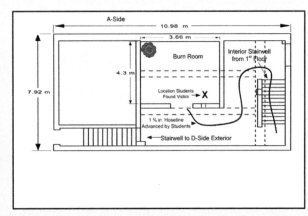

FIGURE 7-B The floor plan of the basement in the non-gas-fired live fire training structure at the Pennsylvania State Fire Academy.
From the NIOSH Fire Fighter Fatality Investigation and Prevention Program, Investigation Report F2005-31.

Incident Report continued

SCBA with integrated personal alert safety system (PASS), and he had a portable radio.

The fatal event occurred during the last evolution. The scenario was to simulate a fire that had not been knocked down enough from the previous evolution.

An instructor waiting to monitor the last crew was in the burn building. He reported good visibility with some smoke. While waiting, he heard the basement door open and saw the victim come out of the burn room and lie on the floor, pulling and moving his bunker gear about. The instructor asked the victim twice if he was okay. The victim answered, "It is hot as hell down there!" The instructor asked the victim if he was okay and if he wanted to go outside. Although the victim responded that he was okay, he was told to go outside. The victim declined, got up, and said that he was fine and would see the other instructor in the basement when the crew arrived. The victim went back down the basement steps. (Note: An estimated 1½ minutes elapsed from the time the victim came up the stairwell until he went back down to the basement.)

It is believed that the victim returned to the burn room and was carrying pallets to add to the fire. In NIOSH's report and following National Institute of Standards and Technology's (NIST) experiments, investigators hypothesized that excessive heat in the burn room caused a catastrophic failure of the lens of the victim's SCBA face piece. The face piece failure was a result of the heat conditions within the burn room and not a manufacturer's problem. The victim dropped the pallets and fell forward. The victim then struggled on the ground and crawled toward the exit.

As the next crew approached the burn room while advancing a hose line, they reportedly heard the victim moaning. They found him struggling on the floor in the right corner of the burn room. One of the candidates declared a mayday by radio, and the candidate serving as the incident commander deployed the rapid intervention crew (RIC). There was initial confusion whether this was part of the scenario or an actual emergency. The crew who found the victim immediately removed him from the burn room to outside, and emergency care was initiated. He was transported initially to a local hospital and then transferred by emergency medical services (EMS) helicopter to a regional trauma/burn unit, where he died 2 days later.

This was the first fatality of an instructor in a permanent live fire training prop, and only the second in such a facility. The fire fighters did not report the PASS device activating, as the victim was still moving. In this case, the PASS device did sound once removed from the victim; due to heat damage it could not be shut off, continuing to sound its alarm 2 days later.

Postincident Analysis: Pennsylvania State Fire Academy

Key Recommendations

- Ensure that two training officers are present with a charged hoseline during the ignition or refueling of a training fire in accordance with **NFPA 1403**.
- Fire departments and training academies should use the minimum fuel load necessary to conduct live fire training (Ref 1403-7.3).
- An instructor shall not serve as an ignition officer for more than one evolution in a row (note, this is not in the report recommendations).

NFPA 1403 Compliant

- Structure in good condition
- Instructors had experience in the fire service and with live fire training
- RIC crew ready
- An ambulance staffed by two EMTs on scene

Other Contributing Factors

- Instructor working alone
- Repeated entries by instructor
- No monitoring of interior temperatures
- No method to rapidly ventilate building
- Basement with no "at grade" exit

After-Action REVIEW

IN SUMMARY

- You must always place students' safety ahead of any performance expectations for the training session.
- The 16 fire fighter life safety initiatives should provide both guidance and motivation to you while you are developing and conducting any training. They highlight the importance of training as a strategy that can help reduce the number of fire fighter line-of-duty deaths.
- As the instructor, you serve as a role model for students. For this reason, you should always reinforce safety policies as part of the teaching process.
- Safety during training begins with safety in the classroom.
- In hands-on training, risk management practices must be applied to limit, reduce, or eliminate risk.
- As an instructor, you must always wear the PPE required for the drill being conducted. The appropriate level of PPE depends on both the potential hazards and your department's SOPs.
- Training should mirror as closely as possible the actual fire-ground operation. Medical care and monitoring, hydration, and cooling/warming materials are just a few of the rehabilitation considerations that are required to be available during live fire training events.
- Knowledge of the department's safety policies and the manufacturer's recommendations for any equipment is the key to keeping students safe during training.
- The example that you set as an instructor sends a far more powerful message to your students than the theories and concepts you present in a lecture or drill.
- You are responsible for inspecting the equipment being used in training to ensure that it meets the required standards prior to conducting the drill. Do not assume that the equipment is in safe and operable condition.
- Live fire training is the pinnacle of all firefighting training, but it is also one of the most hazardous drills.
- NFPA 1403: *Standard on Live Fire Training Evolutions* is the nationally accepted standard for all live fire burn training.
- NFPA 1584: *Standard on the Rehabilitation Process for Members During Emergency Operations and Training Exercises* requires that rehabilitation protocols be practiced at training events that involve strenuous or prolonged operations.
- Instructors are responsible for both the physical and emotional well-being of students while in training.
- Instructors are held accountable for their planning and adherence to standards and SOPs during the training session.

KEY TERMS

Access Navigate for flashcards to test your key term knowledge.

Gross negligence An act, or a failure to act, that is so reckless that it shows a conscious, voluntary disregard for the safety of others.

Hold harmless An agreement or contract wherein one party holds the other party free from responsibility for liability or damage that could arise from the transaction between the two parties.

Ignition officer An individual who is responsible for igniting and controlling the material being burned at a live fire training.

Indemnification agreement An agreement or contract wherein one party assumes liability.

Instructor-in-charge An individual qualified as an instructor and designated by the authority having jurisdiction to be in charge of the live fire training evolution and who has met the requirements of an Instructor II in accordance with NFPA 1041 (NFPA 1403).

Liability Responsibility; the assignment of blame. It often occurs after a breach of duty.

Malfeasance Dishonest, intentionally illegal, or immoral actions.

Misfeasance Mistaken, careless, or inadvertent actions that result in a violation of law.

Negligence An unintentional breach of duty that is the proximate cause of harm.

Safety officer An individual appointed by the authority having jurisdiction as qualified to maintain a safe working environment at all live fire training evolutions (NFPA 1403).

Willful and wanton conduct An act that shows utter indifference or conscious disregard for the safety of others.

REFERENCES

National Fallen Firefighters Foundation. 2013. "16 Firefighter Life Safety Initiatives." Emmitsburg, MD: National Fallen Firefighters Foundation. http://www.lifesafetyinitiatives.com/.

National Fire Protection Association. 2015. *NFPA 1584: Standard on the Rehabilitation Process for Members During Emergency Operations and Training Exercises*. Quincy, MA: National Fire Protection Association.

National Fire Protection Association. 2018. *NFPA 1403: Standard on Live Fire Training Evolutions*. Quincy, MA: National Fire Protection Association.

National Fire Protection Association. 2018. *NFPA 1451: Standard for a Fire and Emergency Service Vehicle Operations Training Program*. Quincy, MA: National Fire Protection Association.

National Fire Protection Association. 2018. *NFPA 1500: Standard on Fire Department Occupational Safety, Health, and Wellness Program*. Quincy, MA: National Fire Protection Association.

National Fire Protection Association. 2018. *NFPA 1582: Standard on Comprehensive Occupational Medical Program for Fire Departments*. Quincy, MA: National Fire Protection Association.

National Fire Protection Association. 2019. *NFPA 1041: Standard for Fire and Emergency Services Instructor Professional Qualifications*. Quincy, MA: National Fire Protection Association.

National Institute for Occupational Safety and Health. 2017. "Preventing Deaths and Injuries of Fire Fighters During Training Exercises." Workplace Solutions, no. 2017-113. https://www.cdc.gov/niosh/docs/wp-solutions/2017-113/pdfs/2017-113.pdf?id=. Accessed October 12, 2018.

CHAPTER PRESENTATION EXERCISE

Instructions: Your course instructor will assign individual or group discussions on the key points and teaching tips of this chapter. You or your group will review the chapter teachings and identify the major learning points from the chapter. You should be able to discuss the points, why they are important, and how and where they apply to your responsibilities at your level of instructor training.

REVIEW QUESTIONS

1. Regarding the 16 Fire Fighter Life Safety Initiatives, which initiative(s) relate to training safety and live fire training? Explain the instructor's role in each one.

2. Name five conditions for which additional instructors should be employed at training sessions.

3. Explain why negligence and misfeasance are "kindred spirits."

CHAPTER 8

Evaluating the Learning Process

NFPA 1041 JOB PERFORMANCE REQUIREMENTS

Note: An asterisk denotes that the 1041 standard contains further information in its annex section.

4.5.2 Administer oral, written, and performance tests, given the lesson plan, evaluation instruments, and evaluation procedures of the AHJ, so that bias or discrimination is eliminated, the testing is conducted according to procedures, and the security of the materials is maintained.

(A) Requisite Knowledge.

Test administration, laws and policies pertaining to discrimination during training and testing, methods for eliminating testing bias, laws affecting records and disclosure of training information, purposes of evaluation and testing, and performance skills evaluation.

(B) Requisite Skills.

Use of skills checklists and assessment techniques.

4.5.3 Grade student oral, written, or performance tests, given class answer sheets or skills checklists and appropriate answer keys, so the examinations are accurately graded and properly secured.

(A) Requisite Knowledge.

Grading methods, methods for eliminating bias during grading, and maintaining confidentiality of scores.

(B) Requisite Skills.

None required.

4.5.4 Report test results, given a set of test answer sheets or skills checklists, a report form, and policies and procedures for reporting, so that the results are accurately recorded, the forms are forwarded according to procedure, and unusual circumstances are reported.

(A) Requisite Knowledge.

Reporting procedures and the interpretation of test results.

(B) Requisite Skills.

Communication skills and basic coaching.

4.5.5* Provide evaluation feedback to students, given evaluation data, so that the feedback is timely; specific enough for the student to make efforts to modify behavior; and objective, clear, and relevant; also include suggestions based on the data.

(A) Requisite Knowledge.

Reporting procedures and the interpretation of test results.

(B) Requisite Skills.

Communication skills and basic coaching.

KNOWLEDGE OBJECTIVES

After studying this chapter, participating in a structured learning environment, and completing assigned assessments, you will be able to:

- Explain the reasons for conducting evaluations of learning. (**NFPA 1041: 4.5.1**, p 185)
- Describe the legal considerations for testing. (**NFPA 1041: 4.5.2**, p 185)
- Describe standard testing procedures. (**NFPA 1041: 4.5.2**, p 186)

- Describe common test administration procedures. (**NFPA 1041: 4.5.2**, pp 186–187)
- Identify considerations for maintaining test security. (**NFPA 1041: 4.5.2**, pp 186–192)
- Describe the methods used to grade evaluations. (**NFPA 1041: 4.5.3**, p 192)
- Explain the purpose of feedback. (**NFPA 1041: 4.5.5**, p 193)
- Identify the information that should be provided to students about their performance on evaluations. (**NFPA 1041: 4.5.5**, p 193)
- Explain the reasons that accurate reporting of evaluations is important. (**NFPA 1041: 4.5.4**, pp 192–193)

SKILLS OBJECTIVES

After studying this chapter, participating in a structured learning environment, and completing assigned assessments, you will be able to:

- Demonstrate methods of administering written, oral, and performance evaluations. (**NFPA 1041: 4.5.2**, pp 186–187)
- Demonstrate how to grade student evaluation instruments. (**NFPA 1041: 4.5.3**, p 192)
- Demonstrate how the results of an evaluation are recorded. (**NFPA 1041: 4.5.2, 4.5.4**, p 192)
- Demonstrate the methods for providing feedback on evaluation performance to students. (**NFPA 1041: 4.5.5**, p 193)

You Are the Fire and Emergency Services Instructor

As a fire instructor currently assigned to your fire training academy, you have been tasked by the academy director to administer and grade the next series of written examinations for the current schedule of classes. Your academy provides training for both recruit and incumbent fire fighters, so the curriculum used is quite broad and includes both classroom and drill-ground instruction.

1. Describe the different methods for administering written examinations.
2. Describe how written examinations are graded and how the results are recorded.
3. Describe methods for providing feedback on written examination results directly to the individual students.

 *Access Navigate for more practice activities.*

Introduction

Testing plays a vital role in training and educating fire and emergency services personnel. A sound testing program allows you to know whether students are progressing satisfactorily, whether they have learned and mastered learning objectives, and whether your instruction is effective. Without a sound testing program, there is no way to assess student progress or to determine which learning objectives have been mastered. In addition, it is impossible to determine the effectiveness and quality of training programs without evaluation.

Fire departments across the country continue to encounter problems with testing programs. Common problems in testing include the following issues:

- Lack of standardized test specifications and test format examples
- Confusing procedures and guidelines for test development, review, and approval

- Lack of consistency and standardization in the application of testing technology
- Failure to perform formal test-item and test analysis
- Inadequate instructor training in testing technology

This chapter is designed to provide the necessary information and training to help overcome these and other common weaknesses, thereby strengthening your testing program.

Legal Considerations for Testing

Testing for completion of training programs, job entry, certification, and licensure is governed by certain legal requirements and professional standards in the United States. These legal requirements and professional standards are readily available and should be important reference documents for anyone involved in the development of test items, construction of tests, administration of tests, analysis and improvement of tests, and maintenance of testing records. Because each state and municipality will have different laws and ordinances, be sure to establish a working knowledge of applicable laws and standards relating to testing and evaluation. As the instructor, you should be aware of laws and standards that apply to some of the following areas:

- Entry-level testing for new candidates for written and physical ability testing
- Certification testing rules and test candidate requirements
- Promotional testing criteria and the evaluation of promotional test results
- Continuing education testing requirements for emergency medical services (EMS), firefighting, and other certifications or licenses
- Recordkeeping, scoring, and test record storage

Among the most important reference documents are *Uniform Guidelines for Employee Selection*, published by the Equal Employment Opportunity Commission; *Standards for Educational and Psychological Testing*, published by the American Psychological Association (American Educational Resource Association, American Psychological Association, and National Council on Measurement in Education, 1999); and *Family Educational Rights and Privacy Act*, published by the U.S. Printing Office and other governmental organizations (U.S. Department of Education, 2011). Each of these references should be available in every fire department's training division library for instructors to study.

Purposes and Types of Tests

Tests should be used for the following three purposes:

1. To measure student attainment of learning objectives
2. To determine weaknesses and gaps in the training program
3. To enhance and improve training programs by positively influencing the revision of training materials and the improvement of instructor performance

The three basic types of tests are written, oral, and performance.

Written Tests

Written tests can be made up of the following types of test items:

- Multiple choice
- Matching (a form of multiple choice)
- Arrangement
- Identification
- Completion
- True/false
- Essay

The multiple choice exam is typically the most common form of written test because it is the most objective and thus the most reliable for testing; however, because there are many different learning styles, students should be given multiple formats in which to demonstrate their competency. The various types of written tests are discussed later in this chapter in the Written Test Items section.

Performance Tests

A **performance test**, also known as a **skills evaluation**, measures a student's ability to perform a task. This test category includes laboratory exercises, scenarios, and job performance requirements (JPRs).

Here is an example of a performance test item:

Perform a daily inspection of self-contained breathing apparatus (SCBA).

Any performance test should be developed in accordance with task analysis information and reviewed by a **subject-matter expert (SME)** to ensure technical accuracy in terms of the actual work conditions, and should be accompanied by a checklist detailing the steps of the skill in order.

Oral Tests

Oral tests, in which the answers are spoken either in response to an essay-type question (oral content)

FIGURE 8-1 Oral tests assess knowledge, skills, and abilities before a performance test.
© Jones & Bartlett Learning. Photographed by Glen E. Ellman.

or along with a presentation or demonstration (oral presentation), are not widely used in the fire service specifically; however, this type of test has a place in technical and emergency services training. Oral tests are given in a structured and standardized manner to determine students' verbal response to assess knowledge, skills, and abilities important on the job (**FIGURE 8-1**). They primarily focus on safety-related skills to be performed. The oral test allows students to clarify answers and you to clarify questions. Oral tests are effective in ascertaining student knowledge and understanding when these aspects are not conveniently measured through other types of tests. An example of an oral test follows:

> *Describe the steps in the proper order that are required to properly engage a fire pump and begin the flow of water through a single 2½-in. port.*

Oral tests used in conjunction with performance tests should focus on critical performance elements and key safety factors. The overhead or directed questioning technique seems to work best when a group of students are preparing to take a performance test. Specific oral test questions can be used when a student is taking a performance test. Oral tests used in conjunction with performance tests are usually not graded, but rather are designed to reinforce key safety factors for students and to ensure critical safety factors are being confirmed by the instructor/proctor.

Standard Testing Procedures

Within a training department, the test development activities should adhere to a common set of procedures. These procedures should be included in standard operating procedure (SOP) or standard operating guideline (SOG) formats. These valuable documents should offer test development concepts, rules, suggestions, and format examples. You should use the procedures and guidelines specified in the SOP/SOG document throughout the test-item development, test construction, test administration, and test-item analysis processes.

Proctoring Tests

You will gain both knowledge and skills in test-item writing and test development when you complete the Instructor II requirements. Now, you should explore the other aspects of a solid testing program. Proctoring, or administering, tests encompasses much more than just being present in the testing environment; it requires specific skills to be performed professionally. Types of tests being proctored require different proctoring skills and abilities.

Proctoring Written Tests

The following are some suggested procedures for proctoring written tests:

- Arrive at least 30 minutes prior to the beginning time for the test to ensure the evaluation area is properly set up and everything is in order.
- Make sure the testing environment is suitable in terms of lighting, temperature control, adequate space, and other related items.
- During the arrival of test takers, double-check those who should be in attendance by checking identification documents and recording their presence. (You may develop a log or other means to document arrival time and departure time.)
- Provide specific written and oral guidance of test-taking rules and behavior guidelines for the testing period. Maintain order in the testing facility, discouraging any activities that might potentially be interpreted as cheating. Discipline anyone who becomes disruptive or violates the rules.
- Remain objective with all test takers. Do not show any form of favoritism.
- Answer questions about the testing process and the test itself.
- Do not answer an individual question about the content of the test unless that information is also shared with the entire group.
- Monitor test takers during the entire period of the test; never leave the testing environment for any reason.
- Require all test-item challenges to be made prior to test takers leaving the room. Advise that their challenge will be noted and addressed after all of the tests are scored.

- Collect and double-check answer sheets or booklets to ensure that all information is properly entered and that any supporting materials are returned before allowing test takers to leave the room.
- Maintain the security of testing materials at all times. Inventory all testing materials and return them to the designated person in the department when you are done with them.

Proctoring Oral Tests

Oral tests are generally given in conjunction with performance tests. Even so, there are preferred procedures that need to be followed to maximize the effectiveness of the oral test:

- Determine the oral test items to be used and the type of assessment technique that will be employed (e.g., overhead, direct).
- Make sure the oral test items are pertinent to the task to be performed.
- Focus on the critical safety items and dangerous steps within the performance.
- If the test taker misses the oral test item, redirect the question to another performer. If you are conducting a one-on-one oral test, then provide the test taker with on-the-spot instruction. Safety knowledge and dangerous task performance steps must be clear and mastered or corrected before performance begins.
- Use a scoring guide for the oral test and record any difficulties and lack of knowledge on the part of the test taker(s).
- If the group or individual has knowledge gaps regarding the critical safety items or dangerous performance steps, *do not* proceed to the performance test. More training and testing are required.
- Make a training record of oral test results using a prepared list of questions and including space for the evaluator to add comments and notes.

> **COMPANY-LEVEL INSTRUCTOR TIP**
>
> It is important to outline the required responses to oral test questions to ensure that proper feedback is given to the student.

Proctoring Performance Tests

Performance test proctoring requires considerably different skills than those needed to proctor the written and oral test. Proctors must have keen observational skills, technical competence for the task(s) being performed, ability to record specific test observations (including deficiency and outstanding performances), ability to foresee critical dangers that may lead to injury to the performer or others, and an objective relationship to the test takers (**FIGURE 8-2**). Following are some suggested procedures, including pre-test preparations:

- Arrive at the test site approximately 1 hour before the performance test to assess the test environment. Make sure all needed tools, materials, and props are present and in good working order.
- Determine whether the performance test will require an oral test before the actual performance of the task. (Obtain oral test items if required.)
- Check the test takers' identification as they arrive and record their presence.
- Review the test procedures and ask for questions from the test takers.
- Verify that all test takers know what will be expected during performance.
- Begin the performance test. Have the students tell you exactly what they will be doing before they undertake each step in the performance. (This is a critical step—it can help prevent an accident or unsafe performance before it happens.) In certification testing, missing safety-related steps or not knowing critical performances that may cause injury or damage to equipment is cause to terminate testing and refer the test taker for more training. This should also be true for the final performance test in a training program that leads to licensure.
- Remain silent and uninvolved in the task performance. Your job is to verify safe and competent performance.
- Record test results using standard protocols. You may provide technical information and critique the performance if your institution or agency permits proctor interaction after the test.
- Provide test results to the person(s) designated to receive them.

Written Test Items

Components of a Test Item

Test items are made up of components that create a testing tool. An important part of the question is the reference material that documents the standard the question is based on and the source of the information in a textbook.

Candidate: _____ Date: _____

ID#: _____

Skill Drill 16-25

Fire Fighter I, 5.5.2

Performing an Accordion Hose Load

Evaluator Instructions: The candidate shall be provided with a fire apparatus, supply hose, and gloves.

Task: Performing an accordion hose load.

Performance Outcome: The candidate shall demonstrate the ability to do an accordion hose load.

Candidate Directive: "Properly perform an accordion hose load."

No.	Task Steps	First Test		Retest	
		P	F	P	F
1.	Determines whether hose will be used for forward lay or reverse lay.				
2.	For forward lay, places male hose coupling in hose bed first. For reverse lay, places female hose coupling in hose bed first.				
3.	Starts hose lay with coupling at front end of hose compartment.				
4.	Lays first length of hose in hose bed on its edge against side of hose bed.				
5.	Doubles hose back on itself at rear of hose bed. Leaves female end extended so two hose beds can be cross connected.				
6.	Lays hose next to first length and brings it to front of hose bed.				
7.	Folds hose at front of hose bed so bend is even to edge of hose bed. Continues to lay folds of hose across hose bed.				
8.	Alternates lengths of hose folds at each end to allow more room for folded ends.				
9.	When bottom layer is completed, angles hose upward to begin second tier.				
10.	Continues second layer by repeating steps used to complete first layer.				

Retest Approved By: _____   Retest Evaluation: _____

Evaluator Comments: _____   Candidate Comments: _____

_____   _____

_____   _____

_____   _____

_____   _____

Evaluator _____ Date _____   Candidate _____ Date _____

Retest Evaluator _____ Date _____   Retest Candidate _____ Date _____

FIGURE 8-2 A skills evaluation sheet can be a useful tool in evaluating a performance test.

© Jones & Bartlett Learning.

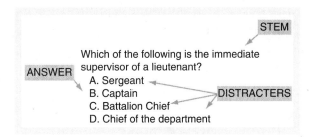

FIGURE 8-3 Multiple choice test items are made up of three components: the stem, the answer, and the distracter(s).
© Jones & Bartlett Learning.

Fire department synthetic rope, new or used, should be kept

A. warm, dry, and in a sunlit place.
B. cool, moist, and in a sunlit place.
C. neatly coiled and stored out of sunlight.
D. damp and in a dark place.

FIGURE 8-4 An example of a properly formatted multiple choice test item.
© Jones & Bartlett Learning.

Multiple Choice Test Items

In a multiple choice test item, the part of the question that asks the information is known as the stem (**FIGURE 8-3**). The choices are made up of a correct answer and distracters. The multiple choice test item is the most widely used in objective testing. This type of item also contributes to high test **validity** and **reliability** estimates. The multiple choice test item is a favorite among test-item developers because it comes nearer to incorporating the important qualities of a good test than any other type of test item. These qualities include content validity, reliability, objectivity, adequacy, practicality, and utility. Ease of grading of the test items and providing immediate feedback to students make multiple choice items extremely flexible for the instructor. The multiple choice test item is so versatile that it can be used to measure almost any cognitive information. The disadvantage is that it may take several well-written test items to cover the learning objective properly. Another disadvantage is the common occurrence of misleading or poorly written distractors. **FIGURE 8-4** provides an example of a properly formatted multiple choice test item.

While the multiple choice question is the most popular type of test item, it is essential to determine the testing method that best ensures competency related to the learning objective rather than to simply fall back on habit.

Matching Test Items

Matching test items are actually another form of multiple choice testing. This type of question is limited in use, but functions well for measuring such things as knowledge of technical terms and functions of equipment. Care should be taken during preparation of matching test items because all information must be factual. Typically, there are no more than four items to match with five choices when providing matches. **FIGURE 8-5** provides an example of a properly formatted matching test item.

Directions: Match the terms listed in column A with the definition provided in column B.

Column A	Column B
1. Oxidation 2. Conduction 3. Convection 4. Radiant heat	A. Rapid oxidation with heat and light B. Oxygen combining with other elements C. Heat energy carried by electromagnetic waves D. Heat transfer by circulation through a medium E. Heat transfer by direct contact

FIGURE 8-5 An example of a properly formatted matching test item.
© Jones & Bartlett Learning.

The matching test item is particularly useful when a question requires multiple responses or when there are no logical distracters. The matching test item is very useful for low-level cognitive information measurement.

Arrangement Test Items

The arrangement test item is an efficient way of measuring the application of procedures for such things as disassembly and assembly of parts, start-up or shutdown, emergency responses, or other situations where knowing a step-by-step procedure is critical. An arrangement test requires the student to place things in the correct order, sometimes using photos or diagrams. These test items lose strength and efficiency steadily as time lags between the paper-and-pencil solution and the actual performance of the procedure. You should plan to administer an arrangement test shortly before an opportunity to perform the actual procedure, disassembly, or assembly.

One disadvantage of arrangement test items is inconsistent grading. A student may miss the order of one step and thus miss the order of all of the steps that follow, even though the remaining steps may be in the

Directions: (*This is a two-part question worth 3 points.*)
The six steps of the basic method that should be used in fire or explosion investigations appear out of order below. **First,** number the steps in the proper order by placing numbers (1–6) in the blanks beside the steps. **Next,** select the answer that matches yours from choices A–D.

_____ Conduct the investigation.
_____ Collect and preserve evidence.
_____ Receive the assignment.
_____ Analyze the incident.
_____ Prepare for the investigation.
_____ Report findings.

A. 1, 4, 5, 2, 3, 6   B. 2, 4, 5, 1, 3, 6
C. 6, 3, 4, 5, 1, 2   D. 3, 4, 1, 5, 2, 6

Answer: D

FIGURE 8-6 A properly formatted arrangement test item.
© Jones & Bartlett Learning.

Directions: (*This is a two-part question worth 5 points.*)
First, label the tools depicted below by placing the number in the blanks provided.
Then, choose the answer below (A–D) that matches yours.

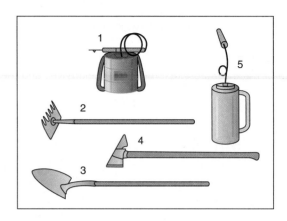

____ McLeod
____ Pulaski
____ Back pack pump
____ Shovel
____ Drip torch

A. 3, 5, 4, 2, 1
B. 2, 4, 1, 3, 5
C. 5, 3, 1, 2, 4
D. 4, 3, 1, 2, 5

Answer: B

FIGURE 8-7 An identification test item.
© Jones & Bartlett Learning.

correct sequence after the error. Another potential issue is key steps that are included in the SOP or SOG being left out of the test item. The test item should be developed in agreement with actual procedures used on the job if such procedures are required. In cases where it is important to test procedures that are longer than 10 steps, consider grouping them at key points in the procedure. Knowledge of emergency procedures, fire drills, triage, or other related procedures where knowing exactly what to do under critical conditions can be measured with high reliability using an arrangement test item. **FIGURE 8-6** provides an example of a properly formatted arrangement test item that includes a multiple choice component.

Identification Test Items

The identification test item is essentially a selection-type test using a matching technique. Its major advantage over the matching test item is the ability of the student to relate words to drawings, sketches, pictures, or graphs. These test items do require reasoning and judgment, but basically focus on the ability to recall information. Identification test items can also be considered to be arrangement test items for assembly- and disassembly-type tasks.

Identification test items provide excellent content validity and contribute to high test reliability. With today's computer publishing capabilities, digital cameras, and other technologies, however, identification questions often can be easily developed using your own fire equipment and apparatus features as the basis for the identification test item. Using this technique makes the test item more job related and makes it much easier to transfer required knowledge from the instructional setting to the job. **FIGURE 8-7** provides an example of a properly formatted identification test item.

Completion Test Items

The completion test item is easy to develop. Completion test items tend to be verbatim from the technical materials, thus they tend to encourage memorization and reinforce rote learning. These test items are appropriate for entry-level positions because they measure low-level cognitive knowledge. The best practice for developing and using completion test items is to follow specific procedures and to develop two test items for every one item that is expected to be included in the final group of test items.

Completion test items are of two types: supply and selection. The supply type requires the student being tested to supply the response that completes the test-item statement. The selection type requires the student being tested to select from a list of responses for completing the test item. The supply type primarily tests the recall capability of the student, whereas the selection type focuses on recognition and analysis.

True/False Test Items

True/false test items have been controversial for years. Test item developers have experimented with numerous ways to make true/false, right/wrong, and yes/no responses more valid and reliable. The major problem with two-answer selection tests such as the true/false is the guessing factor. A student who marks a response without reading the question has a 50 percent possibility of marking the correct response. Application of two or three rules for taking a true/false test could get the student a higher score than just by guessing alone. **FIGURE 8-8** is a typical true/false test item.

A more complex true/false test item can be developed that uses a multiple choice approach to the test item. **FIGURE 8-9** provides an example of a complex true/false test item.

Directions: Circle either "T" for true or "F" for false.

A forward lead-out takes place with a supply line being dropped at the fire area and the engine proceeding to a water supply.   T   F

Answer: F

FIGURE 8-8 Typical true/false test item.
© Jones & Bartlett Learning.

Directions: Read the following statements regarding diversity programs and select your answer from choices A–D.

Statement 1: A diversity program creates an environmental and cultural change in an organization.

Statement 2: Achieving a culturally diverse organization is ethically and managerially a worthwhile accomplishment.

Statement 3: A diversity program is a single effort intended to integrate an organization to meet government mandates.

A. Statement 1 is true; statements 2 and 3 are false.
B. Statements 1 and 2 are true; statement 3 is false.
C. Statement 1 is false; statements 2 and 3 are true.
D. All three statements are true.

Answer: B

FIGURE 8-9 A complex true/false test item.
© Jones & Bartlett Learning.

Essay Test Items

Modern test developers do not consider the **essay test** to be an objective form of testing. This position is based on research in which the major weaknesses of this testing method were examined. Some of the weaknesses that affect the objectivity of essay tests are listed here:

- Writing essay questions requires a lot of the instructor's time. Initial essay questions must be refined over several uses before they will function in a reasonably reliable manner.
- Students require much more time to prepare for and write answers to essay questions. Long tests tend to be fatiguing to students and can sometimes encourage lengthy responses containing extraneous information.
- The length of the student response tends to cloud the actual knowledge required by the question with extraneous knowledge closely related but not directed to the question. Students can "beat around the bush" without ever really addressing the question.
- Poorly prepared essay test items have been shown to reinforce negative learning. This means that the answer the instructor receives tends to support the wrong interpretation of facts and conditions, even among students considered to be outstanding.
- The order in which test papers are graded affects the grading. The first papers graded tend to receive lower grades than those graded last. This is especially true when tests are long and are graded one exam at a time. Grading one question at a time for all students tends to improve the consistency of grading on an essay test.
- Grading essay tests is time-consuming.
- Grading varies widely, even among instructors who are recognized SMEs.
- The "halo effect" influences the grading practices of some instructors. The more they know about a student, either favorable or unfavorable, the more the grade can be affected.

Research has revealed that an instructor can develop an objective test (questions based on objectives) in approximately the same amount of time that it takes to develop an essay test. However, the objective test will achieve higher content validity, comprehensiveness, discrimination, and reliability than is possible with an essay test.

Putting together a valid testing instrument may be one of the most challenging parts of the entire learning process. This process requires developing good questions where a student who knows the material is able to provide the correct response, whereas a student who does not know the material is not able to provide the correct response. It also should give those students who know the material the best chance to get more correct answers than those who know much less.

To accomplish this goal, take each learning objective and create questions that demonstrate competency for that learning objective. Use various styles of questions to discern the level of comprehension. For example, writing a scenario and having students apply the material concepts is much more difficult than developing questions where a student can recite rote information from the training materials.

Cheating During an Exam

With evaluations or exams, there is a risk of cheating. How cheating is addressed should be based on department policy developed by human resources personnel, the agency's legal advisor or attorney, and the chief of the department. Cheating by an individual reflects on the character of that individual and, in many departments, can affect employment status, raises, or promotions. Guidelines related to cheating should be established and communicated clearly before the testing occurs.

In general, a policy that addresses cheating will contain procedures to follow if a student has been observed cheating. In most cases, if you observe a student cheating, you should ask the student for an explanation of his or her conduct. If you determine that the student was indeed cheating, ask the student to leave the test site entirely or give a verbal warning and continue with testing. Again, the actions taken should follow department policy.

You must document what was observed, what the student said, and what you allowed the student to do next, ideally confirmed by a third-party witness. This information should be passed to a supervisor, who will review what happened and then follow up as necessary with additional interviews of the instructor and the student involved. The supervisor will determine what happens from this point. If it is determined that the student did cheat, then department policy will address the consequences. If it is determined that the student did not cheat, then arrangements should be made to allow the student to take the test again. Regardless of the consequences, the department should have a written policy to address how to handle possible cheating during an examination. Determine whether your organization has a policy on how to handle suspicions of cheating and follow it to the letter. Failure to do so is unfair to the student and can get you into trouble.

> **COMPANY-LEVEL INSTRUCTOR TIP**
>
> The proper handling of student information, including test grades, may be covered under laws that apply to you and your department. Know your responsibilities and play it safe by maintaining strict confidence of all student information.

Grading Student Oral, Written, and Performance Tests

The proper and accurate grading of oral, written, and performance tests is an essential function of the instructor. Students will desire feedback on their performance; likewise, instructors will want to know how successful their instructional efforts were. Class answer sheets or skills checklists will need to be carefully evaluated against master answer keys and then the answer keys will need to be properly secured for course validity purposes. Graded exams should be stored properly so that completed exams cannot compromise future offerings of the same test and the confidentiality of test scores remains in place. Elimination of bias in the grading process is necessary, especially when dealing with exam question types that do not have an objective format basis in their structure. After the exams are graded and scores are recorded, do not hesitate to secure the completed exam sheets and answer keys.

Reporting Test Results
Confidentiality of Test Scores

After the completion of the testing process, student scores must be maintained for various reasons. These test results should be protected using strict security measures. Electronic results should be password protected, and hard copies should be kept safe and secure. Test results should be released only with permission of the student who has completed the testing process.

When you release test scores to a class, you should do so on an individual basis. This sharing of the scores should be done in a private session, on a one-on-one basis, or in a personal letter to each student. The practice of posting scores on a bulletin board with Social Security numbers attached to them as identifiers is unsatisfactory and illegal.

Providing Evaluation Feedback to Students

Evaluation of examinations closes with feedback on the testing results with the students. This information must be considered confidential and should be completed in a timely manner using the evaluation data available to you. All feedback should be specific enough for the students to understand the following points:

- The overall score or grade
- Objectives that were missed and those that were met
- Skills that they did not complete to the learning objective standard
- What they may do to modify their behavior and performance in the future
- Any additional study areas or retesting requirements on missed items or failed skill evaluations

Grading scales should be part of the scoring process. When receiving feedback, the student may want access to score tables or percentage breakdowns for letter grades. A written policy should exist that deals with challenges to examination scores or practical skill failures to guide you on which actions should be taken during such a challenge.

Web-Based Training and Testing

In recent years, there has been a steady and significant increase in the use of computers for training and testing purposes. Many colleges and universities encourage students to take a portion of their program off campus, preferably online. Fire departments and EMS have made great strides in incorporating distance learning technology and continue to do so.

There are many options available for putting testing online 24 hours a day, 7 days a week, through computer hardware and professionally developed test banks. One of the primary benefits of online learning and testing is the convenience for people who work full-time jobs. An employer can also extend training and testing time into the available leisure time of willing employees. In addition to the convenience factors, the increasing costs of fuel, hotels, worker salaries, and other overhead costs indicates that web-based certification is here to stay.

Training BULLETIN

JONES & BARTLETT FIRE DISTRICT
TRAINING DIVISION
5 Wall Street, Burlington, MA, 01803
Phone 978-443-5000 Fax 978-443-8000

Instant Applications: Evaluating the Learning Process

Drill Assignment

Apply the chapter content to your department's operation, its training division, and your personal experiences to complete the following questions and activities.

Objective

Upon completion of the instant applications, fire and emergency services instructor students will exhibit decision making and application of job performance requirements of the instructor using this text, class discussion, and their own personal experiences.

Suggested Drill Applications

1. Review a recent written examination and compare the student responses to the answer key. Identify any test questions that were too easy (every student answered correctly) and ones that had poor success (high percentage of failures).

2. Review your local policy or practice on the posting of test scores and student completion records.

3. Develop a strategy to improve student performance based on a poor evaluation result.

4. Review the Incident Report in this chapter and be prepared to discuss your analysis of the incident from a training perspective and as an instructor who wishes to use the report as a training tool.

Incident Report: NIOSH Report F2003-28

Drill Assignment

Review the information in this incident report and prepare a practice presentation for class delivery at the direction of your instructor. Your presentation should include a summary of the incident facts and a review of the NFPA 1403 compliance and noncompliance findings. Use an outline to organize your thoughts. You may be evaluated on your communication skills during your presentation.

Port Everglades, Florida—2003

Note: This was the first U.S. fatality in a permanent live fire training structure.

A live fire training exercise was being conducted in a prop constructed from shipping containers to represent a seafaring vessel (**FIGURE 8-A**). This was the first live fire experience for the students in a class leading to state certification as Fire Fighter II. The students were all new employees of the same fire department.

Three evolutions took place, with no breaks in between the evolutions to allow the all-metal structure to cool. The participants were instructed not to crawl over the metal grating and to avoid holding the handrails, because the metal structure had gotten very hot and there was a possibility of getting burned through their gloves. They were told that failure to complete this evolution or getting injured would mean termination from the department. There was no preburn briefing or walkthrough.

Some members of the third squad received burn injuries during their rotation, and an instructor left during the evolution claiming he had problems with his gloves.

The fire fighter who later died was a member of the fourth squad of students performing the evolution on this particular day. Five students, three instructors, and an observer with a thermal imaging camera entered the enclosed structure on the second level. As in previous evolutions, the students were sent in one by one with the intention being that they would not be able to see or hear one another until arriving at the fire box. According to statements, the instructors did not monitor the students' movement or encourage their progress. Students followed the hose line through a series of three watertight hatches, and then crossed an open-grated catwalk over the gas-fueled fire in the engine room. At the end of the grated catwalk, a combustible materials fire burned in a corner. They then proceeded down a ladder and through the simulated engine room and into the "fire box." The fire box had a raised hearth, similar to a flashover simulator, and the fire was fueled primarily by wood pallets. The students gathered in the room and took turns operating a nozzle in various patterns. They were told to avoid getting water on the fire.

The fire was knocked down, and an instructor in the burn room directed the crew to remain in the extremely hot area as the fire regained intensity. A "dead man" switch operated an open vent in the room, but it had been disabled so that it remained closed.

The instructor in the fire room was quoted by several trainees as saying the environment was too hot. He instructed the group to hurry up and "get out now." One set of students turned around to exit using the same route through which the group members had entered the room. Now out of sequence, the safety officer led the students up the ladder and over the metal grating in a very high heat environment. Meanwhile, two instructors exited through a side door directly from the fire box to an on-grade side exit, without advising the instructor-in-charge.

There are conflicting reports as to what occurred at the exit, but it was apparent that one student was missing. It is believed that he lost track of the hose, and ended up in a chase where heated gases and smoke were venting. Handprints on the walls indicated that he was trying to find a way out. While a search for him ensued, he collapsed. After seeing the downed student's

FIGURE 8-A Maritime training prop at Port Everglades, Florida.
© Jones & Bartlett Learning.

Incident Report continued

personal protective equipment (PPE), one officer entered to retrieve him, without protective clothing or SCBA, but had to abandon his attempt due to the heat. Another fire fighter wearing proper protective clothing and SCBA removed the victim to the outside. The victim's personal alert safety system (PASS) device did not sound while he was inside the structure; it is believed he was mobile until just before he was found.

The victim was unresponsive, with no pulse or respirations. After initial treatment by instructors, a medical rescue unit arrived and the victim was transported to a local trauma center, where he was pronounced dead. He reportedly had severe burns on both hands and sloughing of the skin to both knees and hands. He was described as cyanotic from the neck up. Several members of the same crew received second-degree burns to their hands and knees while they were attempting to exit the structure. One student lost consciousness after exiting and was initially cared for by other students. He and three other students were transported to the hospital.

Two separate investigations were conducted by the state fire marshal—one for criminal violations and an administrative investigation of state training codes. No criminal charges were filed, but the matter went to the state attorney. The fire department was cited by the state fire marshal for numerous code violations, and the reports cited 36 specific findings including almost total failure to follow NFPA 1403.

Postincident Analysis — Port Everglades, Florida

NFPA 1403 Noncompliant

- No written, preapproved plan
- No previous live interior firefighting training
- Safety crews did not have specific assignments
- No preburn walkthrough for the trainees
- No designated safety officer
- No communication plan
- No emergency medical services on site
- Not all instructional personnel had specific live fire training or instructor certification
- Instructors did not monitor the trainees' movement
- Noncompliance with NFPA 1402 and 1403 (both required by state code)
- No temperature monitoring
- Multiple fires inside structure burning simultaneously, including polypropylene rope and other nonorganic materials
- Lead instructor determined the environment was excessively hot, but evolution was not terminated; also failed to identify and correct safety hazards

NFPA 1403 Compliant

- Three to four instructors per squad of five trainees

Other Contributing Factors

- No emergency plans in place
- No rapid intervention crew (RIC) assignment
- Command structure unknown to students and staff
- The fire box vent had been rendered unusable

After-Action REVIEW

IN SUMMARY

- Testing plays a vital role in training and educating fire and emergency services personnel.
- Testing for completion of training programs, job entry, certification, and licensure are governed by certain legal requirements and professional standards in the United States.
- Training and testing leading to hiring, promotion, demotion, membership in a group such as a union, referral to a job, retention on a job, licensing, and certification are all covered under the *Uniform Guidelines for Employee Selection* (Equal Employment Opportunity Commission).
- Tests should be developed for three purposes:
  - To measure student attainment of learning objectives
  - To determine weaknesses and gaps in the training program
  - To enhance and improve training programs by positively influencing the revision of training materials and the improvement of instructor performance
- Oral tests used in conjunction with performance tests should focus on critical performance elements and key safety factors.
- Any performance test should be developed in accordance with task analysis information and reviewed by subject matter experts to ensure technical accuracy in terms of the actual work conditions.
- Test development activities should adhere to a common set of procedures. These procedures should be included in standard operating procedure or standard operating guideline formats.
- Proctoring tests involves much more than simply being in the testing environment; it requires specific skills to be performed professionally.
- Performance test proctors must have keen observational skills, technical competence for the task(s) being performed, ability to record specific test observations, ability to foresee critical dangers that may lead to injury to the performer or others, and an objective relationship to the test takers.
- Performance testing is the single most important method for determining the competency of actual task performance.
- The multiple choice test item is the most widely used in objective testing.
- The arrangement test item is an efficient way to measure the application of procedures for such things as disassembly and assembly of parts, start-up or shutdown, emergency responses, or other situations where knowing a step-by-step procedure is critical.
- A major advantage of the identification test item over the matching test item is the ability of the student to relate words to drawings, sketches, pictures, or graphs.
- A major problem with two-answer selection test items, such as the true/false question, is the guessing factor. A student who marks a response without reading the question has a 50 percent possibility of marking the correct response.
- When cheating is suspected, the instructor must document what was observed, what the student said, and what the instructor allowed the student to do next.
- The proper grading of oral, written, and performance tests is an essential function of the instructor. Students will desire feedback on their performance, and instructors will want to know how successful their instructional efforts were.
- Release of test scores to a class should proceed on an individual basis. It should be done in a private session, on a one-on-one basis, or in a personal letter to each student.
- Evaluation of examinations closes with feedback on testing results being delivered to the students. This step must be considered confidential and should be completed in a timely manner using the evaluation data available to the instructor.
- Computer or web-based testing is increasingly common in fire and EMS organizations.

KEY TERMS

Access Navigate **for flashcards to test your key term knowledge.**

Essay test A test that requires students to form a structured argument using materials presented in class or from required reading.

Oral test A test in which the answers are spoken in essay form in response to direct or open-ended questions. They may accompany a presentation or demonstration.

Performance test A test that measures a student's ability to do a task under specified conditions and to a specific level of competence. Also known as a skills evaluation.

Reliability The characteristic that a test measures what it is intended to measure on a consistent basis.

Skills evaluation A test that measures a student's ability to do a task under specified conditions and to a specific level of competence. Also known as a performance test.

Subject-matter expert (SME) An individual who is technically competent and who works in the field for which test items are being developed.

Validity The documentation and evidence that supports the test item's relationship to a standard of performance in the learning objective and/or performance required on the job.

Written test A test that may be made up of several types of test items, such as multiple choice, true/false, matching, essay, and identification questions.

REFERENCES

American Educational Resource Association, American Psychological Association, and National Council on Measurement in Education. 1999. *Standards for Educational and Psychological Testing*. Washington, DC: American Psychological Association.

Equal Employment Opportunity Commission. *Uniform Guidelines for Employee Selection*, 29 C.F.R. Part 1607. Washington, DC: Equal Employment Opportunity Commission.

National Fire Protection Association. 2018. *NFPA 1403: Standard on Live Fire Training Evolutions*. Quincy, MA: National Fire Protection Association.

National Fire Protection Association. 2019. *NFPA 1041: Standard for Fire and Emergency Services Instructor Professional Qualifications*. Quincy, MA: National Fire Protection Association.

National Fire Protection Association. 2019. *NFPA 1402: Standard on Facilities for Fire Training and Associated Props*. Quincy, MA: National Fire Protection Association.

U.S. Department of Education. 2011. *Family Educational Rights and Privacy Act*. Washington, DC: Family Policy Compliance Office.

CHAPTER PRESENTATION EXERCISE

Instructions: Your course instructor will assign individual or group discussions on the key points and teaching tips of this chapter. You or your group will review the chapter teachings and identify the major learning points from the chapter. You should be able to discuss the points, why they are important, and how and where they apply to your responsibilities at your level of instructor training.

REVIEW QUESTIONS

1. Describe the various types of test questions and discuss their main advantages and disadvantages.
2. What are the three main purposes for which tests are used, and what are the three basic types of tests?
3. Describe five of the important guidelines for proctoring written tests.

SECTION 2

Fire and Emergency Services Instructor II

CHAPTER **9** **Instructional Development**

CHAPTER **10** **Instructional Delivery**

CHAPTER **11** **Evaluation and Testing**

CHAPTER **12** **Program Management and Training Resources**

CHAPTER 9

Instructional Development

NFPA 1041 JOB PERFORMANCE REQUIREMENTS

Note: An asterisk denotes that the 1041 standard contains further information in its annex section.

5.3.2* Create a lesson plan, given a topic, learner characteristics, and a lesson plan format, so that learning objectives, a lesson outline, course materials, instructional technology tools, an evaluation plan, and learning objectives for the topic are addressed.

(A) Requisite Knowledge.

Elements of a lesson plan, components of learning objectives, instructional methodology, student-centered learning, methods for eliminating bias, types and application of instructional technology tools and techniques, copyright law, and references and materials.

(B) Requisite Skills.

Conduct research, develop behavioral objectives, assess student needs, and develop instructional technology tools; lesson outline techniques, evaluation techniques, and resource needs analysis.

KNOWLEDGE OBJECTIVES

After studying this chapter, participating in a structured learning environment, and completing assigned assessments, you will be able to:

- Describe how a Fire Service Instructor II creates a lesson plan. (**NFPA 1041: 5.3.2**, pp 203–214)
- Describe the components of a learning objective. (**NFPA 1041: 5.3.2**, pp 204–208)
- Explain techniques for eliminating bias from instructional material. (**NFPA 1041: 5.3.2**, p 203)
- Describe the components of a job performance requirement (JPR). (**NFPA 1041: 5.3.2**, p 210)
- Explain the relationship between JPRs and learning objectives. (**NFPA 1041: 5.3.2**, pp 210–211)
- Explain copyright considerations for instructors and fair use exceptions. (**NFPA 1041: 5.3.2**, p 215)

SKILLS OBJECTIVES

After studying this chapter, participating in a structured learning environment, and completing assigned assessments, you will be able to:

- Create a lesson plan that includes learning objectives, a lesson outline, instructional materials, instructional aids, and an evaluation plan. (**NFPA 1041: 5.3.2**, pp 203–215)

You Are the Fire and Emergency Services Instructor

As a newly assigned Instructor II at your department's training academy, you have been tasked with developing a lesson plan for an upcoming class of recruit fire fighters. As you begin to develop the lesson plan, you will have to first identify the different components of a successful lesson plan and then determine which of those components will be included in your plan. As a first step, you should determine the learning objectives necessary for your specific lesson plan. To get started, provide responses to the following:

1. What are the different components of a successful lesson plan?

2. Describe the development of learning objectives using the A (Audience), B (Behavior), C (Condition), and D (Degree) method.

Access Navigate for more practice activities.

Introduction

One of the major functions at the Instructor II level is the development of a lesson plan, which in many cases is part of a larger overall curriculum program of a training division or agency. The importance and value of a well-developed lesson plan cannot be overstated because it is the basic road map for an instructor to present important knowledge and skills to the fire fighters and responders who serve their communities.

This chapter will delve into how to develop a functional lesson plan. It is important to remember that a lesson plan could be a stand-alone one-subject plan or be part of an overall curriculum that is established based on the needs of an agency. The components that make up a lesson plan are covered in detail in Chapter 5, *Using Lesson Plans*.

The Lesson Plan Components

It is beneficial to review the various aspects of a lesson plan before delving into the creation of one. Here is a summary of the parts of a lesson plan:

- Lesson title: The lesson title indicates what a particular lesson plan contains and the information about the topic that is to be taught.
- Level of instruction: The level of instruction identifies the target audience level of understanding for the lesson material. A component of the level of instruction is the identification of any course prerequisites.
- Method of instruction: This is the method in which the lesson plan will be delivered (e.g., lecture, demonstration, group discussion).
- Learning objectives: Fundamental to all lesson plans, learning outcomes or learning objectives must define the applicable knowledge and skills that the lesson intends to teach. The Instructor II will use job performance requirements (JPRs) to develop the learning objectives in the **ABCD method**: A (Audience), B (Behavior), C (Condition), and D (Degree), described later in this chapter.
- References/resources: Fire and emergency services instructors who are not subject matter experts may need to refer to additional references or resources to obtain further information on these topics.
- Instructional materials: A well-developed lesson plan will identify the type of instructional materials to be used in the delivery of the lesson plan. Instructional materials are tools designed to help you present the lesson plan to your students.
- Lesson outline: The lesson outline is the main body of the lesson plan; it comprises four main elements: preparation, presentation, application, and evaluation.

- Lesson summary: The lesson summary simply summarizes the lesson plan, reviewing and reinforcing the main points. The lesson summary plays an important role in the overall lesson, allowing the instructor to enhance the application step by asking summary questions on key objective and lesson points.
- Assignment: A typical component of a lesson plan, the assignment could be a homework activity that allows the student to apply the content from the lesson plan.

Creating a Lesson Plan

As a Fire and Emergency Services Instructor II, you are responsible for creating lesson plans. Depending on the subject, the creation of a lesson plan can take anywhere from several hours to several weeks. Regardless of the size of the lesson plan, the ultimate goal is to create a document that any instructor can use to teach the subject and ensure that students achieve the learning objectives.

Many fire departments have lesson plan templates for the Instructor II to use as a starting point. Such a standard format makes it easier for all instructors in the department to understand the lesson plan and ensure consistency in training. If readily accessible and available, it may be easy to access one style of lesson plan provided by a publisher and use that as the template for the training program. Consistency may be achieved if a variety of instructors all use the same lesson plan format to write or teach from.

An important facet of lesson plan development is awareness of diversity and eliminating possible bias in the instructor material or student handouts. All instructors need to be aware of the changing environment in which today's fire and emergency services operates and recognize the need to be inclusive and supportive of all those who choose to be a part of this challenging profession. Work to exercise good judgment in the development of the plan and in the selection of images and any other material you will use in a presentation. Remember to always be mindful of and eliminate any bias based on race, sex, national origin, color, religion, or sexual orientation. Forms of bias could be found in handouts that use gender specific terms such as *he* or *she* or stereotyping a race, culture, or religion in either wording or in images. Developing lesson plans and training materials that eliminate any potential bias should be a fundamental goal and desired outcome.

The creation of materials that are free from your own preferences or bias can be a challenge. To guard against this, do extensive research to learn about best practices or what other material might exist on a topic area. Once you have collected information from multiple sources, check your own agency's policies to ensure there is no conflict, and from there, develop a best practices lesson plan that is as bias-free as possible. For example, when developing a lesson plan for raising a ladder, the plan should outline the proper steps following your department's standard operating procedures rather than how you prefer to raise a ladder.

In addition to lesson material, remember that all students have particular needs in the classroom that may require you to adjust or modify part of the lesson plan at some point. When you develop your plan, it should be detailed and complete enough to achieve the identified goals and desired outcomes, yet flexible enough that it can be adjusted to meet students' needs on the fly. You should be prepared to repeat information or to perform demonstrations in different ways to ensure each student receives and absorbs the information.

Instructor- and Student-Centered Learning

There are many schools of thought on how to assist students in the learning process. Two of the most common methods are **instructor-centered learning** and **student-centered learning**.

Examples of instructor-centered learning include direct teaching such as lecture, the use of slides, and other presentation material where the student remains a passive listener and the instructor is the sole source of knowledge (**FIGURE 9-1**). Assessing instructor-centered learning can be accomplished with quizzes or other testing tools or a take-home assignment.

Examples of student-centered learning include tests of memorization and recall, polling, summarizing, and other interactive lecture methods, as well as simulated clinical activities, such as patient assessments. A key element to student-centered learning is that students must be active participants in the process. The instructor's role in student-centered learning is to ensure all students are learning. Assessments for student-centered learning could be from in-class presentations or the completion of a project.

Regardless of the model, the instructor remains the lead person in the learning process, but as an instructor, you need to understand that each method has a place in the classroom. Instructor-centered learning is

FIGURE 9-1 An example of instructor-centered learning is a lecture with the use of presentation material, where the student is a passive listener and the instructor is the sole source of knowledge.
© Jones & Bartlett Learning. Photographed by Glen E. Ellman.

appropriate for new recruits or the introduction of a new tool being adopted by the department, whereas student-centered learning might be more effective in teaching senior paramedics in a continuing education class.

Determining the Learning Outcomes

The first step of lesson plan development is to determine the purpose of the course. What are students expected to achieve as a result of taking the class? What should the students take away after attending this training session? Often, this desired outcome is obvious, because you are teaching a class to prepare students to perform a certain job or skill. For example, if you were to develop a lesson plan for a class to train fire fighters to drive a fire engine, you would start by listing the JPRs for a fire engine driver; however, on many other occasions, the learning objectives are not that clear.

It is very difficult to develop a lesson plan when there is not a clear path and purpose for the lesson. Many instructors have been in the unhappy position of being told to teach a certain class, such as one dealing with workplace diversity or safety, without clear direction on the intended outcomes of the class. Although the person requesting the class may have a general idea of what the class is intended to accomplish, he or she might not know the specific learning outcomes that the Instructor II needs in order to develop a lesson plan.

For example, the fire chief may want to improve fire fighter safety through a training class on safety but unless given specific learning outcomes, the Instructor II cannot develop a lesson plan to "improve fire fighter safety." When placed in this position, the Instructor II should attempt to clarify the fire chief's vision of improving fire fighter safety. Would he like all fire fighters to understand the chain of events that leads to an accident and to know how to break that chain so that an accident is avoided? Or does the fire chief expect all fire fighters to don their structural firefighting protective equipment properly within a time limit? A clear learning outcome can be written to achieve that goal. Whenever you are asked to develop a lesson plan for a class, start by clarifying the intended outcome of the class with the person who requests the class.

Developing the Learning Objectives

Once you have a clear outcome for a class, you should work backward to develop the learning objectives. As described earlier in this chapter, learning objectives can be written using the ABCD method (**TABLE 9-1**). The ABCD method is described in Chapter 5, *Using Lesson Plans*, but is mentioned here as a review.

Audience

The first component of the ABCD method is the *audience*, which should describe the students who will take the class. If the lesson plan is being developed specifically for a certain audience, the learning objectives should be written to indicate that fact. For

TABLE 9-1 Learning Objectives: Using the ABCD Method

Method Step	Question	Example
Audience	Who	The fire fighter
Behavior	Will do or know what	Will perform a one-person ladder raise
Condition	Using which equipment or resources	Using a two-section, 24-foot ground ladder
Degree	How well or to which level	So that the proper climbing angle is set and all safety precautions are observed

© Jones & Bartlett Learning.

example, if an Instructor II is writing a lesson plan for a driver training class, the audience would be described as "the driver trainee" or "the driver candidate." Both of these terms indicate that the audience consists of individuals who are learning to be drivers. If the audience is not specifically known or if it includes students from mixed backgrounds or various departments, such as seen in a regional delivery, the audience part of the learning objective could be written more generically; i.e., "the fire fighter" or even "the student." but still meet the objective or JPR. When developing a lesson plan for use in a regional delivery where the level of knowledge of the students is not known, you need to follow the objectives/JPRs for the desired outcomes. In some cases, the material might be a refresher for some of these students while for others, it is all new information; however, since all of the students are learning together, they all will obtain the same level of knowledge in the end. This type of lesson plan can be a challenge. You don't want any students to feel lost under an onslaught of new information; at the same time, you want to maintain the attention of the students who are staying up to speed.

Behavior

The *behavior* part of the learning objective should be specified using a clearly measurable action word, which allows the evaluation of the student's achievement of the learning objective. Another important consideration is the level to which a student will achieve the learning objective. This level is most often determined using Bloom's Taxonomy (Bloom, 1956), a method to identify levels of learning within the cognitive domain (discussed further in Chapter 2, *The Learning Process*) (**TABLE 9-2**).

Cognitive Domain Objectives

For the fire service, the most commonly used levels when developing *cognitive* learning objectives are (in

TABLE 9-2 Bloom's Taxonomy of Educational Objectives

	Complete Objective	Parts	
Remembering (Knowledge)	Ability to assemble terms or facts, memorize material, and recall information	Identify the wraparound short backboard device.	List Define Recall State Recite
Understanding (Comprehension)	Ability to use knowledge and interpret or translate the information (i.e., understand the meaning of information)	Recognize situations where the wraparound short backboard is indicated.	Explain Summarize Describe Restate Interpret
Applying (Application)	Ability to use knowledge and comprehension to apply information to new situations	Illustrate situations of poor application of the wraparound short backboard.	Solve Illustrate Apply Put into practice
Analyzing (Analysis)	Ability to break down information into components and determine how each component affects understanding (i.e., looking at elements and the relationships)	Analyze situations when a wraparound short backboard is or is not indicated.	Organize Analyze Compare Contrast

(continued)

TABLE 9-2 Bloom's Taxonomy of Educational Objectives (Continued)

	Complete Objective	Parts	
Evaluating (Evaluation)	Ability to make judgments, critiques, or appraisals of the information	Following the completion of the scenario, evaluate the use of the wraparound short backboard device on an entrapped vehicle crash victim.	Evaluate Judge Defend Justify
Creating (Synthesis)	Ability to reassemble the information in a new manner using abstract thinking and creativity	Describe a scenario that illustrates other uses of the wraparound short backboard device.	Design Hypothesize Discuss Devise

© Jones & Bartlett Learning.

order from simplest to most difficult) knowledge, comprehension, and application, defined as follows:

- *Remembering (Knowledge)* is remembering facts, definitions, numbers, and other specific items.
- *Understanding (Comprehension)* is displayed when students clarify or summarize important points.
- *Applying (Application)* is the ability to solve problems or apply the information learned to situations.

Higher levels of application and understanding occur when the learning objectives are written at the following levels:

1. *Analyzing (Analysis)* is the ability to break down information into components and the ability to know which components affect understanding.
2. *Evaluating (Evaluation)* is the ability to make judgments, critiques, or appraisals of the information.
3. *Creating (Synthesis)* is the ability to reassemble information in a new manner using abstract thinking and creativity.

In 2001, Dr. David Krathwohl and several other educational theorists modified the original Bloom's Taxonomy of Educational Objectives to expand its scope (Anderson et al., 2001) (**FIGURE 9-2**). Several levels were revised:

- *Knowledge* became *Remembering*
- *Comprehension* became *Understanding*
- *Application* became *Applying*
- *Analysis* became *Analyzing*
- *Evaluation* became a Level 5 task *Evaluating*
- *Synthesis* was moved to the highest level and became *Creating*

An Instructor II must determine which level within the cognitive domain is the appropriate level for the student to achieve for the lesson plan. For example, if you are developing objectives for a class on portable extinguishers, the following objectives could be written for each basic level:

- Remembering (Knowledge): "The fire fighter trainee will identify the four steps of the PASS method of portable extinguisher application." For this objective, the student simply needs to memorize and repeat back the four steps of pull, aim, squeeze, and sweep. Achievement of this very simple knowledge-based objective is easily evaluated with a multiple choice or fill-in-the-blank question.
- Understanding (Comprehension): "The fire fighter trainee will explain the advantages and disadvantages of using a dry-chemical extinguisher for a Class A fire." This objective requires the student first to identify the advantages and disadvantages of a dry-chemical extinguisher, and then to select and summarize those that apply to use of such an extinguisher on a Class A fire. This higher-level objective may be evaluated by a multiple choice question but is better evaluated with a short answer question.
- Applying (Application): "The fire fighter trainee, given a portable fire extinguisher scenario, shall identify the correct type of extinguisher and demonstrate the method for using it to extinguish the fire." This is the highest level of objective because it requires the student to recall several pieces of information and apply them correctly based on the situation. This type of objective is often evaluated with scenario-based questions that may be answered with multiple choice items or short answers.

Psychomotor Domain Objectives

For the fire service, the most commonly used levels when developing *psychomotor* learning objectives are

COGNITIVE DOMAIN	PSYCHOMOTOR DOMAIN	AFFECTIVE DOMAIN
REMEMBERING (KNOWLEDGE) Recognition and recall of facts and specifics EXAMPLES: Define, describe, list, state	**IMITATION** Observes skills and attempts to repeat them EXAMPLES: Assemble, build, connect, couple, repeat	**RECEIVING** Listening passively; attending to EXAMPLES: Ask, name
UNDERSTANDING (COMPREHENSION) Interprets, translates, summarizes, or paraphrases given information EXAMPLES: Convert, infer, rewrite	**MANIPULATION** Performs skills by instruction rather than observation EXAMPLES: Transmit, arrange, re-create	**RESPONDING** Complies to given expectation; shows interest EXAMPLES: Answer, recite
APPLYING (APPLICATION) Processes information in a situation different from original learning context EXAMPLES: Demonstrate, relate, produce	**PRECISION** Reproduces a skill with accuracy, proportion, and exactness; usually performed independently of original sources EXAMPLES: Modify, demonstrate	**VALUE** Displays behavior consistent with single belief or attitude; unforced compliance EXAMPLES: Complete, explain, justify
ANALYZING (ANALYSIS) Separates whole into parts; clarifies relationships among elements EXAMPLES: Diagram, outline, illustrate	**ARTICULATION** Combines more than one skill in sequence with harmony and consistency EXAMPLES: Combine, coordinate, develop, modify	**ORGANIZING** Committed to set of values as displayed by behavior EXAMPLES: Integrate, adhere
EVALUATION (EVALUATING) Makes decisions, judges, or selects based on criteria and rationale EXAMPLES: Compare, contrast, justify, summarize		
CREATING (SYNTHESIS) Combines elements to form new entity from original one EXAMPLES: Compile, compose, design	**NATURALIZATION** Completes one or more skills with with ease; requires limited physical or mental exertions EXAMPLES: Design, specify, manage	**CHARACTERIZING** Behavior consistent with values internalized EXAMPLES: Qualify, modify, perform

FIGURE 9-2 Revised matrix for the new taxonomy.

Adapted from Miami-Dade Community College revision of Bloom, B. S. (1956). Taxonomy of Educational Objectives, Handbook I: The Cognitive Domain. New York: David McKay Company; Dave, R. H. (1970). Developing and Writing Behavioral Objectives. Tucson, AZ: Educational Innovators Press; and Krathwohl, D., B. S. Bloom, and B. B. Masia. (1964). Taxonomy of Educational Objectives, Handbook II: Affective Domain. New York: David McKay Company.

(in order from simplest to most difficult) imitation, manipulation, precision, articulation, and naturalization, explained as follows:

- *Imitation* is when the student observes the skill and attempts to repeat it.
- *Manipulation* is performing a skill based on instruction rather than on observation.
- *Precision* is performing a single skill or task correctly.
- *Articulation* is when the student combines multiple skills.
- *Naturalization* is when the student performs multiple skills correctly all the time.

Action verbs associated with the psychomotor domain include the following:

- Demonstrate
- Practice
- Apply
- Perform
- Display
- Show
- Assemble

Affective Domain Objectives

The *affective* domain deals with an individual's expressed interests, ambitions, and values. These kinds of emotional behaviors are essential to the overall learning experience but may not be readily visible during initial learning. The taxonomy of the affective domain identifies five levels of understanding as follows:

- *Receiving* is paying attention and displaying a willingness to learn.
- *Responding* is displaying an acknowledged behavior within the learning experience and participating when given an opportunity.

- *Valuing* is showing active involvement, passion, or commitment toward a topic.
- *Organization* is accepting a new value as one's own and setting a specific goal.
- *Characterization* is comparing and contrasting one's own values to others and using the new value.

Action verbs associated with affective domain objectives include the following examples:

- Accept
- Participate
- Share
- Judge
- Attempt
- Challenge

Condition

Moving on to the "C" of ABCD method, the *condition* describes the situation in which the student will perform the behavior. Specific equipment or resources must be listed in the objective; these are the "givens" necessary to demonstrate the skill or knowledge. For example, the phrases "in full protective clothing, including SCBA" and "using water from a static source such as a pool or pond" would tell both the instructor and the student what needs to be present to complete the behavior. You might notice that many skill sheets have a specific area inside their template that lists the equipment needed for part of the skill sheet. Before teaching the lesson, the instructor must check whether all equipment or resources needed for the performance of the objective are in working order and whether the student can properly operate all equipment or resources.

Degree

The degree may also be known as the *standard* because it describes how well the behavior must be performed. Both the student and the evaluator need to know the criteria against which the student is being measured. Percentage scores, "without errors," and "within a designated time" are all examples of the degree of performance that objectives should contain. A reference to a skill sheet during manipulative performance will guide both the student and instructor in how to approach proper completion and evaluation of the skill.

There is no one correct format for determining which level or how many learning objectives should be written for a lesson plan (**TABLE 9-3**). Typically, a lesson plan will contain knowledge-based (cognitive) learning objectives to ensure that students learn all of

TABLE 9-3 Examples of Formatted Objectives by Domains and Levels

Duty Area	Domain Level	Cognitive	Domain Level	Psychomotor	Domain Level	Affective
Fire fighter	Remembering	A fire fighter trainee, given a written exam, will identify the tools needed to perform vertical ventilation on a peaked roof according to J&B *Fundamentals of Fire Fighter Skills*, 3rd edition.	Manipulation	The fire fighter, given a K-950 rotary saw and an assortment of blades, will follow manufacturer's directions to change the wood blade to a metal blade.	Receiving	Basic fire fighter academy students will name the safety precautions to be taken when using a ground ladder during vertical ventilation.
Officer	Creating	Company officers will be given video clips of five different fire scenes and will explain the type of ventilation required based on smoke conditions according to department standard operating procedures (SOPs).	Articulation	As the officer of a crew with three experienced fire fighters, and given PPE, ladders, and tools, conduct vertical ventilation on a pitched roof so that all barriers are removed and structural integrity is not compromised within 4 minutes.	Organizing	The officer of a crew assigned to ventilation at a structure fire will display adherence to all safety and risk management requirements for the entire crew, according to department policy.

Courtesy of Bryant Krizik.

Job Performance Requirements in ACTION

The lesson plan is the tool used by an instructor to conduct a training session. It is as essential as personal protective equipment (PPE) is to a fire fighter. The lesson plan details the information necessary to present the training session, which includes everything from the title of the class to the assignment for the next training session. In between are the resources needed, the behavioral objectives, the content outline, and the various teaching applications used to complete the training. As an instructor, you must review and practice the delivery of the lesson plan, check the materials needed for the class, and be ready to present the materials. Using the lesson plan, you must present a structured training session by taking advantage of appropriate methods of instruction to engage the students and use a variety of communication skills to complete the learning objectives.

Instructor I	Instructor II	Instructor III
The Instructor I will teach from a prepared lesson plan using appropriate methods of delivery and communication skills to ensure that the learning process is effective. The instructor must understand each component of the lesson plan. It may be necessary to adapt the lesson plan to the needs and abilities of the audience and the teaching environment.	The Instructor II will prepare the lesson plan components and determine the expected outcomes of the training session. The four-step method of instruction should be defined within the lesson plan and all instructional requirements outlined for the presenter's use.	The Instructor III will conduct a training needs assessment and develop a curriculum to meet the training need, including course goals and evaluation strategies.
JPRs at Work	**JPRs at Work**	**JPRs at Work**
Present prepared lesson plans by using various methods of instruction that allow for achievement of the instructional objectives. Adapt the lesson plan based on student needs and specific conditions.	Create and modify existing lesson plans to satisfy the student needs, JPRs, and objectives developed for the training session.	There are no JPRs at this instructor level.

Bridging the Gap Among Instructor I, Instructor II, and Instructor III

A partnership must exist between the developer of the lesson plan and the instructors who will deliver that lesson plan. In many cases, these individuals may be the same person. In other cases, such as in large departments, you may never know who wrote the lesson plan. Your skill in developing a lesson plan that another instructor can use—perhaps long after it was developed—is therefore important. If another person must deliver your lesson plan, however, you must be sure that all components of the lesson plan are clear and concise and that the material and instructional methods match the needs of the students. Your communications skills and knowledge of the learning process will be used at both Instructor I and Instructor II levels in the development and delivery of this content. The Instructor III develops curricula based on a training needs assessment and provides direction to instructors at the other levels on the goal or outcome of the material they are developing or delivering.

the facts and definitions within the class. Comprehension or cognitive objectives are then used to ensure that students can summarize or clarify the material. Psychomotor objectives are used to ensure that the student can perform a task identified in the objective and presented in the lesson plan. Objectives pertaining to the third domain—that is, affective objectives—are often listed with the psychomotor objectives and identify "when" to perform the task. These learning domains are discussed in more detail in Chapter 2, *The Learning Process*.

Converting JPRs into Learning Objectives

Often, an Instructor II needs to develop learning objectives to meet the JPRs listed in a National Fire Protection Association (NFPA) professional qualification standard. Remember that a JPR describes a specific job task, lists the items necessary to complete that task, and defines measurable or observable outcomes and evaluation areas for the specific task. The matching of learning objectives to JPRs occurs when a lesson plan is being developed to meet the professional qualifications for a position such as fire officer, fire instructor, or fire fighter.

The JPRs listed in the NFPA standards of professional qualifications are not learning objectives per se, but learning objectives can be created based on the JPRs. Each NFPA professional qualification standard has an annex section that explains the process of converting a JPR into an instructional objective, including examples of how to do so (**FIGURE 9-3**). By following this format, an Instructor II is able to develop learning objectives for a lesson plan to meet

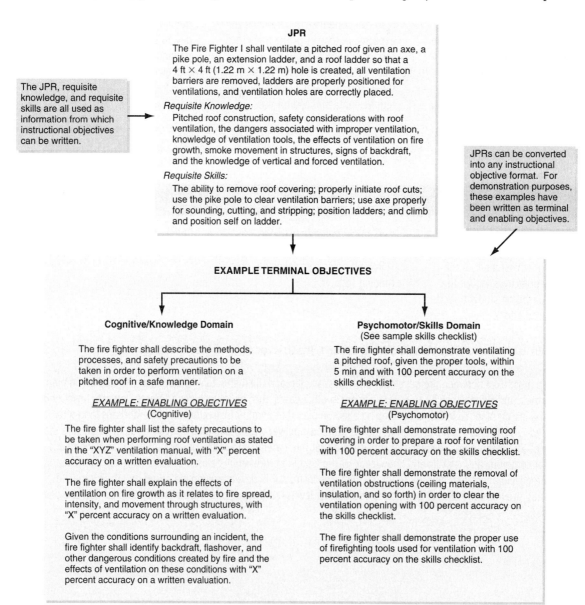

FIGURE 9-3 Converting job performance requirements into instructional objectives.
Courtesy of NFPA.

the professional qualifications for NFPA standards. This process involves the breaking down of a JPR into a **terminal objective** and several **enabling objectives**, including cognitive and psychomotor objectives. Further into the development process, cognitive objectives are written into a lesson plan and include evaluation tools such as test questions, whereas psychomotor objectives are broken down into task steps and made into skill sheets. An Instructor III will use the JPRs to help write course objectives for larger curricula.

Developing the Lesson Outline

After determining the performance outcomes and writing the learning objectives for the lesson plan, the next step is to develop the lesson outline.

One method for creating a lesson outline involves brainstorming the topics to be covered and then arranging them in a logical order (**FIGURE 9-4**). Begin listing all the information that needs to be taught to achieve the learning objectives. Which terms do students need to learn? Which concepts must be presented? Which skills need to be practiced? Which case studies, NIOSH reports, or other real-life examples would demonstrate the need to learn this material?

An effective method for ensuring that the lesson plan covers all of the most current information is to research what material currently exist or can be obtained from a reliable source. Other means to reaching a topic would be the use of the internet, online training resources, publishers, or government sites such as the Learning Resource Center (LRC) at the National Fire Academy. Only use valid sources that will help in the development of your lesson plan. When using other sources make sure you do not violate copyright laws; this is addressed later in this chapter.

Once you have listed all the topics that should be covered in the lesson outline, organize them into presentation and application sections. Arrange the listed topics you will lecture on in a logical and orderly fashion in the presentation section. It may be practical to structure a lesson outline by progressing from the known and working toward the unknown. This is particularly useful in a large course where the instructor aims to ensure that the topics previously covered have been mastered by the students. This establishes a relationship among topics in the larger curriculum. For example, you will need to instruct students on building construction and ladders before you can focus on vertical ventilation. Topics should

---

Pre-Lecture (Preparation Step)
I. You Are the Fire Fighter

Time: 5 Minutes
Small Group Activity/Discussion

Use this activity to motivate students to learn the knowledge and skills needed to understand the history of the fire service and how it functions today.

Purpose
To allow students an opportunity to explore the significance and concerns associated with the history and present operation of the fire service.

Instructor Directions
1. Direct students to read the "You Are the Fire Fighter" scenario found in the beginning of Chapter 1.
2. You may assign students to a partner or a group. Direct them to review the discussion questions at the end of the scenario and prepare a response to each question. Facilitate a class dialogue centered on the discussion questions.
3. You may also assign this as an individual activity and ask students to turn in their comments on a separate piece of paper.

Lecture (Presentation Step)
I. Introduction

Time: 5 Minutes
Slides: 1–6
Level: Fire Fighter I
Lecture/Discussion

A. Training to become a fire fighter is not easy.
 1. The work is physically and mentally challenging.
 2. Firefighting is more complex than most people imagine.
B. Fire fighter training will expand your understanding of fire suppression.
 1. The new fire fighter must understand the roots of the fire service, how it has developed, and the fire service "culture" in order to excel.
 2. This course equips fire fighters to continue a centuries-old tradition of preserving lives and property threatened by fire.

FIGURE 9-4 A sample lesson outline.
© Jones & Bartlett Learning.

II. Fire Fighter Guidelines

Time: 5 Minutes
Slide: 7
Level: Fire Fighter I
Lecture/Discussion

A. Be safe.
  1. Safety should always be uppermost in your mind.
B. Follow orders.
  1. If you follow orders, you will become a dependable member of the department.
C. Work as a team.
  1. Firefighting requires the coordinated efforts of each department member.
D. Think!
  1. Lives will depend on the choices you make.
E. Follow the golden rule.
  1. Treat each person, patient, or victim as an important person.

III. Fire Fighter Qualifications

Time: 30 Minutes
Slides: 8–10
Level: Fire Fighter I
Lecture/Discussion

A. Age requirements
  1. Most career fire departments require that candidates be between the ages of 18 and 21.
B. Education requirements
  1. Most career fire departments require a minimum of a high school diploma or equivalent.
C. Medical requirements
  1. Medical evaluations are often required before training can begin.
  2. Medical requirements for fire fighters are specified in NFPA 1582, *Standard on Comprehensive Operational Medical Program for Fire Departments*.
D. Physical fitness requirements
  1. Physical fitness requirements are established to ensure that fire fighters have the strength and stamina needed to perform the tasks associated with firefighting and emergency operations.
E. Emergency medical requirements
  1. Many departments require fire fighters to become certified at the emergency medical responder (EMR), emergency medical technician (EMT), or higher levels.

IV. Roles and Responsibilities of the Fire Fighter I and Fire Fighter II

Time: 30 Minutes
Slides: 11–17
Level: Fire Fighter I and II
Lecture/Discussion

A. The roles and responsibilities for Fire Fighter I include:
  1. Don and doff personal protective equipment properly.
  2. Hoist hand tools using appropriate ropes and knots.
  3. Understand and correctly apply appropriate communication protocols.
  4. Use self-contained breathing apparatus (SCBA).
  5. Respond on apparatus to an emergency scene.
  6. Force entry into a structure.
  7. Exit a hazardous area safely as a team.
  8. Set up ground ladders safely and correctly.
  9. Attack a passenger vehicle fire, an exterior Class A fire, and an interior structure fire.
  10. Conduct search and rescue in a structure.
  11. Perform ventilation of an involved structure.
  12. Overhaul a fire scene.
  13. Conserve property with salvage tools and equipment.
  14. Connect a fire department engine to a water supply.
  15. Extinguish incipient Class A, Class B, and Class C fires.
  16. Illuminate an emergency scene.
  17. Turn off utilities.

FIGURE 9-4 *Continued*

18. Perform fire safety surveys.
19. Clean and maintain equipment.
20. Present fire safety information to station visitors, community groups, or schools.

B. Additional roles and responsibilities for Fire Fighter II include:
1. Coordinate an interior attack line team.
2. Extinguish an ignitable liquid fire.
3. Control a flammable gas cylinder fire.
4. Protect evidence of fire cause and origin.
5. Assess and disentangle victims from motor vehicle accidents.
6. Assist special rescue team operations.
7. Perform annual service tests on fire hose.
8. Test the operability of and flow from a fire hydrant.
9. Fire fighters must also be prepared to assist visitors to the fire station and use the opportunity to discuss additional fire safety information.

V. Summary

Time: 5 Minutes
Slides: 51–53
Level: Fire Fighter I
Lecture/Discussion

A. Remember the five guidelines: Be safe, follow orders, work as a team, think, and follow the golden rule.
B. Fire fighter qualifications consider age, education, medical and physical fitness, and emergency medical certifications.
C. The roles and responsibilities of Fire Fighter I and Fire Fighter II vary.

Post-Lecture
I. Wrap-Up Activities (Application Step)

Time: 40 Minutes
Small Group Activity/Individual Activity/Discussion

A. Fire Fighter in Action
This activity is designed to assist the student in gaining a further understanding of the roles and responsibilities of the Fire Fighter I and II. The activity incorporates both critical thinking and the application of fire fighter knowledge.

Purpose
This activity allows students an opportunity to analyze a firefighting scenario and develop responses to critical thinking questions.

Instructor Directions
1. Direct students to read the "Fire Fighter in Action" scenario located in the Wrap-Up section at the end of Chapter 1.
2. Direct students to read and individually answer the quiz questions at the end of the scenario. Allow approximately 10 minutes for this part of the activity. Facilitate a class review and dialogue of the answers, allowing students to correct responses as needed. Use the answers noted below to assist in building this review. Allow approximately 10 minutes for this part of the activity.
3. You may also assign these as individual activities and ask students to turn in their comments on a separate piece of paper.
4. Direct students to read the "Near Miss Report." Conduct a discussion that allows for feedback on this report. Allow 10–15 minutes for this activity.

Answers to Multiple Choice Questions
1. A, D
2. B
3. B
4. D

B. Technology Resources
This activity requires students to have access to the Internet. This may be accomplished through personal access, employer access, or through a local educational institution. Some community colleges, universities, or adult education centers may have classrooms with Internet capability that will allow for this activity to be completed in class. Check out local access points and encourage students to complete this activity as part of their ongoing reinforcement of firefighting knowledge and skills.

Purpose
To provide students an opportunity to reinforce chapter material through use of online Internet activities.

Instructor Directions
1. Use the Internet and go to **www.Fire.jbpub.com**. Follow the directions on the Web site to access the exercises for Chapter 1.
2. Review the chapter activities and take note of desired or correct student responses.
3. As time allows, conduct an in-class review of the Internet activities and provide feedback to students as needed.
4. Be sure to check the Web site before assigning these activities, as specific chapter-related activities may change from time to time.

FIGURE 9-4 *Continued*

II. Lesson Review (Evaluation Step)

Time: 15 Minutes
Discussion
Note: Facilitate the review of this lesson's major topics using the review questions as direct questions or overhead transparencies. Answers are found throughout this lesson plan.

A. Name some of the physical fitness requirements established for fire fighters.
B. What requirements do the fire fighter qualifications focus on?
C. How do the roles and responsibilities of the Fire Fighter I and II differ?

III. Assignments

Time: 5 Minutes
Lecture

A. Advise students to review materials for a quiz (determine date/time).
B. Direct students to read the next chapter in *Fundamentals of Fire Fighter Skills* as listed in your syllabus (or reading assignment sheet) to prepare for the next class session.

FIGURE 9-4 *Continued*

---

be presented in order starting from the basic and then moving on to the more complex. Ensure that the topics flow together and that the presentation does not contain any gaps that might confuse a student. If you identify a gap, you may need to create a new topic to bridge it.

In the application section of the lesson, list the topics that require students to apply the information learned in the presentation section. Most often the topics in the application section will be activities or skills practice. If the lesson does not include actual hands-on activities, the application should at least consist of discussion points for you to talk about with the students to ensure the information in the lecture was learned and can be applied.

Many lesson outlines use a two-column format. The first column contains the actual outline of the material to be taught. If this lesson outline is to be used by experienced instructors, a simple outline of the material may suffice. For less experienced instructors (or to ensure consistency among multiple instructors), the outline may be more detailed. The second column of the lesson outline contains comments or suggestions intended to help an instructor understand and present the lesson outline. It is also a good practice to indicate in the second column which learning objectives are being achieved during the presentation or application sections. This information is especially helpful when you are developing a lesson plan to teach an established curriculum that uses a numbering system to identify learning objectives.

> **DEPARTMENT TRAINING OFFICER TIP**
>
> Guided or indirect discussion methods allow students to be actively involved in the lesson presentation. The more involved students are in a lesson, the better the learning experience becomes for the students.

Identifying the Instructional Materials

Once the lesson outline is developed, all instructional materials needed to deliver instruction should be identified and listed in the lesson plan. The list should be specific so that the exact instructional aid or teaching tool can be identified. For example, if the lesson plan is a fire safety lesson for children that incorporate a video as an instructional aid, simply listing "fire safety video" in the lesson plan does not provide enough information for the instructor. Instead, give the exact title of the video, such as "*Sparky Says: Join My Fire Safety Club*, by the NFPA." This information will allow any instructor who uses the lesson plan to obtain the correct instructional aid. Instructional materials also include handouts, computer projectors, computers, manikins, or fire hose to be used for hose evolutions.

Often the inclusion of one instructional aid creates a need for more instructional materials. For example, if a lesson plan lists a video as an instructional aid, the instructional materials would need to be revised to include a monitor. To ensure that you have all instructional resources available to deliver the lesson plan, conduct a resource needs analysis of your current resources by asking the following types of questions to determine precisely what you need:

- Are additional informational resources needed to present the learning objectives to students—for example, a handout describing your department's standard operating procedures (SOPs)?
- Are supplies needed to make props or demonstrations?
- Is there access to a curriculum or do we need to develop slides, videos, or other instructional media?
- Is equipment needed for the activities or skills practice?
- Is equipment needed to ensure student safety?
- Are the correct instructional tools/technology available to deliver the lesson plan?
- If the necessary tools and technology are not available, where can you find them?

Copyright and Public Domain

The first chapter of this text addressed copyright law. Recall that the Copyright Act of 1976 divides work, such as written words and photographs, into two general categories: those protected by the law and those considered to be part of the public domain. The Copyright Act determines specific time frames pertaining to copyrighted material. Work published before 1923 is considered to be in the public domain. Work published between 1923 and 1978 has a 95-year copyright protection from the date of actual publication. For work published after 1977, the author, artist, or photographer has a copyright for life plus 70 years. Violation of the copyright law can result in liability, with the creator of the work being eligible to recover any damages suffered or lost profits.

Commonly, the training materials provided to fire and emergency service employees include copyrighted content. Remember that the Copyright Act does not prohibit the content's use outright or require that permission be sought before it is used, and the fact that some material is protected by a copyright does not necessarily prohibit you from using it in your training, because a "fair use" exception exists relative to the protection afforded by copyrights. Fair use depends on the nature of the copyrighted work and how much of the work is copied. If your intended application of the material does not constitute fair use, you may still be able to use the content in a learning environment by obtaining permission from the copyright holder.

It is important to remember that whether a work is used under the fair use exception or with permission, the user must always credit the source who holds the copyright to the work.

Writing the Evaluation Plan

The evaluation plan is the final part of the lesson plan. Each part of the evaluation plan should be directly tied to one or more learning objectives. When writing this component of the lesson plan, simply describe the evaluation plan—do not provide the actual evaluation. In other words, the lesson plan could indicate that the evaluation plan is a 50-question multiple choice test, but it should not list the actual test questions. The test questions should be a separate document that is securely kept and made available to the instructors presenting the lesson plan. When the evaluation plan lists skills performance tests, these documents should be included with the instructional materials and distributed to students so they can prepare for skills testing. This step is covered in greater detail in Chapter 8, *Evaluating the Learning Process*.

DEPARTMENT TRAINING OFFICER TIP

When creating an evaluation tool for a lesson, make sure the evaluation tool links back to the initial learning objectives outlined at the beginning of class.

Training BULLETIN

JONES & BARTLETT FIRE DISTRICT
TRAINING DIVISION
5 Wall Street, Burlington, MA, 01803
Phone 978-443-5000   Fax 978-443-8000

Instant Applications: Lesson Plans

Drill Assignment

Apply the chapter content to your department's operation, training division, and your personal experiences to complete the following questions and activities.

Objective

Upon completion of the instant applications, fire and emergency services instructor students will exhibit decision making and application of job performance requirements of the instructor using the text, class discussion, and their own personal experiences.

Suggested Drill Applications

1. Using the ABCD method of developing learning objectives, write several learning objectives for an upcoming training session on vehicle extrication skills such as door removal, roof removal, and safety zones.

2. Research and identify a prepared lesson plan for conducting a training session on one of the vehicle extrication skills listed in question 1. From that lesson plan, determine what you may need to do to adapt it to your own organization and identify any instructional materials you have at your disposal that would be beneficial for the delivery of that lesson plan.

3. Review the Incident Report for this chapter and determine what other training topics, besides those mentioned in the summary of the report, would be beneficial for fire suppression personnel to obtain that would have benefit for incidents such as the one described.

Incident Report: NIOSH Report # F2011-15

Drill Assignment

Review the information in this incident report and prepare a practice presentation for class delivery at the direction of your instructor. Your presentation should include a summary of the incident facts and identify the NFPA standard(s) that could apply to the incident. Use an outline to organize your thoughts. You may be evaluated on your communication skills during your presentation.

Pinckneyville, Illinois—2011

On June 17, 2011, a 22-year-old male paid-on-call fire fighter received fatal injuries when he was struck by bricks and falling debris during an exterior wall collapse at a commercial structure fire. Crews worked using defensive operations for about 45 minutes attempting to extinguish the fire in the 96-year-old brick and masonry structure that housed an antique store with living quarters located in a rear addition. The victim and another fire fighter were moving a 35-foot aluminum ground ladder away from the Side D (east) wall of the structure when the top part of the exterior wall collapsed. No other fire fighters were injured in the collapse.

The structure involved in this incident was one of three structures that occupied a city block near the center of town. The fire building occupied approximately half of the block on the eastern side. This Type III, ordinary construction structure was believed to have been built in 1915 with multiple additions and renovations extending the structure to the rear (north and west). The original building was a two-story structure approximately 22 feet high with a flat roof. The exterior walls consisted of three-course brick and masonry construction. The front of the structure faced a city sidewalk that was covered by an overhead awning made of wood and metal. The roof and second floor were supported by wooden joists embedded into the west and east (B- and D-side) walls.

A steel beam running from front to back also supported the second-floor joists. A one-story addition had been added in multiple stages using brick and concrete blocks for the exterior walls. The fire is believed to have originated in the rear of the structure due to an undetermined cause. The exterior walls on both the fire structure and the adjacent theater contained star-shaped anchor plates. Anchor plates were used for structural reinforcement on buildings in the 18th, 19th, and early 20th centuries. These anchor plates were typically made of cast iron and were used as tie plates serving as the washers for tie-rods on brick or other masonry-based buildings. The tie-rod and plate assembly serves to brace the masonry wall against lateral bowing.

The victim joined the fire department in July 2010 and had received training on bloodborne pathogens and Heartsaver AED operations. He had completed Module A of the IFSI basic fire fighter training and was working toward IFSI Fire Fighter I certification. Postincident investigation revealed the department had limited training on structure collapse hazards with two of the key recommendations resulting from the investigation being that the department train all firefighting personnel on the risks and hazards related to structural collapse and on understanding the effects of master streams on structural degradation.

Postincident Analysis Pinckneyville, Illinois

Contributing Factors

- 96-year-old brick masonry structure degraded by fire burning for over 45 minutes.
- Fire fighters with limited experience entered collapse zone to move ground ladder.
- Entering collapse zone in close proximity to master stream directed onto roof.
- Limited visibility at side and rear of structure may have obscured signs of pending collapse.
- Limited training on structure collapse hazards.

After-Action REVIEW

IN SUMMARY

- The importance and value of a well-developed lesson plan cannot be overstated; it is the basic road map for an instructor to present important knowledge and skills to the fire fighters and responders who serve their communities.
- An important consideration in today's emergency services is making sure to avoid potential bias in the lesson material we develop for use in training our responders. Being aware of the increasing diversity of today's emergency responders will help you exercise good judgment in the development of the lesson plan.
- Instructor-centered instruction can include direct teaching such as lecture, the use of slides, and other presentation material where the student remains a passive learner and the instructor is the sole source of knowledge.
- Student-centered instruction can include methods such as small- and large-group activities, student portfolios or journals, and the use of class time for simulation clinical activities such as patient assessments.
- The first step of lesson plan development is to determine the learning outcomes. It is very difficult to develop a lesson plan when the learning objectives are not clearly stated.
- For the fire service, the most commonly used levels when developing *cognitive* learning objectives are (in order from simplest to most difficult):
  - Remembering
  - Understanding
  - Applying
- For the fire service, the most commonly used levels when developing *psychomotor* learning objectives are (in order from simplest to most difficult):
  - Imitation
  - Manipulation
  - Precision
  - Articulation
  - Naturalization
- The taxonomy of the *affective* domain identifies five levels of understanding:
  - *Receiving* is paying attention and displaying a willingness to learn.
  - *Responding* is displaying an acknowledged behavior within the learning experience and participating when given an opportunity.
  - *Valuing* is showing active involvement, passion, or commitment toward a topic.
  - *Organization* is accepting a new value as one's own and setting a specific goal.
  - *Characterization* is comparing and contrasting one's own values to others and using the new value.
- Often, an Instructor II needs to develop learning objectives to meet JPRs listed in a NFPA professional qualification standard. A JPR describes a specific job task, lists the items necessary to complete that task, and defines measurable or observable outcomes and evaluation areas for the specific task.
- Once the lesson outline is developed, all instructional materials needed to deliver instruction should be identified and listed in the lesson plan.
- The evaluation plan is the final part of the lesson plan. Each part of the evaluation plan should be directly tied to one or more learning objectives.
- When you develop your plan, it should be detailed and complete enough to achieve the identified goals and desired outcomes, yet flexible enough that it can be adjusted to meet students' needs on the fly.

KEY TERMS

Access Navigate for flashcards to test your key term knowledge.

ABCD method Process for writing lesson plan objectives that includes four components: audience, behavior, condition, and degree.

Enabling objective An intermediate learning objective, usually part of a series that directs the instructor on what he or she needs to instruct and what the learner will learn in order to accomplish the terminal objective.

Instructor-centered learning Instructional technique where the student is a passive listener and the instructor is the sole source of knowledge, such as a lecture with the use of presentation material.

Student-centered learning Educational methodologies that focus on student engagement and require students to be active, responsible participants in the learning experience (NFPA 1041).

Terminal objective A broader outcome that requires the learner to have a specific set of skills or knowledge after a learning process.

REFERENCES

Lorin W. Anderson, Krathwohl, David R., Airasian, Peter W., Cruikshank, Kathleen A., and Mayer, Richard E. (Eds.). 2001. *A Taxonomy for Learning, Teaching, and Assessing: A Revision of Bloom's Taxonomy of Educational Objectives.* Boston: Allyn & Bacon.

Bloom, Benjamin S. 1956. *Taxonomy of Educational Objectives, Handbook I: The Cognitive Domain.* New York: David McKay Company.

National Fire Protection Association. 2019. *NFPA 1041: Standard for Fire and Emergency Services Instructor Professional Qualifications.* Quincy, MA: National Fire Protection Association.

CHAPTER PRESENTATION EXERCISE

Instructions: Your course instructor will assign individual or group discussions on the key points and teaching tips of this chapter. You or your group will review the chapter teachings and identify the major learning points from the chapter. You should be able to discuss the points, why they are important, and how/where they apply to your responsibilities at your level of instructor training.

REVIEW QUESTIONS

1. Provide a few examples of both instructor-centered instruction and student-centered instruction, explaining the difference between the two types.

2. Name the three most common levels used in the fire service when developing *cognitive* learning objectives.

3. Which of the domains deals with an individual's expressed interests, ambitions, and values?

CHAPTER 10

Instructional Delivery

NFPA 1041 JOB PERFORMANCE REQUIREMENTS

Note: An asterisk denotes that the 1041 standard contains further information in its annex section.

5.2.6 Evaluate instructors, given an evaluation tool, authority having jurisdiction (AHJ) policy, and objectives, so that the evaluation identifies areas of strengths and weaknesses, recommends changes in instructional style and communication methods, and provides opportunity for instructor feedback to the evaluator.

(A) Requisite Knowledge.
Personnel evaluation methods, supervision techniques, AHJ policy, and effective instructional methods and techniques.

(B) Requisite Skills.
Coaching, observation techniques, and completion of evaluation records.

5.4.2 Conduct a class using a lesson plan that the instructor has prepared and that involves the utilization of multiple teaching methods and techniques, given a topic and a target audience, so that the lesson is delivered in a safe and effective manner and the objectives are achieved.

(A) Requisite Knowledge.
Student-centered learning methods, discussion methods, facilitation methods, problem-solving techniques, methods for eliminating bias, types and application of instructional technology tools, and evaluation tools and techniques.

(B)* Requisite Skills.
Facilitate instructional session, apply student-centered learning, evaluate instructional delivery, and use and evaluate instructional technology tools, evaluation techniques, and resources.

5.4.3* Supervise other instructors and students during training, given a specialized training scenario, so that applicable safety standards and practices are followed and instructional goals are met.

(A) Requisite Knowledge.
Safety rules, regulations, and practices; the incident management system; and leadership techniques.

(B) Requisite Skills.
Conduct a safety briefing, ability to communicate, and implement an incident management system.

KNOWLEDGE OBJECTIVES

After studying this chapter, participating in a structured learning environment, and completing assigned assessments, you will be able to:

- Match instructional methods to learning objectives and student characteristics. (**NFPA 1041: 5.4.2**, p 223)
- Describe safety considerations for supervising instructors. (**NFPA 1041: 5.4.3**, pp 225; 230; 233–234)
- Describe the instructor evaluation process. (**NFPA 1041: 5.2.6**, pp 229–230)
- Describe the role of providing feedback to, and receiving feedback from, the fire and emergency services instructor following an evaluation. (**NFPA 1041: 5.2.6**, pp 231–232)
- Explain the contents of a safety briefing. (**NFPA 1041: 5.4.3**, pp 233–234)

SKILLS OBJECTIVES

After studying this chapter, participating in a structured learning environment, and completing assigned assessments, you will be able to:

- Demonstrate transitioning between methods of instruction to facilitate effective learning. (**NFPA 1041: 5.4.2**, pp 224–225)
- Demonstrate basic coaching and motivational techniques. (**NFPA 1041: 5.4.3**, p 226)
- Perform an evaluation of a fire and emergency services instructor using an existing evaluation instrument. (**NFPA 1041: 5.2.6**, p 230)
- Supervise other instructors and students during training. (**NFPA 1041: 5.4.3**, pp 225–233)
- Conduct a safety briefing for students and instructors. (**NFPA 1041: 5.4.3**, pp 233–234)

You Are the Fire and Emergency Services Instructor

Recently, a private contractor was awarded the construction bid for a project entailing the installation of several thousand feet of underground sewer pipe, manholes, and catch basins in a new subdivision in your community that will require that some areas be tunneled rather than excavated. In preparation, the fire chief has approached you and asked you to develop a confined space awareness-level training program that emphasizes the safety concerns for confined spaces and safe response practices. Considering the different nature of this hazard and the required training, do the following:

1. Explain how to adjust the presentation of the training and still meet the objectives of the lesson plan while transitioning from a classroom setting to visiting construction sites.

2. Describe communication techniques that will improve your presentation, especially considering the low-frequency high-risk nature of these incidents.

Access Navigate for more practice activities.

Introduction

As a Fire and Emergency Services Instructor II, one of your primary functions is to develop a lesson plan, but in many cases, you may be expected to present your newly developed lesson plan. At the Instructor II level, you are in a position to teach when needed but also function in an area of mid-management between frontline instructors and a training division chief.

In addition to creating a lesson plan and, on occasion, instructing a class, a person operating at the Instructor II level is a supervisor to those teaching within a training bureau or division. The responsibility to supervise other instructors is an important one that includes monitoring their preparation, classroom conduct, and presentation skills. When an instructor you are supervising receives a complaint, you are the one who must address the complaint and then resolve it. There are times when you will lead an instructional team, perhaps teaching a recruit academy or an 18-month paramedic program; in that role, supervision becomes a critical function as the team leader.

Conducting a Training Session: Methods of Instruction

All classroom or drill field sessions require the instructor to use a **method of instruction** that is designed to share knowledge and material to the students in your class. In many cases, the method of instruction is identified on the lesson plan to help the instructor understand which presenting methodology is to be used. There are different methods of instruction and each can be applied to different types of materials as well as different types of students. Each of the various student demographics and categories will respond differently to the material according to how they are motivated and the method(s) of instruction used. At the Instructor II level, you are the person who decides which method of instruction is most appropriate to deliver the material in the lesson plan.

Chapter 5, *Using Lesson Plans*, discusses formatting of a lesson plan and placeholders in the lesson plan that developers can use to identify the methods of instruction they have determined to be most appropriate for the material. The same chapter also discusses the parts of a lesson plan and gives guidance on how lessons are constructed, considering the method of instruction to be used. A fundamental point to the use of a lesson plan is that it not only provides a road map to achieving the learning objectives of a course, but it will help protect the learner and the instructor from possible bias being introduced into the classroom. As the instructor, when you present material, you give a part of yourself to the students, but at the same time, you need to be mindful of following a lesson plan versus pushing your own agenda. A lesson plan can help in minimizing that possibility.

Chapter 3, *Methods of Instruction*, identifies several factors that determine which method of instruction to use:

- The experience and educational level of the students
- The classroom setting or training environment
- The objectives of the material to be taught
- The number of students to be taught
- The physical condition of the student; i.e., tired or alert
- Attitude about the training session or topic

As an Instructor II, understanding the various methods of instruction is a fundamental skill that you should master. Any level of instructor should be able to adapt a lesson plan to the learning environment as well as to the audience in order to make it valuable. All instructors will perform an audience analysis by assessing the demographics of the audience, and then adapt the lesson plan accordingly. Consideration should also be given to the time of day, lesson topic, and the knowledge level of the students.

The following is a quick recap of the most common and effective methods for teaching new and in-service personnel:

- **Lecture:** This method is the most common and familiar method of fire and emergency service instruction. An illustrative lecture allows the instructor to use a visual aid such as slides or videos to present lesson material.
- **Demonstration/skill drill:** This type of learning is used when the instructor demonstrates the use of a new tool or demonstrates how to perform a specific task-oriented skill; it is also known as a psychomotor lesson or hands-on method of instruction (**FIGURE 10-1**).
- **Discussion:** This method fosters interaction among the instructor and the students; the instructor's role in this method is more of a facilitator or monitor to ensure the discussion stays on track.

FIGURE 10-1 The demonstration method of instruction.
Courtesy of Alan Joos.

- **Role-play:** Many leadership and management classes feature a senior instructor role-playing a problem employee or angry citizen to allow students to practice handling confrontations.
- **Labs/simulations:** Placing students under pressure in a controlled scenario strengthens information learned in the classroom. Many simulation programs now exist to simulate actual fires and incidents that create a very realistic learning environment
- **Case studies:** Using an incident report that was generated after a line-of-duty death and allowing students to read the report, discuss the material, and then determine ways to apply the knowledge is an effective way to work with a case study.
- **Out-of-class assignments:** Many courses require pre-course work to be completed prior to class; the instructor might then give assignments in a workbook to be completed during the course. If the instructor does use this tool as part of the coursework, the instructor needs to make sure that these assignments are evaluated and that feedback is provided to the students.

Based on the material to be covered and the make-up of the class, most instructors use a combination of these methods while delivering course material. It is critical to observe how the class is reacting to the material you are presenting: are they engaged or bored? Asking questions or checking their watches? Sometimes we need to adjust our methodology during a class to help keep everyone engaged and learning.

Other methods of instruction are becoming prevalent in today's fire service: online learning, independent study, and blended learning. All three of these other methods have a place and role in the learning process. Training can be conducted online with students reviewing lesson material, completing a quiz, and printing out a course attendance certificate. Independent study classes are different in that a student manual is sent to students and on their own time, they read the material, complete a learning assessment, and submit it for grading and receive a certificate of completion in the mail or by email. Blended learning is a combination of studying the content online and then meeting in a classroom with an instructor who engages in hands-on training and reviews material covered in the online portion.

Instructional Techniques and Transitioning

In addition to the basic methods of instruction, there are other areas of instruction that all experienced instructors need to keep in mind, like the various techniques for maintaining students' attention while transitioning between media and lecture or moving from one topic or concept to a new one. During these transitions, it can be easy to lose a student's focus or interest in the lesson material. There are some simple and effective techniques for mitigating this. If a physical transition is going to take place, such as moving from the classroom to the training ground, you can lose students as they go to the coffee station or restroom. To avoid this, establish a clear timeline: direct the students to take a break for 10 minutes and then meet at Engine 31 in the truck bay where class will continue. To avoid losing the students' attention when making a transition from one topic to a new one, give them a quick preview of the new concept that is going to be discussed next. You can conclude where you are at in the lesson with a quick summary, then inform the students that after a 10-minute break, you are will introduce the new topic and build on what was just discussed. These are simple examples of transition that can make a big difference in maintaining focus in the learning environment.

Another area to pay attention to is behavior that can distract from the learning environment. There is a wide range of distracting behavior, some of which requires only a simple reminder to the student, and others that may require special techniques or even intervention. A smartphone is a valuable tool; however, when used inappropriately or too frequently, it becomes a distraction. But students may need only a quick reminder on the policy regarding its use during class. Some classroom behavior can be more challenging. How do you manage a student who takes on the role of the class clown, or a student who just does not want to be in your class and is going to make sure you and everyone around knows it? A useful technique is to directly address the student with a question to help bring his or her focus back to the classroom. Asking a student to share his or her experience with the lesson topic can help to elicit positive engagement but be careful that it does not backfire by providing the student with the opportunity to take the discussion in another direction. If you challenge a student, be aware of the door you are opening; it might be difficult to close once it has been opened. One advantage we have as instructors is the opportunity to simply talk to the disruptive student outside the classroom in a one-to-one environment. You can empathize with disruptive students on their frustrations while still reminding them of the purpose of the training. Comment on the value of their time and explore ways to recruit their participation. In the most extreme situations, you might have to ask the student to leave and follow up with his or her supervisor to address the issue.

Remember that an important component of the learning process is to assess the learning process during class, which can be done by asking questions. Effective questioning techniques serve to bring students' attention back into the classroom, as well as act as a refresher for students who might be struggling with the information. Many lesson plans will contain built-in questions the instructor can ask the students to determine whether they are truly understanding what is being taught. Chapter 3, *Methods of Instruction*, discusses this in greater depth.

DEPARTMENT TRAINING OFFICER TIP

A series of well-designed questions can springboard a discussion that reveals how much your students are taking in from the lesson.

Supervising Other Instructors

In the role of a supervisor, the senior instructor will determine who will assist in teaching the course, schedule the facility and equipment for the course, and provide leadership to those assigned instructors. Supervision at this level falls into important areas that will help in providing a safe learning experience for everyone involved. First, know your instructors and those you are training. It may be that you were once alongside them as a responder or fire fighter, but you are now in a position of providing direction, supervision, and leadership. Second, follow your agency's policies in all that you do. There is no excuse for intentionally disobeying an agency policy with the expectation of someone's life being in jeopardy. Always follow department policies regarding rehabilitation, fuel loading of props, instructor-to-student ratio, weather conditions on the training field, and classroom conduct. Last, lead by setting an example in all that you do. Being an example is the most powerful tool a supervisor can use in leading others. While on the training ground, if full protective equipment is being expected of the students, then you should wear full protective equipment also—there is no excuse for not setting the correct and proper example.

Managing fellow instructors requires leadership skills and the ability to recognize training scenarios that require specific supervision relating to safety and standard practice. You must ensure that applicable standards, standard operating procedures (SOPs), and departmental practices are used in the development and execution of training sessions that involve increased hazards to students. All instructors have different styles of presenting materials, different backgrounds, and different experiences that they bring to the training session. As an Instructor II, you should be familiar with these characteristics of instructors when supervising training delivery. Allow for individual styles and experiences in the training sessions, but make sure that all objectives, safety-related guidelines, department SOPs, and other performance expectations are covered. As a senior instructor, it is your responsibility to ensure that all policies and safety guidelines are followed by those you are supervising. For example, at no time in any situation should horseplay or hazing or bullying be allowed in the learning environment, on the training ground, or even during breaks. The fire fighters we are tasked to train should look to us as instructors and know that they are respected and protected by everyone on the department.

The process of supervising other instructors during the training process can be influenced by such factors as the material being presented and the qualifications of the instructors, the skill level of the students, and the number of students involved in the training process. A group of senior fire fighters reviewing ropes and knots being taught by one instructor will not need much supervision. In contrast, an instructor teaching advanced self-contained breathing apparatus (SCBA) skills to a group of new recruits might need additional help and guidance from other instructors and perhaps someone in charge of the entire training session to make sure groups rotate and the class stays on time.

The quality of instructional delivery and ability of the students to learn or perform a new skill are two of the criteria that may be used to evaluate training success. Too often in fire service history, instruction has fallen short of this goal and students and instructors have been injured or killed in poorly constructed or poorly supervised training sessions. Likewise, if the quality of the instruction is substandard, then the learning outcomes will be poor—a shortcoming whose effects can persist long after the student has left the classroom or drill ground.

As a supervisor, there are various skills and techniques to help you shape an instructor who presents lesson material in a safe manner; for example:

- **Coaching:** Coaching is a method of directing, instructing, and training a person in how to perform a new skill or method. As an Instructor II you can show a new instructor how to set a classroom or skill station on the training field, and coaching is a great technique that allows the person being taught the opportunity to do and then receive feedback on their performance.
- **Mentoring:** The process of mentoring is more in depth than coaching and provides the opportunity for a supervising instructor to spend more one-on-one time with a new instructor and help them excel in learning new or additional skills. Mentoring also involves allowing the new instructor to mirror or assist the senior instructor as they do they work and thus, the new instructor is given the opportunity to see the job performed in real time.
- **Challenging:** The process of challenging a new instructor and offering them new opportunities that might push them out of their comfort zone and help them gain new skills and confidence.
- **Delegation:** Another important skill that an Instructor II can use to develop a new instructor is that of delegation. By selecting the right person and matching them with the right project, timeline, and benchmarks to accomplish the project; clear expectations of the desired outcome; and budget (if appropriate), a new instructor can excel and develop new skills and confidence.

This not a complete list but does provide the Instructor II supervisor with basic tools to supervise new instructors, senior instructors, and students with a positive and productive learning environment.

Evaluating the Instructor

Fire and emergency services instructor evaluations may be a requirement for instructors' promotion. For certification at the Fire Officer I level under NFPA 1021: *Standard for Fire Officer Professional Qualifications*, you must obtain Fire Service Instructor I level certification to become a Fire Officer I. Fire and emergency services instructor evaluations may also be required to obtain state or national certification, as most states support a certification process for instructors. The National Fire Academy and other institutions require initial and periodic evaluation of instructors. In some cases, fire service personnel may have to go through (and pass) an instructor evaluation process before they are allowed to instruct students. This evaluation may be done by a supervisor, a peer, or a student, or the instructor may do a self-evaluation. Each evaluator puts his or her own knowledge and experience to work to complete the evaluation appropriately.

Formative Evaluation

Typically, the **formative evaluation** process is conducted for the purpose of improving the instructor's performance by identifying his or her strengths and weaknesses. For new instructors, this might happen several times during their first few classes as they are being developed. This formative evaluation may be done by either a supervisor or another instructor for the purpose of professional development.

The formative evaluation process is intended to assess the instructor's delivery during a class presentation, and the results are reviewed with the instructor to help him or her improve or enhance teaching skills and classroom performance. For this reason, it is often perceived as less threatening than other evaluations. A typical formative evaluation is designed to improve the instructor's abilities by identifying individual skills or deficiencies and allowing for the development of improvement plans.

A formative evaluation incorporates criteria that may allow the evaluator to review and observe the overall performance of the instructor by relating that performance to the expected outcomes of the training session. The instructor's ability to meet these criteria may be measured by the instructional methods used during the training, such as visual aids, transitioning between visual aids and the lesson plan, student participation, and the application of a student testing form to the learning objectives of the class. The evaluator may also participate in observation of student performance and take notes on the instructor's interactions with the students. Much of formative evaluation centers on instructional technique, classroom presence and conduct, and the delivery methods used by the instructor. This information is then compared to student success rates.

Summative Evaluation

A **summative evaluation** of the instructor occurs at the end of the class and could include the students' achievements through testing and performance on the final testing process. Many applications of summative evaluations take place at the end of the training process by providing students an opportunity to provide feedback on the instructor's performance. In some agencies, a summative evaluation is used for the purpose of making personnel decisions about the instructor's job status, merit pay, and promotion, and could affect his or her ability to teach other courses. However, a supervising instructor needs to be cognizant of a student's reasons for being in the class; precautions are necessary when considering information from this type of evaluation form based on a student who might not have wanted to be in class in the first place. In many agencies, the summative evaluation is conducted by someone other than the instructor who taught the course, to ensure the students feel comfortable in being honest on the evaluation form and to remove the possibility and perception that the evaluation has no value.

Student Evaluation of the Instructor

Students are in a position to rate the increased knowledge they have gained and their classroom experience as it relates to a particular instructor. They may provide valuable information about other factors not readily observed during an evaluation by the supervisor during a classroom observation session. Areas such as the instructor's punctuality, use of audiovisual equipment, and typical classroom demeanor that the students have observed during the class can be learned from an evaluation tool.

Some departments include an evaluation form for students to complete at the conclusion of the course. Different delivery schedules and multiple instructors delivering training over a long period of time may require evaluation forms to be distributed earlier, however, so that students can complete the evaluation while specific instructor characteristics are still fresh in their minds. A course evaluation typically includes a section that allows students to evaluate the instructor, classroom setting, instructional material, handouts or audiovisual material, and ability of the material taught to meet their needs. Developing class evaluation forms is addressed in more detail in Chapter 11, *Evaluation and Testing*.

Some students may view the evaluation as an opportunity to take a shot at the instructor because of a negative experience during the class. As a supervising instructor, if you observe one or two negative reviews of an instructor, you should not dismiss those reviews but take the opportunity to visit with the instructor and discuss the reasons for the comments. In many cases, the instructor will remember an incident that led to the negative comments, and this could become an opportunity to counsel the instructor on how to avoid a similar situation in future classes. If several students voice the same complaint, it is likely that there is an issue with the instructor and additional training or counseling might be necessary to correct the instructor's behavior and or conduct.

Other students may go through the motions of completing the evaluation form without giving it much thought or consideration. In such cases, a department can account for the lack of concrete feedback by evaluating the students participating in the class—that is, students can be compared in terms of participation, increase in knowledge, and skill performance level. This information can then be used when reviewing the students' evaluations for the instructor and the course.

Job Performance Requirements in ACTION

Evaluation of a fire and emergency services instructor and the content of a presentation serves as a form of quality assurance. These measurements of the instructor's skills and impressions of how the course material helped students understand and apply the learning objectives are an important part of an instructor's professional development. A student review of the instructor's attributes is one method of performing an evaluation. Another strategy is to have a senior instructor conduct an in-person review of the course delivery.

Instructor I

All instructor levels should develop an appreciation for how course and instructor evaluations benefit the instructional process. At the Instructor I level, you should consider all feedback provided in an evaluation as a measure of professional development and work to improve your performance based on any constructive criticisms provided while building on positive attributes.

Instructor II

Evaluations of instructor performance are a type of personnel evaluation, and their construction and use should reflect the intent of any evaluation—namely, to improve the performance of the person being evaluated. The Instructor II will develop class and instructor evaluations and conduct evaluations of other instructors in an effort to improve both the instructor and the course elements.

Instructor III

Evaluation strategies performed by the Instructor III include the development of an ongoing instructor evaluation plan. This may be similar to the annual performance review used for any other level of responsibility, such as fire fighter or company officer, with the criteria in this case focusing on how well the instructor performed his or her duties during that period. Evaluation strategies are covered in Chapter 12, *Program Management and Training Resources*.

JPRs at Work

The National Fire Protection Association (NFPA) does not identify any JPRs that relate to this chapter. Nevertheless, the Instructor I should understand how to utilize information provided during the evaluation process.

JPRs at Work

Develop evaluation forms and evaluate instructors in an effort to improve instructor skills and the quality of the course materials by providing feedback on class presentation elements.

JPRs at Work

There are no JPRs for the Instructor III.

 Bridging the Gap Among Instructor I, Instructor II, and Instructor III

Personnel who meet the Instructor I JPRs will be able to use their experiences in being evaluated when they transition to Instructor II by recalling the elements of the evaluations that they were able to put to use so as to improve their own performance. When conducting evaluations, the Instructor II should be professional and constructive in identifying areas for improvement. Individual strengths and weaknesses should be highlighted, and both instructor levels should use this step in the instructional process in a positive manner. The Instructor III will acquire these evaluations and incorporate their results into an ongoing evaluation plan. Monitoring performance of instructors is a key part of maintaining an effective training program.

The Evaluation Process

NFPA 1041 references departmental policies and procedures as one of the criteria for an evaluation. The standard requires that the department or authority having jurisdiction (AHJ) use qualified individuals as fire service instructors. NFPA 1041 also expects that the instructor will have the capability and skill level to demonstrate how to perform the skills included in the training session appropriately. The instructor's level of competency must be identified in a policy developed and enforced by the AHJ. In addition, this policy must include a method for verifying the qualifications and competency of the instructor.

A policy for instructor evaluation includes many components. It should identify who does the evaluation and ensure that the evaluation remains confidential. Confidentiality allows the evaluator to be critical without embarrassing the instructor being evaluated. A timetable needs to be established for evaluations, so that instructors are evaluated initially and then periodically. The policy should determine a process for practical instruction as well as presentations. It should address the instructor's preparations to deliver the material, the arrangement of the classroom, and the readiness of any equipment. The instructor should have reviewed the lesson plan and adapted it to the audience. If the coursework will address practical evolutions, the instructor should wear the same level of protection as is required for students.

The policy should require that the instructor follow the lesson outline; keep the class moving; use appropriate equipment; finish material delivery in the allotted time frame; and deliver a summary. The policy must outline positive traits expected of the instructor, including being unbiased, encouraging open discussion, being clear during the communication process, wearing appropriate attire, remaining flexible, and being honest.

The policy for evaluating an instructor can be a short paragraph stating when evaluations will be done, by whom, and in which form; conversely, a detailed policy may enumerate the entire evaluation process. In either case, the department's policies must be incorporated into the process and then followed as outlined.

Preparation

An important function at the Instructor II level as a supervisor is to evaluate a person at the Instructor I level. One place to observe instructors is in the classroom, because it places them in their work environment. To make an evaluation of an instructor's performance a positive event, start by selecting a topic

FIGURE 10-2 Prior to the evaluation, review the lesson plans.
© Jones & Bartlett Learning.

from the training schedule with which you are familiar and comfortable, so you can focus on observing the instructor's performance rather than being distracted by the course content. Prior to conducting an evaluation, you should to review the instructor evaluation criteria and relevance to the job description; review the lesson plans and supporting material; review the evaluation tool used by your agency; and then schedule the date, time, and location for the evaluation (if deemed appropriate by the AHJ (**FIGURE 10-2**).

Prior to the training session you are going to observe, meet with the instructor to state the purpose of your visit and the reason for your evaluation. Identify the areas that you will be reviewing, including the delivery of the provided lesson plan material, the instructor's ability to apply the material to the class, adherence to department and accepted safety practices, and the instructor's overall presentation skills. Encourage the instructor to present normally and not to involve you in the delivery process. Be sure to arrive early enough to determine whether the instructor is properly prepared. The evaluation standards should be objective, identified prior to the evaluation, focused on the instructor's performance, and part of an improvement program for the instructor and course content, if applicable.

Observation

During the presentation of the lesson, evaluate the instructor from a holistic point of view. Look at his or her entire skills set versus focusing in on one area. By using all sources of information, you will give a more complete analysis of the instructor's performance. During classroom training evaluations, assess the following areas:

- **The instructor's attire:** Is the instructor's attire professional, compliant with policy, and appropriate for the audience?

- **The instructor's demeanor:** Is the instructor professional, approachable, and appropriate for audience? Does he or she appear to be engaging the audience? Does he or she inspire confidence and respect?
- **The presentation:** When evaluating the presentation, note the instructor's ability to relate the topic directly to the students' needs. Note mannerisms, language usage, and comfort level with technology. Do you feel that you are watching a black-and-white, still-picture show or a full-color, interactive presentation?
- **The lesson plan:** One effective method is to observe the classroom presentation while monitoring the lesson plan that the instructor is following. The lesson plan components should be reviewed in terms of how they were used and whether the presentation was delivered in a consistent manner. By using a copy of the lesson plan and matching it with the actual presentation, you can also review how well the instructor followed and used the lesson plan.

During practical or drill-ground evaluations, the overall safety of the students is the major focus of the evaluation. Instructors have the twin responsibilities of presenting training in a safe manner and of teaching safe practices. When reviewing a instructor during practical sessions, consider whether he or she was able to anticipate problems in the delivery of instruction such as noise, weather, or equipment availability.

Certain types of training sessions, such as live fire exercises or complex rescue scenarios, may require the assignment of a safety officer and should always follow NFPA guidelines when applicable. Both the instructor and the students must dress appropriately for activity, wear personal protective equipment (PPE) correctly, and clean or inspect the PPE after heavy use. The best review of a practical training session is to determine how closely the instructor followed the lesson plan, provided hands-on instruction on the training ground, and was able to tie it all together and how it would apply in the real world.

Lesson Plan

The lesson plan is the instructor's road map. It is occasionally appropriate to stray off the road, providing that the side trip benefits the overall experience and enhances student knowledge. The instructor must be familiar with the lesson plan and its learning objectives for the class—a critical point to keep in mind during the evaluation. Effective instructors use the lesson plan as a tool to deliver the desired knowledge. As instructors become comfortable with lesson content, they will appear to use the lesson plan less, but in reality, they are so familiar with the lesson content they can teach the desired content without seeming to read the lesson plan. Effective instructors deliver the lesson plan but do so with ease and in a comfortable manner. Remember that just because the instructor you are evaluating is not constantly looking at the lesson plan does not mean he or she is not following it.

Evaluation Forms and Tools

Because of the diverse reasons for which evaluations are undertaken, a variety of forms are used to document the evaluation process. Instructor evaluation forms or tools vary widely, with some being better than others. When a department adopts a certain tool, that document must be used during your evaluation (**FIGURE 10-3**). Before adopting an evaluation tool, it is prudent to review several different types of forms and tools to determine which content and criteria best fit the needs of the department. If an agency plans to select an evaluation tool that is produced commercially or adopted from another agency, all of the instructors who will be evaluated using the tool should be given the opportunity to provide feedback on which tool to select. Allowing instructors the opportunity to share their opinions on the type of instructor evaluation tool and its content, format, rating system, and how the results will help them can assist in the adoption of the evaluation tool. Knowing the purpose and goal of the evaluation process, as well as which information will be gathered, will ensure that instructors are more willing to participate in the evaluation process.

Once the evaluation of the instructor is complete, it is time for the supervising instructor to complete the evaluation process and paperwork. The saying, "The job isn't over until the paperwork is done," is accurate and is a fundamental step for the supervisor. Thorough and accurate records are essential to maintain credibility and help ensure fairness for everyone involved. By procrastinating in the evaluation process, you will only make it harder on yourself as you become overwhelmed with your other responsibilities. Avoid this by doing the paperwork and forms for an evaluation, following these steps:

1. Review the tool that was used in the evaluation.
2. Have appropriate writing materials.
3. Take notes and/or fill in the appropriate areas on the tool.
4. Review the instructor's performance with him or her.
5. Submit paperwork to the AHJ as directed or required.

Lead Instructor Evaluation Criteria

Please shade the letter that indicates your evaluation of the course/instructor feature using #2 pencil.

Instructor Attributes	Excellent	Very Good	Good	Fair	Poor
Ensure that the daily classroom and practical training evolutions were conducted in a safe and professional manner	(1A)	(1B)	(1C)	(1D)	(1E)
Ensure that all learning objectives were covered with more than adequate resource information	(2A)	(2B)	(2C)	(2D)	(2E)
Coordinate the activities of all students so that the class flow and pace was not interrupted	(3A)	(3B)	(3C)	(3D)	(3E)
The instructor was prepared to present the program	(4A)	(4B)	(4C)	(4D)	(4E)
Provide documentation to your progress in class and in general on a regular basis	(5A)	(5B)	(5C)	(5D)	(5E)
Monitor all class and practical activities and provide a structured learning environment	(6A)	(6B)	(6C)	(6D)	(6E)
Provide a positive role image to you at all times and conduct themselves in a professional manner	(7A)	(7B)	(7C)	(7D)	(7E)
Was accessible and responsive to your individual needs as a student in this program	(8A)	(8B)	(8C)	(8D)	(8E)
Was actively involved in your training and education throughout the program	(9A)	(9B)	(9C)	(9D)	(9E)
Worked to improve program content, policy and procedure	(10A)	(10B)	(10C)	(10D)	(10E)
The instructor is knowledgable about this program and acted accordingly	(11A)	(11B)	(11C)	(11D)	(11E)
The instructor was instrumental in your success in this program	(12A)	(12B)	(12C)	(12D)	(12E)

Instructor(s) Evaluation: Please use the following scale to evaluate the instructor(s).

A = Excellent; clearly presented objectives, demonstrated thorough knowledge of subject, utilized time well, represented self as professional, encouraged questions and opinions, reinforced safety practices and standards

B = Good; presented objectives, knowledgeable in subject matter, presented material efficiently, added to lesson plan information

C = Average; met objectives, lacked enthusiasm for content or subject, lacked experience or background information, did not add to delivery of material

D = Below Average; did not meet student expectations, poor presentation skills, lack of knowledge of content or skills, did not interact well with students, did not represent self as professional

E = Not Applicable; did not instruct in section or course area, do not recall, no opinion

Instructor Name	Excellent	Good	Average	Below Average	Not Applicable
	(13A)	(13B)	(13C)	(13D)	(13E)

FIGURE 10-3 A sample instructor evaluation tool.
Courtesy of Forest Reeder.

Instructor Feedback

After completing the evaluation of the instructor, you should meet with the instructor and let him or her know that you enjoyed being in the class (if that is the case) and that you appreciate the opportunity to participate with the instructor in the evaluation process. Unless you are prepared to meet with the instructor immediately, you should set up a time to discuss your evaluation and provide feedback. In many situations, the instructor may be tired or in a hurry to prepare for another class, so take the time to make an appointment when both of you can sit down in a relaxed environment to foster receptivity and focus (**FIGURE 10-4**). Just as students appreciate quick feedback on test results, so instructors should also have quick access to instructor evaluation tools.

In certain cases, the class may have gone wrong or the instructor may have done a poor job in presenting the lesson plan. In such a situation, you might need to take some time to step back and look for positives in the evaluation before you meet to address negative parts of the evaluation. If the true purpose of the evaluator is to improve the quality and performance of the

FIGURE 10-4 Set up a dedicated time and area for instructor feedback.
© Jones & Bartlett Learning. Photographed by Glen E. Ellman.

instructor, then discussing the evaluation after time for consideration works best for all parties involved.

Evaluation Review

After the evaluation of the instructor is complete and you have reviewed the evaluation form and your notes, you are better prepared to provide feedback to the instructor you evaluated. During your discussion with the instructor, take the time to review your notes in person with the instructor (**FIGURE 10-5**). Specifically, address the instructor's strengths and weaknesses. During the review, strive to show professionalism by maintaining a level of formality appropriate for the situation, thereby sending the message that this is a formal process. If agency policy allows, provide the instructor with copies of completed evaluation forms for his or her reference and professional development. At the very least, a summary of evaluation results should be provided to the instructor.

Make arrangements to meet with the instructor for the evaluation soon after the observation is conducted.

FIGURE 10-5 Review your notes in person with the instructor.
© Jones & Bartlett Learning.

To show respect and add value to the process, set time aside and conduct the review in your office where you can focus on the review with the instructor. Start the session by thanking the instructor for the opportunity to observe the class. Establishing a dialogue with the person you are evaluating will help open the lines of communication, so that when you provide the results of the evaluation, there will be a willingness to listen to what you have to share. All feedback should be delivered in a considerate manner, emphasizing the instructor's positive attributes first, and then addressing any areas that were observed to need improvement. Offer ideas or suggest changes to deal with weaknesses and ask the instructor for ideas he or she thinks might help in dealing with those areas. Do not overwhelm the instructor with criticism; instead, provide constructive feedback. Strive for a balance between positive and negative comments, and set goals for improvement when appropriate. To help provide meaningful direction, provide examples from your own experiences as an instructor or provide a suggestion of where the instructor could find additional examples or material. If the instructor struggled with the use of computer equipment, offer time to review the equipment and provide practical tips on that equipment. In this way, you can turn the evaluation session into a learning session. You may also want to include a follow-up review at a later date. As in any evaluation, make sure to observe review dates and complete reevaluations to maintain credibility of the evaluation process. Finally, always make yourself available to the instructor for advice and guidance.

Another situation you might find yourself in as a supervisor could be when the instructor you observed did an outstanding job and you have limited to no areas noted for improvement. Because even the best instructors can usually improve on some aspect, this should be a rare occurrence; however, avoid searching for issues just so you have something to discuss. There is nothing wrong with telling an instructor they did a great job and challenging them to maintain that high level of performance.

Throughout this instructor evaluation process, you should ensure that the instructor who is being evaluated has the opportunity to give you feedback. The evaluation process should be an open, two-way communication process. By fostering an open dialogue with the instructors, they will be more open to your suggestions and recommendations. Encourage an open environment of communication and listening. Effective supervisors learn the art of active listening and learn to build relationships with their staff, allowing for an evaluation that will be more effective in helping to shape a valuable new member of the instructor cadre.

Supervision During High-Risk Training

One of the most important responsibilities you have at the Instructor II level is supervising other instructors during training that involves a high-risk hazard to either the student or the instructor. Examples of high-risk training include live fire, high angle, confined space, water rescue, and other types of training that could endanger a student or instructor (**FIGURE 10-6**).

The creation of NFPA 1403: *Standard on Live Fire Training Evolutions*, came about from fire fighter fatalities that occurred during live fire training. There are other standards that address how training should be conducted to ensure the safety of both students and instructors. In adding to training standards for specific types of training, the NFPA has also developed standards such as NFPA 1584: *Standard on the Rehabilitation Process for Members During Emergency Operations and Training Exercises*, which addresses the need for rehabilitation during live fire or other training activities.

In addition to being aware of the NFPA standards, all instructors must be aware of and have a working knowledge of their agency's policies for conducting training within their department. Some agencies have very detailed policies that address aspects such as rehabilitation, classroom conduct, recordkeeping, levels of protective equipment required for each training scenario, and what precautions to follow during adverse weather conditions on the training ground. An absence of training policies could expose both instructors and students to unpleasant consequences in the case of a training evolution gone awry. Learn your agency's policies and follow them; if your agency does not have training policies, work on developing them to provide a safer training environment.

Live Fire Training

One of the highest-risk activities performed during the training process is the use of live fire training. This training could take place in an acquired structure, a burn building, or with the use of various propane props (**FIGURE 10-7**). The use of live fire training is critical in preparing fire fighters with the fundamental skills to perform one of their primary missions: to suppress fire in all its various forms or locations.

All instructors participating in live fire evolutions should meet NFPA 1403 and the most current NFPA 1041 with the additional live fire instructor qualifications. NFPA 1403 and 1041 are specific in how a live fire evolution should be conducted with the use of different staffing positions to oversee the safety of those participating in the live fire training. The main person in charge of the live fire training is identified as the **instructor-in-charge**, the senior instructor who directs and supervises other instructors during a live fire evolution. NFPA 1403 and 1041 also identify the need for a **safety officer** and an **ignition officer**.

The safety officer is responsible for intervention in and control of any aspect of the operation that he or she determines to present a potential or actual danger at all live fire training events. The instructor-in-charge is required to plan and coordinate all training activities and is responsible for following NFPA 1403 in its entirety. The ignition officer is responsible for igniting and controlling the material being burned during the training evolution. The ignition officer coordinates all live fire training evolutions with the instructor-in-charge, the safety officer, and the fire control team.

Safety Briefings

For all high-risk training, it is important to conduct a safety briefing for instructors and participants. Prior to

FIGURE 10-6 Live fire training is a prime example of high-risk training.
Courtesy of Alan Joos.

FIGURE 10-7 A structural prop used for live fire training.
© Jones & Bartlett Learning. Photographed by Glen E. Ellman.

conducting an actual live fire evolution, NFPA 1403: *Standard on Live Fire Training Safety Evolutions* specifically states that there must be a safety briefing conducted by the instructor-in-charge with the safety officer. All fire fighters who are going to participate in the live fire training are to attend this briefing and participate in the walkthrough of the structure/burn building or props being used for the training. The briefing should address all safety procedures and provide a forum to address questions. It is critical that everyone involved in the training is given the opportunity to ask questions so that the purpose of the event is clearly understood. Each participant must be assigned to a position or role; unassigned students or participants should remain in a staging area away from the evolution. During the briefing, the training objectives should be clearly stated and the desired outcomes conveyed to all participants so that everyone understands their roles and how to accomplish the training goals. The safety briefing should clearly identify the following:

- The objectives for the training session
- The role of each participant involved in the training
- Specific hazards to watch for
- Specific expectations of the instructors' and participants' conduct during the training evolution

High-risk training should be focused on one topic at a time, rather than attempt to cover several aspects of training all at once. It is also important that all high-risk training be conducted with a zero tolerance for any conduct that could injure others involved in the training.

During all live fire training evolutions, the use of "live" victims is absolutely prohibited and is unacceptable. If a manikin is going to be used during the training, all participants must be made aware of its use and exactly how it will be attired. If a manikin is going to be used as a civilian it should be dressed accordingly. If the manikin is being dressed in fire fighter turn out gear, it must be uniquely marked or colored so as not to be confused with the turnout gear being worn by students or instructors in the live fire evolutions. When using a manikin for training its location does not need to be divulged or shared with the students, but the instructors should be aware of its location.

It is important to adhere to a proper student-to-instructor ratio when conducting training involving live fire. That student-to-instructor ratio is 5:1. In some training sessions involving multiple fire departments, you will need a large pool of instructors to assist you and maintain the 5:1 ratio of crew members to instructors. This 5:1 ratio must be upheld throughout the evolution to ensure the safety of the students and accountability for the instructors. This ratio may need to be even lower, depending on the capabilities of the students, the complexities of the training evolution, or the size of the physical location where the training is taking place.

An effective way to conduct a live fire training evolution is to apply a basic incident command structure with the instructor-in-charge being the incident commander, the safety officer functioning within that role, and the ignition officer being the operations officer. Other support areas should be rehabilitation under the logistics designation, a public information officer to control media inquiries, a water supply officer, and emergency medical services for students and instructors. The use of the incident management system is an effective method for controlling a live fire training event. The fire service uses an incident management system on a daily basis; therefore, transposing its structure and principles for use during live fire training can be a very effective method to manage a high-risk environment.

High-risk training that is conducted on a regular basis should follow the same procedures as listed for a live fire training evolution. High angle, confined space, water, ice rescue, and other types of high-risk training should all be addressed with the same degree of caution and preparation. Supervisory skills that have been identified for live fire training are appropriate for use during any high-risk training and include:

- Using qualified instructors for the training evolution
- Using a command structure that identifies each instructor's role in the evolution and accountability for all participants
- Using an Incident Action Plan (IAP)/training plan for the training evolution with specific training and outcome objectives identified
- Conducting a safety briefing for all participants involved in the training
- Discussion of the communications plan and emergency evacuation signal
- Conducting an "After-Action Review" of the training to discuss pros/cons

Even when you have a training event planned down to the smallest detail, you have fallen short if you fail to define and convey the purpose and goals for the event. Communication during all training activities is critical to a safe training environment and is an important management tool. Whether a training session involves live fire or watching a video in a classroom, there needs to be clear communication among instructors, support staff, and students to ensure a safe learning environment and the attainment of the learning objectives.

Training BULLETIN

JONES & BARTLETT FIRE DISTRICT
TRAINING DIVISION
5 Wall Street, Burlington, MA, 01803
Phone 978-443-5000   Fax 978-443-8000

Instant Applications: Instructional Delivery

Drill Assignment

Apply the chapter content to your department's operation, training division, and your personal experiences to complete the following questions and activities.

Objective

Upon completion of the instant applications, fire and emergency services instructor students will exhibit decision making and application of job performance requirements of the instructor using the text, class discussion, and their own personal experiences.

Suggested Drill Applications

1. Identify the most effective and least effective training delivery techniques you have experienced and consider which would be the most effective for a training session involving a low-frequency high-risk type of event.

2. Consider how interfacing with and possibly training alongside a private contractor would take place and what efforts may need to be emphasized in order to maintain the level of professionalism desired of your organization.

3. Visit the website for your state training organization and determine if any of their offerings would be appropriate for the training mentioned in the Incident Report in this chapter.

4. Determine the most effective methods you have witnessed for dealing with disruptive and unsafe behaviors in the classroom.

Incident Report: NIOSH Report # F2010-31

Drill Assignment

Review the information in this incident report and prepare a practice presentation for class delivery at the direction of your instructor. Your presentation should include a summary of the incident facts and identify the NFPA standard(s) that could apply to the incident. Use an outline to organize your thoughts. You may be evaluated on your communication skills during your presentation.

Tarrytown, New York—2010

On September 6, 2010, a 51-year-old male volunteer fire fighter (victim) died after being overcome by low oxygen and sewer gases while climbing down into a sewer manhole in an attempt to rescue a village utility worker. The utility worker had entered the manhole to investigate a reported sewer problem and was overcome by low oxygen and sewer gases.

The incident occurred behind the fire station in an underground sewer line that ran under the fire station. The local utility company contacted the chief of the village's volunteer fire department and requested that a piece of fire apparatus be moved out of the station so they would not block it in while accessing a manhole. The fire chief responded to the station to move the fire apparatus so it would not be blocked by the utility trucks. The victim and another fire fighter also arrived at the station to assist.

A utility worker entered the manhole behind the station to clear a sewer backup and was overcome by a lack of oxygen and sewer gases and then fell unconscious inside the manhole. The victim then entered the manhole without any PPE to help the utility worker and was also overcome by the low oxygen level and sewer gases. The victim and the utility worker were later removed from the sewer manhole by fire department personnel and transported to a local hospital where they were pronounced dead. The medical examiner reported the cause of death as asphyxia due to low oxygen and exposure to sewer gases.

The victim's department conducts in-house training and participates in joint training sessions with neighboring fire departments. Members also attend training provided by the state office of fire prevention and control. The county department of emergency services offers scheduled fire fighter training (including technical rescue confined space training) throughout the year and that training is made available to all departments in the county through notices emailed to the chiefs in the county and is also posted on their website. The department involved in this incident has since met with the department of emergency services and scheduled fire fighter confined space awareness training.

Postincident Analysis Tarrytown, New York

Key Recommendations

- Ensure that fire fighters are properly trained and equipped to recognize the hazards of and participate in a confined space technical rescue operation.
- Ensure that SOPs regarding technical rescue capabilities are in place and a risk–benefit analysis is performed to protect the safety of all responders.
- Ensure that an effective incident management system is in place that supports technical rescue confined space operations.
- Ensure that a safety officer properly trained in the technical rescue field being performed is on scene and integrated into the command structure.

After-Action REVIEW

IN SUMMARY

- At the Instructor II level, you are in a position to teach when needed but also function in an area of mid-management between frontline instructors and a training division chief.
- All classroom or drill field sessions require the instructor to use a method of instruction that is designed to share knowledge and material to the students in your class.
- To deliver your message effectively, you must know who your audience is.
- Managing fellow instructors requires leadership skills and the ability to recognize training scenarios that require specific supervision relating to safety and standard practice.
- The quality of instructional delivery and ability of the students to learn or perform a new skill are two of the criteria that may be used to evaluate training success.
- The formative evaluation process is typically conducted for the purpose of improving the instructor's performance by identifying his or her strengths and weaknesses.
- A summative evaluation process measures the students' achievements through testing or completion of evaluation forms to determine the instructor's strengths and weaknesses.
- During the evaluation process, evaluate the instructor from a holistic point of view rather than focusing on only a few aspects of his or her presentation. Areas to assess include the instructor's attire, demeanor, presentation style, and how well he or she is following the lesson plan.
- The instructor must be familiar with the lesson plan and its learning objectives for the class—a critical point to keep in mind during the evaluation.
- One of the most important responsibilities you have at the Instructor II level is supervising other instructors during training that involves a high-risk hazard. Examples of high-risk training include live fire, high angle, confined space, water rescue, and other types of training that could endanger a student or instructor.
- In addition to being aware of the NFPA standards, all instructors need to be aware of and have a working knowledge of their agency's policies for conducting training within their department.
- One of the most high-risk activities performed during the training process is the use of live fire training. The use of live fire training is critical in preparing a fire fighter with the fundamental basic skills to perform one of their primary missions: to suppress fire in all its various forms or locations.

KEY TERMS

Access Navigate for flashcards to test your key term knowledge.

Formative evaluation Process conducted to improve the instructor's performance by identifying his or her strengths and weaknesses.

Ignition officer An individual who is responsible for igniting and controlling the material being burned at a live fire training.

Instructor-in-charge An individual qualified as an instructor and designated by the authority having jurisdiction to be in charge of the live fire training evolution and who has met the requirements of an Instructor II in accordance with NFPA 1041 (NFPA 1403).

Method of instruction Various ways in which information is delivered to student, both in a classroom and on the training ground. (NFPA 1041)

Safety officer An individual appointed by the authority having jurisdiction as qualified to maintain a safe working environment at all live fire training evolutions. (NFPA 1403)

Summative evaluation Process that measures the students' achievements to determine the instructor's strengths and weaknesses.

REFERENCES

National Fire Protection Association. 2019. *NFPA 1041: Standard for Fire and Emergency Services Instructor Professional Qualifications.* Quincy, MA: National Fire Protection Association.

National Fire Protection Association. 2015. *NFPA 1584: Standard on the Rehabilitation Process for Members During Emergency Operations and Training Exercises.* Quincy, MA: National Fire Protection Association.

National Fire Protection Association. 2018. *NFPA 1403: Standard on Live Fire Training Evolutions.* Quincy, MA: National Fire Protection Association.

CHAPTER PRESENTATION EXERCISE

Instructions: Your course instructor will assign individual or group discussions on the key points and teaching tips of this chapter. You or your group will review the chapter teachings and identify the major learning points from the chapter. You should be able to discuss the points, why they are important, and how/where they apply to your responsibilities at your level of instructor training.

REVIEW QUESTIONS

1. Name four components of a safety briefing.
2. What is the proper student-to-instructor ratio when conducting training involving live fire?
3. During classroom training evaluations, what are the four main aspects you should assess?

CHAPTER 11

Evaluation and Testing

NFPA 1041 JOB PERFORMANCE REQUIREMENTS

Note: An asterisk denotes that the 1041 standard contains further information in its annex section.

5.5.2 Develop student evaluation instruments, given learning objectives, learner characteristics, and training goals, so that the evaluation instrument measures whether the student has achieved the learning objectives.

(A) Requisite Knowledge.
Evaluation methods, evaluation instrument development, and assessment of validity and reliability.

(B) Requisite Skills.
Evaluation item construction and assembly of evaluation instruments.

5.5.3* Develop a class evaluation instrument, given authority having jurisdiction (AHJ) policy and evaluation goals, so that students have the ability to provide feedback on instructional methods, communication techniques, learning environment, course content, and student materials.

(A) Requisite Knowledge.
Training evaluation methods.

(B) Requisite Skills.
Development of training evaluation instruments.

KNOWLEDGE OBJECTIVES

After studying this chapter, participating in a structured learning environment, and completing assigned assessments, you will be able to:

- Describe the types of written examinations. (**NFPA 1041: 5.5.2**, pp 240–241)
- Describe the role of testing in the systems approach to the training process. (p 244)
- Describe how to develop and analyze student evaluation instruments. (**NFPA 1041: 5.5.2**, pp 240–256)
- Describe how to develop a class evaluation form. (**NFPA 1041: 5.5.3**, pp 256–258)

SKILLS OBJECTIVES

After studying this chapter, participating in a structured learning environment, and completing assigned assessments, you will be able to:

- Demonstrate how to prepare an effective exam for student evaluation. (**NFPA 1041: 5.5.2**, pp 245–256)
- Develop a class evaluation form. (**NFPA 1041: 5.5.3**, pp 256–258)

You Are the Fire and Emergency Services Instructor

After presenting several training sessions at your fire training academy on fire fighter safety and survival, the department training director requests feedback on the success of the program, including individual comprehension and retention rates for each student. Because you are relatively new to the cadre of instructors assigned to the academy, you are somewhat unsure of how to go about providing the results to the director. In preparation for conducting an evaluation of the training, you need to do some research and answer a few questions.

1. How are student evaluation instruments and class evaluation forms developed and analyzed?

2. What type of written examination would be the most appropriate for evaluating the comprehension and retention rates of students who attended the fire fighter safety and survival training?

 Access Navigate for more practice activities.

Introduction

As an Instructor II, one of most important responsibilities you have is to develop student evaluation instruments that will be used to evaluate their performance and abilities. Testing plays a critical role in training and educating fire and emergency services personnel, as explained in Chapter 8, *Evaluating the Learning Process*. A quality testing program allows you to assess student progress and ascertain which objectives students have met.

In testing, fire and emergency agencies typically face problems such as the following:

- Lack of standardized test specifications and test format examples
- Confusing procedures and guidelines for test development, review, and approval
- Lack of consistency and standardization in the application of testing technology
- Failure to perform formal test-item and test analysis
- Inadequate instructor training in testing technology

This chapter will build on the information provided in Chapter 8, *Evaluating the Learning Process*, to help you overcome these common issues and offer guidance for strengthening your testing and evaluation program.

Types and Purpose of Testing

Testing and evaluating students is a process used within fire and emergency services training to validate the training process, to determine what the student has retained, and to demonstrate competency and subject mastery. There are numerous terms that are often used to discuss this process but have at times been confused or used incorrectly. To help in eliminating confusion for instructors and developers, we will use the following definitions:

- Pre-test. This is an evaluation tool use by instructors to determine a student's level of understanding of the material that will be presented in a course. One example often used in the academic environment of this type of test is a placement exam to determine in which level of math to place a student based on his or her performance. Another example could be the feature used in this text called "You Are the Instructor." This is an example of a pre-test for a chapter to help the student to evaluate his or her understanding of content to be covered during a class session, which will also help the instructor know the level of understanding of the students.
- Posttest. This is an evaluation tool that could be the summative assessment tool or could

be used at the end of a chapter to determine understanding of the material.

- **Formative assessments.** This evaluation tool is used during the course of a class to help the instructor determine the level of understanding of the students and for students to determine where they might need additional assistance from the instructor. The formative evaluation tool allows for interaction between the instructor and student to discuss the learning process and how to improve. This type of testing has been referred to as "low-risk" evaluation because this assessment process is meant to be an exchange of ideas without a negative impact on the final grade. Examples include end-of-unit or chapter quizzes, group presentations of a case study, or an in-class oral discussion on a topic.
- **Summative assessments.** This evaluation tool is used at the end of a course. Often summative assessments are considered "high-risk" evaluations because the goal of this level of evaluation is used to measure mastery of the skills and knowledge of the course material. Examples of high-risk evaluations include the following:
  - **Final training test.** Typically, final training tests are used to support the decision to pass or fail a student at the end of the training program. The importance and consequences of these tests can have an effect on whether the fire fighter or medical first responder can qualify for a certification test or licensure examination. These tests *must* be supported with job-relevant documentation and achieve high levels of test reliability. The consequences of failure from both the examinee and training institution's point of view are important and direct.
  - **Certification test.** Certification tests often require examinees to meet certain prerequisite conditions, such as completion of training, specified levels of experience in the job domain, or both. For the student to obtain a job in the domain, he or she may be required to pass the licensure or certification examination. Failure to pass has important and direct consequences. These examinees typically are removed from job consideration or removed from rank-order lists of eligible applicants. These tests must be supported with job-relevant documentation and achieve high levels of test reliability.
  - **Selection tests.** A testing method that determines whether an applicant is accepted or rejected for a particular educational or employment opportunity, a selection test must focus on the "must-know" and "need-to-know" information necessary to perform a job successfully. Job relevance is a major focus of the test. All candidates should have appropriate training or experience prior to taking selection tests. These tests must be supported with job-relevant documentation and must achieve high levels of test reliability. These tests are not intended to be taken by inexperienced or untrained personnel or by the general public.
  - **Promotional tests.** As the name implies, promotional tests are used to select the most qualified individuals from a group of candidates to be promoted to a higher position. Promotional tests must also demonstrate high levels of job relevance. In the fire and emergency medical services (EMS), it is important that the criteria to be measured are based on the standard operating procedures/standard operating guidelines (SOPs/SOGs), the appropriate National Fire Protection Association (NFPA) professional qualification standards, or both. Job relevance must be documented and demonstrated. These tests must achieve high levels of test reliability. Results of these tests are often rank-ordered from the highest score to the lowest. Most organizations then promote from the top of the list to the bottom. It is very important for the list to be updated periodically by using a new series of promotional tests.

Testing in the high-stakes arena requires considerable work, professional guidance, and test-item documentation. It is not wise to undertake the development of high-stakes tests without considerable professional test developer involvement. In addition to understanding how to create a written or skill evaluation sheet, the Instructor II needs to understand the legal considerations for testing and evaluating. The American Psychological Association is one resource that provides guidance on many aspects of test development (American Educational Resource Association, American Psychological Association, and National Council on Measurement in Education, 1999). The Equal Employment Opportunity Commission (EEOC) and other government organizations provide direction for testing and, if available, an agency's legal counsel is a source for areas of testing (EEOC, 2010).

> **DEPARTMENT TRAINING OFFICER TIP**
>
> Evaluations should be objective assessments, without any bias or prejudice. Your attitudes, opinions, or values should not influence your evaluation of a student's performance.

Uniform Guidelines for Employee Selection

Uniform Guidelines on Employee Selection Procedures (EEOC, 1978, Title VII) is viewed by many as the national standard for employee testing and selection. It is primarily focused on employee selection/promotion procedures. Training and testing leading to hiring, promotion, demotion, membership in a group such as a union, referral to a job, retention on a job, licensing, and certification are all covered under these guidelines. Written and performance tests should be job related in such training programs. It is up to the user of the testing materials to make sure that the tests are job related. If the tests are acquired from a professional publisher or a distributor of tests, then that source must be able to provide you with the job validity information. For example, if a question is based on the NFPA's professional standards for a particular topic and the question is current, then it would be defined as containing job-validity information.

As an instructor, you must address job-content validity even for the test items that you personally develop and use on a day-to-day basis. Almost all training in the fire service leads to some sort of certification, pay raise, or potential for promotion; therefore, job-content-related testing should be paramount when using any form of testing.

Test-Item Validity

When developing a test, it is crucial to make sure that each test item actually measures what it is intended to measure. All too often, tests contain items that are totally unrelated to the learning objectives of the course. The term *valid* is used to describe how well a test item measures what the test-item developer intended it to measure. Taking steps to ensure validity will prevent your test items from measuring unrelated information.

Forms of test-item validity are distinguished as follows:

1. Face validity
2. Technical-content validity
3. Job-content validity/criterion-referenced validity

Although this chapter deals primarily with face validity, technical-content validity, and job-content validity, you will encounter other forms during your career as an instructor.

Face Validity

Face validity is the lowest level of validity. It occurs when a test item has been derived from an area of technical information by a **subject-matter expert** who can attest to its technical accuracy and can provide backup evidence. Each level of validity requires documentation and evidence.

Technical-Content Validity

Technical-content validity occurs when a test item is developed by a subject-matter expert and is documented in current, job-relevant technical resources and training materials.

Job-Content/Criterion-Referenced Validity

Job-content/criterion-referenced validity is obtained through the use of a technical committee of subject-matter experts who certify that the knowledge being measured is required on the job. This level of validity is often called *criterion-related validity*. In the case of criterion-referenced validity, you may use professional standards such as the NFPA professional qualification standards that are based on a job and task analysis. This level of validity should be carefully documented with each test item so that anyone can trace the validity information to the specific part of the NFPA standards and the reference material used to develop the question.

Currency of Information

An important aspect of developing evaluation tools is using the most current information in the field of knowledge that a student should know and use. With many areas of knowledge evolving in the fire and emergency services profession, currency of the information is relevant to job performance.

Job Performance Requirements in ACTION

For the learning process to be complete and reach its maximum potential, it needs to be evaluated. Evaluation can come in many forms, ranging from written evaluations and hands-on evaluations to ongoing evaluation during the delivery of training. Numerous legal and ethical considerations exist during the evaluation phase of the learning process, and as the instructor, you must be aware of these considerations at all times.

Instructor I

As part of the delivery of training, the Instructor I will be responsible for administrating written evaluation and practical skills sessions.

JPRs at Work

Administer evaluations of student learning and grade evaluations according to the type of evaluation used. Record evaluation scores and provide feedback to students regarding their performance.

Instructor II

The Instructor II develops evaluation instruments to measure the student's ability to meet the performance objectives.

JPRs at Work

Develop and analyze evaluation instruments to ensure that the student has achieved the learning objectives and that the evaluation is a valid measure of student performance.

Instructor III

There are no JPRs at the Instructor III level for this chapter.

JPRs at Work

Responsible for the overall evaluation plan for a course. Written and skills evaluations will be used to check knowledge and skill ability related to each course objective.

Bridging the Gap Among Instructor I, Instructor II, and Instructor III

The Instructor I will administer the evaluations developed by the Instructor II, observing the legal and ethical principles of test administration. The overall course evaluation plan responsibilities lie with the Instructor III, as does the responsibility for evaluation policy development. The Instructor I must learn to identify the desired outcomes of evaluations so as to improve his or her instructional skills. All levels of instructors will have to make sure that the evaluation measures the stated objectives for the lesson plan and that the evaluation correctly measures the learning process.

Test Item and Test Analysis

Test analysis occurs after a test has been administered. Three questions are usually answered in the posttest analysis:

- How difficult were the test items?
- Did the test items discriminate (differentiate) between students with high scores and those with low scores?
- Was the test reliable? (Were the results consistent?)

If a test is reliable and the test items meet acceptable criteria for difficulty and differentiation, it is usually considered to be acceptable. The process used to make this determination is known as **quantitative analysis**, and it entails the use of statistics to determine the acceptability of a test. If a test is not reliable and the test items do not meet acceptable criteria for difficulty and differentiation, a careful review of the test items should be performed and adjustments made to the test item. This process is known as **qualitative analysis**, and it entails an in-depth research study performed to categorize data into patterns to help determine which test items are acceptable.

The purpose of test analysis is to determine whether test items are functioning as desired and to eliminate, correct, or modify those test items that are not. Bear in mind that it takes the use of a written test question, for example, being given several times to see any type of pattern develops with students selecting the correct answer to determine if a question is valid. The best way to achieve acceptable test analysis results is to ensure validity of test items as they are developed and to follow a standard set of test development procedures as the test is constructed and administered. See Chapter 14, *Program Evaluation,* for more information.

The Role of Testing in the Systems Approach to Training Process

The **systems approach to training (SAT) process** was developed by the U.S. military during the early 1970s and 1980s. Many improvements have been made since the early days of this training approach. The effectiveness and efficiency of the SAT have been well documented in training journals, academic studies, and actual practice by leading businesses and industries in the United States and around the world.

Dr. Robert F. Mager, often referred to as "the father of the performance objective/learning objective," played a key role as a leading researcher during the development of performance-based or criterion-referenced instruction. Dr. Mager introduced the idea of learning objectives that have three distinct parts. First, the learning objective is task-based by using verbs as part of the learning objective (which implies doing something). Sometimes these verbs are referred to as the action part of the learning objective. The second part of the learning objective deals with the condition(s) under which the action is to be performed. Third, the learning objective should contain a standard or measure of competence. An example of this approach is seen with the following learning objective:

> *Given a self-contained breathing apparatus (SCBA), the fire fighter will don the SCBA and place it in operation within 1 minute.*

All forms of testing and training should be based on learning objectives. In technical training, whether the tests are written, oral, or performance based, they must be based on the learning objective.

Within any training program, testing serves three purposes:

1. Measure student achievement of learning objectives.
2. Determine weaknesses and gaps in the training program.
3. Enhance and improve training programs by positively influencing the revision of training materials and improving instructor performance.

In a performance-based training program, the goal is for all students to master all learning objectives. This mastery is evaluated or measured using performance tests and safety-related oral tests. If this is the case, then what purpose does the written test serve in a performance-based training program? Written tests are used to take a "snapshot" of students' knowledge at predetermined points throughout a course of instruction. They may take the form of a pre-course exam used to determine a student's level of knowledge upon entering a course, a midcourse test or formative exam to determine a midpoint level of knowledge, and a final exam or summative exam. These snapshots provide information on how well the student is progressing. If a student is not progressing satisfactorily, additional instruction can be provided to bring the student up to the required knowledge level. Testing then allows you to provide needed assistance before the student's lack of knowledge becomes critical. Lack of knowledge mastery is often a primary reason for poor performance or nonperformance of a learning objective and tasks on the fire ground.

Test-Item Development

The fire service is committed to safety and productivity through improved training programs and courses. Testing, as an important part of any approach to training, comprises two activities:

- The preparation of test questions using uniform specifications
- A quantitative/qualitative analysis to ensure that the test questions function properly as measurement devices for training

The development of objective test items requires the application of specific technical procedures to ensure that the tests are both valid and reliable. **Validity** means the test items (questions) measure the knowledge that the test items are designed to measure, and **reliability** means the test items measure that knowledge in a consistent manner.

The fire service is fortunate that it has the NFPA professional qualification standards, which provide an excellent basis for criterion-referenced training and testing. Make sure you carefully consider using the technical references in these standards as part of your own validity evidence and documentation. A test item has content validity when it is developed from a body of relevant technical information by a subject-matter expert who is knowledgeable about the technical requirements of the job. Remember that validity is the ability of a test item to measure what it is intended to measure, and reliability is the characteristic of a test that measures what it is intended to measure on a consistent basis. In other words, a valid and reliable test is one that measures what it is supposed to measure each time it is used.

To develop a test comprehensively, it is important to have a group (bank) of test items. The test-item bank can be organized on a course basis, by standard reference, by job title, by duty, or in any other way that seems sensible and logical. Such a database of test items permits a test developer or instructor to alternate test questions or produce different versions of a test for use between classes or among students within a class. This balancing of test versions is important in today's legal environment and must be applied on a procedural basis to reduce the possibility of any discrimination occurring as a result of the testing program.

To begin building statistically sound tests, test items are written by subject-matter experts, who should be technically competent and actually working in the job for which the test items are being developed. The reason for this level of competence is readily apparent, given that changes are occurring at a more rapid rate than ever in the fire service. This initial activity provides the first level of validity: face validity. From this point, test items are analyzed on the basis of their individual performance in the test using test-item analysis techniques and data collected from responses of students taking the test.

Once data are collected from approximately 30 uses of a test item, the first test-item analysis can be completed. The test-item analysis data provide specific information for establishing a quality test-item bank.

Test Specifications

Test specifications permit instructors and course designers to develop a uniform testing program consisting of valid test items as a basis for constructing reliable tests. Writing and developing test items is difficult for instructors. Indeed, it is difficult to write a test item the first time that is both content valid and reliable. Thus, most valid test items and reliable tests are the result of refinement through use. Test-item writers and test developers must follow basic rules or specifications to create a valid and reliable test. This section presents and briefly explains the test specifications, which are intended as a guide for developing test items and tests with initial technical content validity and testing reliability.

Written tests are designed to measure knowledge and acquisition of information. These tests have limitations, however. One major limitation is that testing knowledge and information does not ensure that a student can *perform* a given task or job activity; it simply means that the student has the knowledge *required* for performance. Skill development—that is, the ability to perform tasks or task steps—is obtained through actual performance. It may be achieved during evolutions on the training ground, and performance then documented via a **skills evaluation** during and at the end of training.

A simple analogy using golf explains the difference between testing for knowledge and information and actual performance of a task. For instance, a person can read numerous books, watch films, view videos, attend golf tournaments, and perform other information-gaining activities. By doing these things in the areas of golf, the person could probably make an acceptable score on a written test about golf. However, even with all this information and a perfect score on the written test, the person probably could not actually play a game of golf and score under 100 after 18 holes (**FIGURE 11-1**).

This concept can be applied to the many tasks necessary to score well in any course. The basic skills associated with any task must be practiced and

FIGURE 11-1 Obtaining knowledge and applying knowledge are separate skills.
Courtesy of Alan Joos.

kept sharp for the student to be a competent performer. Of course, these same conditions apply to being a public service professional. Whether the job is in firefighting, policing, administration, supervision, or training, success is based on the ability to apply information to achieve levels of professional competence. A balance of must-know and need-to-know information and competent performance is the foundation of performance-based learning. The following specifications, examples, and test development criteria will help ensure that instructors, subject-matter experts, and curriculum developers prepare written test items, construct written tests, and develop performance tests that measure what they should measure and function reliably over a long period of time.

Written Test

The written test-item type to be developed is determined by two considerations:

- The knowledge requirement to be evaluated
- The content format of the resource material

As you research and identify passages of relevant information in the resource material, ask yourself which type of written test item the material might support. For example, one passage full of terms and definitions might support developing a matching question, while another passage containing a scenario or other complex situation might support multiple choice questions. You may want to note this observation somewhere in the resource material as you go through the review and research phase. One technique that works well is to use a different-colored highlighter for each type of test item. With this system, when you return to write test items, the analysis for the type of test item is already complete.

Complete a Test-Item Development and Documentation Form

Test-item development and documentation forms serve many purposes other than the most obvious—recording the test item. The form is also useful to do the following:

- Connect planning/research with test-item development efforts
- Record pertinent course-related information
- Record the source of test-item technical content
- Provide a record of format and validity approval
- Record the learning objective and test-item number for banking

By documenting the required information on the form, you are using it to accomplish all of these purposes.

Collectively, the information you place in the blanks of test-item development and documentation forms will serve to link the identifiers of the program or course and the reference with the test item. Each blank is important to proper identification, so fill in each one carefully with the appropriate information.

A continuation sheet may be used when developing a test item that is too long to fit on a single form.

The **reference blank** is where the current job-relevant source of the test-item content is identified. In this blank, you should supply enough information to pinpoint the specific location within the resource material where you found the information you are using to develop the test item. Entries might appear as follows:

Starting at the top of page one of the Test-Item Development and Documentation form, complete it as follows:

- **Program or Course Title.** In this blank, put the title of the program or course for which you are writing a test item. An example would be "Fire Officer I."
- **Learning Objective Title.** In this blank, place the learning objective you are working on in preparing test items. An example would be "Fire-Ground Operations."

Components of a Test Item

Test items are made up of components that create a testing tool. The part of the question that asks the information is known as the *stem* (**FIGURE 11-2**).

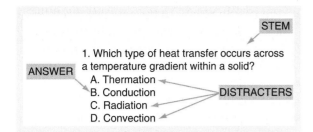

FIGURE 11-2 Test items are made up of three components: the stem, the answer, and the distracter(s).
© Jones & Bartlett Learning.

What is the correct angle for positioning a ladder against a structure?

A. 60°
B. 65°
C. 75°
D. 80°

FIGURE 11-3 An example of a properly formatted multiple choice test item.
© Jones & Bartlett Learning.

The choices are made up of a correct answer and distracters. An important part of the question is the reference material that documents the standard the question is based on and the source of the information in a textbook.

Selection-Type Objective Test Items

The various selection-type objective test items were described in an earlier chapter; however, each type is briefly described here as a quick review.

Multiple Choice Test Items

The multiple choice test item, which is the most widely used in objective testing, also contributes to high test validity and reliability estimates. It incorporates the important qualities of a good test, which are content validity, reliability, objectivity, adequacy, practicality, and utility. The most important advantage of the multiple choice test item is that it can be used to measure higher mental functions, such as reasoning, judgment, evaluation, and attitudes, in addition to knowledge of facts. The versatility of this test item allows it to be used to measure almost any cognitive information; the disadvantage is that it may take several well-written test items to cover the learning objective properly. **FIGURE 11-3** provides an example of a properly formatted multiple choice test item.

Preparing Multiple Choice Test Items

Suggested guidelines to keep in mind when preparing multiple choice test items include the following:

- Select one and only one correct response for each item. The remaining distracters should be plausible, but wrong, to serve as distracters for the correct response.
- Construct the same number of responses for each test item. Four responses are preferable. More than four response choices may be used and sometimes are necessary, but they do not improve the reliability of the test.
- Responses having several words should be placed on one or two lines with space separating them from other responses. If responses consist of numbers or one word, arrange two or more of them on the same line, uniformly placed.
- Provide a line or parentheses on the right or left margin of the page for students to write their responses. This placement permits ease of test taking for students and encourages use of a grading key for more rapid and accurate test grading. Disregard this step if you will be using a machine-scannable answer sheet.
- Change the position of the correct response from test item to test item so that no definite response pattern exists.
- Review the test item to ensure proper grammar and punctuation in the stem and answer and distracters.
- Avoid wording such as "which of the following is *incorrect*" or "all of the following are correct *except*." These are considered negative questions and, generally, are no longer considered acceptable formats.
- Avoid questions with "all of the above" or "none of the above" as the answer. In many cases, it is a giveaway because that is the correct answer.
- Prepare clear instructions to the student on how to mark the correct answer on the answer sheet.

Matching Test Items

Matching test items, which are another form of multiple choice testing, is limited in use, but works well for measuring knowledge of technical terms and

Directions: Match the tool in Column A with the best example in Column B.

Column A	Column B
1. Striking tool	A. K-Tool
2. Prying tool	B. Axe
3. Cutting tool	C. Sledgehammer
4. Lock tool	D. Wedge bar
	E. Hux bar

FIGURE 11-4 An example of a properly formatted matching test item.
© Jones & Bartlett Learning.

functions of equipment. Since all information must be factual, it is important to take care during preparation of matching test items.

Although this type of test item is relatively easy to develop, it requires more space on the test. The matching test item is particularly useful when a question requires multiple responses or when there are no logical distracters. It is also very useful for low-level cognitive information measurement. **FIGURE 11-4** provides an example of a properly formatted matching test item.

Preparing Matching Test Items

Suggestions for preparing matching test items include the following:

- Check to make sure that similar subject matter is used in both columns of the matching test item. Do not mix numbers with words, plurals with singulars, or verbs with nouns.
- Use up to four items in Column A and up to five choices for the match in Column B. This prevents students from earning credit simply by applying the process of elimination.
- Make sure there is only one correct match for each of the items to be matched.
- Arrange statements and responses in random order.
- Check for determiners or subtle clues to the answers.
- Provide a line or parentheses beside each response statement for the student to indicate the answer. Arrange the parentheses or lines in a column for ease of marking and scoring. This step is not required if you are using a machine-scannable answer sheet.
- Prepare directions to the students carefully and observe their following of instructions to ensure that the instructions function as intended.
- Attempt to keep all matching test items on one page.
- When preparing the draft of questions, follow the format guide to expedite typing, reproduction, response by the student, and rapid grading.

Arrangement Test Items

Knowledge of emergency procedures, fire drills, triage, or other related procedures where knowing exactly what to do under critical conditions can be measured with high reliability using an arrangement test item. Plan to administer an arrangement test shortly before an opportunity to perform the actual procedure, disassembly, or assembly.

A disadvantage of arrangement test items is inconsistent grading. A student may miss the order of one step and consequently miss the order of all of the steps that follow. Another potential issue is key steps that are included in the SOP or SOG being left out of the test item. The test item should be developed in agreement with actual procedures used on the job if such procedures are required. The test item is not effective if the procedure is much longer than 10 steps. In cases where it is important to test procedures that are longer, consider grouping them at key points in the procedure. **FIGURE 11-5** provides an example of a properly formatted arrangement test item.

Preparing Arrangement Test Items

Suggestions for preparing arrangement test items include the following:

- Make sure that arrangement test items developed for procedures are based on current procedures. Remember that arrangement test items should relate to emergency procedures or to things that must be performed in a certain order.
- Have one or more subject-matter experts review the test items for technical accuracy.
- Make sure the steps of the procedure are disordered and that steps with clues to the correct order are separated.
- State exactly what credit will be given for correct sequencing of all steps.

Directions: *(Note: This is a two-part question.)*
There are six steps for the proper use of a fire extinguisher and they appear below but are out of order. **First**, number the steps in the proper order by placing numbers (1–6) in the blanks beside the steps. **Next**, select the answer that matches yours from choices A–D.

_____ Exit the room.
_____ Pull the pin on the extinguisher.
_____ Sweep at the base of the fire.
_____ Notify occupants/fire department.
_____ Aim extinguisher nozzle at base of fire.
_____ Squeeze handle of extinguisher and discharge content.

A. 6,2,5,1,3,4
B. 5,1,4,6,2,3
C. 6,2,3,1,5,4
D. 6,1,4,5,3,2

FIGURE 11-5 A properly formatted arrangement test item.
© Jones & Bartlett Learning.

- Include cautionary statements with the procedure to be rearranged; for example: (Caution: This is a two-step process worth 5 points) or (Caution: Reorder these procedures from the point of view of the Initial Company Officer).

Identification Test Items

Recall that the identification test item is essentially a selection-type test using a matching technique. It also can be considered to be arrangement test items for assembly- and disassembly-type tasks. Its major advantage over the matching test item is the ability of the student to relate words to drawings, sketches, pictures, or graphs. Identification test items do require reasoning and judgment but focus on the ability to recall information.

Identification test items provide excellent content validity and contribute to high test reliability. With computer publishing capabilities and digital cameras, identification questions can often be rather easily developed using your own fire equipment and apparatus features as the basis for the identification test item. This approach makes the test item more job related and makes it much easier to transfer required knowledge from the instructional setting to the job.
FIGURE 11-6 provides an example of a properly formatted identification test item.

Directions: (*This is a two-part question worth 5 points.*)
First, label the tools depicted below by placing the number in the blanks provided.
Then, choose the answer below (A–D) that matches yours.

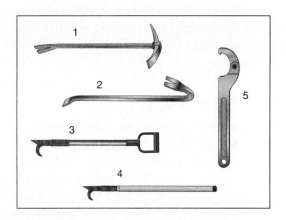

____ Halligan too
____ Spanner wrench
____ Closet hook
____ Crowbar
____ Pike pole

A. 3, 5, 4, 2, 1
B. 1, 5, 3, 2, 4
C. 5, 3, 1, 2, 4
D. 4, 3, 1, 2, 5

Answer:

FIGURE 11-6 An identification test item.
© Jones & Bartlett Learning.

Preparing Identification Test Items

Suggestions for preparing identification test items include the following:

- Select graphic, art, or other object for inclusion in the question.
- Place the object into a drawing or illustration software.
- Identify the parts or items you want identified. (Caution: Do not specify more than seven or eight items to be labeled, or even fewer items if the object will be cluttered by arrows or letters. You may choose to make two identification test items using the same graphic and reduce the number of responses from eight to four on each test item.)
- Develop the desired alternatives A–D, with only one answer being the correct identified object.
- The identification test item may give a point value of more than 1 for the correct answer.

(A rule of thumb is to award one point for every two correct responses.)
- Have a subject-matter expert review the technical accuracy of the question and alternatives. (Be prepared to provide your validity evidence and documentation.)
- Use the test item several times with the target population to make sure it measures what was intended.

True/False Test Items

The major problem with two-answer selection tests such as the true/false is the guessing factor. Without reading the question, a student has a 50 percent possibility of marking the correct response. Application of two or three rules for taking a true/false test could get the student a higher score just by guessing alone. **FIGURE 11-7** is a typical true/false test item.

A more complex true/false test item can be developed that uses a multiple choice approach to the test item. **FIGURE 11-8** provides an example of a complex true/false test item.

Directions: Circle either "T" for True for "F" for False.

Scene safety for EMS responders only applies to incidents such as vehicle accidents or technical rescue incidents:

T   F

FIGURE 11-7 A typical true/false test item.
© Jones & Bartlett Learning.

Directions: Read the following statements regarding fire fighter standards and select your answer from choices A–D.

Statement 1: The only standard a fire fighter needs to use is NFPA 1001.
Statement 2: The only standard a fire officer needs to use is NFPA 1500.
Statement 3: The NFPA standards are applicable based on which level of fire fighter is being trained.

A. Statement 1 is true, statements 2 and 3 are false.
B. Statement 1 and 2 are true, statement 3 is false.
C. Statement 1 and 2 are false, statement 3 is true.
D. All three statements are false.

Answer: C

FIGURE 11-8 A complex true/false test item.
© Jones & Bartlett Learning.

Developing True/False Test Items

When developing true/false text items, follow these guidelines:

- True/false test items do not cover the subject matter in depth because of the limited application possible in most technical information or courses. Use of true/false test items should be limited to factual information at the basic level of an entry-level course, such as a course for Fire Fighter I candidates. There is little discernible value for the true/false test item in technical instruction when student knowledge and performance measurement are required for safe and efficient operation of equipment and emergency task performance. For this reason, true/false questions should be avoided in operations and maintenance programs.
- The complex true/false test item is much more suited to technical courses and information. These items should be developed from a specific area of knowledge. Statements should be brief and technical in nature. This type of question does require higher-level cognitive skills.
- If used in a multiple choice format, this test item can be objectively scored.

Essay Test Items

Although there are several weaknesses that affect objectivity (see Chapter 8, *Evaluating the Learning Process*), **essay test** items can be improved to the point that some inherent weaknesses are diminished. The strength of the essay test is that the student must recall facts from acquired knowledge and present these facts logically and in writing. Writing answers provides the student with an opportunity to express ideas and attitudes, interpret operational situations, and apply knowledge gained to an individual solution. Outlining a response to an essay question and marking the key points so that they may be emphasized in the answer is a valuable approach to improve reliability.

Formatting Essay Test Items

It is important to word the essay question properly and to develop a preliminary outline of the key points expected in the response to the question. This exercise helps to determine whether the question will actually elicit the intended responses. The outline also serves as a guide to grading student responses and assigning relative point values to the key points in an answer before the test is given.

A hierarchy of words can be used to develop essay test items, which will provide a range from easy to difficult. For instance, questions—in ascending order of difficulty—may require the student to "describe," "explain," "discuss," "contrast," and "evaluate." That is, asking the student to *describe* an event in plant operations is much easier than asking the student to *evaluate* the same event and *contrast* that event with the implications for plant safety. In the essay test, some of each of these types of questions should be asked so that the relative difficulty of the test is maintained. The easier questions (*describe*, *explain*) should be presented earlier in the test.

Preparing Objective Essay Test Items

Suggested guidelines for preparing objective essay test items include the following:

- Review the learning objectives. Test items should be developed considering the learning objectives and using them as the basis for testing.
- Make a list of major points contained in the body of the course material that are to be covered by the test. This list should focus on major points, rather than on trivial details. Check the list once it is completed to confirm that it is comprehensive in its coverage of the subject matter, keeping in mind that an essay test contains only a few test items and that it requires more time for the student to answer and for you to grade.
- Arrange the list of major points in an order similar to that used when presenting the information or assigning reading material for student outside study.
- Identify the major points that you want the student to describe. These major points should be covered in a less difficult essay question for students to begin with on a test. The use of a relatively easy question for the first item will get students ready for the more difficult questions that follow.
- Next, prepare a question or two requiring the student to describe and explain a key point or major concept in the subject matter. Again, this approach takes the student to the next level of difficulty in an essay exam.
- Prepare the most difficult essay items last and place them toward the end of the exam. These questions usually require the student to contrast, compare, and evaluate a situation or complex set of facts. It is a good testing practice to use the earlier questions in an exam to form the basis for these contrasts, comparisons, and evaluations. Placing difficult questions at the end of an exam has the added value of permitting the well-prepared and knowledgeable student an opportunity to pull the facts together and express personal views in terms of the facts. It is not uncommon for students to expand their knowledge well beyond the level expected by the instructor and to offer plausible solutions that may not have been presented previously. These unexpected results may be rewarded with extra credit.
- Take the test. Answer the questions in terms of the course material presented, assigned reading, and lab and shop activities. Once you have prepared your answers, prepare a detailed outline of each essay test-item response. Determine the amount of credit for each key point in the outline. Determine in advance where extra credit will be awarded for answers that clearly excel, and determine the number of points to be deducted for major omissions in test-item answers. The outlines then become the scoring guide and key to help ensure consistent grading from question to question and from one exam to another.
- When essay exam questions are completed, review them once again. While reviewing the exam, it is of extreme importance to keep in mind the exact response behavior desired from the students and the specific abilities and knowledge that students should possess relative to course material covered by the exam.

Grading Essay Test Items

A crucial activity necessary to make an essay exam more objective is the use of a scoring key with associated criteria for consistent grading. Following are a few suggestions to make the grading process more reliable:

- Fold the exam cover sheets back or cover the test takers' names before beginning any grading activity. This permits the application of the grading criteria as anonymously as possible.
- Grade one question at a time on all exam papers before moving to the next question. This tends to keep the focus on the grading key and the response at hand.
- If the exam has an effect on a student's career or promotion, it is good professional practice to have another qualified person give a second grade to the exam. An average of the two grades provides more objectivity.
- Make sure that you provide some flexibility in the grading key or outline for giving

additional credit for answers that produce results beyond those expected when the test was developed. Penalties for omission of key points must also be deducted from the test score. Make sure, however, that the omission penalties are uniformly applied to all questions and exams.

- The *whole method*, which can improve grading of exams, begins with a preliminary reading of the exam papers. During the reading process, a judgment is made of each exam in terms of four groupings: excellent, good, average, below average. Each exam is then sorted into one of these preliminary groups. Once the grouping is completed, grading is accomplished in the same manner as described previously. A shift from group to group of one or more exam papers is not uncommon when the scoring is completed.

- Another grading tool that will assist the evaluator in grading an essay test is the use of a rubric which has defined areas and a rating scale. By using a rubric, the evaluator will be more consistent in grading, plus the criteria will help remove some of the subjective traps that an evaluator could fall into (**FIGURE 11-9**).

Research Paper Grading Guidelines

Name _____
Date _____
Title _____
Final Grade _____

_____ **Format** *(25 points max., 1-5 points each sub-area)*
　　_____ APA Style
　　　　- *Use of written paper guideline*
　　　　- *Correct format*
　　_____ Title Page
　　　　- *Layout (-2 pts)*
　　　　- *Information (-3pts)*
　　_____ Margin
　　　　- *1" throughout document (-1pt ea.margin)*
　　_____ Reference Page
　　　　- *References used in body of paper (-2pt)*
　　　　- *Layout (-3 pts incorrect)*
　　_____ Parenthetical Documentation
　　　　- *References in body match reference page (-2pt)*
　　　　- *Correct documentation (-3pt)*

_____ **Content** *(25 points max., 1-5 points each sub-area)*
　　_____ Current/Correct Information
　　　　- *Material is relevant*
　　　　- *Material is correct*
　　　　- *Material is current*
　　_____ Topic Narrowly Defined
　　　　- *Topic is focused*
　　　　- *Distinguish main point from supporting material*
　　_____ Originality in Choice of Topic
　　　　- *Topic is original*
　　　　- *New insight into topic*
　　　　- *Topic is insightful*
　　_____ Clear Understanding of Material
　　　　- *Writer confident in approach*
　　　　- *Readers questions are answered*
　　　　- *Writer has "learned" something*
　　_____ Illustrations and Examples
　　　　- *Illustrations/examples apply to research paper (-3pt if none)*
　　　　- *Correct reference(s) for illustrations*

_____ **Grammar & Mechanics** *(25 points max., 1-5 points each sub-area)*
　　_____ Sentence Construction
　　　　- *Transitions from point to point*
　　　　- *Use of slang (hazmat, FDC, etc.)*
　　_____ Spelling
　　　　- *Misspelled words (-3pt 2+ words)*
　　　　- *"Spell check" errors ("right" vs. "write") (-2pt 2+ words)*
　　_____ Recognizable Thesis
　　　　- *Able to identify thesis in introduction*
　　　　- *Thesis re-stated/reinforced in conclusion (-3pt if not done)*
　　_____ Readability (Run-on Sentences, Comma Splices, Sentence Fragments)
　　　　- *Sentences make it hard to breathe*
　　　　- *Sentences are choppy*
　　_____ Non-Gender Specific
　　　　- *His/her instead of just his, her*

_____ **Research/Writing Process** *(25 points max., 1-5 points each sub-area)*
　　_____ Assignment Understood and Followed Precisely
　　　　- *Turned-in on time (-2pt late)*
　　　　- *Length of paper (minimum 5 pages of text, plus cover and reference pages) (-3pt if not 5 pages)*
　　_____ Research Material
　　　　- *Enough references to support thesis*
　　　　- *Minimum 3 different sources (-3pt)*
　　_____ Research/Writing Effort
　　　　- *Time spent on the paper*
　　　　- *References applicable to topic*
　　_____ Attention to Detail
　　　　- *Overall appearance of paper*
　　　　- *Coffee stains, hand-written*
　　_____ Plagiarism
　　　　- *Reference used for everything that did not originate in the mind of the writer*

Comments: _____
_____
_____

FIGURE 11-9 An example of a rubric.
Courtesy of Alan Joos.

Remember—be as objective as possible in the use of the grading criteria. Application of the criteria for grading will improve the objectivity of those applying the grade. An essay test is difficult to effectively write and grade.

Preparing an Essay Test-Item Grading Key

It is important to develop a preliminary outline of key points expected in the response to the question (**FIGURE 11-10**). While helping to determine whether the question will actually elicit the intended response, this exercise also assists the instructor in assigning relative point values to the key points in an answer before the test is given.

In outline format, list the key points that are expected in the answer to the essay test item. For each key point in the outline, assign a weighted point value. These values may vary in weight depending on their degree of importance to the overall answer. Key points of a critical nature or those pertaining to safety-related activities and information should be given higher weighted values. Because of the importance of grading essay test items, the methods for increasing objectivity should be considered.

Performance Testing

Performance or skills testing is the single most important method for determining the competency of actual task performances. For this reason, the major emphasis in performance-based instruction is the validity and reliability of the performance tests. The focus on student performance is the critical difference between traditional approaches to instruction and the use of performance-based instructional procedures. Performance must be evaluated in terms of an outside criterion derived from a job and task analysis. The justification for training program knowledge and skill development activities centers on the concept that the activities are derived from, and will be applied to, training for tasks performed on the job. The final demonstration of job knowledge and skill application in fire and EMS training programs is the completion of specific on-the-job tasks.

The development of a performance test skill sheet takes as much effort and ability as creating a multiple choice written exam question. The skill being evaluated needs to be valid and referenced to the appropriate NFPA JPR or other applicable standard such as an SOP. When creating a skill sheet to be used for evaluation, make sure to do the following:

- Clearly identify what is being evaluated in the objective or performance statement.
- Clearly identify what tools, equipment or other resources will be available or allowable during the performance of the skill.
- Identify a time element if the skill is to be performed in a set time frame. If using a time frame you need to justify how that time element was determined.
- Identify the steps necessary to successfully complete the skill. If the skills need to be performed in a certain order, that needs to be clear to the student and evaluator in the criteria.
- Identify the rating or scoring system and what the criteria are for passing the skills. Rating systems can be points awarded for each step with a minimum passing total or pass/fail criterion.
- Treat skill evaluation sheets with the same level of security as written test banks (**FIGURE 11-11**).

Question
State four precautions for handling or working with caustic soda, including the appropriate first-aid actions, in the event contact with caustic occurs.

Grading Key	Point Value
1. Hot water must be used when dissolving caustic soda. Eyes, face, neck, and hands should be protected.	1
2. Wherever caustic soda is stored, unloaded, handled, or used, abundant water should be available for emergency use in dissolving or diluting and flushing away spilled caustic.	1
3. General first aid is of prime importance in case of caustic coming in contact with the eyes or skin. Prolonged application of water to the affected area at the first instant of exposure to caustic is recommended.	2
4. a. If even minute quantities of caustic soda (in either solid or solution form) enter the eyes, the eyes should be irrigated immediately and copiously with water for a minimum of 15 minutes.	2
b. The eyelids should be held apart during the irrigation for contact of water with all the tissues of the surface of the eyes and lids.	1
c. If a physician is not immediately available, the eye irrigation should be continued for a second period of 15 minutes.	2
d. No oils or oily ointments should be applied unless ordered by the physician.	1
Total Point Value	**10**

FIGURE 11-10 An example of an essay test-item grading key.
© Jones & Bartlett Learning.

SA203:	Assist a rescue operation so that procedures are followed, rescue items are recognized and retrieved, and the assignment is completed.
Reference:	NFPA 1001, 2019 Edition: 5.4.2(A)(B)
Condition:	Given SOPs for the AHJ, necessary rescue equipment for the given scenario, and an assignment based on the scenario (confined space, hazardous material, rope rescue, or vehicle extrication).
Competence:	Determined based on the completion of the following:

Complete the following:	First Attempt		Second Attempt	
	PASS	FAIL	PASS	FAIL
Correctly identify requested tool or equipment requested by Ops Team				
Safely retrieve tool/equipment				
Carry tool/equipment in safe manner				
Complete task in allotted time				

Time: 8:00 minutes

FIGURE 11-11 An example of a skill sheet format.
© Jones & Bartlett Learning.

Test Generation Strategies and Tactics

There are many ways to put together tests. You can do it yourself or use a computer or web-based test-item bank. Doing it yourself is generally referred to as "instructor-made tests." The primary problem with instructor-made tests, even though they may be valid and reliable, is the tendency to use the same test—over and over again. The word gets out about the test content and the answers. Computer or web-based testing is rapidly gaining favor in fire and EMS contexts. Many educational publishers and other private organizations, for example, provide test-item banks with their publications. Although these tools are certainly helpful, you must carefully analyze such test banks to make sure the items in them are properly valid and meet the needs of your agency or purpose.

Commercial testing software can also be used to generate a written exam. The testing software allows you to generate a randomized test by entering how many questions you want on the test and the software will generate a test. This method of test-item selection removes any potential bias of the instructor.

Remember that most testing in fire and EMS training environments leads to certification, selection for a job, pay raises, and promotions. If you are testing for any of these reasons, use only test banks that have been rigorously validated and that are known to produce reliable tests. The question then becomes, "How do I know?" A checklist is provided here to help you determine the quality of a test-item bank no matter who is providing the product (**FIGURE 11-12**).

Of course, there are many more questions you can add to this checklist. Cost is one of the possible criteria to be added, although it should not be the primary concern. The critical emergency tasks that fire and EMS personnel perform can open them up to costly legal challenges. Training and testing programs are the likely targets for litigation if a suit for liability is filed.

Computer and Web-Based Testing

Distance learning is now widely present in conjunction with most colleges, universities, and 2-year institutions. Fire and EMS organizations are likely to increase the accessibility of distance learning to their personnel, given that it has the potential to reduce costs for education and training for both the student and the department. There are many options on the market, each with a wide range of testing formats (**FIGURE 11-13**). Fire and EMS departments also have many options available for putting testing online through professionally

Questions to Be Answered by Test Bank Provider	Provider 1	Provider 2	Provider 3
Do you claim to have valid and reliable test items in your test banks?	Yes __ No __ Notes:	Yes __ No __ Notes:	Yes __ No __ Notes:
Are documentation and data available for determining test-bank validity and test reliability?	Yes __ No __ Notes:	Yes __ No __ Notes:	Yes __ No __ Notes:
Do your test banks comprehensively cover NFPA Professional Qualification Standards? If yes, do you have cross-reference tables to document the extent of coverage?	Yes __ No __ Notes:	Yes __ No __ Notes:	Yes __ No __ Notes:
How long have you been providing test banks? Is this your primary business focus?	Yes __ No __ Notes:	Yes __ No __ Notes:	Yes __ No __ Notes:
How many customers do you have, and can you provide user names and telephone numbers?	Yes __ No __ Notes:	Yes __ No __ Notes:	Yes __ No __ Notes:
Do you provide technical support for your test banks and software? If yes, then how?	Yes __ No __ Notes:	Yes __ No __ Notes:	Yes __ No __ Notes:
Is there regularly scheduled training for the test banks and software? If yes, how often and at what locations?	Yes __ No __ Notes:	Yes __ No __ Notes:	Yes __ No __ Notes:
Do you have technically competent staff members in testing technology? If yes, will you provide detailed résumés of their qualifications?	Yes __ No __ Notes:	Yes __ No __ Notes:	Yes __ No __ Notes:
What are your revision and updating strategies for your test bank?	Yes __ No __ Notes:	Yes __ No __ Notes:	Yes __ No __ Notes:
Do you provide qualified expert witnesses supporting your test bank in case of lawsuits? If yes, what are the costs for the service?	Yes __ No __ Notes:	Yes __ No __ Notes:	Yes __ No __ Notes:

FIGURE 11-12 A test bank validation checklist.
© Jones & Bartlett Learning.

Question 1

Not yet answered
Marked out of 1.00
⚑ Flag question
✱ Edit question

Current NFPA standards require an SCBA EOSTI to sound when cylinder pressure drops to _____ percent of capacity.

Select one:
- A. 30
- B. 20
- C. 25
- D. 35

FIGURE 11-13 An example of an online quiz from a learning management system.
© Jones & Bartlett Learning.

developed test banks. Some of these organizations are already conducting online and web-based certification testing programs. Given the currently high price of fuel, hotels, worker salaries, and other costs, web-based certification is a growing trend.

A significant advantage when using valid and reliable computer-based test banks with large numbers of questions is the reduced risk of test compromise. Maintaining large numbers of test items in a test bank can help ensure that no two tests will ever be alike. For instance, if a test bank contains 1000 test items and you randomly generate a new 100-question-test each time you need one, the chances of getting exactly the same test are negligible.

If an agency uses online testing, test security and proctoring remain the same and should follow the same agency policies for traditional testing. A proctor should be present to ask that students check-in with photo identification to ensure the correct person is testing. In places where online testing is being used, the student must first receive a letter or mail with a student code that is unique to him or her and must be provided with proper identification, so the site proctor knows the correct person who will be testing. The proctor will make sure the student does not bring or access some type of reference material or access online information with a smart phone or other device during the test. In many cases, the proctor will collect the cell phone or mobile device and return it to the student on completion of the test. The testing environment should be free from other distractions, disruptions, or visitors in the testing area where the computers are located. It is important to maintain the integrity of the testing environment and to ensure that online testing is not compromised, which can affect the value of this modern technology.

Remember, in fire and EMS organizations, you are dealing with situations that require rapid responses and high levels of technical knowledge. Providing your students with the best means for learning and performing is your primary challenge. However, use caution when committing to new technology: bookshelves and storage cabinets are filled with equipment and software that were purchased with good intentions but remain unused. Dedicate time to research the currently available and evolving options. Learn about the strengths and potential shortcomings of the leading products so that you can make informed recommendations to your supervisors. In researching the available systems and programs, you soon will narrow down the options to the types that will best meet your needs. It is easy to become overwhelmed by all the options for hardware and learning systems, causing you to spend more of your budget than is needed on the wrong solutions, leading to frustration. Realistic consideration of your budget and a thorough assessment of the software and equipment will ensure that you obtain the desired return on your investment.

Developing Class Evaluation Forms

One of the skills an Instructor II needs to cultivate is the ability to develop a class evaluation form or instrument. An *instrument* is something that is used to measure performance, so this term is more appropriate when considering the process of developing something to measure performance. The most commonly used evaluation instrument is a fill-in-the-blank form or a bubble form that requires the use of a #2 pencil. Some departments may provide you with a class evaluation form, whereas others may have you develop your own. Whatever its source, the evaluation form should cover specific topics, such as the training environment, instructional methods, communication techniques, course content, and student material (**FIGURE 11-14**).

Remember to update the content of instructor and course evaluation forms when newer content or content specific to a particular course is needed. If an existing form does not fit your course, change it to reflect the rating areas on which you want feedback.

Illinois Fire Chiefs Foundation
End-of-Course Evaluation Form

Course Title: _____ Course Location: _____

Course Evaluation: Please shade the letter that indicates your evaluation of the course/instructor feature using #2 pencil.

Course Feature	Excellent	Very Good	Good	Fair	Poor
1. AV material/course materials	(1A)	(1B)	(1C)	(1D)	(1E)
2. Course organization/flow/delivery rate	(2A)	(2B)	(2C)	(2D)	(2E)
3. Observance to safety procedures and practices	(3A)	(3B)	(3C)	(3D)	(3E)
4. Hands on experience (if applicable to course)	(4A)	(4B)	(4C)	(4D)	(4E)
5. Time per subject	(5A)	(5B)	(5C)	(5D)	(5E)
6. Evaluation tools	(6A)	(6B)	(6C)	(6D)	(6E)
7. Will this course help you with your professional development?	(7A)	(7B)	(7C)	(7D)	(7E)
8. Did the course meet your expectations?	(8A)	(8B)	(8C)	(8D)	(8E)
9. Would you recommend this course to others?	(9A)	(9B)	(9C)	(9D)	(9E)
10. Overall impression of course	(10A)	(10B)	(10C)	(10D)	(10E)

Add comments on specific areas on reverse

FIGURE 11-14 An end-of-course survey.
Courtesy of Illinois Fire Chiefs Foundation.

Evaluating the learning environment is important to student success. Was the room too hot or cold? Was the classroom lit properly? Did the audiovisual equipment work? Were the chairs and tables comfortable? Was the learning environment free from distractions? These are questions that affect students' learning environment, which has a direct impact on students' ability to learn. When used properly, student evaluations of the learning environment can be used to justify additional funds for classroom improvements or upgrades to audiovisual equipment.

Evaluation forms should also include areas for evaluating the course material. Did the material meet the students' needs? Did it meet the learning objectives? Did the training material match NFPA standards, the department's SOPs, and the department's own mission statement? What was the quality of the student material? Did the student manual match what the instructor was teaching? Did the audiovisual material enhance the learning process? If the course was offered to other agencies or on a regional or national basis, the form might also include questions on the registration process or the advertisement of the course.

The class evaluation form should also include questions regarding the instructor:

- Did the class start and end on time?
- Did the instructor know the course material?
- Did the instructor identify the learning objectives?
- Was the instructor dressed appropriately?
- How well did the instructor meet the needs of the class?
- Did the instructor communicate the material in a manner that enables the student to learn?

In addition to the questions asked on the evaluation form, the format and the manner in which the form is completed represent an important part of the instrument. In general, all evaluation instruments follow a similar format and can be broken down into three areas:

- The heading area, where the students fill in the date, course title, location of course, or other agency-specific information so the evaluation form can be tied to a course and instructor
- The question area, which asks questions related to the following topics:
  - Course content, learning objectives, and course expectations
  - Learning environment, facility, and classroom temperature
  - Whether student expectations were met
  - Instructor performance and conduct
- An open section where students can write in their thoughts and opinions about the course

A rating scale is a common element on an evaluation form and is used to evaluate the various aspects of the course in the question area. The scale is usually 1–5, with 1 indicating the respondent strongly agrees with a statement about the course and 5 indicating the respondent strongly disagrees with the statement. It is very important when administering such a form to make sure the students understand which end of the scale indicates "strongly agrees" or "strongly disagrees." When developing an evaluation form for use in your agency, you need to make sure the questions are asked in a consistent format. Improperly designed evaluation tools will mix questions with the desired responses changing from "agree" to "disagree" for every other question. This becomes confusing to the student and, in frustration or unintentionally, they may mark the wrong response. This, in turn, could have a negative impact on a course or instructor.

Many forms are scanned by a reader or scanner that is attached to a computer and uses software that gathers the information from the form and provides a printout of the results. These types of instruments use a bubble-form format, wherein students fill in or darken a bubble indicating their opinion per the rating scale (**FIGURE 11-15**). The resulting data can be viewed one evaluation at a time or by assigning a value to each response to provide statistical averages. Optical scanning device readers can computerize these results and make the statistics easier to interpret.

A potential drawback to reliable evaluation occurs when a student evaluation form asks questions that students cannot or do not judge reliably. For example, many times these evaluations may indicate that the class should be shorter when the lesson plan actually calls for a longer presentation.

In addition to the format of the questions, the number of questions asked on an evaluation instrument needs to be considered. The inclusion of too many questions or redundant questions could cause the student to become frustrated, which could then prompt the student to disregard his or her true feelings or opinions and replace them with a hurried attitude and just "bubbling in spots."

The way in which the information gained from an evaluation form is used is just as important as the evaluation form itself. If a department wants to improve the quality of its programs, then it must act upon the information obtained from the evaluation process. If improvements to the classroom are necessary, then the department can budget for those improvements. If the course material is not meeting students' needs, then revisions to the lesson plan may be necessary. If there are issues with the instructor, they can be addressed, as well. Over the long term, a well-structured evaluation

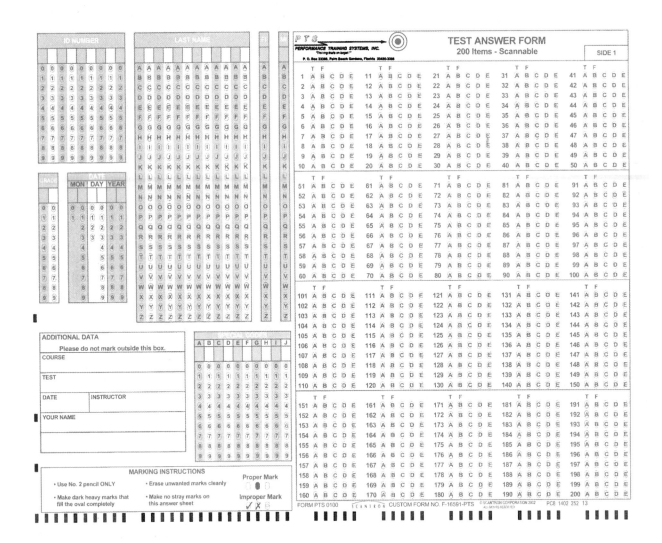

FIGURE 11-15 A sample of a bubble sheet evaluation form.
Courtesy of Ben Hirst, Performance Training Systems, Inc.

program will improve the learning environment, the quality program material, and highly trained and professional cadre of instructors.

Regardless of how the evaluation form was obtained (purchased, borrowed, or developed in-house), the form needs to be reviewed by the instructors who will be evaluated by it, the supervising instructors to make sure the form addresses the needs of the agency, and the training chief because all activities and actions fall back on the leadership of the department and training division. A course or instructor evaluation process is only as important and valuable to an organization as the degree to which the department's administration supports it. If needs or problems are found during an evaluation process, whether they involve an instructor or a poor classroom environment, progressive and effective chiefs are willing to fix problems and improve their department because ultimately, they know they are responsible for the safety and well-being of their fire fighters, officers, and instructors.

Training BULLETIN

JONES & BARTLETT FIRE DISTRICT
TRAINING DIVISION
5 Wall Street, Burlington, MA, 01803
Phone 978-443-5000   Fax 978-443-8000

Instant Applications: Evaluation and Testing

Drill Assignment

Apply the chapter content to your department's operation, training division, and your personal experiences to complete the following questions and activities.

Objective

Upon completion of the instant applications, fire and emergency services instructor students will exhibit decision making and application of job performance requirements of the instructor using the text, class discussion, and their own personal experiences.

Suggested Drill Applications

1. Determine where written examinations have been used within your organization and evaluate their relative level of effectiveness by speaking to department members that have participated in the exam process.

2. Search the internet for university, technical college, or trade school websites that describe the testing services they offer and any policies, procedures, or best practices they may describe.

3. With input from your shift members, develop a class evaluation form that can be used by your department to evaluate the effectiveness of your in-house training programs and solicit feedback from the participants.

Incident Report: NIOSH Report # F2012-27

Drill Assignment

Review the information in this incident report and prepare a practice presentation for class delivery at the direction of your instructor. Your presentation should include a summary of the incident facts and identify the NFPA standard(s) that could apply to the incident. Use an outline to organize your thoughts. You may be evaluated on your communication skills during your presentation.

Beaumont, Texas—2012

On September 15–16, 2012, a 46-year-old male volunteer captain ("Captain") participated in a training course called Smoke Diver. Smoke Diver provides instruction in the advanced use of a SCBA and advanced fire fighter survival skills. The course was taught by instructors from a component of a state fire fighter's association at a private nonprofit fire fighter training facility. Participants performed such drills as air consumption, victim rescue, and self-rescue in a variety of situations while wearing personal protective equipment (PPE) that consisted of full turnout gear and SCBA. The training lasted approximately 12 hours on September 15, 2012, and approximately 10 hours on September 16, 2012. The heat index was 92.2 degrees Fahrenheit (°F) on September 15, 2012, and 90.3°F on September 16, 2012. Rehabilitation (rehab), in an unshaded area, occurred several times each day and consisted of doffing SCBA, partially removing turnout gear, resting, and drinking fluids.

On the first day of training, the Captain completed all the drills without incident. However, two trainees required evaluation and treatment by the on-scene paramedics: one for a medical condition exacerbated by heat stress, and the other for heat syncope treated on-scene with intravenous (IV) fluids. In addition, one student quit the program on the first morning because of training difficulties, and two additional students quit at the end of the day because they reportedly considered the training unsafe.

The next day, the Captain completed the drills throughout the day and was nearing the end of the last training drill known as the "tower." The tower drill consisted of descending six floors of a concrete tower with each floor having an obstacle/drill. On the second floor the Captain had made his way through the "entanglement" simulation and exited the confined space box, when for unclear reasons, he re-entered the confined space room. An instructor monitoring the room noted that the Captain suddenly stopped moving. The instructor found the Captain unresponsive and called a mayday.

The Captain was carried down the interior stairs and outside the tower. Crew members removed the Captain's gear while the on-scene paramedics and emergency medical technicians began an assessment. The Captain was found to be in cardiac arrest; cardiopulmonary resuscitation (CPR) and advanced life support (ALS) were begun as a transport ambulance was requested. The transport ambulance arrived 6 minutes later, and en route to the emergency department (ED) the ambulance crew found the Captain's temperature to be 107.9°F (tympanic [ear] membrane). ALS continued as cooling treatments began with cool IV fluids and ice packs applied to the Captain's skin. The Captain regained a pulse, but remained unconscious as the ambulance arrived at the ED.

In the ED the Captain's core (rectal) temperature ranged from 104.4°F to 106.6°F. The Captain was hospitalized, and additional treatments for hyperthermia (cold IV fluids, fans, mist, cooling blanket, and ice packs) were administered. Treatment continued for over 24 hours, but complications of heatstroke developed, including rhabdomyolysis, acute renal insufficiency, acute respiratory failure, and hypoxic encephalopathy. On September 17, 2012, a brain scan showed results consistent with brain death. After consulting family members, the attending physician pronounced the Captain dead, and life support machines were turned off. The autopsy report listed the cause of death as hyperthermia. National Institute for Occupational Safety and Health (NIOSH) investigators concluded that the Captain's hyperthermia was caused by exertional heatstroke following heavy physical exertion in full PPE and severe environmental conditions with insufficient rehab.

According to the Smoke Diver coordinators, this was the first fatality in their 17-year history. However, numerous trainees in previous courses suffered from heat-related illness (HRI) that required emergency medical assistance either on-scene or in the ED. All 10 trainees in this course interviewed by NIOSH investigators reported symptoms consistent with mild to moderate HRI such as feeling hot, tired, fatigued, exhausted, nauseated, and having headaches. In addition to the Captain, two other trainees suffered heat syncope and heat exhaustion. The individual with heat syncope that occurred on September 15, 2012, was treated on-scene and recovered. The individual with heat exhaustion occurring on September 16, 2012, was transported to the ED and was subsequently hospitalized for 2 days.

Postincident Analysis Beaumont, Texas

Key Recommendations

- Recommendation #1: Ensure a comprehensive rehabilitation program complying with NFPA 1584: *Standard on the Rehabilitation Process for Members During Emergency Operations and Training Exercises* is in place and operating.

- Recommendation #2: Ensure that all organizations that use the training facility comply with the facility's heat stress program.

- Recommendation #3: Strengthen the facility's heat stress program by implementing the following recommendations before, during, and after all training courses at the facility.

- Recommendation #4: Require safety officers to review the planned training exercise and require their presence on all technically difficult/challenging training activities.

- Recommendation #5: Ensure that fire fighters, including training instructors, are trained in situational awareness and personal safety and accountability.

- Recommendation #6: Ensure that training maze props used in SCBA confidence training have adequate safety features such as emergency egress panels, emergency lighting, ventilation, and a temperature monitoring system to measure the ambient temperature inside the maze.

- Recommendation #7: Ensure PASS devices remain on during SCBA drills to signal if a fire fighter is lost or becomes unresponsive.

- Recommendation #8: Ensure that training facility participants are equipped with radios and properly trained in mayday standard operating guidelines and survival techniques.

After-Action REVIEW

IN SUMMARY

- As an Instructor II, one of most important responsibilities you have is to develop student evaluation instruments that will be used to evaluate their performance and abilities.
- Testing is used in many different ways within the fire and emergency services profession. Regardless of the approach, testing must follow rules and policies of the agency administering the exam.
- Testing in the high-stakes arena requires considerable work, professional guidance, and test-item documentation. It is not wise to undertake the development of high-stakes tests without considerable professional test developer involvement.
- The term *valid* is used to describe how well a test item measures what the test-item developer intended it to measure. Taking steps to ensure validity will prevent your test items from measuring unrelated information.
- Forms of test-item validity are distinguished as follows:
  - Face validity
  - Technical-content validity
  - Job-content validity/criterion-referenced validity
- Test analysis occurs after a test has been administered. Three questions are usually answered in the posttest analysis:
  - How difficult were the test items?
  - Did the test items discriminate (differentiate) between students with high scores and those with low scores?
  - Was the test reliable? (Were the results consistent?)
- If a test is reliable and the test items meet acceptable criteria for difficulty and differentiation, it is usually considered to be acceptable.
- The Systems Approach to Training (SAT) process was developed by the U.S. military during the early 1970s and 1980s. The effectiveness and efficiency of the SAT have been well documented in training journals, academic studies, and actual practice by leading businesses and industries in the United States and around the world.
- In technical training, whether the tests are written, oral, or performance based, they must be based on the learning objective.
- The fire service is committed to safety and productivity through improved training programs and courses. Testing, as an important part of any approach to training, comprises two activities:
  - The preparation of test questions using uniform specifications
  - A quantitative/qualitative analysis to ensure that the test questions function properly as measurement devices for training
- The written test-item type to be developed is determined by two considerations:
  - The knowledge requirement to be evaluated
  - The content format of the resource material
- Ease of grading of the test items and providing immediate feedback to students make multiple choice items extremely flexible for the instructor.
- Remember that most testing in fire and EMS training environments leads to certification, selection for a job, pay raises, and promotions. If you are testing for any of these reasons, use only test banks that have been rigorously validated and that are known to produce reliable tests.
- One of the skills an Instructor II needs to cultivate is the ability to develop a class evaluation form. Some departments may provide you with a class evaluation form, whereas others may have you develop your own.

KEY TERMS

Access Navigate for flashcards to test your key term knowledge.

Essay test A test that requires students to form a structured argument using materials presented in class or from required reading.

Face validity A type of validity achieved when a test item has been derived from an area of technical information by an experienced subject-matter expert who can attest to its technical accuracy.

Job-content/criterion-referenced validity A type of validity obtained through the use of a technical committee of job incumbents who certify the knowledge being measured is required on the job and referenced to known standards.

Qualitative analysis An in-depth research study performed to categorize data into patterns to help determine which test items are acceptable.

Quantitative analysis Use of statistics to determine the acceptability of a test item.

Reference blank Where the current job-relevant source of the test-item content is identified.

Reliability The characteristic that a test measures what it is intended to measure on a consistent basis.

Skills evaluation A test that measures a student's ability to do a task under specified conditions and to a specific level of competence. Also known as a performance test.

Subject-matter expert A person who is technically competent and who works in the job for which test items are being developed.

Systems approach to training (SAT) process A training process that relies on learning objectives and outcome-based learning.

Technical-content validity A type of validity that occurs when a test item is developed by a subject-matter expert and is documented in current, job-relevant technical resources and training materials.

Validity The documentation and evidence that supports the test item's relationship to a standard of performance in the learning objective and/or performance required on the job.

REFERENCES

American Educational Resource Association, American Psychological Association, and National Council on Measurement in Education. 1999. *Standards for Educational and Psychological Testing.* Washington, DC: American Psychological Association.

Equal Employment Opportunity Commission. 1978. *Uniform Guidelines on Employee Selection Procedures.* Washington, DC: Equal Employment Opportunity Commission. 29 C.F.R. Part 1607.

National Fire Protection Association. 2018. *NFPA 1001: Standard for Fire Fighter Professional Qualifications.* Quincy, MA: National Fire Protection Association.

National Fire Protection Association. 2019. *NFPA 1041: Standard for Fire and Emergency Services Instructor Professional Qualifications.* Quincy, MA: National Fire Protection Association.

CHAPTER PRESENTATION EXERCISE

Instructions: Your course instructor will assign individual or group discussions on the key points and teaching tips of this chapter. You or your group will review the chapter teachings and identify the major learning points from the chapter. You should be able to discuss the points, why they are important, and how/where they apply to your responsibilities at your level of instructor training.

REVIEW QUESTIONS

1. Explain the differences between a high-stakes test and a low-stakes test.

2. What is the systems approach to training (SAT) process, and what is its significance?

3. What do *qualitative* analysis and *quantitative* analysis determine, and how?

CHAPTER 12

Program Management and Training Resources

NFPA 1041 JOB PERFORMANCE REQUIREMENTS

Note: An asterisk denotes that the 1041 standard contains further information in its annex section.

5.2.2 Assign instructional sessions, given authority having jurisdiction (AHJ) scheduling policy, instructional resources, staff, facilities, and timeline for delivery, so that the specified sessions are delivered according to AHJ policy.

(A) Requisite Knowledge.
AHJ policy, scheduling processes, supervision techniques, and resource management.

(B) Requisite Skills.
Select resources, staff, and facilities for specified instructional sessions.

5.2.3 Recommend budget needs, given training goals, AHJ budget policy, and current resources, so that the resources required to meet training goals are identified and documented.

(A) Requisite Knowledge.
AHJ budget policy, resource management, needs analysis, sources of instructional materials, and equipment.

(B) Requisite Skills.
Resource analysis and preparation of supporting documentation.

5.2.4 Gather training resources, given an identified need, so that the resources are obtained within established timelines, budget constraints, and according to AHJ policy.

(A)* Requisite Knowledge.
AHJ policies, purchasing procedures, and budget.

(B) Requisite Skills.
Records completion.

5.2.5 Manage training recordkeeping, given training records, AHJ policy, and training activity, so that all AHJ and legal requirements are met.

(A) Requisite Knowledge.
Recordkeeping processes, AHJ policies, laws affecting records and disclosure of training information, professional standards applicable to training records, and systems used for recordkeeping.

(B) Requisite Skills.
Records management.

KNOWLEDGE OBJECTIVES

After studying this chapter, participating in a structured learning environment, and completing assigned assessments, you will be able to:

- Describe how to schedule instructional sessions. (**NFPA 1041: 5.2.2**, pp 267–268)
- Describe the selection of instructors for specific instructional sessions. (**NFPA 1041: 5.2.2**, p 276)
- Describe the types of training records necessary to document instructional sessions. (**NFPA 1041: 5.2.5**, pp 276–277)
- Describe record and report retention and storage considerations. (**NFPA 1041: 5.2.5**, pp 276–277)

- Describe the budget process, the creation of a bid, and budget management. (**NFPA 1041: 5.2.3**, pp 278–282)
- Describe the procedures for recommending budget items to meet training goals. (**NFPA 1041: 5.2.3**, pp 282–285)
- Explain purchasing considerations including documentation associated with purchases. (**NFPA 1041: 5.2.4**, pp 282–287)
- Explain considerations when evaluating resources for a program. (**NFPA 1041: 5.2.4**, pp 282–287)

SKILLS OBJECTIVES

After studying this chapter, participating in a structured learning environment, and completing assigned assessments, you will be able to:

- Demonstrate the ability to schedule training. (**NFPA 1041: 5.2.2**, pp 267–275)
- Select instructors for specific assignments. (**NFPA 1041: 5.2.2**, p 276)
- Demonstrate completion of a training record report form. (**NFPA 1041: 5.2.5**, pp 276–277)
- Prepare a recommendation for a budget item. (**NFPA 1041: 5.2.3**, pp 282–285)
- Use procedures to acquire training resources. (**NFPA 1041: 5.2.4**, pp 282–287)

You Are the Fire and Emergency Services Instructor

As the newest Instructor II assigned to the Fire Training Academy, you are tasked with making sure that the training records are properly completed and filed. At the end of the most recent recruit training class, the director of the training academy requested a report on the actual number of hours spent completing the job performance requirements (JPRs) contained within NFPA 1001: *Standard for Fire Fighter Professional Qualifications*, the standard that identifies the minimum JPRs for career and volunteer fire fighters whose duties are primarily structural in nature.

1. Describe the types of training records necessary to document instructional sessions and the methods used to preserve such records.
2. Discuss the importance of proper recordkeeping and possible pitfalls encountered while maintaining records.
3. Describe the records and reports that are required and the legal implications associated with such recordkeeping.

 Access Navigate for more practice activities.

Introduction

Training and education are tools used by fire departments to improve efficiency in their operations and fire fighter safety. Managing a training team in any type of fire department is a staff-level function often fulfilled by a department training officer. The training officer may have a rank position within the department and must possess the managerial skills to accomplish the many job tasks assigned to the training team. The majority of these management- and supervisory-level functions are performed by the Instructor II or higher, but an Instructor I could also be responsible for handling these duties on occasion. If you wish to be a leader of a training program, you must develop or improve your administrative and leadership skills.

A fire and emergency services training program is a critical part of the effort to reduce fire fighter injuries and line-of-duty deaths. Lack of training, failure to train, inadequate training, and improper training are contributing factors to many injuries and fatalities. Fire fighter injuries and fatalities could be prevented through a successfully managed training program. Much is at stake and much relies on the quality of the training program.

Training records and reports are also becoming more important for use as evidence in liability lawsuits brought against an organization's fire-ground activities. The instructor who is assigned the responsibility for managing the training division may be referred to as a training officer, director of training, training program manager, chief of training, or similar title designating a specific level of job performance above that of an instructor who delivers a lesson plan. For the purposes of this chapter, the title of training officer (TO) will be used to identify the person responsible for these functions of budget needs matching them with the training needs.

The job description of the TO covers all four of the main responsibility areas identified by NFPA 1041: *Standard for Fire and Emergency Services Instructor Professional Qualifications*:

- Program management
- Instructional development
- Instructional delivery
- Evaluation and testing

This chapter covers the program management component.

Scheduling of Instruction

As a Fire and Emergency Services Instructor II, you must take into consideration the many aspects of your department's operation when developing a training schedule. A comprehensive schedule includes all job areas and covers elements of both initial and ongoing training. A training schedule must balance the various regulatory requirements with the training sessions needed to help fire fighters deliver safe and effective service. Regulatory requirements that affect the department's training schedule include everything from emergency medical services (EMS) recertification requirements to hazardous materials training. Ongoing skill and knowledge retention and refresher training programs are an important part of the scheduling process. In part-time, paid-on-call, or volunteer organizations, the scheduling and delivery of training can be an extremely challenging task. The amount of time available for delivering training is often compromised by members' occupations and family commitments in such departments.

Developing a Training Schedule

An **agency training needs assessment** should be performed at the direction of the department administration. This is covered in greater detail in the Instructor III chapters that follow. The agency training needs assessment helps you identify any regulatory compliance matters that must be included in the training schedule. It may also help you identify the topics in which the department wants their personnel to be trained. The following methods may be used to determine which topics need to be given priority over others:

- Identify skills and knowledge necessary for safety and survival.
- Review the skills and knowledge necessary for delivery of the department's mission statement.
- Examine the subject areas tied directly to the job description of each job title or function.

Every department has specific regulatory issues to address in its schedule, but some issues have a universal impact. Although these areas require interpretation locally, certain types of training requirements are usually specified by many regulatory agencies. Consider the Insurance Service Office (ISO)—it has numerous training requirements that cover initial training for new fire fighters and vehicle operators and ongoing training for fire officers. Although technically not a regulatory agency for fire departments, the ISO may influence the operations of many departments because property classification ratings can be improved by development and implementation of an effective training plan. A good technique is to contact your local ISO representative to have a discussion on the training requirements that are part of the ratings survey.

The Occupational Safety and Health Administration (OSHA) has specific training requirements that should be considered, ranging from respiratory protection training to infectious disease protection training. Some states are considered OSHA states, meaning that enforcement authority for OSHA requirements is with the state. In OSHA states, public entities fall within the authority of the state OSHA. Because OSHA has different levels of enforcement power and authority in each state, you should contact your local OSHA representative to discuss which areas of your program are affected by OSHA training requirements. Fines or other sanctions could be levied against your department if it fails to meet OSHA requirements.

Consider other regulating commissions or agencies that may have training requirements, such as your local insurance carrier or risk management departments, EMS state or licensing agent, National Fire Protection Association (NFPA), or the State Fire Marshal or statewide certification entity. These agencies may have significant requirements for both entry-level and ongoing training for your students.

Knowledge of these and other regulatory agency requirements is a key part of the scheduling puzzle. Each agency may require training to be conducted at specific intervals or to last for a certain number of hours. Become familiar with each of these areas and develop ways to cover as many areas as possible by combining resources and sessions to meet the varied requirements.

Part of the Instructor II's job description is to keep current with the requirements of the regulatory agencies that affect your department. As new laws and standards are passed, new or different requirements in your department training policy must be included to comply with those new laws and standards. Best practices learned from case studies and other local requirements may require the development of a department training policy. High-risk areas of training, such as live fire training or special operations training, should enforce a student-to-instructor ratio of 5:1, with this requirement being included within the department's written training policy. The department training policy should also specify safety rules during training, the scope of coverage of the training program, and many other areas of administrative responsibility.

Types of Training Schedules

After establishing a clear vision of what a department's training needs are, the Instructor II who is developing the training must determine how that training will be administered. Establishing a training schedule can be challenging because of the many unknowns that are inherent with the fire and emergency services profession.

The two types of training that occur most often in the fire service are formal training programs and in-service training. At the most basic level, in-service training or on-duty training, often referred to as an **in-service drill**, takes up the majority of a training schedule (**FIGURE 12-1**). An in-service drill can comprise a single station drill or involve all on-duty companies. In-service drills can be scheduled to run on specific days or they can take place only when a specific training need is identified. In-service drills are the most common delivery method of training. For the purposes of this chapter, in-service drills are considered to be conducted while fire fighters are available on duty and in service to respond to incidents, whereas formal training courses take place outside the structured workday.

In-Service Drills

In-service drills, or in-service training, typically take place at a specified time and location. All types of fire departments use some form of in-service drills. There is very little time to take units out of service to train. Instead, apparatus and personnel must be kept in some form of readiness in the event that an emergency occurs, including during training. On the schedule, some departments identify the topic and level of coverage for the session, whereas others simply state that the company officer will select a topic from a list provided by the training office, or the company officer will create an in-service drill from a National Institute for Occupational Safety and Health (NIOSH) report or recent incident.

In-service training can be broken into different levels, described as follows.

Skill and Knowledge Development

Skill and knowledge development occurs when new approaches to firefighting operations are introduced to the department (**FIGURE 12-2**). These sessions expand on previously learned information or skill levels to increase the fire fighters' ability to do their job. This level of training is used when a new method, piece of equipment, or procedure is implemented. All new equipment should be used in training by all fire fighters prior to placing that equipment into service. This approach may take a considerable amount of time to accomplish but will pay off on the fire ground.

Skill and Knowledge Maintenance

Skill and knowledge maintenance is one of the goals of a comprehensive training program (**FIGURE 12-3**). Both skills and knowledge degrade over time if not used. In a life-safety profession, you cannot afford not to function at your highest level of ability. Skill and knowledge maintenance training is used to develop performance baselines for core duties and functions so as to establish company-level standards and to measure individual skills and weaknesses. All company members should be able to perform basic tasks within a reasonable amount of time. An incident commander makes incident assignments knowing how long it takes to stretch a hose line or ventilate a roof, and if the company is untrained or unprepared to meet that standard, other fire-ground assignments may suffer.

Skill and Knowledge Improvement

Skill and knowledge improvement is necessary when individual weaknesses become apparent. These drills

JBL Fire Department
Monthly Training Brief
January 2019
Wear It: Survive Today & Retire Healthy
PPE   Seat Belts   Safety Vests   Hearing & Eye Protection

Weekly Skill Drill Summary

Subject	Date(s)	FH Entry Code	Location	Skill Description
Tie Figure Eight Knot	Dec 30–Jan 5	1WSD1	Stations	Tie a handcuff knot on partner and self
60 Sec. SCBA Donning	Jan 6–12	1WSD2	Stations	Don SCBA from floor in 60 sec or less
Operate Power Saws	Jan 13–19	1WSD3	Stations	Operate and maintain all power saws
Set-up PPV and Smoke Ejector	Jan 20–26	1WSD4	Stations	Assemble RIC equipment
2 Person Salvage Cover Throws	Jan 27–Feb 2	1WSD5	Stations	Perform salvage cover throws and folds

Company Level Training

Subject	Date(s)	FH Entry Code	Locations	Assignments/Info
Daily Quick Drill	Daily	DQD	Station Roll Call	Assigned to Acting Officer/Driver. List subject covered in notes on FH.
Company Readiness Training-Daily	Daily	CRTD	Stations	Daily company tool, equipment, and apparatus training.
Daily Driver Training	Daily	DRIV DAILY	Stations	Daily pre-trip driving inspection, routine and emergency responses experience.
Company Readiness Training-Weekly	Saturdays	CRTW	Stations	Weekly company tool, equipment, and apparatus training (Saturdays).
Tactical Walk Through Drill	Mondays	TWT	Company Officer Selected	Tactical walk through for first-due operational information.
Daily Physical Fitness Training	Daily	FITS	Stations	Strength, flexibility, cardio, and aerobic capacity fitness training.
Tool Assignments	Jan 2–4	TOOLS1	Stations	Define standard tool assignments by company type and assignment.
MAP Book/CAD Training	Jan 9–11	MAPS	Station 1 0900 Station 3 1330	Introduction to new map book and grids.
Fire Behavior	Jan 16–18	BEHAVIOR1	Station 1 0900 Station 3 1330	Introduction to fire behavior research.
RIC – FF Through the Floor	Jan 23–25	1RITOPSFLOOR	TBD	Rescue FF who has fallen through the floor.
Ice Rescue	Jan 30–Feb 1	ICE	TBD	Ice rescue evolutions.

Weekly Tactical Training

Building Name	Date(s)	FH Entry Code	Assignments/Info
Polo Inn	Jan 6–12	2SIMPOLO	Review tactical walk through and complete tactical simulation.
Lee Manor Nursing	Jan 13–19	2SIMLEEMNR	
Stonecrest Condos	Jan 20–26	2SIMSTONECRST	

Pump Operator Training

Subject	Date(s)	FH Entry Code	Locations	Assignments/Info
Apparatus Operators Intro	Jan 9–11	2PUMPINTRO	Station 1 0900 Station 3 1330	Review of apparatus daily, weekly, and maintenance procedures. Response policy overview.

HazMat Operations Training

Subject	Date(s)	FH Entry Code	Locations	Assignments/Info
Not Scheduled This Month				

Technical Rescue Awareness Training

Subject	Date(s)	FH Entry Code	Locations	Assignments/Info
Not Scheduled This Month				

Officer Training

Subject	Date(s)	FH Entry Code	Locations	Assignments/Info
Officer Training Session	TBD	TBD		Self-study activities and terminology/SO preview. Assignments via email groups.
Acting Officer Training	TBD	TBD		
Incident Command Training	TBD	TBD		

EMS Training

Subject	Date(s)	FH Entry Code	Locations	Assignments/Info
EMS-Paramedic CE	Jan 8, 15, 22		Station 1 1330 Station 3 0900	TBA
EMS-EMT Basic CE	Jan 8, 15, 22		Station 1 0900 Station 3 1330	TBA

Special Teams Training
See Training Calendar on Shared Calendars for Special Teams Drill Information

FIGURE 12-1 A sample training brief.

are used in the case of errors or poor performance, or when other undesirable outcomes have been observed. A company may be able to stretch a hose line and place the nozzle in a ready position effectively, but if the pump operator is unable to have the desired gallons per minute (gpm) and nozzle pressure arrive at the nozzle, the entire company will fail in its assignment. This is a signal for the company officer to train with the pump operator, using instructors, to improve the entire company's skill level.

The in-service drill is the type of training that is most vulnerable to cancellation, delay, or reduction in participation due to emergency response. Consequently, the instructor should have a **supplemental training schedule** available in case some type of mitigating factor requires a change in the original training session (**TABLE 12-1**). This should be a published program. The resources needed to accomplish the training should be on standby and ready to use when unplanned events occur.

FIGURE 12-2 Skill and knowledge development occurs when new approaches to firefighting operations are introduced to the department.
Courtesy of Forest Reeder.

FIGURE 12-3 Skill and knowledge maintenance is one of the goals of a comprehensive training program.
© Jones & Bartlett Learning. Photographed by Glen E. Ellman.

Formal Training Courses

Certain types and levels of training require the use of a set curriculum matched to a directed number of hours or evolutions. Training to NFPA professional qualification standards using JPRs may require that students participating in a course attend a predefined-hour program to meet all of the JPRs. This type of training schedule affects the use of the department classroom or training center, and such a program could possibly overlap with the use of the same facilities for in-service training. Formal training courses may also have specific needs related to the facility—for example, audiovisual equipment, props, or appliances—that a department may have to budget for or acquire from other sources. An example of a formal training course could be a Fire Officer I course, based on the NFPA 1021 *Standard*

TABLE 12-1 A Supplemental Training Schedule

Supplemental Training Schedule: Available Topics			
Fire Suppression	**Rescue Tools**	**Hazardous Materials**	**EMS**
Nozzles and streams	Air bags	*Emergency Response Guidebook*	Vital signs
Extinguishers	Glass removal tools	Terminology	Patient movement
Self-contained breathing apparatus (SCBA) donning	Hydraulic tools	Decontamination	Documentation
Ladder carries	Hand tools	Spill containment	Initial patient surveys

© Jones & Bartlett Learning.

for Fire Officer Professional Qualifications, which could include certification testing conducted by an outside certifying agency.

Other types of training also must be developed when creating a comprehensive training schedule. Special areas of training to consider include the following:

- Technical rescue teams
- Special operations teams
- Hazardous materials teams
- EMS
- Vehicle operator training

The following areas of department operations may overlap with a training program and should be considered part of in-service training:

- Vehicle and equipment inspection and maintenance
- Hose testing
- Ladder testing
- Pump testing
- Daily/weekly/monthly equipment checks

Other areas within a department that must be considered within the framework of training program management include other staff-level functions within the organization. Keeping the concept of training commensurate with duty means that the training needs of staff—personnel such as fire investigators, EMS, or other staff assignments—must be included in the development of your training plan. Consider the functions of fire-cause determination personnel, fire inspection personnel, public education teams, dispatch and communication personnel, and apparatus mechanics when developing your training plan.

Each of these areas may take up substantial amounts of available time and require the coordination of resources and time management. These areas may also require specialized resources, instructors, or equipment that must be considered when creating the schedule.

Develop a good working relationship between instructors and operational team leaders. These team leaders have backgrounds and expertise that will help in the delivery of instruction. When you work with these team leaders to prepare lesson plans, schedule sessions, and make sure division objectives are incorporated into the course, your training program will be more successful.

Scheduling for Success

To meet the needs of the department and its fire fighters, the schedule you develop must be consistent, easy to understand, and clear (**FIGURE 12-4**). To accomplish this goal, you must have access to, and fully understand, the department's training policies. A department's training policy or standard operating procedure (SOP) that relates to training may identify who must train, how often, where, and who will provide the training. If this information is already known, the scheduling chore is almost complete.

Subject coverage in a curriculum-based training program should be sequential. When looking at your subjects, consider any prerequisite learning and skill levels. Ask yourself which topic needs to be covered before the next one can be covered. For example, it would be advisable to instruct on ladders before teaching a class on vertical ventilation to a group of new fire fighters. Students should learn and perform skills in a sequential order to maximize their understanding as well as to improve their safety and survival.

Important information in a training policy or SOP that will help you develop a training schedule includes the following elements:

- Who must attend the training session
- Who will instruct the training session
- Which resources are needed for the training session

Information specific to the training session may also be used to help identify the scope and level of the training session. Students appreciate knowing the subject of the training session before they walk into the classroom. Highly motivated students may complete reading assignments or research the topic before the training takes place.

Fire officer development includes the ability to prepare a crew for a training session by practicing skill sets before the drill (**FIGURE 12-5**). Advanced polices also detail the responsibilities of those who attend, supervise, and participate in the training. Recordkeeping processes are included in this policy as well. Information about the training session itself includes the following points:

- What the training subject(s) is(are)
- Whether it is a practical (hands-on) or classroom type of session
- Whether there are any pre-training assignments
- Where the training will take place

A standardized process must be used to schedule training sessions. As stated earlier, the schedule must have some flexibility so that you can take advantage of unforeseen opportunities or overcome obstacles in the delivery of training. For example, you might schedule a block of training on hose deployment when a local

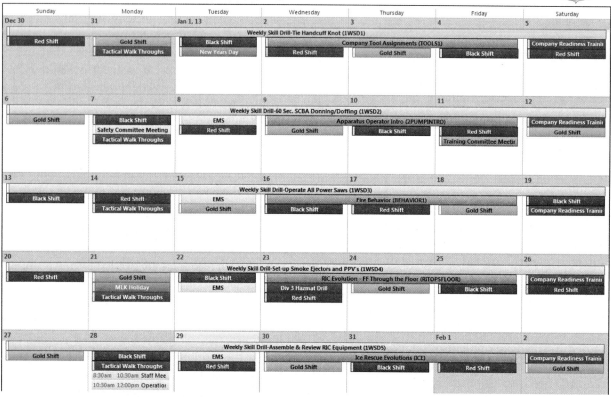

FIGURE 12-4 Training schedules should be easy to follow and clearly list the training topics along with other pertinent information.

FIGURE 12-5 Fire officer development includes the ability to prepare a crew for a training session by practicing skill sets before the drill.
© Jones & Bartlett Learning. Photographed by Glen E. Ellman.

developer calls you and offers the use of a structure scheduled for demolition. If the property is suitable for use, it may present a great opportunity to enhance your training program through the use of the acquired structure. You might also want to consider covering your hose deployment drills during the training performed at this structure, along with the multitude of other areas you can train on.

Obstacles can range from weather complications to the scheduled vacation time of students who need training. Additionally, new equipment may be purchased that needs to be trained on before it is put into service, with this consideration taking precedence over other scheduled activities. In addition to the creation of the schedule, a process for the publication, distribution, and delivery of the training

Job Performance Requirements in ACTION

Most fire departments have a training division, often led by a training officer or fire officer designated to complete the management-related functions tied to training. Recordkeeping, scheduling of classes/training, and compliance are some of the functions that must be attended to on a regular basis, regardless of the size of the department. Resources must be managed, and budgets must be prepared and monitored. Instructors need to be assigned and facilities secured to ensure that instruction and student experiences are of the highest quality. Some of these functions are related to typical fire officer management and organizational principles, but in the area of training program management, many special considerations must be learned and applied to manage the training team. The guidance for many of these functions is agency policies that, when followed, protect the instructor at every level from problems both internal and external.

Instructor I

Teamwork within the training team is as important as it is in the fire company. As the training program is a division within the fire department, proper management requires the Instructor I to act as part of the team and to complete routine administrative tasks in addition to the delivery of training.

Instructor II

Managing the resources, staff, facilities, and budget of a training program requires sound management and leadership skills. Many fire officer JPRs will be put to work while completing these functions.

Instructor III

Working as part of the training team, focus on policy development and good communication flow between instructors on the training team. Supervision and evaluation of instructors and the program are key duty areas.

JPRs at Work

Function as part of the training team by delivering training, scheduling single instructional sessions, managing resources, and ensuring the safety of students during training. Creating and maintaining records, reports, and proper documentation of the learning process while following agency policy will be a large part of the Instructor I's responsibilities.

JPRs at Work

Schedule and acquire the resources necessary to conduct training sessions and assign qualified instructors while administering a budget for these functions. Ensure the safety of the students and instructors by administering department policies and applicable standards while developing training materials and requesting resources to delivery training.

JPRs at Work

Create the policies and procedures for the management of instructional resources, staff, facilities, records, and reports for the training program. He or she will have knowledge of the methods used to evaluate equipment and resources used for training delivery.

Bridging the Gap Among Instructor I, Instructor II, and Instructor III

The Instructor I can increase his or her ability to work as part of the training team by accepting additional responsibilities, such as scheduling a single instructional session and other management-related functions. The Instructor II can serve in a mentoring position and also reduce his or her workload by utilizing the Instructor I to assist in completing routine tasks. This prepares the Instructor I for increased management responsibilities and helps to fully develop the training team and utilize each member to his or her full potential. An Instructor III will develop processes, policies, and practices that will support the delivery of course goals, agency objectives, and the overall quality of the training program.

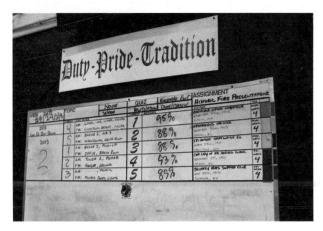

FIGURE 12-6 Schedules should be posted and visible to all members to ensure consistent delivery of training and proper resource management.
Courtesy of Forest Reeder.

schedule must exist so that all who are affected by it are informed about the training-related activities of the period (**FIGURE 12-6**).

One of the challenges in scheduling instruction in the volunteer world of emergency response is the demand placed on time. Career responders have the opportunity to schedule training while on duty and meet the requirements for recertification or learning new equipment. For responders who volunteer their time to provide emergency services within their communities, training time competes with work, family, and other pressures that can challenge any department. As an Instructor II or TO for a volunteer department, you still need to meet all required training hours for OSHA or EMS, but you end up doing so in smaller blocks of time. Training sessions for volunteer departments might be broken up over several nights or on the weekends. A Fire Officer I course that might have taken a week at a full-time agency could take several weeks to complete, but in the end the same knowledge, skills, and competencies are delivered and the volunteer responder can obtain the same level of proficiency.

Master Training Schedule

A useful method to organize the multiple demands for schedule coverage is a system capturing all the required training areas within a single forum. A simple table form, called a **master training schedule**, may suffice to begin this process (**FIGURE 12-7**). It can serve as the starting point in the development of your in-service training program schedule. When a period of training (typically one month in the year) is planned, you can go to this table to identify the topics that must be included into the schedule.

Here are the steps required to develop a master training schedule for your department:

1. Complete an agency needs assessment of the regulatory agencies that require training on specific topics (OSHA, ISO, NFPA, local agency/authority).
   a. Determine the frequency with which the training must be included.
   b. Determine the type of training that is needed.
      i. Initially upon hire
      ii. Ongoing after hire
      iii. Classroom or hands-on
      iv. Individual, company, or multi-company
      v. Mutual aid training
      vi. Training on special hazards or topics
2. Identify other areas of training that are necessary to train your students on the skills and knowledge necessary to keep them safe at emergencies.
   a. SCBA skills
   b. Fire fighter survival techniques
   c. Minimum company standards or other proficiency-based training
3. Identify ongoing activities that affect the training schedule so that you do not overload the crews with too many activities within a period.
   a. Equipment testing
   b. Fire prevention activities
   c. Apparatus testing
   d. Applicable recertification
4. Create a table that allows you to list the information compiled in steps 1–3 in one column on individual rows and create at least 12 additional columns to represent each month of the year.
5. Insert a checkmark or other indicator of when each type of training needs to be completed into the number of month columns required to meet the standard.
   a. If SCBA needs to be trained on quarterly, insert an X in one month of each quarter.
   b. If protective clothing needs to be donned, doffed, and cleaned semiannually, insert an X in each of two 6-month periods.
6. Once all elements have been added, balance out your coverage to allow for an even distribution of the content to be covered each month.
7. Develop categories for other types of training by subject matter areas and distribute them throughout the calendar to give your program even coverage.
   a. Examine the subject area coverage in the training materials to establish a well-rounded training plan.
   b. Make sure that all of the JPRs of a given job level have been covered.
8. Review and revise the schedule on a regular basis, preferably annually.

JBL Fire Department
Master Training Schedule

Core Training Sessions	JAN	FEB	MAR	APR	MAY	JUN	JUL	AUG	SEP	OCT	NOV	DEC
Protective Clothing Inspection	X						X					
Officer Training/Acting Officer Training	X	X	X	X	X	X	X	X	X	X	X	X
Minimum Company Standards	Under Development and Initial Training
SCBA Module A (D/D, Care, Maint., Fit Testing)		Don Doff									Fit Test	
SCBA Module B (Emergency Procedures)			X									
SCBA Module C (Air Management Training)					X							
SCBA Module D (Practical Applications)											X	
Engineer Recertification & Testing		Hydraul. 6–8		Drive 17–19					18–20 25–27			
HazMat Operations Refresher		X								X		
Pump Operator Refresher (Modules)	Maint. & Driving		Mod 1 & 2		Relay	S-Pipe	Foam	Def.			Skids 2½"	Written
Technical Rescue Team Support	Ice			Con Sp		Trench				Vert		
IDOL Compliance Session (BBP/LOTO/RTK)				X								
COMPANY TRAINING												
Company Skill Evolutions (Weekly Skill Drill)	X	X	X	X	X	X	X	X	X	X	X	X
Multi-Company Level Evolutions			X			X					X	
EQUIPMENT TESTING												
Pump							X	X				
Hose				X	X							
Ground/Aerial Ladders									X	X		
SAFETY TRAINING												
Occupational Health & Safety Programs	Mod 1	Mod 2	Mod 3	Mod 4	Mod 5	Mod 6	Mod 7	Mod 8	Mod 9	Mod 10	Mod 11	Mod 12
Trends, Current Events, LODD Analysis	X	X	X	X	X	X	X	X	X	X	X	X
SPECIAL TRAINING PROGRAMS												
EMS Training (Locally Assigned Medic/EMT)	X	X	X	X	X	X	X	X	X	X	X	X
Technical Rescue Team (Local Team)	X	X	X	X	X	X	X	X	X	X	X	X
HazMat Team (Local Team @ Tech Level)	X	X	X	X	X	X	X	X	X	X	X	X
ONGOING TRAINING PROGRAMS												
Min. Company Standards (Individual and Co) Daily Quick Drills Safety Audits	X	X	X	X	X	X	X	X	X	X	X	X
Self Survival and RIT Skill Development	X	X		X	X		X		X	X	X	
Weekly Tactical Walk Through & Exercise	Every Monday afternoon assigned by Company Officers

FIGURE 12-7 Sample master training schedule.

Once the master training schedule has been created by the Instructor II, it should be approved by the training division chief (if applicable) and the agency chief or department head. There are benefits from having these approvals: first, the chief knows what the training division is doing; second, leadership approvals send the message that training is valued and important to the agency and department.

Selection of Instructors

As an Instructor II, one of your duties is to assign other instructors to deliver training. In many agencies, in-service training is conducted by the company officer following the agency's training schedule. On other occasions, training is delivered by members of a training division, or line personnel are assigned on a temporary basis to assist in delivering training to on-duty responders. Regardless of the reason, the selection of instructors to teach a course must be considered an important duty and one that should be taken seriously.

When selecting instructors to deliver a course, ask yourself the following questions: What are their qualifications? What are their values concerning fire service safety? What do their previous course evaluations say about them? Despite your best efforts to keep your training safe, you cannot be everywhere at all times; therefore, you must rely on other instructors to help supervise the drill.

Instructors who are not competent in or confident with the subject being presented should not be placed in a situation where they are expected to teach that material to others. It is important to assess the motivation of your subordinate instructors as well. An instructor who has a propensity to show off or try to prove that he or she is superior to students is an accident waiting to happen. Often the focus of such an instructor is not on the students or their safety. Know your staff, their attitudes toward safety, the way they operate on the fire ground, their motivations for being instructors, and their teaching methods. All support personnel for the drill, including those who simply assist in prop construction, should be included in this assessment, because this information is essential to conduct training safely.

In addition to which instructors to assign a training session, determine how many instructors are needed to safely deliver and conduct the session. Two factors should be considered in making this decision: the *type* or *risk level* of the training and the *skill level* of the students. Remember that training can be classified based on risk; for example: high risk (such as live fire training evolutions), medium risk (such as ladder training), or low risk (such as connecting a hose to a fire department connection). Also consider the skill level of those being taught. High-risk training of new recruits should generate different concerns from high-risk training involving senior experienced fire fighters. The selection process should follow department polices that might also direct which instructors should be used during a training session.

Other areas of planning include assembling the equipment that may be needed for the training session. All instructors involved in training are responsible for inspecting all equipment being used in the training to ensure that it meets the required standards prior to the training or drill session. Do not assume that the equipment is in safe and operable condition. If your department does not have the equipment available to make the training safe, then it needs to find alternative equipment—perhaps by borrowing items from a neighboring fire department, a local fire equipment vendor or the state fire training academy. Acquisition of the appropriate equipment should be arranged prior to the drill. As an instructor, you are responsible for protecting students from physical and emotional harm in the classroom and on the drill ground.

Part of the scheduling process includes securing the necessary facilities for a training session to be conducted. For in-service staff, making sure the training tower or field is open is an important first step in conducting a training session. If you are the Instructor II that is planning and coordinating a course, in addition to selecting other instructors you also need to make sure you have access to the training ground, classroom, or other facilities you might need. Using a checklist to help you plan and to ensure that all resources are available is a simple but effective method to coordinate an upcoming training session.

Record Management

One of the most important aspects of Instructor II responsibilities is the management of training records and a records management system. In today's world, where litigation is all too common, it is important to remember that one of the first lines of defense is always proper and thorough recordkeeping. These records should be constructed contemporaneously and maintained for a period of time and in accordance with local and state laws and agency policy.

The recordkeeping process starts at the beginning of each training activity. For example, when you are organizing a training evolution, the materials you prepare should set forth the purpose and objectives of the training, the activities to be performed, the equipment to be used, and the end goals of the training. You should have available any sources that were relied upon in creating the evolution and should confirm that the training complies with all policies and procedures of your fire department.

Recordkeeping is vital to the smooth operation of the instructional process. In addition, recordkeeping pertaining to certifications, permits, and licenses is needed to ensure the continued operation of the fire department. Failure of the instructor to keep track of EMS recertification or continuing education, for example, could impair the staffing of the fire department.

If you are unsure of what should be included in a training record, NFPA 1401: *Recommended Practice for Fire Service Training Reports and Records* (Appendix D) is a good place to start. Within this standard are the areas that need to be covered and addressed in a training record that will document the following: date of training/course, who attended, who participated, who the instructor was, what material was taught, where the training was conducted, and any problems that occurred during the training session.

Additional information that should be part of the training records could include a copy of the lesson plan, method of delivery, student scores, and evaluations with feedback on the instructor and material. Knowing your agency's policy on what should be gathered and maintained is important to the recordkeeping process.

Confidentiality

The thoroughness and accuracy of the records you keep is fundamental in mounting a sound defense against litigation. In cases where an injury occurs to a participant in or an observer of a training session, thorough documentation is imperative, including who was present and what each person observed. Chapter 1, *Today's Fire and Emergency Services Instructor*, emphasizes the importance of diligent recordkeeping.

Remember that state and federal laws (Privacy Act of 1979) prohibit the disclosure of certain information. Failure to abide by the **confidentiality** provisions required by law or your department could lead to liability or adverse action against you.

The following records must be kept secure and confidential:

- Personnel files: Usually include information such as date of birth, Social Security number, dependent information, and medical information.
- Hiring files: Include test scores, pre-employment physical reports, psychological reports, and personal opinions about the candidate.
- Disciplinary files: Any report or document about an individual's disciplinary history and related reports.

Maintain records for all training sessions for a period of at least 5 years, or longer as required by law or agency policy. Often, lawsuits are filed long after the event occurred; therefore, the only concrete protection for the fire department or you as the instructor might be the records and documentation you have created and kept on file.

Other Considerations for Records

Each state has freedom of information laws that govern public access to government records, documents, and other information kept by government bodies. The purpose of the Freedom of Information Act (FOIA) is to promote government accountability and the public's right to know. Chapter 1, *Today's Fire and Emergency Services Instructor*, further discusses federal and state laws.

Generally, members of the public can make a FOIA request through designated public officials. Depending on the applicable state law, there may be exceptions to the disclosure of certain documents, or the government body releasing the documents may be required to redact personal information from certain documents. In addition, government entities may be required to retain public records for a specific period of time. In most cases, state record retention laws will dictate the minimum length of time public records must be retained by government entities. Review your state law and consult with your attorney for more specific information about the requirements of FOIA, the exceptions to FOIA disclosure, and the record retention requirements in your state.

If the training will involve participants from other fire departments, additional documentation should be obtained for those fire fighters. Moreover, your fire department should enter into a **hold harmless** and **indemnification agreement** with any other department or agency involved in the program. Such an agreement provides that your fire department will not be held monetarily or legally liable for injuries or conduct of the other department or agency's employees. This document is necessary to protect your fire department from individuals over whom you have no control. As a representative of the hosting fire department, you should also verify that all outside participants are medically cleared to participate in the training, have met the necessary prerequisites, and possess the proper credentials. While this approach might seem somewhat burdensome, the protection it affords is priceless.

The critical point in all of this information is that, as the supervising instructor, you are aware of which laws pertain to you and your organization regarding recordkeeping, storage, and retention. If you are unclear on those details, research and become familiar with them before an incident occurs and you are placed in a position of uncertainty.

Budget Development and Administration

The budget process of a fire department can be very confusing, if not overwhelming. Understanding the budget development process and your role in specifying and purchasing resources and materials, as well as justifying the funding requirements of a training project or program, are crucial to successful training program management.

> **DEPARTMENT TRAINING OFFICER TIP**
>
> Developing budgets is a great way to demonstrate what the training division does and how it adds value not only to the department, but to the community as a whole. For this reason, you should approach this process by viewing it as an opportunity. Know the goals and be able to outline the direct relationship between the attainment of those goals and your need for the requested resources. Be articulate in your justification by being focused and concise but also thorough. Don't fluff the numbers to give negotiation room, but don't sell yourself short.

Introduction to Budgeting

A **budget** is an itemized summary of estimated or intended revenues and expenditures. **Revenues** are the income of a government from all sources appropriated for the payment of public expenses and are stated as estimates. **Expenditures** are the money spent for goods or services and are considered appropriations that allow the expenditure authority. Every fire department has some type of budget that defines the funds that are available to operate the organization for a particular period of time, generally 1 year. The budget process is a cycle that follows these steps:

1. Identification of needs and required resources
2. Preparation of a budget request
3. Local government and public review of requested budget
4. Adoption of an approved budget
5. Administration of approved budget, with periodic review and revision
6. Close out of budget year

Budget preparation is both a technical and a political process. The funds that are allocated to the fire department define which services the department is able to provide for that year. The technical part relates to calculating the funds that are required to achieve different objectives, whereas the political part is related to elected officials making the decisions on which programs should be funded among numerous alternatives.

Budget Terminology

Most municipal governments use a base budget in the financial planning process. The **base budget** is the level of funding that would be required to maintain all services at the currently authorized levels, including adjustments for inflation and salary increases. A built-in process assumes that the current level of service has already been justified and that the starting point for any changes in the budget should be the cost of providing the same services next year. Budgets can be increased or decreased from the base level due to a variety of reasons. When an approved budget needs to be changed from the base budget, a modification is introduced to address any supplement budget requests.

There are other types of budgets such as program, performance, or zero-based budgets. Each of these types has an appropriate place and use in agencies. The key is that you need to know and understand the type of budget system used within your agency. Understanding the difference between a capital expense and an operational expense will affect how you develop a budget and which budget is used to pay salaries or purchase a projector for the training division.

Two types of budgets that all instructors are aware of are operating budget and capital budget. An **operating budget** is what is used to run the agencies day-to-day in the delivery of services to their community. This budget covers everything from uniforms, to toilet paper, to fuel for the apparatus, to salaries of the fire fighters and paramedics. The **capital budget** comprises purchases that have a value that would extend over several years of use, for example: a fire engine, SCBA, ambulance, or a training tower with a burn chamber for live fire training.

Budget Preparation

Successful budgeting requires justifications of the amount of money being requested. A budget justification is useful in defending your position when requested money is allocated for a specific line item or category. A training division may be an example of a line item in a budget document. When developing the line item for the training division, the justification can be based on previous expenses in this area, the agency training needs assessment, and the forecasting of the training needs for the upcoming period (**FIGURE 12-8**). Note that the *Modifications and Justification* column shown in this figure includes just

Category/Budget Code	Justification Explain the goal or target for each line item category	2016-17 Requested	2016-17 Actual	2017-18 Requested	2017-18 Modifications and Justification
Contingency 7100-01	Unforeseen expenses related to training program	1000.00	1000.00	1000.00	Status quo, no changes
Outside Seminars 7100-02	FF attendance at certification programs and seminars	17,500.00	15,000.00	20,000.00	Increase fire officer training course work for officers
Fire Fighter I Academy 7100-03	1 new full-time member to attend academy	3000.00	3000.00	6000.00	2 hires on schedule for retirements in 2013
Hosted Courses 7100-04	Expenses for in-house hosted classes; coffee, pop, snacks	1000.00	1000.00	1000.00	Status quo, no changes
Text/Publications/Video 7100-05	Purchase of updated manuals, magazine subscriptions	8500.00	7000.00	9000.00	Library updates are needed for stations 2 and 3
Drill Site Maintenance 7100-06	Upkeep of training center props and buildings	30,500.00	25,000.00	30,500.00	Upkeep and replacement of burn tower linings and prop construction
Supplies/Maintenance 7100-07	Smoke fluid, office supplies, repairs of equipment	8,000.00	5375.00	5000.00	LP fuel, smoke fluid and repair to training equipment
Part-Time Instructors 7100-08	Part-time instructional staff wages	35,000.00	25,000.00	35,000.00	Increase post-academy training program hours
Instructor Seminars 7100-09	Attendance at local, state and national seminars, conf.	15,500.00	7000.00	10,000.00	Conference price increases and travel expenses.
TOTALS	Support of Division of Training Mission and Assignments	120,000	89,375	$117,500	

FIGURE 12-8 Training budget and justification columns.
Courtesy of Forest Reeder.

a synopsis of the justification, and it is likely that a more in-depth justification will be needed.

The Budget Cycle

Every department has a budget cycle, which is typically 12 months in length. The budget document describes where the revenue comes from (input) and where it goes (output) in terms of personnel, operating, and capital expenditures. Annual budgets usually apply to a fiscal year, such as starting on July 1 and ending on June 30 of the following year. Thus, the budget for fiscal year 2018, referred to as FY18, would start on July 1, 2018, and end on June 30, 2019. Some agencies use a true calendar-year budget process beginning on January 1 and ending on December 31. Another example of a fiscal budget year is the federal government, which has a fiscal budget year that runs October 1 through September 30. Knowing which type of budget calendar is important to managing a training program.

In many cases, the process for developing the FY18 budget would start in 2017, a full year before the beginning of the fiscal year (**TABLE 12-2**). The timeline in Table 12-2 shows that there is a long period from the time you make a budget request to the time the money is available to spend. As a consequence, all training needs for the forthcoming fiscal year may not be apparent to you during the budget development process. Effective budget management requires you to anticipate as much as possible when preparing the budget and to be able to develop contingency plans for unforeseen events. Some actions and purchases may have to be cancelled or delayed when others take priority.

Mandated continuing training, such as hazardous materials and cardiopulmonary resuscitation (CPR) recertification classes, is also factored into the operating costs portion of the budget. That figure includes the cost of the training per fire fighter per class. The cost to pay another fire fighter overtime to cover the position during the mandated training could also be included. As new mandates are established, finding the money to cover these expenses can prove very difficult. New mandated training is one of the most difficult challenges for you to budget and justify.

Capital Expenditures and Training

Capital expenditures refer to the purchase of durable items that cost more than a threshold amount and last for more than one budget year. Local jurisdictions

TABLE 12-2 Fiscal Year 2018 Timeline

August–September 2017	Fire station commanders, section leaders, and program managers submit their FY18 requests to the fire chief. Their concentration is on proposed new programs, expensive new or replacement capital equipment, and physical plant repairs. Depending on how the fire department is organized, a station commander might request funds for a new storage shed or a replacement dishwasher for the fire station at this point. Documentation to support replacing a fire truck or purchasing new equipment, such as an infrared camera or a hydraulic rescue tool, would be submitted, and a proposal for a special training program would be prepared. In most fire departments, these requests are prepared by fire officers throughout the organization and submitted to an individual who is responsible for assembling the budget proposals.
September 2017	The fire chief reviews the budget requests from fire stations and program proposals from staff to develop a prioritized budget "wish list" for FY18. This is the time when larger department-wide initiatives or new program proposals are developed. For example, purchasing a new radio system or establishing a decontamination/weapons of mass destruction unit would be part of the fire chief's wish list.
October 2017	The city's budget office distributes FY18 budget preparation packages to all agency or division heads. Included in each package are the application forms for personnel, operating, and capital budget requests. The budget director includes specific instructions on any changes to the budget submission procedure from previous years. The senior local administrative official (mayor, city manager, county executive) also provides specific submission guidelines. For example, if the economic indicators predict that tax revenues will not increase during FY18, the guidelines could restrict increases in the operating budget to 0.5 percent or freeze the number of full-time equivalent positions.
November 2017	Deadline for agency heads to submit their FY18 budget requests to the budget director.
December 2017	The budget director assembles the proposals from each agency or division head within the local government structure. Each submission is checked to ensure that it complies with the general directives provided by the senior local administrative official. Any variances from the guidelines must be the result of a legally binding agreement, settlement, or requirement or have the support of elected officials.
January 2018	Local elected officials receive the preliminary proposed budget from the budget director. For some budget items or programs, two or three alternative proposals may be included that require a decision from the local elected officials. Once the elected officials make those decisions, the initial budget proposal is completed. This is the "Proposed Fiscal Year 2018 Budget."
February or March 2018	The proposed budget is made available to the public for comment. Many cities make the proposed budget available on an internet site, inviting public comment to the elected officials. In smaller communities, the proposed budget may show up in the local newspaper.
March or April 2018	Local elected officials conduct a public hearing or town meeting to receive input from the public on the FY18 proposed budget. Based on the hearings and on additional information from staff and local government employees, the elected officials debate, revise, and amend the budget. During this process, the fire chief may be called on to make a presentation to explain the department's budget requests, particularly if large expenditures have been proposed. The department may be required to provide additional information or submit alternatives as a result of the public budget review process.
May 2018	Local leaders approve the amended budget. This becomes the "Approved" or "Adopted" FY 2018 budget for the municipality.
July 1, 2018	FY19 begins. Some fire departments immediately begin the process of ordering expensive and durable capital equipment, particularly items that have long lead times for delivery. Many requests for proposals are issued in the early months of the fiscal year.

October 2018	Informal first-quarter budget review. This review examines trends in expenditures to identify any problems. For example, if the cost of diesel fuel has increased and 60 percent of the motor fuel budget has been spent by the end of September, there will be no funds remaining to buy fuel in January. This is the time to identify the problem and plan adjustments to the budget. Amounts that have been approved for one purpose may have to be diverted to a higher-priority account.
January 2019	Formal midyear budget review. The approved budget may be revised to cover unplanned expenses or shortages. In some cases, these changes represent a response to a decrease in revenue, such as an unanticipated decrease in sales tax revenue. If the money is not coming in, expenditures for the remainder of the year might have to be reduced. Sometimes, revenues exceed expectations and funds become available for an expenditure that was not approved at the beginning of the year.
April 2019	Informal third- and fourth-quarter review. Activity within the third quarter can be reviewed and projections for the remaining quarter can be made. Final adjustments in the budget are considered after 9 months of experience. Year-end figures can be projected with a high degree of accuracy. Some projects and activities may have to stop if they have exceeded their budget, unless unexpended funds from another account can be reallocated.
Mid-June 2019	The finance office begins closing out the budget year and reconciling the accounts. No additional purchases are allowed from the FY18 budget after June 30th.

© Jones & Bartlett Learning.

differ on the amount and the time period that these items are supposed to last. Items such as hydraulic rescue tools, apparatus, and SCBA are examples of equipment purchased as capital items. Each of these items requires training when delivery takes place, and additional costs may be incurred to prepare both instructors and fire fighters to train on the new equipment. A Fire Officer III may be responsible for assigning values and depreciation rates to fixed assets and for developing policies that define these items.

The training budget should not be approached any differently than any other part of the budget document and process in a fire department. Budgets must be prepared, justified, and managed throughout the life of the budget cycle. You must be able to understand the budget process used by your department and be able to perform an administrative review of your budget area. When budgets need to be cut or reduced, administrations may seek out areas that can be trimmed by reducing activities or resource purchases. In the training area, you might lose funding for new textbooks or curriculum packages, for class fees and seminar registrations, or for the purchase of a much-needed training prop. Capital expenditures seem to be targets for cutting during the budget process because generally, they are large ticket items, so they appear to be an easy target to cut. However, this is being shortsighted when considering the overall improvement and growth of a training division. The need for a training tower or live fire prop is just as great as for a new fire engine because without properly trained fire fighters, that fire truck has limited use.

DEPARTMENT TRAINING OFFICER TIP

When purchasing training resources, first be sure you understand your department's purchasing policy in relation to bid requirements.

Training-Related Expenses

Expense areas that may be included in the training line item include the following:

- Instructors' salaries, benefits, and expenses
- Student expenses for class attendance
- Tuition, overtime if applicable, transportation, lodging, and meals
- Textbooks and other publications
- Office supplies
- Equipment and training aid maintenance
- Course and seminar fees
- Facility maintenance and improvements
- Capital expenditures
- Subscriptions and memberships
- Contracts

- Educational assistance programs
- Tuition reimbursement
- Grant programs

Ongoing budget maintenance activities should take place to monitor the training division's expenditures to date as part of your fiscal responsibilities. Your performance may be measured in part on your ability to operate within the financial constraints of your program. Good budget management starts with an understanding of the department's budget process and ends with you following the purchasing procedures and tracking of expenses related to the program on a regular basis.

Acquiring and Evaluating Training Resources

Purchasing of resources and products used in the instructional area is a management skill. With shrinking budgets, trial and error does not suffice as a means of selecting training resources. Understanding of the purchasing responsibility and your authority in this process must be evident and documented.

Resource management—whether of equipment, materials, facilities, or personnel—requires a full understanding of the training goals of the department. Purchasing new training equipment every year may not be an option for every department. Often, frontline equipment that has been replaced by newer equipment may be designated for training use. This practice helps with the resource allocation process but can require additional resource management skills on behalf of the training division.

Training Resources

Training resources can be defined as any equipment, materials, or other resources that are used to assist in the delivery of a training session. They can be either consumable or reusable. Consumable items can be used only once, such as a student workbook that will be written in. Reusable items can be refilled or recharged, such as smoke fluid used for a smoke-generating machine used in SCBA training. Prop structures may have certain durable components, while other parts are disposable (such as a ventilation prop) (**FIGURE 12-9**).

Books and curriculum packages may have reusable sections in the form of presentation slides and reproducible handout materials. As noted throughout this text, you as the instructor must identify the types of resources needed to conduct a training session each time you develop or prepare to instruct from a lesson plan (**FIGURE 12-10**). The equipment and

FIGURE 12-9 An instructor should carefully explain all steps of an evolution before it is performed, using props and equipment in a controlled environment in which it is easy to see and hear.
Courtesy of Nathan Flowers.

materials that are usually identified in a lesson plan and are considered resources include the following items:

- Materials that the students need to complete learning objectives
- Materials that the instructor needs to present, apply, and evaluate the learning objectives
- Props, aids, equipment, or other resources necessary to complete the session

Preparing for an effective training session requires you to review the availability, usability, and reliability of each resource. All equipment must be checked for proper operation and safety and be ready to use for the duration of the training session. Enough fuel, batteries, blades, paper, flipchart paper, or whiteboard markers need to be available for the entire training session. Have a backup plan in case the resources you identified are not available or do not work properly.

Hand-Me-Downs

When a power saw is taken off duty as frontline equipment and given to the training division to use in training, the instructor is able to teach a forcible entry session without worrying about creating wear and tear on in-service equipment. The downside of this practice

Instructor Guide
Lesson Plan

Lesson Title: Use of Fire Extinguishers

Level of Instruction: Fire Fighter I

Method of Instruction: Demonstration

Learning Objective: The student shall demonstrate the ability to extinguish a Class A fire with a stored-pressure water-type fire extinguisher. (NFPA 1001, 4.3.16)

References: *Fundamentals of Fire Fighter Skills*, 4th Edition, Chapter 7

Time: 50 Minutes

Materials Needed: Portable water extinguishers, Class A combustible burn materials, Skills checklist, suitable area for hands-on demonstration, assigned PPE for skill

Slides: 73–78*

Step#1 Lesson Preparation:
- Fire extinguishers are first line of defense on incipient fires.
- Civilians use for containment until FD arrives.
- Must match extinguisher class with fire class.
- FD personnel can use in certain situations, may limit water damage.
- Review of fire behavior and fuel classifications.
- Discuss types of extinguishers on apparatus.
- Demonstrate methods for operation.

FIGURE 12-10 A lesson plan example showing resources needed for training evolution.
© Jones & Bartlett Learning.

is the reason why the power saw was taken off frontline service in the first place. Was it difficult to start? Did it have broken parts? Did it leak fuel or oil? Was it obsolete? If the answer to any of these questions is "yes," then you must prepare for the impact of that issue.

A good practice may be to have in-service equipment ready to use at the training site so that if the hand-me-down does not work, the flow of the training session is not interrupted. For example, that old rotary cut-off saw that has been given to the training division should be backed up by the new saw that was purchased to replace it. It is also important to make sure that any operational differences between the two models are highlighted. Notify the proper company officers or shift commanders before removing in-service equipment from apparatus for use in training. It may be hard to justify a delayed response to an incident because the company had to repack hose before they were able to respond on a fire run.

Reserve apparatus may assist in delivery of your training session, but you must also be sure to identify any differences between what students would normally use in the performance of their duties and the equipment being used for the training. Make sure that significant differences in safety equipment, SOPs, and other dynamic factors are spelled out and understood during the training session (**FIGURE 12-11**). It would be great to allocate duplicate resources to the training

FIGURE 12-11 Reserve apparatus and in-service company.
Courtesy of Gonzales Fire Department.

division that are the same as the frontline equipment, but few departments are able to afford such luxuries.

Purchasing Training Resources

All resources used in the delivery of training should be identified during the construction of a lesson plan. Books, projection equipment, training aids, computer resources, and even straw or hay used to generate smoke in a burn tower—that is, all of the resources identified in the lesson plan—need to be considered within the budget process.

The next step would be to conduct a needs analysis of what materials are on hand and what might need to be purchased. Regardless, if the material to be used during training is disposable (such as pallets) or reusable (such as a manikin), knowing what is on hand helps in the budget process. A well-kept equipment record will help in identifying what inventory is available and what will need to be purchased. Inventory control is an important part of budget management and often is overlooked during the scheduling process. Instructors need to track what materials have been used during training sessions, such as bales of hay or rubber gloves. Many agencies build their training budget based on what training needs to occur during the next fiscal year and understanding what resources are needed to successfully conduct the training.

When given the budgetary approval to purchase training resources, you must follow all departmental guidelines. Solicitation of bids for resources costing more than a stated amount of money is often a requirement that you have to observe, for example. Many departments are required to solicit bids from competitive vendors in an effort to get the best price for a product. Textbooks, smoke fluid for smoke generators, and training foams are all examples of materials that could have different price points from vendors. Other materials may be identified as single-source products that are manufactured or distributed by only one vendor. In that case, many agencies do not require multiple bids for a resource since there are no comparable products in the marketplace; therefore, you may not have to go through the bid process. However, you may need to justify the use of a single-source vendor, proving that no suitable alternatives are available.

Evaluating Resources

Instructors who are responsible for evaluating resources for purchase must observe and follow all purchase policy requirements. Before the submission of a request to purchase a resource, you must evaluate the quality, applicability, and compatibility of the resource for the department. Many high-quality resources have been developed in recent years, whereas some time-proven resources still dominate in certain market areas. When considering and evaluating resources for purchase, consider the following questions:

1. How will this product/resource be used in your department?
   - This is a form of audience analysis, as discussed Chapter 5, *Using Lesson Plans*.
   - Consider both the audience aspect and the instructor(s) who will use the resource. Will it require training for the instructors who will use it? Will it require new learning skills of the students?
2. Which standard(s) is the product/resource developed in conjunction or compliance with?
   - This question may be very important when considering adoption or reference of a textbook or training package. Certain state training agencies use specific references for certification training.
3. Is this a "turnkey" product—meaning it is ready to use as is—or does the instructor have to adapt the product to local conditions?
4. Is this product/resource compatible with your agency procedures and methods?
   - Consider the content-specific issues of the product/resource. Are these the methods you use to operate, or are there significant differences in the equipment or methods of the product/resource?
5. Does the product/resource fit within budgetary restrictions?
6. Are there advantages to purchasing larger quantities of the product/resource?
7. Can you get references from vendors on the product/resource that you are considering, or do you have to do your own investigation within your local area or statewide instruction networks?
8. Can the product/resource be available for your use within the prescribed timeline within which you are working?
9. Are there any local alternatives, such as your own design of a product/resource or identification of other local agencies that might be willing to share the use of the product/resource?
10. Is manufacturer demonstration, product expert, or other developer assistance available to the instructor about the product/resource? For example:
    - On-site assistance
    - Phone support
    - Website information
    - Product literature
    - Suggested user guidelines or instructions

Experienced instructors may develop a resource evaluation form to help justify this process and validate the selection process (**TABLE 12-3**). A simple checklist of these questions may assist in your decision making. Effective and efficient training program management requires the development and use of some standardized practice to evaluate products and resources. In today's budget-conscious departments, doing your homework before submitting budget requests can pay off by getting your resources/products purchased quickly (**FIGURE 12-12**).

TABLE 12-3 Resource Evaluation Form

Criteria	Vendor/Product 1 Rating (1 = poor to 5 = excellent)	Vendor/Product 2 Rating (1 = poor to 5 = excellent)	Vendor/Product 3 Rating (1 = poor to 5 = excellent)
Ease of use within our department ■ Instructor ■ Student			
Amount of modifications needed for our department			
Standard compliance ■ NFPA ■ OSHA ■ American National Standards Institute (ANSI) ■ Other			
Cost of product/resource			
Other similar products already in use by dept.			
Quantity discount available			
References			
Availability			
Vendor support			
Local alternatives			
Other			

© Jones & Bartlett Learning.

PURCHASE ORDER

FIGURE 12-12 Purchase request form.

1.	PRICE		CONTACT PERSON	
	VENDOR NAME		PHONE	
	ADDRESS		FAX	
	CITY, STATE, ZIP		EMAIL	
2.	PRICE		CONTACT PERSON	
	VENDOR NAME		PHONE	
	ADDRESS		FAX	
	CITY, STATE, ZIP		EMAIL	
3.	PRICE		CONTACT PERSON	
	VENDOR NAME		PHONE	
	ADDRESS		FAX	
	CITY, STATE, ZIP		EMAIL	

VENDOR RECOMMENDED		TOTAL AMOUNT BUDGETED	
OTHER PROJECT COSTS		ONGOING OPERATING COSTS	

THIS PURCHASE AND OTHER PROJECT COSTS ARE EXPECTED TO REMAIN WITHIN THE BUDGET: (CHECK BOX)	☐ YES	☐ NO

PLEASE MARK ALL BOXES BELOW THAT APPLY TO THIS PURCHASE:

☐	APPROVAL OF LOWEST RESPONSIBLE BIDDER	☐	PROFESSIONAL SERVICES CONSULTING
☐	EMERGENCY PURCHASE	☐	SOLE SOURCE SUPPLIER
☐	EQUIPMENT STANDARDIZATION	☐	TECHNICAL NATURE OF ITEMS MAKES COMPETITION IMPOSSIBLE
☐	JOINT GOVERNMENT PURCHASING PROGRAM	☐	OTHER (PLEASE EXPLAIN BELOW)
EXPLANATION:			

FOR PURCHASES OVER $10,000

A FORMAL REQUEST FOR A RESOLUTION HAS BEEN SUBMITTED TO LEGAL: (CHECK BOX)	☐ YES	☐ NO
THIS ITEM HAS BEEN PLACED ON CONSENT AGENDA: (CHECK BOX)	☐ YES	☐ NO

DATE	REQUESTED BY	DEPARTMENT HEAD	FINANCE DIRECTOR	CITY MANAGER

FIGURE 12-12 Continued

Training BULLETIN

JONES & BARTLETT FIRE DISTRICT
TRAINING DIVISION
5 Wall Street, Burlington, MA, 01803
Phone 978-443-5000   Fax 978-443-8000

Instant Applications: Program Management

Drill Assignment

Apply the chapter content to your department's operation, training division, and your personal experiences to complete the following questions and activities.

Objective

Upon completion of the instant applications, fire and emergency services instructor students will exhibit decision making and application of job performance requirements of the instructor using the text, class discussion, and their own personal experiences.

Suggested Drill Applications

1. Develop a training schedule by reviewing the JPRs for NFPA 1001: *Standard for Fire Fighter Professional Qualifications*, select a JPR and then determine which subjects within the JPR are best suited for scheduling in the classroom and which are best suited for the drill grounds, taking into consideration predictable weather conditions and other factors that would influence the schedule.

2. Demonstrate the completion of a department-specific training record report form. If your department does not have such a form, search the internet for examples and determine which would be best for the training model/schedule used by your department.

3. Review the Incident Report for this chapter and be prepared to discuss your analysis of the incident from a training perspective as well as how such incidents can be addressed through training, including the frequency of such training sessions and any other ancillary topics that need to be included for training that involves topics that are typically high-risk low-frequency events.

Incident Report: NIOSH Report # F2011-12

Drill Assignment

Review the information in this incident report and prepare a practice presentation for class delivery at the direction of your instructor. Your presentation should include a summary of the incident facts and identify the NFPA standard(s) that could apply to the incident. Use an outline to organize your thoughts. You may be evaluated on your communication skills during your presentation.

Cambridge, Minnesota

On May 23, 2011, a 35-year-old male volunteer fire fighter (victim) died after falling from a rope he was climbing after the conclusion of a ropes skills class. The department was conducting a ropes and mechanical advantage haul systems training session that consisted of classroom and practical skills intended to provide the fire fighters rope skills. The drill had concluded and the students were in the process of breaking down the drill site and putting the equipment away. The victim and two fire fighters were standing in front of the tower ladder when the victim decided to climb one of two suspended ropes in an attempt to access the other suspended rope. The victim was climbing up a rope that had been used to demonstrate rope haul systems and attempted to grab another rope out of his reach. The victim likely lost his grip on the rope and fell to the asphalt pavement striking his head. Emergency medical aid was administered by fellow fire fighters and he was transported to a local hospital, where he died from his injuries.

When planning for practical skill training, fire departments should ensure that a sufficient number of instructors are available to deliver the required training and maintain the safety discipline of the students. The proper student-to-instructor ratio when teaching practical skills is recommended to be 5:1. The instructor must pay attention to detail when planning and conducting drills where students may face a risk of injury. When conducting practical drills, many variables come into play and many things can go wrong, resulting in an injury to a student. One way the instructor can help minimize that possibility is to pay special attention to every detail of the drill. Mentally walking through the evolution during the planning session and drawing on personal experiences or case reports of actual incidents can help develop your awareness of what might potentially go wrong.

Postincident Analysis Cambridge, Minnesota

Contributing Factors

- Lack of a safety officer
- Lack of proper personal protective equipment
- Student-to-instructor ratio

Key Recommendations

- Recommendation #1: Fire departments should ensure that a qualified safety officer (meeting the qualifications defined in NFPA 1521) is appointed in practical skills training environments.
- Recommendation #2: Fire departments should ensure that minimum levels of personal protective equipment are established for practical skills training environments (as defined in NFPA 1500).
- Recommendation #3: Fire departments should ensure that sufficient instructors or assistant instructors are available for the number of students expected to participate in practical skills training evolutions (NFPA 1403).

After-Action REVIEW

IN SUMMARY

- A training schedule must balance the various regulatory requirements with the training sessions needed to help fire fighters deliver safe and effective service. Regulatory requirements that affect the department's training schedule include everything from EMS recertification requirements to hazardous materials training.
- After establishing a clear vision of what a department's training needs are, you must determine how that training will be administered. Establishing a training schedule can be challenging because of the many unknowns that come with the fire and emergency services profession.
- The two types of training that occur most often in the fire service are formal training programs and in-service training.
- The in-service drill is the type of training that is most vulnerable to cancellation, delay, or reduction in participation due to emergency response. Consequently, you should have a supplemental training schedule available in case some type of mitigating factor requires a change in the original training session.
- A standardized process must be used to schedule training sessions. The schedule must have some flexibility so that you can take advantage of unforeseen opportunities or overcome obstacles in the delivery of training.
- A useful method to organize the multiple demands for schedule coverage is a system capturing all the required training areas within a single forum. A simple table form called a master training schedule can serve as the starting point in the development of your in-service training program schedule.
- In many agencies, in-service training is conducted by the company officer following the agency's training schedule. On other occasions, training is delivered by members of a training division.
- Understanding the budget development process and your role in specifying and purchasing resources and materials, as well as justifying the funding requirements of a training project or program, are crucial to successful training program management.
- Budget preparation is both a technical and a political process. The technical part relates to calculating the funds that are required to achieve different objectives, whereas the political part is related to elected officials making the decisions on which programs should be funded among numerous alternatives.
- Most municipal governments use a base budget in the financial planning process. The base budget is the level of funding that would be required to maintain all services at the currently authorized levels including adjustments for inflation and salary increases.
- Budgets can be increased or decreased from the base level due to a variety of reasons. When an approved budget needs to be changed from the base budget, a modification is introduced to address any supplement budget requests.
- Successful budgeting requires justifications of the amount of money being requested.
- Every department has a budget cycle, which is typically 12 months in length. The budget document describes where the revenue comes from (input) and where it goes (output) in terms of personnel, operating, and capital expenditures.
- Capital expenditures refer to the purchase of durable items that cost more than a threshold amount and last for more than 1 budget year. Local jurisdictions differ on the amount and the time period that these items are supposed to last.
- Your performance may be measured in part on your ability to operate within the financial constraints of your program. Good budget management starts with an understanding of the department's budget process and ends with you following the purchasing procedures and tracking of expenses related to the program on a regular basis.
- Purchasing of resources and products used in the instructional area is a management skill. With shrinking budgets, trial and error does not suffice as a means of selecting training resources. Understanding of the purchasing responsibility and your authority in this process must be evident and documented.

- Instructors who are responsible for evaluating resources for purchase must observe and follow all purchase policy requirements. Before the submission of a request to purchase a resource, you must evaluate the quality, applicability, and compatibility of the resource for the department.

KEY TERMS

Access Navigate for flashcards to test your key term knowledge.

Agency training needs assessment A needs assessment performed at the direction of the department administration, which helps to identify any regulatory compliance matters that must be included in the training schedule.

Base budget The level of funding that would be required to maintain all services at the currently authorized levels including adjustments for inflation and salary increases.

Budget An itemized summary of estimated or intended revenues and expenditures.

Capital budget A budget that is used to purchase an item that would last for several years, such as a fire engine.

Confidentiality The requirement that, with very limited exceptions, employers must keep medical and other personal information about employees and applicants private.

Expenditures Money spent for goods or services that are considered appropriations.

Hold harmless An agreement or contract wherein one party holds the other party free from responsibility for liability or damage that could arise from the transaction between the two parties.

Indemnification agreement An agreement or contract wherein one party assumes liability from another party in the event of a claim or loss.

In-service drill A training session scheduled as part of a regular shift schedule.

Master training schedule Form used to identify and arrange training topics by the number of times they must be trained on or by the type of regulatory authority that requires the training to be completed.

Operating budget A budget that is used to pay for the day-to-day expenditures such as fuel, salaries or fire prevention flyers.

Revenues The income of a government from all sources, which is appropriated for the payment of public expenses and is stated as estimates.

Supplemental training schedule Form used to identify and arrange training topics available in case of a change in the original training schedule.

REFERENCES

National Fire Protection Association. 2019. *NFPA 1041: Standard for Fire and Emergency Services Instructor Professional Qualifications.* Quincy, MA: National Fire Protection Association.

National Fire Protection Association. 2014. *NFPA 1021: Standard for Fire Officer Professional Qualifications.* Quincy, MA: National Fire Protection Association.

National Fire Protection Association. 2015. *NFPA 1521: Standard for Fire Department Safety Officer.* Quincy, MA: National Fire Protection Association.

National Fire Protection Association. 2018. *NFPA 1403: Standard on Live Fire Training Evolutions.* Quincy, MA: National Fire Protection Association.

National Fire Protection Association. 2018. *NFPA 1500: Standard on Fire Department Occupational Safety, Health, and Wellness Program.* Quincy, MA: National Fire Protection Association.

National Fire Protection Association. 2019. *NFPA 1001: Standard for Fire Fighter Professional Qualifications.* Quincy, MA: National Fire Protection Association.

National Fire Protection Association. 2019. *NFPA 1401: Recommended Practice for Fire Service Training Reports and Records.* Quincy, MA: National Fire Protection Association.

CHAPTER PRESENTATION EXERCISE

Instructions: Your course instructor will assign individual or group discussions on the key points and teaching tips of this chapter. You or your group will review the chapter teachings and identify the major learning points from the chapter. You should be able to discuss the points, why they are important, and how/where they apply to your responsibilities at your level of instructor training.

REVIEW QUESTIONS

1. What would be examples of external demands placed on an agency's training schedule?
2. List the information to include in a training policy or SOP.
3. Explain why it is critical to maintain complete training records and when the recordkeeping should begin in the training process.

SECTION 3

Fire and Emergency Services Instructor III

CHAPTER **13**   **Program Development**

CHAPTER **14**   **Program Evaluation**

CHAPTER **15**   **Program Administration**

CHAPTER 13

Program Development

NFPA 1041 JOB PERFORMANCE REQUIREMENTS

Note: An asterisk denotes that the 1041 standard contains further information in its annex section.

6.2.8 Present evaluation findings, conclusions, and recommendations to AHJ administrator, given data summaries and target audience, so that recommendations are unbiased, supported, and reflect AHJ goals, policies, and procedures.

(A) Requisite Knowledge.
Statistical analysis and AHJ goals.

(B) Requisite Skills.
Presentation skills and report preparation following AHJ guidelines.

6.3.1 Definition of Duty.

Plans, develops, and implements comprehensive programs and curricula.

6.3.2 Conduct an authority having jurisdiction (AHJ) needs analysis, given AHJ goals, so that instructional needs are identified and solutions are recommended.

(A) Requisite Knowledge.
Needs analysis, gap analysis, instructional design process, instructional methodology, learner characteristics, instructional technologies, curriculum development, facilities, and development of evaluation instruments.

(B) Requisite Skills.
Conducting research and needs and gap analysis, forecasting, and organizing information.

6.3.3 Design programs or curricula, given needs analysis and AHJ goals, so that the goals are supported, learner characteristics are identified, audience-based instructional methodologies are utilized, and the program meets time and budget constraints.

(A) Requisite Knowledge.
Instructional design, instructional methodologies, learner characteristics, principles of student-centered learning and research methods.

(B) Requisite Skills.
Technical writing and selecting course reference materials.

6.3.4 Write program and course outcomes, given needs analysis information, so that the outcomes are clear, concise, measurable, and correlate to AHJ goals.

(A) Requisite Knowledge.
Components and characteristics of outcomes, and correlation of outcomes to AHJ goals.

(B) Requisite Skills.
Technical writing.

6.3.5 Write course objectives, given course outcomes, so that objectives are clear, concise, measurable, and reflect specific tasks.

(A) Requisite Knowledge.
Components of objectives and correlation between outcomes and objectives.

(B) Requisite Skills.
Technical writing.

6.3.6 Construct a course content outline, given course objectives, and reference sources, so that the content outline supports course objectives.

(A) Requisite Knowledge.
Correlation between course objectives, instructor lesson plans, and instructional methodology.

(B) Requisite Skills.
Technical writing.

6.5.4 Develop a program evaluation plan, given AHJ policies and procedures, so that instructors, course components, program goals, and facilities are evaluated, student input is obtained, and needed improvements are identified.

(A) Requisite Knowledge.
Evaluation methods and AHJ goals.

(B) Requisite Skills.
Construction of evaluation instruments, technical writing.

6.5.3* Develop a course evaluation plan, given course objectives and AHJ policies, so that objectives are measured and AHJ policies are followed.

(A) Requisite Knowledge.
Evaluation techniques, AHJ constraints, and resources.

(B) Requisite Skills.
Decision making and technical writing.

KNOWLEDGE OBJECTIVES

After studying this chapter, participating in a structured learning environment, and completing assigned assignments, you will be able to:

- Define training needs analysis. (**NFPA 1041: 6.3.2**, pp 297–298)
- List reasons that a needs analysis may be conducted. (**NFPA 1041: 6.3.2**, pp 298–299)
- Identify common sources of data for a needs analysis. (**NFPA 1041: 6.3.2**, pp 298–299)
- Explain ways to analyze data collected during a needs analysis. (**NFPA 1041: 6.3.2**, pp 298–301)
- List considerations for reporting results of a training needs analysis. (**NFPA 1041: 6.3.2, 6.2.8**, p 299)
- Explain how to describe and summarize data from an analysis. (**NFPA 1041: 6.2.8**, p 299, 301)
- Explain considerations for reviewing and interpreting data. (**NFPA 1041: 6.3.2, 6.2.8**, p 299, 316)
- Explain how units of study go together to create a curriculum. (**NFPA 1041: 6.3.3**, pp 301–302)
- Differentiate between course and program outcomes and explain how they work together. (**NFPA 1041: 6.3.4**, p 302)
- Identify a sequence for a curriculum or program to meet the agency training goals. (**NFPA 1041: 6.3.3**, pp 303–307)
- Identify components of program outcomes. (**NFPA 1041: 6.3.4**, p 302)
- Explain how to create program and course outcomes to meet an identified need and goal. (**NFPA 1041: 6.3.4**, p 302)
- Differentiate between goals, outcomes, and objectives. (**NFPA 1041: 6.3.5**, pp 301–302)
- Identify parts of course objectives. (**NFPA 1041: 6.3.5**, p 303)
- Explain methods to make objectives measurable. (**NFPA 1041: 6.3.5**, pp 302–304)
- Identify course content to meet specified objectives (**NFPA 1041: 6.3.6**, pp 301–306)
- Explain considerations for sequencing of topics in a course outline. (**NFPA 1041: 6.3.6**, p 306)
- Identify methods of evaluating courses. (**NFPA 6.5.3**, pp 310–313)

SKILLS OBJECTIVES

After studying this chapter, participating in a structured learning environment, and completing assigned assessments, you will be able to:

- Create a plan for training needs analysis that includes a gap analysis, job and task analysis, and needs assessment. (**NFPA 1041: 6.3.2**, pp 298–301)
- Gather and analyze data to draw conclusions about needs. (**NFPA 1041: 6.3.2**, pp 298–301)
- Report results of needs assessment. (**NFPA 1041: 6.3.2, 6.2.8**, pp 297–301)
- Create a plan to evaluate a course against specified objectives. (**NFPA 1041: 6.5.3, 6.5.4**, pp 310–313)
- Create a program or curriculum outline to meet goals. (**NFPA 1041: 6.3.3**, pp 302–307)
- Create measurable course and program outcomes correlated to specified goals. (**NFPA 1041: 6.3.4**, pp 301–307)
- Convert a job performance requirement (JPR) into a valid learning objective. (**NFPA 1041: 6.3.5**, pp 305–306)
- Write properly constructed learning objectives to support the course goal and intended outcome. (**NFPA 1041: 6.3.3, 6.3.5**, pp 303–306)

You Are the Fire and Emergency Services Instructor

You recently received a promotion to assistant director of the fire academy. Your new boss, the director, has informed you that your first assignment is to prepare a report for the department chiefs and the city council on the curriculum of the academy and how successful the academy is. The director has also asked you to decide what courses will be offered next year and prepare a justification of your choices. As you pour coffee into a mug, which you are now certain is undersized given the enormity of your new tasks, you ponder the following questions:

1. How will you determine whether the academy is successful?
2. How will you decide what courses to offer?
3. How can you justify what you decide about the success or failure of the program and the courses you want to offer?

Access Navigate for more practice activities.

Introduction

Planning and developing an instructional program begins with determining which training is necessary and who needs the training. Each fire and emergency services instructor must develop a system to plan instruction, just like an elementary or secondary school teacher does. The duration of the training is determined by the goal of the program, the number of objectives, and the number of participants. You may find yourself planning instruction as short as a few minutes (e.g., a quick drill or tabletop discussion), a semester-long course run at a local college, and every combination in between. Ultimately, instruction needs to be centered on specific outcomes that benefit the user—that is, the student.

Few curricula are developed to benefit the instructor or the agency hosting the event. Training and learning are user centered, a principle that must always be observed. NFPA 1041: *Standard for Fire and Emergency Services Instructor Professional Qualifications*, charges the Instructor III with planning, developing, and implementing comprehensive programs and curricula. To plan instruction properly, first a training need must be identified, and then a clear, student-centered goal must be established. The course development and program management model identifies the training needs assessment as the center of the development and management process.

Training Program Development

A training program is a systematic approach to organizational and personnel development. The activities, classes, and courses that make up a program are all designed to achieve the same ultimate organizational goal, that of ensuring the members in your department are performing at the highest level possible and are prepared to continue to do so in the future. In the fire service, this means department members have the skills and knowledge to do their current tasks in a manner that exceeds community and departmental expectations while preparing for the changes that the future inevitably brings.

To achieve this goal, the training program must be carefully designed to meet the existing and the upcoming needs of the department, the community, and individual staff members. The process of planning, developing, and implementing a comprehensive training program starts with assessing the training needs. **FIGURE 13-1** is a checklist for a step-by-step approach that covers the main steps or activities of the development process. In some cases, steps may be omitted if information or material exists and is ready to use to meet course intent.

Training Needs Analysis

The first step in creating a successful training program is to assess the needs. A **training needs analysis (TNA)**

- ❏ Conduct a training needs analysis.
    - ❏ Identify a training need.
    - ❏ Define the audience.
- ❏ Present findings for approval or authorization.
- ❏ Design the program.
    - ❏ Modify an existing curriculum.
        - ❏ Review available curricula and references.
    - ❏ Develop course goals.
    - ❏ Write course objectives.
    - ❏ Construct a content outline.
    - ❏ Develop delivery materials.
        - ❏ Lesson plans
        - ❏ Media
        - ❏ Student resources
        - ❏ Class activities
        - ❏ Practice environments
        - ❏ Evaluation instruments
        - ❏ Delivery schedule
        - ❏ Documentation process and recordkeeping
    - ❏ Select instructional staff.
        - ❏ Determine qualifications.
        - ❏ Schedule instructors and support.
    - ❏ Deliver the program.
        - ❏ Monitor a pilot course.
        - ❏ Modify the program as indicated.
    - ❏ Evaluate the program.
        - ❏ Student evaluation
        - ❏ Program evaluation

FIGURE 13-1 A course development checklist.
© Jones & Bartlett Learning.

is a process where research and evaluation are done to determine what should be included in a training program. This first step is often overlooked or ignored when people think they already know what they need to do. In reality, creating a training program without conducting a needs analysis is like wandering around in a strange city hoping to find things you plan to see. It can be fun if you have lots of time and lots of money, but unless you have unlimited resources, it's not likely to produce satisfactory results before you must go home.

A basic needs analysis will answer questions including the following:

- What training is needed?
- Why is the training needed?
- Who needs the training?
- What outcomes is the training expected to achieve?

A thorough needs analysis will also accomplish the following:

- Identify gaps between what members need to know and what they do know.
- Identify training that is outdated or no longer relevant.
- Identify gaps in training pathways designed to provide career advancement for members.
- Provide assessment of existing programs.
- Evaluate how the training be provided.
- Estimate how much the training will cost.

Note that the terms *needs analysis*, *needs assessment*, and *front-end analysis* are often used interchangeably but they are technically different. A needs analysis and a front-end analysis include the process of analyzing, meaning examining the component parts and their functions together and apart. A needs assessment is a part of an analysis, specifically, it is the narrower process of identifying and quantifying needs.

Parts of an Analysis

Several different types of assessment and analysis can be included in a needs analysis. The most common types for the fire and emergency services are gap analysis, job and task analysis, and needs assessment. Each of these is a different way of examining data to answer, as follows:

- A gap analysis examines the knowledge, skills, and abilities (KSAs) personnel have now versus KSAs they need now and in the future to identify and define the gap between the two. This is the most common type of needs analysis done.
- A job and task analysis provides a breakdown of the basic elements of a job and tasks, in other words, the KSAs. The fire service is fortunate to have National Fire Protection Association (NFPA) standards that contain this information at its fingertips.
- A needs assessment examines the needs of the organization and individual members as they relate to the organizational goal. In some senses, it is finding out what the organization and the people within it want and need and sorting out the difference between wants and needs. A needs assessment is useful for addressing the need for training to prepare an organization for change. Needs assessments can be simple or complex. An example of a large-scale, complex needs assessment is the NFPA's *Fire Service Needs Assessment Survey* that is published periodically. For a very small program with a very specific issue, the assessment might be done in a week or a day.

Other types of assessments include performance problem assessments that consider a specific issue and whether it is a training issue or something else, and learner assessments that focus on characterizing a learner.

Data Gathering

Once a decision is made about what tools will be used, sources of data for the analysis must be identified.

Data for a fire department training needs analysis can come from a variety of sources. Common sources of data include the following:

- NFPA standards, Occupational Safety and Health Administration (OSHA) standards, and other standards
- Legal mandates
- Testing records
- Run reports
- Municipal statistical data
- Accident reports
- Municipal long-range plans
- Surveys
- Interviews
- Job observation
- Focus groups
- Literature reviews
- Community demographics and reports
- NFPA reports
- Incident critiques
- Department mission and long-range plans
- Current training program

Select sources that make sense based on the information sought.

TRAINING PROGRAM MANAGER TIP

When assessing your agency's training needs based on external requirements, find out exactly what is required to ensure your agency is in compliance with those requirements.

Analysis and Report

Once the data is collected, it then must be analyzed to make decisions. Combine the data to develop a picture of where the training program is right now. Make sure that an understanding of all parts of the existing system is developed. Critical aspects include who is trained, what they are trained in, when training is delivered, where training is delivered, and why training is delivered, as well as other details that create a detailed overview of the current training program.

Next create the same detailed overview of where the department wants to be with training now and in the future. Again, pay attention to the who, what, when, where, and why. Include the identified organizational goals.

Once both things are understood, the next step is to analyze the gap between where the program is now and the program that is desired. Identify any methods of closing the gap that have been uncovered during the analysis. For example, suppose a gap was discovered between existing training to respond to confined space calls and the desired level of training given that a new factory with several confined spaces is being built in the jurisdiction. During the analysis, information on existing training programs may have been researched. If it was noted that the department down the street has a standard confined space training program that they run twice a year and they are willing to allow your members to attend, this should be captured in the analysis report.

Create a report that explains the findings and the research methods. A good report will include the following elements:

- Statement of purpose or statement of the issues
- Detailed description of the analysis that was undertaken
- Review of the data
- Discussion of the ways the data were interpreted
- Explanation of the gaps that were found
- Discussion of options that were identified
- Conclusions
- A mention of anything that needs further research

Some reports may also include a cost–benefit analysis and recommended solutions to any issues evaluated. Reports should be accurate, timely, and present necessary information in an understandable format. Remember the purpose of the report is to explain the needs analysis, how it was conducted, the findings, and the recommendations. If lengthy, detailed, or technical information is included, it is advisable to provide an executive summary for the reader. If the report is intended for an internal audience, it is appropriate to use fire service terminology and acronyms. If, however, the report is to be submitted to external organizations, a city council, or the media, these terms must be explained in detail to avoid confusion, misinterpretation, and frustration. The document should be through, yet succinct. Avoid interjecting personal opinion and editorializing. Keep the report as objective as possible. Any written report carries with it the potential of becoming the subject of an oral presentation to a fire chief, a city council, or a community organization. The Fire and Emergency Services Instructor III should prepare for a presentation that summarizes the report, suitable for diverse audiences. Prepare a list of potential questions and practice the language, tone, and delivery of the appropriate answers. If a topic is subject to even the slightest chance of controversy, discuss the questions and responses with the fire chief or a senior director before presenting it. By this point in an Instructor III's career, the ability to prepare the oral presentation should be fairly routine.

Job Performance Requirements in ACTION

Planning, developing, and implementing comprehensive programs and curricula is the major instructional development duty area for the Instructor III. Although other instructors will be responsible for delivery and use of the materials, the instructor who develops any curriculum used by the department must complete an agency needs analysis and conduct research while designing the program. Knowledge of which agency goals are addressed by the course is essential, and modification of existing curricula allows for program and course goals to be met. When writing course objectives, keep in mind that you are creating the "big picture" outcomes rather than individual learning objectives. These high-level skill applications require a large amount of time and research to create usable and effective programs.

Instructor I

The Instructor I functions at the delivery level of course materials. Make sure you are familiar with all course elements and are well prepared to deliver the training. Review prepared application exercises and make sure you are ready for questions and possible performance problems.

Instructor II

As an Instructor II, one of your first responsibilities is to modify materials from existing curricula to accommodate student characteristics and even time schedules. Review class rosters and available time to break down larger courses into manageable class segments. Make sure you are able to ensure consistency and continuity of material flowing from class to class.

Instructor III

The Instructor III develops the entire curriculum package, including course goals, course objectives, lesson plans, media, and evaluation materials. As the course is developed, ensure that you are selecting the right instructors to deliver the material and that they have adequate time to review and prepare. Conduct a pilot course with a small audience of content experts, if possible, to work out any problems that may occur before finalizing the product.

JPRs at Work

There are no JPRs at the Instructor I level; however, all of the curricula produced by the Instructor III may eventually be delivered by the Instructor I.

JPRs at Work

There are no JPRs at the Instructor II level; however, all of the curricula produced by the Instructor III may eventually be modified by the Instructor II.

JPRs at Work

Discuss the methods of training needs assessment. List the steps in the curriculum development model. Discuss the importance of properly stated program and course goals. Describe the content of properly constructed learning objectives. Explain the purpose of modifying existing curricula. Write a clear, concise, and measurable program or course goal to address training needs assessment gaps identified by the developer. Write properly constructed learning objectives to support the course goal and intended outcome.

Bridging the Gap Among Instructor I, Instructor II, and Instructor III

All levels of the instructional delivery team play varying roles in the curriculum process. While the Instructor III develops the material, the Instructor I and II will make the lesson plans to achieve the course goals. The outcome-based system of instructional design revolves around input and constant evaluation of program material to strive toward making all programs more effective and efficient. Communications skills within the instructional delivery team will be vital in ensuring the success of the course goal. Those persons who adapt and modify materials must communicate the reasons why such changes are necessary. The designer should seek input and actively participate in the delivery process.

As a reminder, four simple steps for preparing an oral report based on a written report are as follows:

- Get their attention: Find the hook that makes the subject important to the audience. Use the *anticipatory set* to help them attend to the upcoming information.
- Statement of interest: Briefly explain why the audience should be interested in the topic. What makes this important to them? Remember the importance of the cognitive, affective, and psychomotor domains of learning.
- Organize the details: Using skills as an educator, make sure the information is presented in a systematic, logical progression, especially if there are many details to be included. Even if the information makes sense to you, the listener may be overwhelmed if the report does not follow a reasonable path. Practice presenting the report to someone not familiar with the information first. When finished with the practice session, ask the listener what they took away from the presentation. This will serve as a litmus test for your delivery.
- Call to action: At the close of the presentation, remember the importance and value of a direct and impactful conclusion. For a report, it is best to ask the audience to take some specific action(s) that relate directly to the statement of interest.

It may or may not be appropriate to ask the audience for questions regarding the report. This decision will depend of the forum, the type of audience, or the judgment of the fire chief.

The completed training needs analysis provides the necessary information to enable the creation of new parts of programs and modification of existing ones to meet the needs of the department.

Designing Training Programs

Training programs have multiple layers and tracks by their nature. The process of developing programs, courses, and classes is called instructional design. Instructional design includes all the steps of developing, delivering, and evaluating a training program. It begins with the needs analysis and progresses through evaluation and an ultimate return to a needs analysis to continue the cycle. For this text, we will review the basic steps of the general instructional design process; however, it is important to know that many variations of models for instructional design exist. The most common include the ADDIE model (analyze, design, develop, implement, and evaluate), and the National Fire Academy's (NFA) five-step model of determining training needs, design, development, administration, and evaluation. You may find it useful to formally review these methods as well and adapt them to create your own strategy for developing program and course content.

Program Goals and Program Structure

The group of courses and classes that make up a **program** when considered all together are sometimes called the **curriculum** (*curricula* is the plural form of curriculum). The curriculum is the entirety of a program of study. Fire department training programs have a curriculum that covers a wide variety of subjects including areas such as firefighting, EMS, and rescue. Curriculum goals, that is, the overall program goals, should be identified and set out in writing. These are the highest-level goals and focus on meeting the organizational and personnel needs throughout training. A program goal can be viewed as the outcome of an entire series of courses, classes, drills, or training sessions conducted over a period of time. Having well-defined program goals sets the stage for a well-designed program.

It may be understood that any course delivered is part of a department training program and the program itself is self-evident. In contrast, certification entities, college or university course developers, and fire academies may have to identify which courses fall under particular program areas and the curriculum may contain multiple programs. In these cases, specific program goals will be necessary for each program area. A fire academy or college delivery system may break its program delivery into areas of responsibility such as the following:

- Firefighting programs
- Hazardous materials programs
- Officer training programs
- Emergency medical care
- Special operations team programs
- Public education and code enforcement programs

In this example, a program goal is defined for each program in the curricula and individual courses are developed to meet the program goals.

In addition to program goals, the architecture or framework of the program should be identified. It is useful to create a map of sorts of the entire training program and the relationship between the courses and classes. Identify which courses or KSAs are prerequisites to other courses, which courses have multiple parts, and which classes are stand-alone classes. This is useful to the training officer doing program development, as well as the department members looking to understand how to progress toward a career goal. Think of it like blueprints to build a program. Use the results of the needs analysis to determine where the program requires new development or adjustment to meet the organizational goals.

Program Outcomes

Program outcomes are the way program goals are translated into actions. A program outcome is written in objective form and provides a clear, concise, and measurable way to assess progress toward a goal. Like individual class objectives, overall program objectives must be measurable. While very similar and often used interchangeably, outcomes are different from objectives. Objectives are the smaller steps to achieving a learning outcome. **Learning outcomes** are the measurable effects of a program on the learners. For example, a basic fire academy program goal may be to train new fire fighters to become successful members of the department. An outcome for that program might be "Eighty percent (80%) of the students achieve certification as a Fire Fighter I." Programs can have one or more goals and each goal can have one or more outcomes that support it. When the program is evaluated a large part of the evaluation will be an assessment of whether the programs are meeting the stated outcomes. For further discussion of learning objectives associated with a lesson plan, see Chapter 9, *Instructional Development*.

Program Evaluation

Evaluating a training program should be part of the ongoing and constant improvement goal of any training division. Program evaluation may consider the following aspects of a program:

- Performance
- Reliability or consistency
- Facilities and equipment
- Reputation of the program
- Conformance with standards
- Response to needs of the students and the community at large

The department training officer should establish a system for reviewing key course and program components on a regular basis. Among the items that should be reviewed as part of a program evaluation plan are the following:

- Administrative details
  - Costs
  - Paperwork processing
- Instructor performance
- Student performance
- Course components and features
  - Media
  - Textbook
  - Tests, quizzes, and evaluations
  - Time allotments
  - Homework and study assignments
- Facilities and equipment used in course
- Written and verbal comments by students for course improvements

Program evaluation tools are designed to assess the performance of the courses as a group and individually. They identify strengths and weaknesses and assess student attitudes. Program evaluations are also designed to look internally at the organization's objectives (e.g., How did the program measure up? Were the goals met?) and externally at students' opinions and community-wide impressions (e.g., Did the program meet your expectations? What did you like or dislike about the program?) in order to determine the overall degree of excellence within the program. Program evaluations are a continual process; a key aspect of a program assessment is to provide ongoing evaluation throughout the entirety of the course, not just at the end of the course. As part of the program design, an evaluation plan should be created that details how the program will be evaluated, provides time frames for evaluation, and gives outcomes that the evaluation process is expected to generate.

TRAINING PROGRAM MANAGER TIP

Developing an evaluation plan should not be confused with developing evaluation items such as test questions or practical skill sheets. The former process is a management function that usually takes place after a course or program is completed; as part of this process, the instructor—who may be the developer of the program—evaluates the effectiveness and efficiency of the tools used for the course.

Designing Courses

The basic design of a **course** will include taking the course goal and outcomes and creating a framework including course objectives, course outlines, and student evaluation methods. A program is made up of one or more courses with stated goal and course outcomes. Each course must then have course objectives that are the task-specific steps to achieving the outcomes. Content in the course must cover the material needed for the students to achieve the outcomes and the students should be evaluated in ways designed to measure their achievement of the outcomes.

Writing Course Objectives

A Fire and Emergency Services Instructor III must be able to create objectives for courses that are one or more classes long. In addition, an Instructor III must be able to evaluate objectives proposed by instructors under their supervision. Course objectives should always be written in terms of the student rather than the instructor's performance or actions. Objectives that describe instructor behavior or list a classroom activity are not useful for guiding student learning. For example:

- Students will watch a video about tying knots.
- Students will be given the opportunity to practice tying knots.

These examples only tell us what the instructor was supposed to do: show a video and provide practice time.

Likewise, objectives that are unmeasurable or overly vague are not useful. For example:

- Students will understand knot tying.
- Fire fighters will appreciate the importance of safety when using ropes.
- The fire fighter will know how to use rope.

These are too vague and unmeasurable. How would one objectively measure appreciation or understanding? Similarly, "know how to use rope" is so vague that it would seem everyone coming into class already knows how to use rope, albeit perhaps not for the purpose the class design envisions.

An objective that is focused on the learner tells what the learner must be able to do. For example:

- Given a piece of rope, students will tie a bowline knot within 60 seconds.
- Students will be able to list the advantages and disadvantages of synthetic fiber ropes.
- The fire fighter will identify an overhand safety knot, half hitch, clove hitch, figure eight knot, bowline knot, water knot, and sheet bend with 90 percent accuracy.

A good way to remember this is that we want to write SMART objectives:

- **S**pecific
- **M**easurable
- **A**ssignable/Achievable
- **R**elevant
- **T**ime-bound

Writing objectives for a single class session was introduced in Chapter 9, *Instructional Development*, as part of creating a lesson plan.

Many different methods for writing learning objectives have been proposed. One method commonly used in the fire service is the ABCD method (Heinich, Molenda, and Russell, 1998). ABCD stands for **a**udience (Who?), **b**ehavior (What?), **c**ondition (How?), and **d**egree (How much?). Learning objectives do not always need to be written in that order, however.

The audience aspect of the learning objective describes who the students are. If the lesson plan is being developed specifically for a certain audience, the learning objectives should be written to indicate that fact. In contrast, if the audience is not specifically known or includes individuals with varied backgrounds, the audience part of the learning objective could be written generically, such as by referencing "the fire fighter" or even "the student."

Once the audience has been identified, the behavior is listed. The behavior must be an observable and measurable action. A common error in learning objective writing is using words such as *know* or *understand* for the behavior. Is there really a method for determining whether someone understands something? It is better to use the words *state*, *describe*, or *identify*—these are actions that you can see and measure. It is much easier to evaluate the ability of a student who is identifying the parts of a portable fire extinguisher than to evaluate how well the student understands the parts of a portable fire extinguisher.

The condition describes the situation in which the student will perform the behavior. Items that are often listed for the conditions include specific equipment or resources given to the student, personal protective clothing or safety items that must be used, and the physical location or circumstances for performing the behavior. For example, consider these fragments of objectives:

- "... in full protective equipment, including self-contained breathing apparatus (SCBA)."
- "... using the water from a static source, such as a pond or pool."

The last part of the learning objective indicates the degree to which the student is expected to perform the behavior in the listed conditions. With what percentage of completion is the student expected to perform the behavior? Total mastery of a skill would require 100 percent completion—this means perfection, without missing any steps. Knowledge-based learning objectives are often expected to be learned to the degree stated in the passing rate for written exams, such as 70 percent or 80 percent. Another degree that is frequently used is a time limit. This can be included on both knowledge and skill learning objectives.

ABCD learning objectives do not need to contain all of the parts in the ABCD order. Consider this example:

> In full protective equipment, including SCBA, using a bowline knot, the fire fighter trainees will hoist a ventilation fan to a second-floor window in less than 3 minutes.

In this example, the audience is "the fire fighter trainees." The behavior is "hoist a ventilation fan." Hoist is an observable and measurable action. The conditions are "full protective equipment, including SCBA," "using a bowline," and "to a second-floor window." These describe the circumstances for the hoist. The degree is "less than 3 minutes." The fire fighter trainees must demonstrate the ability to perform the behavior to the proper degree to meet the learning objective successfully.

Strictly speaking, well-written learning objectives should contain all four elements of the ABCD method. Shortening learning objectives should be done only when the assumptions for the missing components are clearly stated elsewhere in the lesson plan. It is not very likely that a learning objective would omit the behavior component because this component is the backbone of the learning objective.

Level of Achievement

Another important consideration is the level to which a student will achieve the learning objective. This is most often determined by applying Bloom's Taxonomy (Bloom, 1965), a method used to identify levels of learning within the cognitive domain. For the fire service, the three lowest levels in Bloom's Taxonomy are commonly used when developing learning objectives. In order from simplest to most difficult, these levels are knowledge, comprehension, and application. Knowledge is simply remembering facts, definitions, numbers, and other data. Comprehension is displayed when students clarify or summarize the important points. Application is the ability to solve problems or apply the information learned in situations.

To use this method, an Instructor III who is developing objectives must determine which level within the cognitive domain is the appropriate level for the student to achieve the course objectives. For example, if an Instructor III is developing objectives for a portable extinguisher class, the following objectives could be written for each level:

Remembering (Knowledge): "The fire fighter trainee will identify the four steps of the PASS method of portable extinguisher application." For this objective, the student simply needs to memorize and repeat back the four steps of pull, aim, squeeze, and sweep. This is a very simple knowledge-based objective and easily evaluated with a multiple choice or fill-in-the-blank question.

Understanding (Comprehension): "The fire fighter trainee will explain the advantages and disadvantages of using a dry-chemical extinguisher for a Class A fire." This objective requires the student first to identify the advantages and disadvantages of a dry-chemical extinguisher, and then to select and summarize those that apply to use on a Class A fire. This is a higher-level objective. Such an objective may be evaluated by a multiple choice question but is better evaluated with a short answer type question.

Applying (Application): "The fire fighter trainee, given a portable fire extinguisher scenario, shall identify the correct type of extinguisher and demonstrate the method for using it to extinguish the fire." This is considered the highest level of objective, because it requires the student to recall several pieces of information and apply them correctly based on the situation. Such an objective is often evaluated with scenario-based questions that may be answered with multiple choice items or short answers.

There is no one correct method used to determine which level or how many learning objectives should be written for a lesson plan. Typically, a lesson plan will contain knowledge-based learning objectives to ensure that students learn all of the facts and definitions within the class. Comprehension objectives are then used to ensure that students can summarize or clarify the knowledge-oriented material. Finally, application objectives are used to ensure students' ability to use the information that was learned in the lesson. As the course goals and course objectives are created, the learning objectives used in the lesson plan will satisfy the course goal.

Converting JPRs into Learning Objectives

Often, a Fire and Emergency Services Instructor III needs to develop learning objectives to meet JPRs listed in an NFPA professional qualifications standard. Recall that this occurs when a course is being developed to meet the professional qualifications for a position such as Fire Officer, Fire Instructor, or Fire Fighter. The JPRs listed in the NFPA standards of professional qualifications are not learning objectives, but learning objectives can be created based on the JPRs. Each NFPA professional qualification standard has an annex section that explains the process of converting a JPR into an instructional objective (**FIGURE 13-2**). By following this format, the

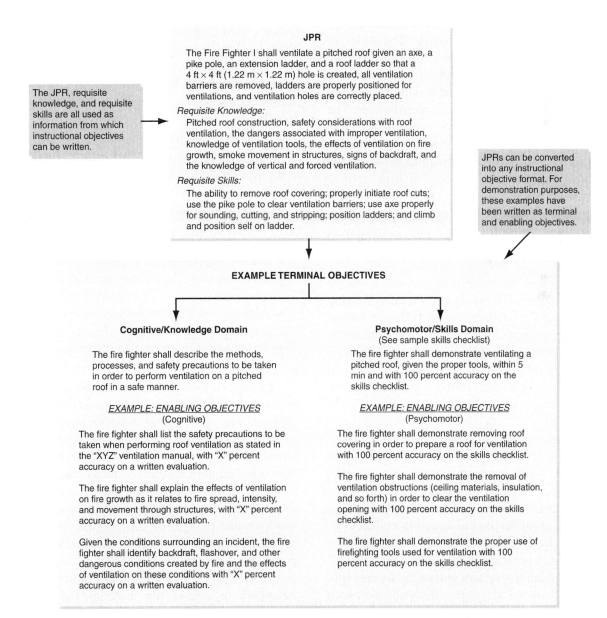

FIGURE 13-2 Converting job performance requirements (JPRs) into instructional objectives.
Courtesy of NFPA.

Instructor III can develop learning objectives for a lesson plan to meet the professional qualifications for NFPA standards.

Course Content Outlines

Once the course goals, outcomes, and objectives are determined, the course content can be outlined. The outline is often called a **syllabus**. The outline should include, at a minimum, the following items:

- Goals
- Outcomes
- Objectives
- Structure
- Reference sources

The course outline should make it clear how the content is related to the objectives. The order of presentation in the outline should be given specific attention to make sure to sequence the instruction so that students can understand it without needing additional knowledge they do not already have. The exact method of sequencing for specific material will depend on the topic and the learner. Some topics are easy to sequence by matching them to the sequence of the job; for example, putting on turnout gear then putting on a SCBA. Other things may need to be taught in a specific order because the first skill is necessary to complete the second skill. For instance, students must learn to tie knots before learning to hoist tools.

Textbooks and reference materials are a critical part of the outline. The information used to develop the course should be clearly identified in a format that allows current and future users of the outline to locate and review the publications or other sources. This ensures that courses are created with valid and reliable information and instructors teaching the course can review the information relied on by the course designer.

Additionally, it is useful to include the method of student assessment in the outline. As part of course development, many instructors generate correlation sheets that show visually each goal connected to the related outcomes connected to the related objectives and attached to the method of assessing student mastery of the objectives.

Syllabi can be generic if little is known about the ultimate user or delivery system or they can be very specific and customized to a particular application such as a course for a specific fire company or station.

A detailed course outline should specify the length of the course and the material it will cover, identify the resources necessary, and determine the methods of instruction. The developer should consider the following factors in completing this process:

- Time available and required to deliver the material
- Experience and knowledge of the audience
- Order of delivery (which material has to be taught before other material is introduced)
- Relationships between objectives and units
- Research of resources used in the course
- Availability of instructors and instructional resources
- Environmental considerations
- Deadlines on completion of the course
- Methods of instruction

Developers will have to modify the decision-making process used to determine course content every time they develop or modify a course. Many factors play into the decision making, the most prevalent being time constraints. A breakdown of the tasks that are undertaken to create the content outline might look like this:

- Determine available and required time to deliver the course.
- Identify the audience and audience delivery factors.
- Assemble the course into units (days or hours) to deliver.
- List the objectives for each unit.
- Determine the logical sequence of related tasks.
- Determine and locate the resources needed for each unit.
- Identify qualified instructors needed to deliver each unit.
- Develop the lesson plan, instructional media, and activities.

If the course developed is new, a **pilot course** should be used, during which students should be made aware that they are participating in a pilot or first-run program. This approach allows the developer to adapt to course flow determinations. During the pilot course, the course content outline will be adjusted according to the rate and flow of the information to the students. Functional groupings of information can be implemented to organize the information being presented and can also be used in the creation of lesson plans, typically at the level of the larger heading areas. As a course developer, you should estimate the time necessary to deliver the material and determine which activities, such as student applications, are necessary to make the learning effective for the learner. Once this breakout is created, you have developed a delivery schedule for the course; this matching of time estimates and activities will also help keep you on task to complete material on time. Many pilot courses do not fully represent the final course product, since much feedback is exchanged between students and instructors in the pilot program to improve the final product.

Conducting Research and Identifying Resources

Conducting research into current issues and trends and obtaining up-to-date statistics are part of the developer's responsibility to ensure that all objectives reflect current practices and are relevant to the topic. The internet brings the firefighting world to each developer's desktop and allows for personnel in even the busiest of departments to learn from one another. Adult learners prefer to be able to relate learning to where and how they will use it. Case studies and current events allow for that to happen if they are used in the right way. Instructors must make sure they blend these types of discussions into the lecture or demonstration, but must also be cautious that they do not result in "Monday morning quarterbacking" of incidents and operations. It is easy for a class to become sidetracked on these types of discussions unless the instructor keeps the group on task.

When creating a plan for a new program and conducting research, be conscious of the legal applications and copyright information that must be adhered to by the developer when citing and referencing materials researched or used in the development of class material. Photos, lesson plans, PowerPoint presentations, articles, books, and other studies and reports, while beneficial in enhancing the quality of the class, need to be used and referenced in the correct manner.

The peer support network established by many developers is another great resource. Instructors are often willing to share their work and will help you when you are stuck for resources or ideas. The NFA has a free, subscription-based resource network titled TRADE (Training Resource and Data Exchange). Developed by Chief Robert Bennett (retired), this network allows developers and others to post "help needed" messages and announcements into a large e-mail listing that goes to several thousand users each week (**FIGURE 13-3**). When a developer needs resources, it is remarkable to see the outpouring of support that the TRADE list supplies after just one e-mail listing.

State and local training bulletins and manuals may also represent time-saving resources for the developer. Many states have baseline curricula available for use, with most of those materials referencing NFPA standards and state-specific requirements for training. It is a best practice to reference and incorporate these curricula into your instructional planning. The Illinois Office of the State Fire Marshal uses an ad hoc committee process to develop referenced baseline curricula for fire departments to reference and use in the development and delivery of training programs. Objectives, lesson plans, basic slide presentations, and sample evaluation tools are available to departments that have the instructors to develop the program.

Evaluating Instructional Resources

Instructors who are responsible for evaluating resources for use within a program or curriculum should develop or follow an objective-based decision-making process when comparing resources. As part of such an assessment, you must evaluate the quality, applicability, and compatibility of the resource for the training program. Many high-quality resources have been developed in recent years, whereas some time-proven resources still hold sway in certain market areas. When evaluating resources for use within your curriculum, consider the following questions:

1. How will this product/resource be used in your department? Consider this question to be a form of audience analysis. Consider both the audience and the instructors who will use it. Will this material require training for the instructors who will use it? Will it require new learning skills of the students?
2. Which standard(s) is the product/resource developed in conjunction or compliance with? This question may be very important when considering adoption or reference of a textbook or training package. Certain state training agencies use specific references for certification training.
3. Is this a "turn-key" product (i.e., ready to use as is), or must the instructor adapt the product to local conditions? Almost all material purchased as part of a curriculum must be adapted to local conditions and student needs. Always check to see if the product allows you to edit all components.
4. Is this product/resource compatible with your agency procedures and methods? Consider the content-specific issues of the product/resource. Does it include the methods you use to operate, or are there significant differences in the equipment or methods used or advocated by the product/resource?
5. Does the product/resource fit within your budgetary restrictions? Every agency has some sort of

Special Notice: December 7, 2018

Free Webinar: American Wood Council Presentation: Pre-planning and Suppression of Buildings Under Construction (BCD232)

CONTACT: AWC Education
Date: Tuesday, December 11, 2018
Time:
1:55PM-3:30PM Eastern
12:55PM-2:30PM Central
11:55AM-1:30PM Mountain
10:55AM-12:30PM Pacific

This program is designed to provide background and information to fire departments that may experience the construction of large area buildings in their community. Many fire departments have limited experience in planning and response to these complex buildings. This requires a thorough understanding of fire and building code provisions as well as the proper use of NFPA 241, Standard for Safeguarding Construction, Alteration, and Demolition Operations and NFPA 1620, Standard for Pre-Incident Planning.

Learning Objectives:
1. Identify risks & hazards on constructions sites. Learn the leading causes of fires in structures under construction.
2. Apply model codes and standards that pertain to safety precautions during construction and pre-incident planning.
3. Identify procedures and methods of pre-incident planning from the moment a building is contemplated.
4. Develop strategies and tactics to suppress a fire on a construction site of a large area building.

Instructor
RAYMOND O'BROCKI, CBO, is the Manager of Fire Service Relations for the American Wood Council. Before that he was the Chief Building Official for the City of Rockville, MD. He retired as the Assistant Fire Chief the Baltimore City Fire Department in 2013. He was appointed fire marshal for Baltimore City in 2008. During his tenure as fire marshal, Baltimore City recorded the three lowest annual fire fatality totals in its history. O'Brocki has served on the Maryland State Child Care Advisory Council, Maryland State Fire Code Update Committee, State Fire Marshals Legislative Working Group in Annapolis and the steering committee for the Mid-Atlantic Life Safety Conference. He has served on the NFPA Urban Fire Safety Task Force and has presented at the National Fire Academy. He is a graduate of the University of Baltimore School of law and a licensed attorney.

Equivalencies: 1.5 Hour of Instruction = 0.15 Continuing Education Units (CEU) = 1.5 Professional Development Hours (PDH) = 1.5 Learning Units (LU)

The PDF of the presentation will posted 24 hours before the webinar, and can be accessed here.
View System Requirements

FIGURE 13-3 An example of the type of resource found on the NFA's resource network, TRADEnet.
Courtesy of U.S. Fire Administration/FEMA.

budgetary oversight, and the instructor should be prepared to complete a budget justification for the material.

6. Are there advantages to purchasing larger quantities of the product/resource? Many vendors will discount certain training materials when they are purchased in larger quantities. Allowances for shipping or enhanced vendor support are also opportunities for comparison of materials.

7. Has your research regarding the product/resource that you are considering included both references from vendors and your own investigation within your local area or statewide instruction networks? You should seek out users who are not necessarily on the list provided by vendors. This may take some extra work, but instructor associations and chiefs groups may have users who have varied opinions on the material being evaluated.

8. Will the product/resource be available for your use within the prescribed timeline within which you are working? Some publications and resources may become backordered due to shortages and demand. Make sure that the vendor selected has a sufficient quantity of the material ready to ship to meet your timetable.
9. Are there any local alternatives, such as your own design of a product/resource or identification of other local agencies that may be willing to share the use of the product/resource? Some fire academies have lending libraries that may be available for your use if your budget demands require this approach.
10. Are manufacturer demonstrations, product experts, and other types of developer assistance related to the product/resource available to the instructor? Consider the availability of onsite assistance, phone support, website information, product literature, and suggested user guidelines or instructions.

Also consider the level to which the resource was written. You may want to compare the level of writing, the level of objectives, and any correlations that may have been done on the curriculum (**TABLE 13-1**). Networking with other developers may provide insight into the benefits and shortcomings of any curriculum being considered for adoption.

Modifying Courses

The Instructor III is also able to review and modify a course, making basic or fundamental changes to it. Such changes may include changing the outcomes,

TABLE 13-1 Resource Evaluation Form

Criteria	Vendor/Product 1 Rating (1 = poor; 5 = excellent)	Vendor/Product 2 Rating (1 = poor; 5 = excellent)	Vendor/Product 3 Rating (1 = poor; 5 = excellent)

Ease of use within our department
    Instructor
    Student			

Amount of modification needed for our department			

Standard compliance
    NFPA
    OSHA
    Other			

Cost of product/resource			
Similar products already in use by department			
Quantity discount available			
References			
Availability			
Vendor support			
Local alternatives			
Other			

rewriting learning objectives, or even modifying the content of the lessons. Reasons a developer may modify an existing program include the following:

- Update of information
- Changes in students' ability or knowledge in the current program
- Outdated references to equipment, technology, or techniques
- Application of an individual department or organization's policy and procedures
- Safety improvements
- Changes in applicable standards
- Audience analysis factors
- Time constraints
- Available resources

Many publishers release prepackaged curriculum, including instructor support materials in a format that allows the user to modify and adapt those materials to his or her specific needs. This approach represents a tremendous time and money saver for the user, and significantly reduces the amount of time necessary to create and prepare material. Modification is always necessary and recommended to ensure compliance with local practice and procedures.

Course Evaluation Plans

A course evaluation plan will evaluate many facets of course delivery. The course or class evaluation forms filled out by the students who were the direct users of the materials are often the first component of the course or program evaluation plan. Instructors, agency policies, and procedures relating to the course; course components; and facilities are all often covered by such student-provided evaluations. Test results can be part of the overall evaluation as well and should be included in the evaluation plan. The overall evaluation plan should encompass the following items:

- Course evaluations completed by the students in the class. Course features should be evaluated for ease of use, amount of work, time to complete work, and application of the content related to the job.
- Instructor evaluations identifying the strengths and weaknesses of each instructor in the course.
- Test results that have been properly analyzed based on the following criteria.
  - Reliability
  - Discrimination
  - Percentage scores against the pass/fail rate established
- Agency policies and procedures.
  - Attendance requirements
  - Testing and grading policies
  - Student and instructor codes of conduct
- Facilities and equipment.
  - Up-to-date
  - Comfortable and accessible to all students
  - Equipment used was safe, serviceable, and contemporary
- Safety analysis.
  - Injuries and accident data
  - Adherence to safety policies

Do not confuse course evaluation with student evaluation. Students are assessed, tested, or evaluated to determine whether they met the course objectives. Did the students learn to tie a bowline knot? Course evaluation is an examination of how well the course met the outcomes and goals of the organization or agency.

Gathering Feedback

In addition to statistical evaluation, student and instructor course evaluations are a helpful tool for assessing whether a program or course has achieved its established goals. Responses to objective questions posed to students and faculty can identify how the organization is meeting its goals (**FIGURE 13-4**).

FIGURE 13-4 An important component of course evaluations is student feedback.
© Jones and Bartlett Learning Publishers. Photographed by Kimberly Potvin.

Course evaluations should contain the following key elements:

- Student opinions
- Instructor viewpoints
- Course coordinator/administration views
- Institute/facilities

The survey could be a comprehensive form that asks questions about each of these four categories, or it can be specifically geared to just one category. This second approach is more time consuming, but it provides better information.

Surveys

The survey is an aspect of the evaluation strategy. With the development of the evaluation strategy comes the development of specific surveys—for example, geared toward students or toward instructors—that will be used to determine the effectiveness of the course and program.

Rating Scales

Several rating scales may be used in surveys. The most commonly used scales are the numerical rating scale and the graphic rating scale. The numerical scale is the simplest type. The following is an example:

5 = Outstanding

4 = Above average

3 = Average

2 = Fair

1 = Poor

For example:

The primary instructor's preparation of course material was:

1 2 3 4 5

The quality of the audiovisual materials was:

1 2 3 4 5

A program administrator can quickly tally this kind of data and identify the common trends and opinions expressed in the survey (**FIGURE 13-5**). For example, if two surveys were administered, one designed for students and the other designed for instructors, it would be appropriate to assess the same objectives for both surveys. The responses to the same objective from the two different perspectives can then be compared. The numerical rating scale would make this type of comparison very easy to conduct because specific values are listed on the rating scale.

The graphic rating scale is not as neatly defined as the numerical version. It uses a line graph to plot the respondents' answers to a specific question. With this rating scale, it is possible to obtain values that are between a given value (e.g., 3.3, 4.5, 2.3). Instead of locking a respondent into a specific number, a specific value assessment for a question can be attained. Although this rating scale is much more difficult to use for tallying purposes, the information it provides better reflects the respondents' actual opinions for the question.

A variation on the two previously mentioned rating scales is the multiple choice rating scale. It is written using the same format as a multiple choice examination; that is, specific answers are pre-identified. The respondent selects the answer that best matches his or her opinions. On a typical form, four answers are pre-identified and a fifth answer is left blank. When respondents believe that the pre-identified answers do not accurately reflect their opinion, they can write in a unique response. The following is an example of a multiple choice survey item:

The audiovisual materials that were used during the program were:

- Poor: Most were old and outdated. They did not enhance the course material.
- Fair: A few of the audiovisual materials were outdated, but overall they complemented the lesson material.
- Good: The materials were current and enhanced the lesson materials.
- Excellent: The materials greatly enhanced the course materials.
(Write in the response on the attached form.)

The responses to a multiple choice survey are very easy to tally. If necessary, the questions can be reworded to allow a comparison between two different categories—for example, students and instructors. The preselected responses allow a respondent who is not familiar with educational concepts to respond to the question. The downside to using preselected responses is that the survey author points a respondent toward a conclusion that may not totally reflect the individual's opinion. The respondent often selects an answer simply because he or she does not want to write in a response. For this reason, care must be used when analyzing the results of a multiple choice survey.

Survey Questions

The key to a successful survey is the way the questions are written on the survey form and interpreted by the person completing the survey. Program evaluation questions are based on accepted educational concepts or on an accepted curriculum. In other words, the goals for a program become the evaluation mechanism for the program.

Fire Officer Training Program
QUALITY ASSURANCE REPORT SUMMARY
COURSE NAME: Fire Service Instructor I

Course Evaluation

Course Feature	Number of Surveys	Excellent (5)	Very Good (4)	Good (3)	Fair (2)	DNA (0)
1. AV material / course materials / textbook	20	18	2			
2. Course organization / flow / delivery rate	20	16	2	2		
3. Observance to safety procedures and practices	20	14	3			3
4. Hands-on experience *If applicable to course*	20	19	1			
5. Time per subject *Add comments on specific areas on reverse*	20	12	4	4		
6. Evaluation tools	20	16	2	2		
7. Will this course help you with your professional development?	20	19	1			
8. Did the course meet your expecations?	20	18	2			
9. Would you recommend this course to others?	20	18	2			
10. Overall impression of course	20	17	2	1		

Instructor(s) Evaluation

A = Excellent 5.0; clearly presented objectives, demonstrated thorough knowledge of subject, utilized time well, represented self as professional, encouraged questions and opinions, reinforced safety practices and standards.

B = Good (4.0); presented objectives, knowledgeable in subject matter, presented material efficiently, added to lesson plan information

C = Average (3.0); met objectives, lacked enthusiasm for content or subject, lacked experience or background information, did not add to delivery of material

D = Below Average (2.0); did not meet student expectations, poor presentation skills, lack of knowledge of content or skills, did not interact well with students, did not represent self as professional

E = Not Applicable; did not instruct in section or course area, do not recall, no opinion

Instructor Name	Number of Surveys	Excellent (5)	Good (4)	Average (3)	Below Average (2)	Not Applicable	Evaluation Score
Chief Smith	20	20					5.0
Capt. Brown	20	15	5				4.5
Bill Engine	20	5	12	3			3.8

LIST STUDENT COMMENTS IN THIS SPACE

- Good class, lots of practice speeches helped my confidence
- Too much PowerPoint® on day one
- Chief Smith's insight was very valuable and his evaluations of the speeches were very helpful.
- Thanks to all instructors who helped out
- Should get reading assignments before class
- Doing workbook assignments and preparing practice speeches as homework was a lot when I was on shift.
- Looking forward to Instructor II, opened my eyes to instructor's job

FIGURE 13-5 Survey results can be tallied to identify trends in responses.

Courtesy of Illinois Fire Chiefs Educational and Research Foundation.

Each question covers a specific goal. The question is to be written in simple English. Just like written examination questions, a program evaluation question should be a complete sentence whenever possible. Remembering these concepts can increase the validity of the survey.

Survey Format

The following basic guidelines are used by most surveys. First, the overall survey should appear simple, not overwhelming. Many surveys are too complex in their appearance. The respondent should be able to look at the survey and readily identify the information being requested. In addition, the questions must be written to the point. There is no need for fancy terms or lengthy sentences. The questions for a survey should be based on accepted educational objectives or accepted standards.

Second, only one rating scale should be used. This consistency will lessen the respondent's confusion as to which rating scale is being used for a particular section. The survey designer should write the rating scale at the top section of each page of the survey. If the rating scale has the scale built into each question—for example, in graphic type—then the respondents have all the information they need right in front of them.

Third, because subjectivity can influence a survey, the respondent might want remain anonymous. Thus, a survey should consider the value of not requesting the name of the respondent.

A sound survey reflects the clarity of the survey's directions. Students or faculty members often complete surveys with no monitor present to clarify a particular question. The directions to a survey must contain enough details so that anyone who completes the survey can do so without difficulty. Poor directions result in poor information because the respondent, if unsure or confused, may provide inaccurate information. This outcome invalidates some of the information gathered from the survey.

The use of these guidelines not only helps in writing a survey, but can also lead to improved validity for the survey. A survey is constructed to assess the opinions of a group of individuals. From this survey, key programming decisions will be made. If a survey has inaccurately assessed the individuals, then the decisions based on its results may be made in error. Care must be used when developing, using, and evaluating with a survey. The information attained from a survey should be helpful, not harmful.

A survey can only collect opinions from the respondents. Often, this information is not enough to base program changes on.

Monitored Performance Evaluations

A performance evaluation is a tool that can provide more in-depth information than is possible to collect via a survey alone. The categories used in such evaluations are the same as those for surveys. The difference between a performance evaluation and a survey is that a trained evaluator physically monitors each group during a performance evaluation. The monitor uses prepared objectives-based checklists. As with the survey approach, these objectives reflect the broad goals of the educational institute. As with any educational evaluation, the information that is obtained through this method needs to be closely examined. The performance evaluation is a tool that, when used in conjunction with a survey, can provide answers to objectives not addressed by the survey.

The use of a performance evaluation versus the use of a survey differs somewhat, as can be seen in the following example. Suppose students provide their impressions of the instructor's performance on a survey. However, they may not be able to fairly or objectively assess the abilities of an instructor, nor do they know what should and should not be taught to a class. To obtain this information, an educationally oriented evaluator needs to monitor various class sessions. The instructor can be evaluated by this impartial evaluator, and the information collected in this manner can then be compared with the students' impressions.

Performance Assessment Checklists

Checklists may be written either based on the main goals for a program or to reflect accepted standards of performance. Both of these elements are important when developing instructor, administrative, and facility checklists. The instructor, administrative, and facility checklists are written in nearly the same format; the significant differences among them reflect the objectives being assessed.

The performance evaluation for the instructor is nothing more than a modified survey (**FIGURE 13-6**). With this type of tool, a trained evaluator measures the performance of an instructor against specific educational objectives. Most checklists consist of the following components:

- Basic demographic data
- Date(s) of the evaluation
- Evaluator(s) name
- Rating scale
- Specific objectives
- General comments
- Strengths and weaknesses

National Fire Academy Course Evaluation Form

Part 2 of 4 - Course Instructors / Overall Training

Considering the instructors for this course, to what extent do you agree that...
1=Strongly disagree, 2=Disagree, 3=Unsure, 4=Agree, 5=Strongly agree, NA=Not Applicable
Instructor: John Smith

	1	2	3	4	5	NA
knew the material well.	○	○	○	○	○	●
regularly clarified course and assignment expectations.	○	○	○	○	○	●
encouraged independent thinking from students.	○	○	○	○	○	●
fostered a collaborative "team-based" learning experience.	○	○	○	○	○	●
supplemented course with helpful experience.	○	○	○	○	○	●
answered students' questions clearly.	○	○	○	○	○	●
presented engaging lectures.	○	○	○	○	○	●
led the learning process without dominating it.	○	○	○	○	○	●
exhibited a positive attitude toward students.	○	○	○	○	○	●
conducted the class in a professional manner.	○	○	○	○	○	●
worked well with co-instructor.	○	○	○	○	○	●
is worth recommending to others.	○	○	○	○	○	●

Considering the instructors for this course, to what extent do you agree that...
1=Strongly disagree, 2=Disagree, 3=Unsure, 4=Agree, 5=Strongly agree, NA=Not Applicable
Instructor: Richard Karlsson

	1	2	3	4	5	NA
knew the material well.	○	○	○	○	○	●
regularly clarified course and assignment expectations.	○	○	○	○	○	●
encouraged independent thinking from students.	○	○	○	○	○	●
fostered a collaborative "team-based" learning experience.	○	○	○	○	○	●
supplemented course with helpful experience.	○	○	○	○	○	●
answered students' questions clearly.	○	○	○	○	○	●
presented engaging lectures.	○	○	○	○	○	●
led the learning process without dominating it.	○	○	○	○	○	●
exhibited a positive attitude toward students.	○	○	○	○	○	●
conducted the class in a professional manner.	○	○	○	○	○	●
worked well with co-instructor.	○	○	○	○	○	●
is worth recommending to others.	○	○	○	○	○	●

Is there particular feedback you have regarding the instructors for this course?

FIGURE 13-6 A sample course instructor and overall training experience rating survey that is completed online at the end of a course.
Courtesy of FEMA.

Considering the training overall, to what extent do you agree that this training experience...
1=Strongly disagree, 2=Disagree, 3=Unsure, 4=Agree, 5=Strongly agree, NA=Not Applicable

	1	2	3	4	5	NA
will help me do my current job better.	○	○	○	○	○	●
was consistent with my department's training expectations.	○	○	○	○	○	●
will be useful for a department the size of mine.	○	○	○	○	○	●
will be applicable to my future work.	○	○	○	○	○	●
included material helpful to my department's prevention efforts.	○	○	○	○	○	●
will help reduce the fire-related risks in my community.	○	○	○	○	○	●
provided sufficient opportunities for networking.	○	○	○	○	○	●
provided information my department can use when responding to an all-hazards and/or terrorist event.	○	○	○	○	○	●
is worth recommending to others.	○	○	○	○	○	●

How satisfied are you with NFA's...

classroom and learning facilities.	○ Satisfactory	○ Unsatisfactory	○ N/A
Learning Resource Center (LRC).	○ Satisfactory	○ Unsatisfactory	○ N/A
dormitory rooms.	○ Satisfactory	○ Unsatisfactory	○ N/A
student center facilities.	○ Satisfactory	○ Unsatisfactory	○ N/A
dining facilities.	○ Satisfactory	○ Unsatisfactory	○ N/A
course registration and administrative details.	○ Satisfactory	○ Unsatisfactory	○ N/A
workout/weight room facilities.	○ Satisfactory	○ Unsatisfactory	○ N/A

[Save For Later] [Back] [Continue]

The Save for Later button will allow you to save answers already provided and come back to the form at a later time (provided the evaluation period hasn't expired) to complete it.

FEMA Form 064-0-5

Paperwork Reduction Act Notice | OMB No. 1212-0065

Last Reviewed: December 28, 2006

U.S. Fire Administration, 16825 S. Seton Ave., Emmitsburg, MD 21727
(301) 447-1000 Fax: (301) 447-1346 Admissions Fax: (301) 447-1441

FIGURE 13-6 *Continued*

- Tally column
- Didactic presentation
- Skill presentation

For administration checklists, objectives oriented to assess customer satisfaction are often used. Either the administration is seen as responsive to a customer's needs (e.g., students, faculty, or the community) or support provided to the customer is perceived as lacking. These business concepts are appropriate objectives to utilize when assessing the performance of the administration because any educational program inevitably has bureaucratic aspects. For example, how the paperwork is handled often influences the quality of instruction. How can paperwork affect a program? Perhaps handouts should be made but are forgotten. Or perhaps audiovisual equipment that was supposed to be in the classroom is missing. Maybe the classroom is too small for the size of the class. The list goes on. All of these details are administrative in nature, yet all influence the quality of the program. Because of their impact, administrative staff members should be assessed in the same manner as other program components are.

The administrative assessment being referred to here is an overall assessment—one that is not targeted at a specific individual's performance. Such an assessment should focus on those administrative tasks that directly affect the classroom programs.

Interpreting Evaluation Results

In ideal circumstances, a committee consisting of faculty, administration, and members from the community at large should review the program material. This committee should look at the areas of concern that have been identified in any of the assessments. For example, a negative response to a question about the adequacy of the audiovisual materials used in the classroom is a potential area of concern. Cross-checking the various surveys—for example, student opinion and instructor surveys—may provide additional information about the area of concern. In addition to negative responses, positive achievements should be reviewed. For example, if a program goal was "Students will achieve a score of 80 percent or higher on the state certification examination," and the class average was 82.4 percent, then this outcome is a positive achievement. The people involved in the program should be given praise for meeting this goal.

Once the committee has reviewed both the positive and negative aspects of the program, then a final report needs to be written. In this report, findings related to each assessment area are summarized. A discussion of the positive and negative aspects of the program is then initiated.

The report concludes with a plan of action, or program recommendations. This plan of action specifies which parts of a program need to be changed to correct a problem or to improve the educational experience. As stated in the first section of this chapter, to have a quality program, assessment and reassessment of the program are required. This means rewriting objectives, policies, and procedures; fixing broken equipment and buying new equipment; and redefining the assessment process so that these improvements can be assessed in the next program.

Training BULLETIN

JONES & BARTLETT FIRE DISTRICT
TRAINING DIVISION
5 Wall Street, Burlington, MA, 01803
Phone 978-443-5000   Fax 978-443-8000

Instant Applications: Program Development

Drill Assignment

Apply the chapter content to your department's operation, training division, and your personal experiences to complete the following questions and activities.

Objective

Upon completion of the instant applications, fire and emergency services instructor students will exhibit decision making and application of job performance requirements of the instructor using the text, class discussion, and their own personal experiences.

Suggested Drill Application

1. Step back and take an outsider's look at your department and its culture and attitude toward training.

2. From that outsider's point of view, conduct a SWOT analysis of your department's training culture and identify your agency's strengths, weaknesses, opportunities, and threats (SWOT).

3. From this analysis/assessment, identify one area you considered to be a high priority that, if changed, would change the training culture of your department.

Incident Report

Drill Assignment

Review the information in this Incident Report and prepare a practice presentation for class delivery at the direction of your instructor. Your presentation should include a summary of the incident facts and identify the NFPA standard(s) that could apply to the incident. Use an outline to organize your thoughts. You may be evaluated on your communication skills during your presentation.

On January 22, 2015, a 49-year-old male career fire fighter participated in air management training during his 24-hour shift as part of the fire department respiratory protection program. The fire fighter, wearing full turnout gear and SCBA (on-air) completed one evolution lasting about 10 minutes without incident. After removing his SCBA and turnout gear, he sat on the engine's tailboard and told his battalion chief that he "was just a little winded." About 5 minutes later, the fire fighter stated that his "chest was hurting." Dispatch was notified as crewmembers noted the fire fighter's clammy skin and a weak and irregular pulse. An electrocardiogram revealed changes consistent with an acute myocardial infarction (heart attack), and the fire fighter was transported to the local hospital's emergency department (ED). In the ED, an acute heart attack was confirmed and the fire fighter was taken emergently to the cardiac catheterization lab for coronary angiography and angioplasty. While in the cardiac catheterization lab, the fire fighter suffered cardiac arrest. Despite resuscitation efforts for over 20 minutes, the fire fighter died at 2145 hours.

Postincident Analysis

Key Recommendations

- Recommendation # 1: Provide annual medical evaluations to all fire fighters consistent with *NFPA 1582: Standard on Comprehensive Occupational Medical Program for Fire Departments* to identify fire fighters with risk factors for coronary heart disease (CHD).
- Recommendation # 2: Perform symptom-limiting exercise stress tests (ESTs) on fire fighters at increased risk for CHD and sudden cardiac events.
- Recommendation # 3: Discontinue routine exercise stress tests on asymptomatic fire fighters with no risk factors for CHD.
- Recommendation # 4: Perform candidate and member physical ability evaluations.
- Recommendation # 5: Discontinue routine PSA testing.
- Recommendation # 6: Conduct annual respirator fit testing.
- Recommendation # 7: Install diesel exhaust source capture systems in fire stations.

After-Action REVIEW

IN SUMMARY

- Planning and developing an instructional program begins with determining which training is necessary and who needs the training. Each instructor must develop a system to plan instruction. The duration of the training is determined by the goal of the program, the number of objectives, and the number of participants.

- Training and learning are user centered, a principle that must always be observed. NFPA 1041: *Standard for Fire and Emergency Services Instructor Professional Qualifications* charges the Instructor III with planning, developing, and implementing comprehensive programs and curricula.

- To plan instruction properly, first a training need must be identified, and then a clear, student-centered goal must be established. The course development and program management model identifies the training needs assessment as the center of the development and management process.

- A training program is a systematic approach to organizational and personnel development. The activities, classes, and courses that make up a program are all designed to achieve the same ultimate organizational goal, that of ensuring the members in your department are performing at the highest level possible and are prepared to continue to do so in the future.

- The process of planning, developing, and implementing a comprehensive training program starts with assessing the training needs.

- A training needs analysis is a process where research and evaluation are done to determine what should be included in a training program.

- Once a decision is made about what tools will be used, sources of data for the analysis must be identified. Data for a fire department training needs analysis can come from a variety of sources.

- The process of developing programs, courses, and classes is called instructional design. Instructional design includes all the steps of developing, delivering, and evaluating a training program. It begins with the needs analysis and progresses through evaluation and an ultimate return to a needs analysis to continue the cycle.

- The group of courses and classes that make up a program when considered all together are sometimes called the curriculum. Fire department training programs have a curriculum that covers a wide variety of subjects including areas such as firefighting, EMS, and rescue.

- Program outcomes are the way program goals are translated into actions. A program outcome is written in objective form and provides a clear, concise, and measurable way to assess progress toward a goal.

- Evaluating a training program should be part of the ongoing and constant improvement goal of any training division.

- A program is made up of one or more courses with stated goal and course outcomes. Each course must then have course objectives that are the task-specific steps to achieving the outcomes. Content in the course must cover the material needed for the students to achieve the outcomes and the students should be evaluated in ways designed to measure their achievement of the outcomes.

- A Fire and Emergency Services Instructor III must be able to create objectives for courses that are one or more classes long. In addition, an Instructor III must be able to evaluate objectives proposed by instructors under their supervision.

- Once the course goals, outcomes, and objectives are determined, the course content can be outlined. The outline should include, at a minimum, the following items:
  - Goals
  - Outcomes
  - Objectives
  - Structure
  - Reference sources

- Instructors who are responsible for evaluating resources for use within a program or curriculum should develop or follow an objective-based decision-making process when comparing resources. As part of such an assessment, you must evaluate the quality, applicability, and compatibility of the resource for the training program.
- Do not confuse course evaluation with student evaluation. Students are assessed, tested, or evaluated to determine whether they met the course objectives. Course evaluation is an examination of how well the course met the outcomes and goals of the organization or agency.
- A performance evaluation is a tool that can provide more in-depth information than is possible to collect via a survey alone. The categories used in such evaluations are the same as those for surveys. The difference between a performance evaluation and a survey is that a trained evaluator physically monitors each group during a performance evaluation.
- In ideal circumstances, a committee consisting of faculty, administration, and members from the community at large should review the program material. This committee should look at the areas of concern that have been identified in any of the assessments.

KEY TERMS

Access Navigate for flashcards to test your key term knowledge.

Course A unit of instruction. In college or high school, a course usually lasts one term or semester. In the fire service, a course may be a single-topic course, for example a confined space course that consists of four classes of 4 hours each.

Curriculum The complete body of study offered. A college will have a curriculum for each major it offers. (*Curricula* is the plural form of curriculum.)

Learning outcomes Statements of what a student can do as a result of completion of a program or course of study.

Pilot course The first offering of a new course, which is designed to allow the developers to test models, applications, evaluations, and course content.

Program A program of instruction comprises the fully body of teaching in a particular area. A fire academy may divide their curricula into several program areas such as firefighting, officer development, and technical rescue.

Program outcomes The measurable outcomes of an assessment process intended to determine whether program goals are being met.

Syllabus An outline of a course that identifies course objectives, course material, assignments, and the assessment process.

Training needs analysis (TNA) A process of identifying organizational training and development needs.

REFERENCES

Bloom, Benjamin S. 1965. *Taxonomy of Educational Objectives, Handbook I: The Cognitive Domain.* New York: David McKay Company.

Heinich, Robert, Michael Molenda, and James D. Russell. 1998. *Instructional Media and the New Technologies of Education.* 6th ed. Upper Saddle River, NJ: Prentice Hall College Division.

National Fire Protection Association. 2019. NFPA 1041: *Standard for Fire and Emergency Services Instructor Professional Qualifications.* Quincy, MA: National Fire Protection Association.

National Fire Protection Association. 2018. NFPA 1403: *Standard on Live Fire Training Evolutions.* Quincy, MA: National Fire Protection Association.

National Fire Protection Association. 2018. NFPA 1582: *Standard on Comprehensive Occupational Medical Program for Fire Departments.* Quincy, MA: National Fire Protection Association.

CHAPTER PRESENTATION EXERCISE

Instructions: Your course instructor will assign individual or group discussions on the key points and teaching tips of this chapter. You or your group will review the chapter teachings and identify the major learning points from the chapter. You should be able to discuss the points, why they are important, and how/where they apply to your responsibilities at your level of instructor training.

REVIEW QUESTIONS

1. Several different types of assessment and analysis can be included in a needs analysis. Name the three most common types for the fire and emergency services and what defines each.

2. A basic needs analysis will answer four basic questions. What are they?

3. An objective that is focused on the learner tells what the learner must be able to do. You can remember this by writing the SMART objectives. What does the acronym represent?

CHAPTER 14

Program Evaluation

NFPA 1041 JOB PERFORMANCE REQUIREMENTS

Note: An asterisk denotes that the 1041 standard contains further information in its annex section.

6.5.2 Develop a system for the acquisition, storage, and dissemination of evaluation results, given authority having jurisdiction (AHJ) goals and policies, so that the goals are supported and so that those affected by the information receive feedback consistent with AHJ policies and federal, state, and local laws.

(A) Requisite Knowledge.
Record keeping systems, AHJ goals, data acquisition techniques, applicable laws, and methods of providing feedback.

(B) Requisite Skills.
The evaluation, development, and use of information systems.

6.5.5 Analyze student evaluation instruments, given test data, objectives, and AHJ policies, so that validity and reliability are determined and necessary changes are made.

(A) Requisite Knowledge.
AHJ policies and applicable laws, test validity and reliability, and item analysis methods.

(B) Requisite Skills.
Item analysis.

KNOWLEDGE OBJECTIVES

After studying this chapter, participating in a structured learning environment, and completing assigned assessments, you will be able to:

- Describe where to find information on the laws regarding management of evaluation data. (**NFPA 1041: 6.5.2**, pp 324–325)
- Explain ethical considerations and recordkeeping for managing evaluation results. (**NFPA 1041: 6.5.2**, pp 324–325)
- Describe recordkeeping systems requirements for managing evaluation results. (**NFPA 1041: 6.5.2**, p 325)
- Describe methods of disseminating evaluation results. (**NFPA 1041: 6.5.2**, p 325)
- Explain test validity, reliability, and item analysis. (**NFPA 1041: 6.5.5**, pp 327–334)

SKILLS OBJECTIVES

After studying this chapter, participating in a structured learning environment, and completing assigned assessments, you will be able to:

- Conduct an analysis of student evaluation instruments to determine if the test items are reliable and valid based on results. (**NFPA 1041: 6.5.5**, pp 327–334)

You Are the Fire and Emergency Services Instructor

As the senior member of your fire academy staff, you have started preparation for an upcoming annual budget cycle presentation. As part of this preparation, you have been asked to provide a detailed report outlining the graduation rates of the recruit fire fighter training program, including the rationale used to determine those rates, the actual performance evaluations that are used, and the validity and reliability of the evaluation process and results.

1. Considering the request, discuss the data that you would want to analyze to determine the answer to this question.

2. Explain the validity, reliability, and item analysis for written tests and the impact they have on the effectiveness and trustworthiness of the program.

3. Explain how management of evaluation records is related to the overall success of the program.

 *Access Navigate for more practice activities.*

Introduction

Evaluation systems, commonly known as tests, come in a variety of forms. Much is at stake for the student as well as the instructor when the evaluation system is used to demonstrate students' competence and knowledge or for certifications. Evaluation systems must be developed to comply with many laws and standards, and the Instructor III will be responsible for understanding these requirements. For more on the various forms of tests, see Chapter 8, *Evaluating the Learning Process*, and Chapter 11, *Evaluation and Testing*.

When designing an evaluation system, many facets of evaluation and testing will require careful attention to both student results and instructor performance. A system to collect acquired evaluation results needs to be established in order to allow the course developer to review success and failure rates on particular test questions and to track trends over time, comparing various classes on the same criteria. In many agencies, these data can be provided by the examining authority, such as a state fire marshal's office, training academy, or other certifying entity. In other cases, the local department will be responsible for data analysis of the results. Many records will be produced as part of testing and evaluation process, and a formal process of result disclosure and storage should be available to guide the instructor on how to handle this information properly. As a senior instructor you need to be aware of federal, state, and local laws that address release, storage, and length of retention of test records. Policies should also be clear on how test results are released to students and to no one else. In addition to student testing, a course and program evaluation plan will help the agency make sure that the goals of a program were met and the course components, instructors, and evaluations met the expectations of the agency.

Acquiring Evaluation Results

Evaluation results should flow from the students who have completed the test, through the instructor who administered the exam, and then to the individual or agency responsible for grading the results. Once test results are graded and data compiled on the success or failure rate on a particular question, skill, or objective, the instructor can analyze those results. A simple flowchart may direct those who come in contact with this information on the next step in the process. Some systems will use an electronic data summary.

Storing Evaluation Results

Examination results and student assignment results should be treated as confidential every step of the way. This confidentiality should be part of your organization's policies and made clear to the student. The physical score sheets, practical skill sheets, and other summary reports and performance documentation may need to be retained for defined periods of time depending on local laws or ordinances. Other considerations related to the storage of evaluation results include the following:

- Evaluation records may need to be made available for inspection by examining or auditing authorities.
- Access to evaluation results should be available in case a challenge to student performance occurs.
- Data should be available for recurring tests of the same nature for the purpose of establishing and identifying trends in the tests' results.

Each fire department should have a policy that states how long evaluation results must be stored and the manner in which they must be kept. It may be acceptable to scan results into a data storage system or to compile the results onto spreadsheets and then destroy the original evaluation forms or checklists. It is a good practice to establish some type of backup system in the event one system fails. NFPA 1401: *Recommended Practice for Fire Service Training Reports and Records* provides guidance on the legal aspects on what type of records should be kept, how to protect the records, determining the effectiveness of a records management system, and the legal aspects of record-keeping. (See Appendix D.)

Dissemination of Evaluation Results

To reemphasize a point made earlier, evaluation results should be kept confidential, and policy should dictate which staff members and administrative staff have access to these records. A best practice in this area may be to have any request for release of evaluation records approved by the fire chief and the student(s) involved. Summary tables that do not include individual names or other identifiable data of the students might be acceptable for the purpose of assisting instructors and staff in reviewing the course participants' progress (**FIGURE 14-1**). It is best not to single out any individual student during a mass release of data to instructors; only key administrators of a program should have full access to individual records.

Students should be granted full access to their evaluation results, but not necessarily to the evaluation tools themselves. Some agencies prepare summary reports of tests administered that show each objective tested, the number of test items for the objective, and how many are correct for that objective. Students can then review these reports as part of their personal exam analysis, without having the actual exam questions.

Training division staff should exercise caution when releasing student scores; gone are the days of posting scores on a bulletin board. If testing results are shared during a class session, do so in a way that will protect each student's individual score. Ideally, you should release the scores in a one-on-one environment that would allow for an open discussion about the results as well as addressing any questions the student might have about the results.

Course Result Summary

Student ID	Building Construction	Ladders	Ventilation	Forcible Entry	Water Supply
1871	88	94	94	80	97
6112	94	96	92	95	100
3809	82	96	88	83	94
4018	97	84	92	93	91
1796	95	90	90	88	96

FIGURE 14-1 Evaluation result summary tables are a resource for instructor and course evaluation that maintains student confidentiality.
© Jones & Bartlett Learning.

Job Performance Requirements in ACTION

Developing an evaluation plan should not be confused with developing evaluation items such as test questions or practical skill sheets. The former process is a management function that usually takes place after a course or program is completed; as part of this process, the instructor—who may be the developer of the program—evaluates the effectiveness and efficiency of the tools used for the course.

Instructor I	Instructor II	Instructor III
The Instructor I who administers tests and course evaluations must do so in a professional manner. Test item security and confidentiality of test results must be maintained.	The Instructor II may develop test items and course evaluation forms that will be used to evaluate student knowledge and skill levels.	The Instructor III is responsible for evaluating the learning process using tests, instructor evaluations, and other performance measures.
JPRs at Work	**JPRs at Work**	**JPRs at Work**
There are no JPRs for the Instructor I for this chapter.	There are no JPRs for the Instructor II for this chapter.	Develop an evaluation plan that reviews test data, course components, facilities, and instructors and manages the results of evaluation measures.

 Bridging the Gap Among Instructor I, Instructor II, and Instructor III

Instructors at all three performance levels will develop, administer, and analyze the evaluation systems created for each course. The Instructor I should adhere to the agency policies and procedures for test administration, while the Instructor II develops test items based on course objectives. The review of all evaluation data becomes the responsibility of the Instructor III, and a comprehensive course review mechanism is used to identify program and content strengths and weaknesses.

Analyzing Evaluation Tools

A great deal of attention has been given to the procedure for developing valid and reliable test items. It is equally important, however, to follow up each administration of the test items with a posttest item analysis. This analysis will allow you to look at each item on the test in terms of its difficulty, or P+ value, and its ability to discriminate; this analysis also allows you to ascertain the reliability of the overall test. You can then use these data to identify potentially weak or faulty items, scrutinize each item to identify the possible cause of a weakness or fault, and make necessary improvements.

Analyzing Student Evaluation Instruments

One of the areas that educators commonly examine when assessing an institute's or program's quality is the written examination scores. A written examination is designed to objectively assess the degree of learning that has occurred. Because it is based on the course curriculum or textbook, the results of this examination are one of the first indicators reviewed. Of course, test results alone are not a sufficient measure of overall program quality. Performance evaluations of the instructors, opinion surveys of the students, and course coordinator reports regarding student grievances should also be considered.

Item Analysis

An **item analysis** is a listing of each student's answer to a particular question. This evaluation tool is useful because examinations may be flawed. Performing an item analysis of the examination can determine whether it measured what it was supposed to, had poorly worded questions, did not contain the correct answer, or had other problems. An item analysis should be done for every examination.

For example, if fewer than 50 percent of the students answered a particular question correctly, the question should be reviewed for problematic sentence structure, misspelling, or lack of a clear answer. If a particular question was answered correctly by almost all students, it should also be reviewed to determine if it is too easy or its answer was given away by another question in the examination.

Item analysis can indicate how well the students understood the course material. The wealth of information provided by the item analysis makes it worth the time it takes to create this kind of analysis.

Item Difficulty

Exam questions should be reviewed for level of difficulty. Building on the information provided by an item analysis, each question should be examined by looking at the number of correct answers. A high item-difficulty score means that it is an easy question. Conversely, a low item-difficulty score indicates a difficult question. Identification of the item difficulty can enable the instructor to improve an examination for future use by linking the item difficulty to the test blueprint. For each question, an item-difficulty score can be identified. When it comes time to modify an examination, the instructor can look at this score and insert a new question with a similar item-difficulty score. This allows the examination blueprint to remain nearly unchanged.

The ideal examination will demonstrate balanced item-difficulty scores. As a general guideline, any item that has an item-difficulty score of less than 0.55 should be reviewed for question validity. At 0.55, almost half the group answered the item incorrectly. An examination that consists of questions with item-difficulty scores ranging between 0.60 and 0.95 has a normal distribution of scores and is considered to be a balanced examination.

Balancing the question difficulty using an item-difficulty analysis offers a clear-cut way to determine the actual item performance for each question. When an examination is used repeatedly, a sense of how a class should perform on the examination develops. This predictability enables the instructor to look at the performance of one class versus another class. The comparison between the two classes allows the instructor to gauge each class's performance for a particular point in its education. Changes can be based on this comparison, if deficiencies are noted between the two classes.

Item difficulty is a powerful tool with which to evaluate each question on an examination. Instructors should respect its usefulness and routinely analyze examination questions in this way. The objective information provided is invaluable in determining the validity of the examination.

A test-item analysis helps determine whether the test items discriminate between respondents with high or low scores and whether the results were consistent. Such an analysis can use data from one administration of a test and formulas for computing item difficulty (P+ value) and discrimination. The analytical objective will be met when the P+ value and discrimination index have been computed for at least 10 items on the test, and each item has been identified as acceptable or needing review.

A posttest item analysis should be approached methodically, using the following process:

1. Gather data.
2. Compute the P+ value.
3. Compute the discrimination index.
4. Compute the reliability index.
5. Identify whether items are acceptable or need review.
6. Revise test items based on posttest analysis.

Data Collection

Immediately following the administration and grading of a test, it is important to collect information on the responses to each item. For each item, you need to know how many times each alternative was selected—that is, whether the alternative was the keyed (correct) response. On a tally sheet, tally each participant's response to each alternative for each test item.

To see how this process works, we will use a multiple choice test item from a 10-item test that was administered to a group of 12 students. This particular test item had four alternatives—A through D—and B was the keyed response (i.e., correct answer). After tallying each participant's response to each test item, you are able to determine how many correct responses were marked and how many incorrect responses were marked for each distracter. Now, as you begin to analyze one test item at a time, isolate the data (**TABLE 14-1**).

Other types of test items will require different tally-sheet formats or provisions for more alternatives than the four (A–D) shown in Table 14-1. The example in **TABLE 14-2** provides space for 10 alternatives.

TABLE 14-1 Isolation of Data

Item Number	Alternatives				Responses	
	A	B	C	D	Right	Wrong
1	1	*8	2	1	8	4

*Bold indicates keyed response.

TABLE 14-2 Test-Item Analysis Tally Sheet

Item Number	Alternatives (Where * indicates the answer)										Responses	
	A	B	C	D	E	F	G	H	I	J	Right	Wrong

Total:

Matching, arrangement, and identification test items require five alternatives; therefore, you will need to provide adequate space on your tally sheet.

Score Distribution

Score distributions may be depicted in many ways. One popular method uses a listing of high to low scores in one column. Then, beside each score, a second column lists the number of students who attained that score on the examination. An example of such a distribution is shown in **TABLE 14-3**. This list of numbers is not easily visualized when it is collected as a random set of numbers. If the numbers are placed onto a chart, however, the resulting graphic provides a visual depiction of the students' scores. Recording the scores on a chart is a common method for showing how well students performed.

Educators like to see a perfect distribution of scores—referred to as a normal distribution—from an examination. In a normal distribution, the majority of the scores cluster around the average score for the examination. The mean, or average score, is positioned in the center of the distribution. In ideal circumstances, 95 percent of the class will have scores within the major area covered by the distribution,

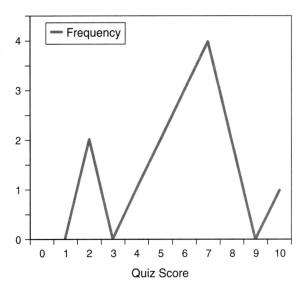

FIGURE 14-2 Quiz distribution curve.
© Jones & Bartlett Learning.

which leaves only 2.5 percent on each side of the curve for extremely high and low scores. **FIGURE 14-2** shows the normal curve.

In practical applications, the normal distribution is not the rule, but rather the exception. Most examination distributions do not conform to the normal distribution. Often, the scores may vary so dramatically that no clustering of the scores is evident. Alternatively, they may be off in one direction, either to the high side or to the low side. In such situations, the distribution is said to be skewed. Two specific types of skews are important to an instructor: a positive skew and a negative skew.

The discovery of a skewed distribution immediately alerts an instructor to a potential problem with an examination. With a positive skew, very few scores appear toward the top of the distribution, and the majority of the scores are below the mean. Interpreting the score distribution chart reveals potential causes of this distribution. Perhaps the examination was too difficult for the majority of the class. To determine whether this perception is accurate, look at the individual student scores. In particular, examine the scores of those students who were performing exceedingly well before the exam and those who were performing poorly before the test. Compare their performance on the examination. If the individuals who were expected to do well on the examination actually did poorly, it could be an indicator that the examination was too difficult.

A negative skew has the opposite effect of a positive skew. In a negative distribution, scores for the majority of students are above the mean. An immediate interpretation is that the examination was too easy. Only a few individuals scored poorly on the examination.

TABLE 14-3 Sample Score Distribution

Score	Score Number of Students
10	1
9	0
8	**2**
7	**4**
6	**3**
5	**2**
4	1
3	0
2	2
1	0
0	0

Note: The main cluster of students falls between 8 and 5, as shown in **bold**.

A negative distribution needs to be subjected to the same kind of investigative review as a positive skew; that is, the instructor needs to review the scores and validate the examination results. If an examination has overevaluated or underevaluated students' knowledge, then it has not fairly represented the amount of learning achieved by that group of students. For this reason, instructors should routinely review the distribution results.

Average Scores

For an instructor to assess the true average score for an examination, the mean, mode, and median scores should be calculated. In a normal distribution, all three should be found on the center line that divides the distribution. In a skewed distribution, each of these values may have a different line location within the distribution. An instructor should use the measurement that best represents the distribution of the students' score as a gauge of the test's effectiveness.

Mean Score

The most commonly used measurement for the average score is the **mean**. It is calculated by adding up all the scores from the examination and then dividing by the total number of students who took the examination:

$$\frac{\text{Total scores}}{\text{Number of students}} = \text{Mean}$$

The mean offers a crude estimate of how the students performed on the examination. With skewed distributions, however, the mean does not reflect all students' scores because just one or two extremely high (or low) scores may raise (or lower) the mean significantly. In other words, the mean will move with the extreme scores and will not reflect the achievement attained by the majority of the students. For this reason, an instructor should not use the mean as the only measurement; the mode and median also need to be assessed.

Mode Score

The **mode** is defined as the most commonly occurring score in the examination. To determine the mode, count the number of student scores for each examination score. The largest number of single examination scores is the mode. If the majority of student scores are clustered in a distribution, the likelihood of the mode appearing in that cluster is high. If the distribution is skewed, the mode may appear anywhere. In a very large distribution, multiple modes may exist.

FIGURE 14-3 shows the mean and the mode for a distribution of exam scores. The mean is the more reflective average score for this particular distribution. More than 11 students attained at least the mean score, whereas only 4 students were at a score of 18. The mode does not reflect all the student scores for this distribution.

Median Score

The **median** score is a more involved measurement than either the mean or the mode. The median score is the score located exactly in the middle of the distribution. It is based on a percentage of the total number of students taking the examination. To calculate the median score, the first step is to create a column called a *cumulative frequency* (cf) by adding up the number of student scores at each level. The top value is equal to the actual number of students who took the examination. Then, each cumulative score is divided by the total number of students taking the examination. The score that is found at the 50 percent indicator is called the median score.

Many educators refer to the C% column as a percentile. Percentiles show student scores based on a 100 percent scale (**TABLE 14-4**). Notice that the median score is between the scores of 15 and 16 in Table 14-4. A statistical method is used for determining the exact score. For the purpose of this example, it is important to note that the mean and the median are nearly identical. Thus, for this distribution, either the mean or the median score would be equally reflective of the actual average score for the distribution.

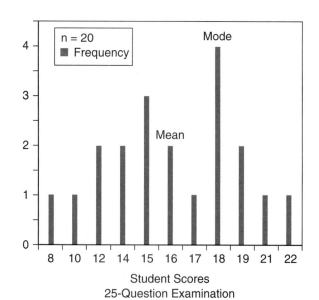

FIGURE 14-3 Mean and mode of score distribution.
© Jones & Bartlett Learning.

TABLE 14-4 Average Scores

Score	Score Frequency	Cumulative Frequency (cf)	C% (Percentile)
22	1	20	100%
21	1	19	95%
19	2	18	90%
18	4 (mode)	16	80%
17	1	12	60%
16	2 (mean)	11	55% (median 15–16)
15	3	9	45%
14	2	6	30%
12	2	4	20%
10	1	2	10%
8	1	1	5%

Identifying the true average for the students' scores assists the instructor in locating the correct average for the distribution. As noted earlier, the mean may be significantly affected by just one or two extraordinarily high or low scores. Finding the true average score requires that all three measures of the distribution—mean, mode, and median—be assessed and compared. At that point, the instructor can select the measurement that best reflects the average score for the distribution.

Computing the P+ Value

The **P+ value** of a test item indicates how difficult that item was for the class taking the test. Your calculation of the P+ value will be useful later on as you revise a test item based on the results of your analysis. Here is the formula to calculate the P+ value of a test item:

$$P+value = \frac{\text{Number right (R)}}{\text{Number taking the test (N)}}$$

$$P+ = \frac{R}{N}$$

Now, let's apply that formula to our example test item. Eight of the 12 respondents answered the question correctly.

$$P+ = \frac{8}{12}$$
$$= 0.666$$
$$= 0.67$$

Thus the P+ value of that item is 0.67.

The P+ value is merely a statistical indicator and should not be used as the sole judgment of the relative merit of a test item. Instead, it should be used in conjunction with the discrimination and reliability indexes.

Computing the Discrimination Index

The **discrimination index** refers to a test item's ability to differentiate (or distinguish) between test takers with high scores on the total test and those with low scores. Several formulas may be used to compute the discrimination index. Our example formula divides the group tested into three subgroups, as equal in number as possible, based on their grades on the overall test. The example test was administered to a group of 12 participants. We arrange their scores in descending order and divide them into three groups (**TABLE 14-5**).

Once the groups have been determined, you can then calculate each test item's discrimination index. Another tally sheet, such as the one shown in **TABLE 14-6**, is useful for this purpose.

In this example, the four best grades constitute the high group and the lowest four grades constitute the low group. We now count the number of respondents

TABLE 14-5 Group Determination

Score	Number of Respondents	Group (12/3 = 4)
100	2	High
90	2	High
80	4	Middle
70	4	Low

Notes: When working with a group size that is not evenly divisible by 3, place equal numbers in the high and low groups and place the remaining numbers in the middle group.
It is important to have equal numbers in the high and low groups.

TABLE 14-6 Test-Item Discrimination Index Tally Sheet

Groups by Test Score	Test-Item Alternatives (Indicate number of students choosing each response choice)										Total Number of Respondents Answering Correctly
	A	B	C	D	E	F	G	H	I	J	
High third	1	3	0	0							3
Low third	0	2	1	1							2

in the high and low groups who answered the item correctly. In the high group, three respondents answered correctly; in the low group, two answered correctly. The number of respondents in the low group who answered correctly is subtracted from the number in the high group who also answered correctly. The difference (1) is then divided by one-third of the total number of respondents (4). The formula for computing the discrimination index follows:

$$D_S = \frac{H - L}{N/3} = \frac{3 - 2}{12/3} = \frac{1}{4} = 0.25$$

where
H = Number of respondents from the high group who answered the item correctly = 3
L = Number of respondents from the low group who answered the item correctly = 2
N = Number of respondents taking the test = 12
Number of groups = 3

The procedure follows these steps:

1. Sort test papers or answer sheets into high, middle, and low groups.
2. Tally all high-group and low-group responses to the item on the tally sheet in the space under "Test-Item Alternatives."
3. Carry the number of correct responses for each group to the right column.
4. Insert the numbers representing the "respondents answering correctly" for the high and low groups into the formula to replace H and L, respectively.

$$\frac{H - L}{N/3} = \frac{3 - 2}{N/3}$$

5. Insert the total number of test respondents to replace N.

$$\frac{3 - 2}{N/3} = \frac{3 - 2}{12/3}$$

6. Complete the calculation to derive the discrimination index for that test item.

$$\frac{3 - 2}{12/3} = \frac{1}{4} = 0.25$$

The index of 0.25 indicates the degree to which the item discriminates between those respondents who obtained high total-test scores and those who obtained low total-test scores. Theoretically, the range of the discrimination index is from –1 to +1.

- An index of +1 results when all respondents in the high group and no respondents in the low group answer correctly.
- An index of –1 results when all respondents in the low group and no respondents in the high group answer correctly.
- If the same number of respondents from both groups answer either correctly or incorrectly, the index is 0 (the exact center of the range).

Here are some assumptions that can be made after studying discrimination-index trends:

- When more respondents from the high group than from the low group answer correctly, the item has positive discrimination.
- When more respondents from the low group than from the high group answer correctly, the item has negative discrimination.
- When the same number from both groups answer the item correctly or incorrectly, there is neutral (zero) discrimination.

Test items that have very low positive discrimination, zero discrimination, or negative discrimination need to receive a quality review to determine their validity. If neutral or negative discrimination occurs with a test item of medium difficulty, you should determine the reason and correct that item before it is used again.

As mentioned earlier, the discrimination index has more meaning when it is used in conjunction with the P+ value (i.e., the measure of difficulty). Our example test item has a P+ value of 0.67 and a discrimination index of 0.25. Thus, although the P+ value is high, the difficulty level and the discrimination index are moderate to low. The answer to the question, "Is this good or bad?" depends on the purpose of the test item. Generally speaking, if both the difficulty level and the discrimination index are in the same range (low–low, medium–medium, or high–high), the test item is doing its job. An intentionally difficult yet valid test item should have a high discrimination index; an intentionally easy test item should have a low discrimination index.

Computing the Reliability Index

The first two indexes are used to rate the difficulty and discrimination characteristics of a single test item. The **reliability index**, by comparison, rates the consistency of results for an overall test. Statisticians use a variety of measurements for determining whether an examination is reliable. They apply these measurements to evaluate whether the examination has consistency among the questions. Through the use of a mathematical formula, a determination of the examination's reliability is made.

One of the formulas most frequently used for this purpose is the Kuder–Richardson reliability formula, which determines the internal consistency of the examination. The range for scores computed from this formula is from 0 to 1. An examination is said to be reliable with a score of 0.60 or higher.

Kuder–Richardson Formula 21

$$\text{Reliability} = \frac{ns^2 - M(n - M)}{ns^2}$$

where
n = Number of test items
s = Standard deviation
M = Mean

Another formula to calculate reliability is the Spearman–Brown formula, also known as the split-half method.

Spearman–Brown Formula (Split-Half Method)

$$r_{11} = 2\left(\frac{1S_a^2 + S_b^2}{S_t^2}\right)$$

where
r_{11} = Reliability of total test
S_a = Standard deviation of the first half of the test (odd-numbered items)
S_b = Standard deviation of the second half of the test (even-numbered items)
S_t = Standard deviation of the whole test

Given that both of the preceding formulas use the **standard deviation**, this value must be calculated before proceeding with the reliability index. The formula for computing standard deviation is as follows:

$$\sqrt{\frac{\Sigma(x - x_2)^2}{n}}$$

where
Σ = Sum of
x = Each score
x_2 = Mean
n = Number of scores

The reliability index computation, using any formula, is a complex and time-consuming process that is best accomplished using a computer and is beyond the scope of this text. Your computer technician should have no trouble implementing the formula.

Ideally, the reliability index should be as close to 1.00 as possible. In reality, anything greater than 0.60 is generally acceptable. If the reliability index for the test does not reach 0.80, you need to look for possible causes, such as the following problems:

- Nonstandard instructions
- Scoring errors
- Non-uniform test conditions
- Respondent guessing
- Other chance fluctuations

Educators also measure an examination's reliability through repeated administrations of an examination. Theoretically, if an examination is given to a group of students one week, and the same examination is given to the same students a week later, the scores should be nearly identical. Although fire and emergency services instructors would not perform this type of educational test in their classes, an instructor should keep track of an examination's performance each time it is administered. Although different students are taking the examination, if the material presented in the course is the same, the scores on the examination should be similar. This can be a factor to consider when evaluating a program.

Instructors should check the reliability of their examinations. Assessing the examination reliability is part of ensuring a valid measurement of a student's learning experience.

Identifying Items as Acceptable or Needing Review

To identify test items as acceptable or in need of review, the first step is to perform a quantitative analysis of the test item. This analysis uses difficulty (P+), discrimination, and reliability data to identify potentially weak or faulty test items. To accomplish the quantitative analysis, compare the test item's statistical data against the following screening criteria:

- Difficulty level (P+ value) between 0.25 and 0.75
- Discrimination index between +0.25 and +1.0
- Reliability of test 0.80 or greater

When a test item fails to meet these criteria, identify the item so that you can later conduct a qualitative analysis of it. During the qualitative analysis, you take a close look at the test item in terms of its technical content and adherence to the format guidelines for its type. Keep in mind that some test items identified as potentially weak or faulty during the *quantitative* analysis may not be found to be weak or faulty when reviewed more closely during the *qualitative* analysis. Some items on each test cover the "core" body of knowledge that everyone must possess. The instructor appropriately emphasizes this information during training, and the test item may then be intentionally written at a low enough difficulty level to ensure that the majority of the trainees "get it right." Therefore, when conducting the quantitative analysis, try to remain objective about the uses of statistical data.

Revising Test Items Based on Posttest Analysis

When you have determined the probable causes of your test-item faults, make changes accordingly. Even after second and third revisions, some items may remain problematic and require additional revision. The analytical item-analysis method does not guarantee immediate success in all cases, but it does promise gradual success over time.

Psychomotor Skills Evaluations

Many fire fighter courses have psychomotor, commonly called practical, skills that are evaluated as part of the testing process. When reviewing testing instruments, it is important not to overlook practical skills tests as well. Many of the same analysis techniques discussed in this chapter are useful for evaluating practical skills assessment instruments to identify possible issues in the practical testing process and make necessary improvements.

Training BULLETIN

JONES & BARTLETT FIRE DISTRICT
TRAINING DIVISION
5 Wall Street, Burlington, MA, 01803
Phone 978-443-5000   Fax 978-443-8000

Instant Applications: Program Evaluation

Drill Assignment

Apply the chapter content to your department's operation, training division, and your personal experiences to complete the following questions and activities.

Objective

Upon completion of the instant applications, fire and emergency services instructor students will exhibit decision making and application of job performance requirements of the instructor using the text, class discussion, and their own personal experiences.

Suggested Drill Applications

1. Research the test administration policy of your state fire service training office or your local technical college fire training program and compare it to your own individual policy and determine if any revisions need to be made to your policy. If you do not have such a policy, create one for proctoring oral, written, and performance-based tests that defines student and evaluator expectations.

2. Review a set of class evaluation forms and identify any features that need modification to better meet the needs of your students.

3. Analyze different student evaluation tools to determine which are the most appropriate for your individual training program and create an implementation plan for incorporating new tools into your program.

4. Review the Incident Report in this chapter and be prepared to discuss your analysis of the incident from a training perspective and as an instructor who wishes to use the report as a training tool.

Incident Report: NIOSH Report #F2010-10

Drill Assignment

Review the information in this incident report and prepare a practice presentation for class delivery at the direction of your instructor. Your presentation should include a summary of the incident facts and identify the NFPA standard(s) that could apply to the incident. Use an outline to organize your thoughts. You may be evaluated on your communication skills during your presentation.

Homewood, Illinois—2010

On March 30, 2010, a 28-year-old male career fire fighter/paramedic (victim) died and a 21-year-old female part-time fire fighter/paramedic was injured when caught in an apparent flashover while operating a hose line within a residence.

Units arrived on scene to find heavy fire conditions at the rear of a house and moderate smoke conditions within the uninvolved areas of the house. A search and rescue crew had made entry into the house to search for a civilian who was entrapped at the rear of the house. The victim, the injured fire fighter/paramedic, and a third fire fighter made entry into the home with a charged 2½-inch (64-mm) hose line. Thick, black rolling smoke banked down to knee level after the hose line was advanced 12 feet into the kitchen area.

While ventilation activities were occurring, the search and rescue crew observed fire rolling across the ceiling within the smoke. They immediately yelled to the hose line crew to "get out." The search and rescue crew were able to exit the structure safely, then returned to rescue the injured fire fighter/paramedic first and then the victim.

The victim was found wrapped in the 2½-inch (64-mm) hose line, which had ruptured, and without his face piece on. He was quickly brought out of the structure, received medical care on scene, and was transported to a local hospital where he was pronounced dead.

Postincident Analysis Homewood, Illinois

Key Recommendations

- Ensure that a complete 360-degree situational size-up is conducted on dwelling fires and others where it is physically possible and ensure that a risk-versus-gain analysis and a survivability profile for trapped occupants is conducted prior to committing to interior firefighting operations.
- Ensure that interior fire suppression crews attack the fire effectively to include appropriate fire flow for the given fire load and structure, use of fire streams, appropriate hose and nozzle selection, and adequate personnel to operate the hose line.
- Ensure that fire fighters maintain crew integrity when operating on the fire ground, especially when performing interior fire suppression activities.
- Ensure that fire fighters and officers have a sound understanding of fire behavior and the ability to recognize indicators of fire development and the potential for extreme fire behavior.
- Ensure that incident commanders and fire fighters understand the influence of ventilation on fire behavior and effectively coordinate ventilation with suppression techniques to release smoke and heat.
- Ensure that fire fighters use their self-contained breathing apparatus (SCBA) and are trained in SCBA emergency procedures.

CHAPTER 14 Program Evaluation

After-Action REVIEW

IN SUMMARY

- Evaluation systems, commonly known as tests, come in a variety of forms. Much is at stake for the student as well as the instructor when the evaluation system is used to demonstrate students' competence and knowledge or for certifications.
- Evaluation systems must be developed to comply with many laws and standards, and the Instructor III will be responsible for understanding these requirements.
- Evaluation results should flow from the students who have completed the test, through the instructor who administered the exam, and then to the individual or agency responsible for grading the results. Once test results are graded and data compiled on the success or failure rate on a particular question, skill, or objective, the instructor can analyze those results.
- Examination results and student assignment results should be treated as confidential every step of the way. This confidentiality should be part of your organization's policies and made clear to the student.
- Always keep evaluation results confidential; policy should dictate which staff members and administrative staff have access to these records. A best practice in this area may be to have any request for release of evaluation records approved by the fire chief and the student(s) involved.
- There is great importance placed on the procedure for developing valid and reliable test items; however, it is also important to follow up each administration of the test items with a posttest item analysis. This analysis will allow you to look at each item on the test in terms of its difficulty, or P+ value, and its ability to discriminate; this analysis also allows you to ascertain the reliability of the overall test.
- One of the areas that educators commonly examine when assessing an institute's or program's quality is the written examination scores. A written examination is designed to objectively assess the degree of learning that has occurred. Because it is based on the course curriculum or textbook, the results of this examination are one of the first indicators reviewed.
- Test results alone are not a sufficient measure of overall program quality. It is also necessary to consider performance evaluations of the instructors, opinion surveys of the students, and course coordinator reports regarding student grievances.

KEY TERMS

Access Navigate for flashcards to test your key term knowledge.

Discrimination index The value given to an assessment that differentiates between high and low scorers.

Item analysis A listing of each student's answer to a particular question used for evaluation.

Mean A value calculated by adding up all the scores from an examination and dividing by the total number of students who took the examination.

Median The score in the middle of the score distribution for an examination.

Mode The most commonly occurring value in a set of values (e.g., scores on an examination).

P+ value Number of correct responses to the test item. Example: P = 67 means that 67 percent of test takers answered correctly.

Reliability index Value that refers to the reliability of a test as a whole in terms of consistently measuring the intended material.

Standard deviation The value to which data should be expected to vary from the average.

REFERENCES

National Fire Protection Association. 2017. *NFPA 1401: Recommended Practice for Fire Service Training Reports and Records.* Quincy, MA: National Fire Protection Association.

National Fire Protection Association. 2019. *NFPA 1041: Standard for Fire and Emergency Services Instructor Professional Qualifications.* Quincy, MA: National Fire Protection Association.

CHAPTER PRESENTATION EXERCISE

Instructions: Your course instructor will assign individual or group discussions on the key points and teaching tips of this chapter. You or your group will review the chapter teachings and identify the major learning points from the chapter. You should be able to discuss the points, why they are important, and how/where they apply to your responsibilities at your level of instructor training.

REVIEW QUESTIONS

1. Why is it important to share with students their test results, and how does this tie into your role as a program administrator?
2. What is gained by understanding the math behind test results?
3. How can analyzing students' performance improve a program, curriculum, and instructor cadre?

CHAPTER 15

Program Administration

NFPA 1041 JOB PERFORMANCE REQUIREMENTS

Note: An asterisk denotes that the 1041 standard contains further information in its annex section.

6.2.2* Administer a training record system, given authority having jurisdiction (AHJ) policy and type of training activity to be documented, so that the information captured is concise, meets all AHJ and legal requirements, and can be accessed.

(A) Requisite Knowledge.

AHJ policy, recordkeeping systems, professional standards addressing training records, legal requirements affecting recordkeeping, and disclosure of information.

(B) Requisite Skills.

Development of records and report generation.

6.2.3 Develop recommendations for policies to support the training program, given AHJ policies and procedures and the training program goals, so that the goals are achieved.

(A) Requisite Knowledge.

AHJ procedures and training program goals, and format for AHJ policies.

(B) Requisite Skills.

Technical writing and decision making.

6.2.4 Select instructional staff, given personnel qualifications, instructional requirements, and AHJ policies and procedures, so that staff selection meets AHJ policies and achievement of AHJ and instructional goals.

(A) Requisite Knowledge.

AHJ policies regarding staff selection, instructional requirements, the capabilities of instructional staff, employment laws, and AHJ goals.

(B) Requisite Skills.

Evaluation techniques and interview methods.

6.2.5 Construct a performance-based instructor evaluation plan, given AHJ policies and procedures and job requirements, so that instructors are evaluated at regular intervals, following AHJ policies.

(A) Requisite Knowledge.

Evaluation methods, employment laws, AHJ policies, staff schedules, and job requirements.

(B) Requisite Skills.

Evaluation techniques, scheduling, technical writing.

6.2.6 Formulate budget needs, given training goals, AHJ budget policy, and current resources, so that the resources required to meet training goals are identified and documented.

(A) Requisite Knowledge.

AHJ budget policy, resource management, needs analysis, sources of instructional materials, and equipment.

(B) Requisite Skills.

Resource analysis and required documentation.

KNOWLEDGE OBJECTIVES

After studying this chapter, participating in a structured learning environment, and completing assigned assessments, you will be able to:

- Explain considerations for creating a records management system, including management of the collection, storage, and use of necessary information. (**NFPA 1041: 6.2.2**, pp 341–342)

- Explain the purposes of policies and procedures. (**NFPA 1041: 6.2.3**, pp 350–353)
- Identify content considerations when writing policy. (**NFPA 1041: 6.2.3**, pp 350–353)
- Identify selection criteria for a position that needs to be staffed. (**NFPA 1041: 6.2.4**, pp 354–357)
- Explain evaluation methods for reviewing and ranking candidates for assignment. (**NFPA 1041: 6.2.4**, pp 354–357)
- Explain legal and ethical requirements that must be considered in the selection process of candidates for instructor positions. (**NFPA 1041: 6.2.4**, pp 355–356)
- Discuss methods of constructing instructor evaluation plans. (**NFPA 1041: 6.2.5**, pp 356–357)
- Explain the purpose of instructor performance evaluation plans. (**NFPA 1041: 6.2.5**, pp 356–357)
- Identify necessary components of an instructor evaluation plan. (**NFPA 1041: 6.2.5**, pp 356–357)

SKILLS OBJECTIVES

After studying this chapter, participating in a structured learning environment, and completing assigned assessments, you will be able to:

- Administer a training record system. (**NFPA 1041: 6.2.2**, pp 341–342)
- Prepare policies for records management including management of the collection, storage, and use of necessary information while meeting legal and organizational requirements. (**NFPA 1041: 6.2.2**, pp 341–342)
- Prepare forms for data collection meeting legal and organizational requirements. (**NFPA 1041: 6.2.2**, pp 341–342)
- Prepare a training program policy to meet an identified goal and recommend its adoption. (**NFPA 1041: 6.2.3**, pp 350–354)
- Create instructor selection guidelines based on specific agency needs and polices. (**NFPA 1041: 6.2.4**, pp 354–356)
- Demonstrate using a policy and procedure to review potential candidates for an instructional assignment. (**NFPA 1041: 6.2.4**, pp 355–356)
- Interview a candidate for an instructional staff position. (**NFPA 1041: 6.2.4**, pp 355–356)
- Select an instructor that meets specified position and organization criteria. (**NFPA 1041: 6.2.4**, pp 355–357)
- Create a plan for an instructor evaluation program based on job requirements and agency policy. (**NFPA 1041: 6.2.5**, p 357)
- Write purchasing specifications for training resources. (**NFPA 1041: 6.2.6, 6.2.7**, pp 361–362)
- Prepare a proposed budget to meet agency training goals. (**NFPA 1041: 6.2.6**, pp 357–362)

You Are the Fire and Emergency Services Instructor

A collaborative effort by your fire department and several of your neighboring departments has resulted in the signing of an agreement forming a joint fire/emergency medical services (EMS) training program designed to provide initial training to recruit fire fighters based on National Fire Protection Association (NFPA) 1001 Fire Fighter I, as well as ongoing and refresher training for incumbent personnel. As part of this program, you have been tasked with heading the committee responsible for the development of the policies and procedures for the agreement.

1. What policies and procedures will be necessary for the management of instructional resources, staff, facilities, records, and reports?
2. What policies will be needed to support the training program? Where might you find examples of the necessary policies?
3. How will staff be selected for the program and what instructional responsibilities will they have?

Access Navigate for more practice activities.

Introduction

As a Fire and Emergency Services Instructor III, you have risen within an organization to senior leadership of the training division or bureau. In a smaller agency, you are the individual who has a passion for training but also understands the need to document the training activities of your members. Regardless of the size of your agency, the need for accurate recordkeeping and ongoing maintenance is a critical skill that requires focus and dedication. The ethical and legal ramifications for accurate training records or the lack of training records can cause any department or agency legal nightmares. There is an expectation from the national, state, and even local level that agencies adhere to recordkeeping requirements for fire and emergency medical providers for training, incident reports, accidents, and injuries. There is no legal standing to say, "We don't keep records in our department."

Training Record Systems

As mentioned throughout this text, training record systems are one of the most important management functions assigned to the training officer. A well-known saying—"If it isn't documented, it didn't happen"—sums up almost every aspect of a fire department training record system. A record system must be developed by the Instructor III, who may function as the training officer or training division chief, to document all aspects of the training program. Most instructors are familiar with the training record report, which is the most common type of training record, but they may not be aware of the many types of training reports that are created to summarize and detail training activities. NFPA 1401: *Recommended Practice for Fire Service Training Reports and Records*, details best practices for these training record systems and provides structure for the management functions of this duty area. An excerpt of NFPA 1401 is included in Appendix D.

Using Training Records

Training records are used for many purposes. In some organizations they are analyzed to identify personnel assignments or to make pay-scale adjustments, whereas in other organizations the number of training sessions attended are tallied to allow for continued assignment to emergency response duties. A formal department policy—created by the training division and approved by the fire chief—should exist to determine the uses, preparation requirements, release of records, and storage of all types of reports.

Compliance

A major use of training records is to ensure regulatory compliance. Many local, county, state, provincial, and national organizations require very specific types of training, and the documentation of those training sessions is often reviewed in audits. The training officer should become very familiar with the highly dynamic world of compliance mandates and should ensure that every effort is made to follow the guidelines or procedures necessary to meet the intent of the agency. In some cases, emergency service agencies have been denied federal and state grants due to the lack of training level documentation. In other cases, fire departments have seen their Insurance Service Office (ISO) rating either increase or decrease as a result of careful examination of training record reports that document the numbers of hours of individual fire fighters' training completed. In many volunteer fire and EMS organizations, training records are used to maintain and document membership status in the organization.

TABLE 15-1 identifies several regulatory agencies and some examples of the types of training records they may require the emergency service agency to create and make available during audits or investigations and for compliance purposes.

Types of Training Records

Many types of training record reports exist throughout the emergency service community, the most common of which is the training record report used to document the ongoing training of the members. Although each agency should create records and reports that meet its own specific needs and criteria, NFPA 1401 does provide some direction on the content and use of training records and reports. At a minimum, training needs to document who attended, what was taught, who was the instructor, hours completed, and what date it occurred.

Electronic or Paper Records and Reports

Efforts to reduce paper use and storage are a basic environmental and space-saving management decision. Training officers should be aware, however, that some agencies still require the permanent storage of certain types of training records, especially ones used to document initial training on occupational health

TABLE 15-1 Training Record Examples per Regulatory Agency

Insurance Service Office	State Certification Boards	Local EMS Providers	OSHA
■ Training record reports ■ Summary of training hours ■ Recruit, pump operator, and officer training reports ■ Mutual aid training record reports ■ Preplan review training ■ Equipment and procedure training	■ Certifications earned by members ■ Hours completed toward certifications ■ Testing record reports	■ Certifications and licenses earned ■ Continuing education hours ■ In-service training records ■ CPR training ■ AED training ■ Policy training	■ Mandated training area documentation ■ Respiratory protection training ■ Hazardous materials training ■ Infection control ■ Confined space and special hazard training ■ Injury and accident reports

AED = automated external defibrillator; CPR = cardiopulmonary resuscitation; OSHA = Occupational Safety and Health Administration.

and safety. It is acceptable to maintain a combination of electronic and paper-based record systems and to provide a clear identification of which types should be kept in which form. This arrangement should be identified in the training policy.

Many software programs specific to emergency service training documentation storage exist and make instant recall and custom reporting easy for the department. Paper-and-pen systems may be easy to use, but they require a lot of file storage space and further management decisions about the length of time for which records must be kept. Some documentation aspects that may need to be decided by local legal counsel include the following:

- The types of records and reports that must be kept in original form
- The length of time that records and reports must be kept before disposal
- Access and permissions to electronic and paper records and reports

The training officer should work with local agencies responsible for reviewing the content of fire and EMS training records, as they may provide guidance or clarification to determine record requirements. For example, the Illinois Office of the State Fire Marshal (OSFM) has made a specific ruling that allows for an electronic daily training record (paperless training record) to be used as long as a hard-copy monthly summary report is created for each member, then reviewed and signed by the member and a qualified instructor (**FIGURE 15-1**). This summary report must be kept in the member's training record jacket and made available for audit and inspection by the OSFM.

Performance Tests, Exams, and Personnel Evaluations

Performance tests, examination score reports, and personnel evaluations related to training may be some of the more confidential types of records and reports the training officer will be responsible for maintaining. These documents may be covered by the Privacy Act of 1979 or other state and local acts or policies that deal with provisions of confidentiality and disclosure. A person's training file may be classified as a personnel file and should be treated as confidential information by the training division. Some learning management systems include features that will electronically score and store examination and quiz score records. As discussed in Chapter 11, *Evaluation and Testing*, performance tests and examination records will include a detailed breakdown of success and failure of the member.

Storage and maintenance of personnel evaluations may not fall directly under the responsibilities of the training officer. Nevertheless, a training officer may be able to provide valuable training-related information to other officers for personnel evaluations by supplying records of training hours, certifications attained, and any performance deficiencies related to training. Local policy will dictate this area of record-keeping responsibility and address the use of this information.

WILLOW SPRINGS FIRE-RESCUE
Monthly Training Record

DECEMBER 2018

Fire Chief Larry Moran

Date	Time	Hours	Class Description
12/04/2018	09:00	03:00	MABAS 10 Haz Mat Training
12/04/2018	13:00	02:00	Christ Hospital Continuing Education
12/05/2018	19:00	03:00	Incident Command System
12/07/2018	08:00	05:00	Communications and Information Management
12/11/2018	09:00	03:00	IMAT Training
12/13/2018	11:00	02:00	MABAS 10 Training Meeting
12/17/2018	09:00	03:00	Fire-ground Communications and Mayday Procedures
12/19/2018	10:00	02:00	MABAS 10 Chief Meeting
12/27/2018	10:00	01:00	District Familiarization
		24:00	**Total Hours**

_____ _____ _____
Member's Signature Training Officer's Signature Chief's Signature

FIGURE 15-1 A sample monthly summary training record report.
Courtesy of Willow Springs Fire Department.

Progress Reports

Much in the same way that school districts track the academic progress of elementary and secondary school students, the training officer will track the progress of fire fighters as they pursue certifications and licenses. Many state certification agencies have specific hour- and content-related requirements that must be documented. A progress report may help track these pursuits by logging hours completed, test scores, and practical skill testing toward the certification or license. In some cases, a certification may require completion of several courses, and a progress record will assist the member and training officer in tracking which additional courses are needed.

During a fire academy, weekly progress reports may be issued back to the employing fire department because the recruit may be attending the academy at some distance from the department (**FIGURE 15-2**). In such a case, it would be important to keep the department up-to-date on the progress of its members in an effort to identify any performance issues early in the training process. Many fire chiefs who are sponsoring fire fighters in fire or EMS distance-learning programs need these updates because these reports are tied to employment status, insurance, and other benefits.

State Certification Records

Each state training entity will have specific requirements for obtaining a certification, including the records that relate to these criteria. It may be a handy and useful practice to keep this portion of the training file separate from other areas for quick access, as it may be more frequently utilized than other training files (**FIGURE 15-3**).

A Typical Training File

At a minimum, the following information should be kept in a state certification record file and in all types of training files in general:

- Master individual training accomplishments
- Certifications received or in-progress
- Dates, hours, locations, and instructors of all special courses
- Monthly summaries of all departmental training

NIPSTA FIRE ACADEMY
Weekly Evaluation Report
January 25, 2018

Candidate Name: Fire Fighter Newguy   Company: 5
Candidate Department: J & B Fire Rescue

Week 2	Monday	Tuesday	Wednesday	Thursday	Friday
Subject	Safety	Fire Behavior	SCBA 2	Ladders	Ladders
Attendance	Present	Present	Late/Excused	Present	Present
Quiz	Orientation	Terms	SCBA 1	Ropes	Tools
Score	100%	97%	100%	90%	88%

Overall Weekly Evaluation		Comments
Discipline	Meets Expectation	
Effort	Needs Work	Step up and complete tasks on time
Tools	Meets Expectation	
Ability	Improving	Showing progress from Week 1
Initiative	Needs Work	Ask for additional assignments
Ladders	Meets Expectation	
SCBA	Meets Expectation	Donning time is currently :48 sec.

Additional Information		Comments
Prepared for each day:	Yes	
Obeys rules and regs:	Yes	
Appearance and hygiene:	Very Good	
Overall attitude:	Good	Respectful to instructors and others
Level of participation:	Needs Work	Waits to be told what to do
Quality of work:	Good	
Ability to work w/others:	Very Good	Is involved in group work
Injuries Reported:	None	
Areas of Concern:	Study habits should improve	
Corrective Action Plan:	Increase use of workbook and online resources	

Additional Comments:
Continue to work on understanding of terminology and tool names/uses. Increase your participation in house duties, classroom and drill ground set-up, and group team-building exercises. Physical fitness level is improving, keep it up. Document additional online study and workbook completion before unit 2 exam.

Upcoming:	Monday	Tuesday	Wednesday	Thursday	Friday
Week 3	Hose 1	Hose 2	Live Fire 1	Ventilation	Live Fire 2

Academy Director: Deputy Chief Drew Smith
Academy Coordinators: Scott Exo   Mike Fox   Ken Koerber

FIGURE 15-2 A sample fire academy progress report.
Courtesy of Northeastern Illinois Public Safety Training Academy.

Required Signatures

Almost every type of training record must identify both the instructor of the course and the person who was instructed. In most circumstances, initials are not acceptable for this purpose on permanent training records and state certification training records. It may be acceptable to use a preprinted roster listing a department roster, shift, company, or unit, but a column should be available for each participating member to sign in as a record of attendance and participation. Certificates of completion for coursework should include the trainee's name, the title of the training

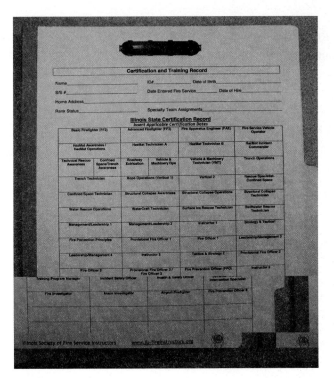

FIGURE 15-3 A sample training record jacket.
Courtesy of Illinois Society of Fire Service Instructors.

session, the dates of attendance, hours completed, and the signature of a lead instructor or program coordinator who is validating all of the information on the certificate is accurate.

Elements of Training Record Information

A standard training record report must be developed and used anytime training takes place in the department (**FIGURE 15-4**). Some information may be preprinted, whereas other portions of the record are a fill-in-the-blank format. NFPA 1401 recommends the areas in **TABLE 15-2** as the minimum areas that should be included on the record. These requirements pertain to both electronic and paper-based training records.

A summary of the objectives and content of the training session should also be included, as should a narrative provided by the instructor in charge.

Additional Information on Training Record

Training record reports should be easy to complete. Nevertheless, the more thorough the information, the more useful the document will be. Additional information that might prove beneficial could include source information used as the basis for training, such as textbook title and edition and any available lesson plan information. If a local policy name and version is used as a reference, then be sure to include the reference number in the record. Some departments use a coding system related to state certification objectives and insert the objective numbers into the record, as the objective number is a known factor in the state program. As distance-learning programs become more common as part of overall training curricula, remember that distance-learning sources should be cited and referenced just like textbooks or other curricula.

Agencies that conduct post-incident analysis or after-action review (AAR) of incidents that have occurred within their agencies find that these post-incident analyses can be a very valuable training tool in identifying areas of below standard performance. Opportunities for improvement can be inserted into the training program from the use of a post-incident analysis, and these opportunities should be noted in the narrative. As a best practice, a specific code can be created in an electronic recordkeeping system to identify the use of post-incident analysis in the training program, which in turn can be tied back to the post-incident report itself.

The delivery method used for training should be identified as part of the training record. Some program evaluations comprise an assessment of the percentages of each method of instruction used. A good balance among the various methods used should be apparent. Common methods of instruction that should be noted in the training record report include the following:

- Lecture
- Demonstration
- Skills training
- Discussion
- Role play
- Labs/simulations
- Case studies
- Out-of-classroom assignments
- Self-study
- Online/video/computer based

If a combination of methods is used in the training session, a breakdown of how much time is spent doing pre-course study such as an online activity, lecture/classroom time, and skills training at the tower, training ground, or sims-lab. All of this should be documented on the training record for this training session. In addition to this information, include documentation of any evaluation of training objectives that takes place. Written tests or quizzes given in class, skills examinations or skill sign-offs, or other tools used to assess the student's understanding or ability to meet the learning objectives should be documented. Maintain score confidentiality by keeping scores separate from the training record; all that is needed on the training record is a notation of the evaluation's completion. This notation

Training Record
Training Attendance Form

Date _____ Station(s) _____ Description _____

Start Time _____ End Time _____ Credit Hours (Total Time) _____

Method of Training: ☐ Classroom ☐ Practical ☐ Self-Directed Certification Credit: ☐ Yes ☐ No

Lead Instructor _____ Additional Instructor(s) _____

Print Name	Dept ID #	Signature	Hours Attended

Objectives: _____

Description of Training (Notes) _____

_____ _____
Instructor Signature   Training Officer Approval

Type of Training (ISO Category)				
Company	Multi-Comp.	Officer	Mutual Aid	Night
Tower Burn	Classroom	Practical	Combo.	Driver

Equipment Used in Training Session	Feet of 1 1/2" Hose Used	Feet of 2 1/2" Hose Used	Supply Hose Used/ft.	Feet of Ladders	Number of Engines	Number of Trucks	Gallons of Water	Number of SCBA	Total Number of Fire Fighters

FIGURE 15-4 A sample training record report.
Courtesy of Forest Reeder.

could take the form of a check box indicating whether an evaluation was completed or as a note in the narrative section of the report.

Types of Training Records

Many types of training records can be created, each with a different purpose. Consider the following NFPA 1401 recommended types of training records (NFPA 1401, Section A.5.3):

- Departmental training record: A report that serves as a permanent record showing all the training fire personnel receive. These reports usually are completed on a yearly basis. Company officers usually make entries on

TABLE 15-2 Minimum Areas for a Training Record Report

Who	Who was the instructor(s)?
	Who was included in the training (individuals, company, multiple companies, or organization)?
What	What was the subject covered?
	Which equipment was utilized?
	What were the stated objectives?
When	When did the training take place?
	■ Date and start/end time
Where	Where did the training take place?
	■ Station, classroom, drill ground, off-site, academy, school
Why	Why did the training take place?
	■ In-service training, certification based, compliance training, professional development, performance improvement

this form; however, training officers in small departments might take on this responsibility.

- Individual special course record: Special schools or courses made available to fire fighters. Their attendance and certificates of completion should be recorded.
- Individual training record: A record containing a chronological history of an individual's progress from the time of entry into the organization until time of separation.
- Progress chart: A record form that provides an accurate and complete picture of all class activities and work accomplished by both the instructor and the students. The chart also shows, at a glance, how the class or program is progressing with respect to calendar or time schedules.
- Certification training record (fire fighter): A worksheet that maintains a record of each fire fighter's progress during the pursuit of certification.
- Educational courses: A documentation of courses taken outside the department, such as at institutions for higher education or the National Fire Academy.
- Vocational courses: A documentation of courses, most of which are provided by state or regional programs in the form of workshops, region or state fire school, or demonstrations outside the department.
- Seminars and other training: A documentation of all seminars, short courses, and other individual or group development meetings attended. A certificate of achievement or certificate of completion obtained in this area of training should be made a part of the individual's training file.
- Periodic company summary: A report showing all other training conducted or attended by the company. The number of hours spent by each individual on each subject should be recorded, and this report should be submitted through proper channels to the officer in charge of training. Such reports should be submitted monthly.
- Chief officers' periodic training summary: A report showing all training conducted by fire companies within a division, battalion, or district. This report serves as confirmation for chief officers that company officers are properly conducting company training, and it can be submitted to the officer in charge of training for inclusion in monthly and annual summaries.
- Group training records and evaluation: Because most fire-ground operations are accomplished by more than one fire fighter, group performance of basic evolutions should be an important part of any training system. A means or method of measuring the effectiveness of the organization's evolutions and how well they prepare fire fighters to make an attack on a fire should be established. The group or company performance standard evaluation report form should be designed to allow for quick and accurate determination of the group's ability to meet minimum basic requirements. The report should also enable the individual group or company to check on the progress of its abilities in carrying out standard operating procedures, and it should recommend areas in which additional training is needed.

Reports for the Training Program

Periodically, the training division must compile and submit reports on the activities, inventory, statistics, and successes of the division. These reports can be used for budget justification, equipment purchasing, and many other useful management functions. Generally speaking, training reports—like other reports submitted to a supervisor or administrative body—are considered technical reports, and specific guidelines for their creation may exist. Reports should be clear and concise and follow specific organizational steps. The reports may include the total number of classes taught, hours of teaching, student contact hours, total number of students attending, and a summary of courses that were offered.

Format for Training Reports

A logical sequence of the information contained should be followed throughout each report, and the same sequence should be used whenever the report is used. With some frequently submitted reports, such as monthly training activity reports, the data fields may simply be updated on a monthly basis once the format is approved by the receiver of the report (**FIGURE 15-5**).

The organization of a report can follow a five-step process that can be helpful in identifying, investigating, evaluating, and solving a problem. These five steps should be completed before the report is written:

1. Determine the purpose and scope of the report.
2. Outline the method or procedure.

JBL Fire Department
Division of Training & Safety

Monthly Training Report Summary January 2018

	Month 2017/2018	Year-to-Date 2017/2018
Total In-Service Training Hours	3127/3377	3127/3377
Total Classes in Period	309/430	309/430
Total Attendees in Period	2061/2791	2061/2791
Acting Officer Training Hours in Period (Lt)	0/48	0/48

In-Service Training/Activities Scheduled

1. Company Tool Assignments
2. Grid Map Book Training
3. Fire Behavior 1
4. Apparatus and Equipment Maintenance
5. Ice Rescue
6. EMT/Paramedic Refresher

Out-of-Department Training or Schools Attended

Leadership I (40 hours)	FF/PM John Williams, Eng Dave Perry, FF/PM Ryan Smith
Ice Rescue (16 hours)	FF/PM John Williams, FF/PM Ed Dillon, FF/PM Mike Russo
Peer Fitness Trainer (40 hours)	FF/PM Mike Johnson, Eng Matt Dowd

Total Out-of-Department Training: 248 hours

Division Related and Attended Meetings & Events
MABAS Division 1 Training Officers Meeting 1-15-17
IFCA Educational Research Foundation 1-17-17
Promotional Assessment Orientation 1-23 & 24, 2017
NWC Mass Casualty Training 1-17-17
Monthly Staff and Operations Mtg 1-28-17

Certifications Received
Ed Collins   Provisional Fire Officer I

_____
Division Chief Name

FIGURE 15-5 A sample monthly training activity report.

3. Collect the essential facts.
4. Analyze and categorize the facts.
5. Arrive at the correct conclusions and make the proper recommendations.

Other Useful Reports

The training officer may be required to submit any number of training reports to a superior officer or regulating body. The frequency and content of these reports are determined by the local authorities and should be included in the overall training policy. Some reports, including those profiling assets assigned to the training division and detailed plans to improve the training division, can be used for budget request purposes. Another useful document is a periodic report on the progress of any probationary fire fighters; this type of a progress report can help identify strengths and weaknesses and document hours of training toward a certification (**FIGURE 15-6**). If a probationary fire

JBL Fire Department
DAILY INSTRUCTOR EVALUATION OF CANDIDATE
Operational Readiness Period – Probationary FF/PM

Date	Evaluator	Company Officer	Company Assigned	Eval. Start	Eval. End

List training modules completed:

Scale	Not Applicable N/A	Unacceptable 1	Needs Work 2	MEETS EXPECTATION 3	Above Average 4	Excellent 5

Candidate Name		
Work Ethic	N/A 1 2 3 4 5	Injuries Received/Reported:
Judgment	N/A 1 2 3 4 5	Areas for Concern:
Teamwork	N/A 1 2 3 4 5	
Time Management	N/A 1 2 3 4 5	
Skill Level	N/A 1 2 3 4 5	
Physical Ability	N/A 1 2 3 4 5	Provide examples of positive and negative performance observed throughout evaluation period.
Communication	N/A 1 2 3 4 5	
Initiative/Motivation	N/A 1 2 3 4 5	
Accepts Guidance	N/A 1 2 3 4 5	
Adaptability/Stress Mgt.	N/A 1 2 3 4 5	
Community Awareness	N/A 1 2 3 4 5	
Empathy	N/A 1 2 3 4 5	
Appearance	N/A 1 2 3 4 5	

Evaluator Signature/Date

Candidate Signature

FIGURE 15-6 Periodic reports on the progress of probationary fire fighters can be helpful.

fighter must be released from service due to poor performance, this report will help justify these difficult but necessary management decisions.

A monthly summary of all activities of the training division is one of the most frequently used reports completed by the training officer. These monthly reports can then be used as the basis to complete an annual report of all activities of the training division at the end of a fiscal or calendar year.

Legal Aspects of Training Records

There are many laws and standards that apply to recordkeeping in a training division. The training officer and department should make it a regular management practice to consult with legal representation about the quality, type, and quantity of information contained within the training record report. An example of how NFPA standards are used to specify training records in different ways can be found by examining NFPA 1403: *Standard on Live Fire Training Evolutions*. NFPA 1403 requires specific records and reports to be completed by the instructor in charge and also requires participants' training reports to be submitted to verify prerequisite training of participants. Failure to follow these guidelines can result in civil lawsuits and criminal charges in the event that a participant or instructor is injured or killed during the training event.

Privacy of records must be maintained and observed over the life of the organization. Training records are considered to be personnel records and, therefore, can be disclosed only with written permission from the student, unless required by law, statute, or court order. The use of Social Security numbers in such documents should be strictly prohibited, as should inclusion of other confidential individual information. Older training record jackets may display and contain Social Security number information, and those files should be retroactively stripped of such information to avoid potential legal issues. No confidential medical information should be kept within the training record. In fact, medical records should be kept separate from all other types of records. Personally identifiable or proprietary information is restricted by several privacy laws.

Records Storage and Retention

A records retention schedule that details the life expectancy of the various records and reports maintained and stored by the training division is required to be created; in fact, creation of this document may be required by state regulatory offices, such as the Secretary of State (**FIGURE 15-7**). Do not dispose of any form of training record or report without first checking the records retention schedule. Computerized forms can be kept forever and backed up on multiple hard drives and other electronic media. Keep in mind that some records must be kept in their original form, including some completion certificates and licenses.

Training Program Policies

Policies serve a number of important functions in a training program. The most important reason policies exist is to document the organization's approach to operations. Policies also define expectations, guide decisions, create accountability, and ensure consistency in an organization or program's operations. Formal training program policy and procedures should be created to help manage the training division activities and the instructors who work and teach in the division.

Policies and procedures are more than simply tools to guide decisions and actions. It is up to the officers and administrators of the training division to ensure that the policies and procedures are communicating the desired message of the fire chief and the division. Policies and procedures are an opportunity to direct organizational communication. The flow of communication is just as important as the accuracy of the information itself. Policies and procedures help shape the organization's culture and should be used to align the organization's mission, priorities, culture, and operational realities.

A person writing good policies and procedures must have a firm understanding of how they are defined and work together. A policy is a statement of principle used to guide decisions and actions. A procedure is the method used to achieve compliance with the principle the policy describes. These two terms relate in that a policy is like a strategy—something specific that must be done—and a procedure is like the tactic used to accomplish the strategy. In other words, a policy is what should be done, and a procedure is how to get it done. For example, in a promotional policy one part might say, "Only candidates passing a three-part assessment will be eligible for promotion." This is a strategy. The procedure or tactic part might read something like this: "The assessment will consist of three parts: (1) written test, (2) in-basket exercise, and (3) a tactical problem simulation." Understanding

APPLICATION FOR AUTHORITY TO DISPOSE OF LOCAL RECORDS
(CONTINUATION SHEET)

APPLICATION NO. 84:156C

PAGE 39 OF 67 PAGES.

ITEM NO.	DESCRIPTION OF ITEMS OR RECORD SERIES	ACTION TAKEN
	Some records contained on this application are subject to state and/or federal audits. If that is the case the retention period will be followed by this symbol *. The audit clause which applies in this case is as follows. Provided audit completed according to Illinois Revised Statutes 1983, Chapter 24, Article 8, Division 8. Some records contained on this application are to be retained permanently. However, the original records may be disposed of if microfilmed in accordance with the regulations and standards of the Local Records Act, and providing the microfilm is retained permanently. If that is the case the retention period will be listed as "Retain permanently." followed by this symbol + . If microfilming is not an option, the retention period will be listed as "Retain permanently."	
195.	SAFETY TRAINING PROGRAM FILE – STATE CERTIFICATION RECORDS Dates: 1974 – Volume: 6 cu. ft. Annual Accumulation: 1 cu. ft. Arrangement: Chronological by year Recommendation: Retain permanently.+	
196.	ENGINEER TRAINING RECORDS Dates: 1980 – Volume: 2 cu. ft. Annual Accumulation: 1/2 cu. ft. Arrangement: Chronological by year Recommendation: Retain permanently.+	
197.	PARAMEDIC TRAINING FILE, DEPARTMENT TRAINING RECORDS (State Certifications) Dates: 1976 – Volume: 2 cu. ft. Annual Accumulation: Negligible Arrangement: Chronological by year Recommendation: Retain permanently.+	
198.	DAILY DRILL RECORDS Dates: 1973 – Volume: 6 cu. ft. Annual Accumulation: 1 cu. ft. Arrangement: Chronological by year Recommendation: Retain 5 years and dispose of.	
199.	PERFORMANCE EVALUTIONS Dates: 1981 – Volume: 2 cu. ft. Annual Accumulation: 1/2 cu. ft. Arrangement: Chronological by year Recommendation: Retain 5 years after termination of employment and dispose of.	

FIGURE 15-7 A sample records retention schedule.
Courtesy of Forest Reeder.

how these terms are defined and relate to each other will make it easier to write and understand policies and procedures.

The policies should follow the same format as other department policies. Therefore, it is incumbent on the Instructor III to understand authority having jurisdiction (AHJ) systems and help to create the policies and directives that affect or support the mission of the training division safely, ethically, and efficiently. The Instructor III should be familiar with the following common types of fire service policies and procedures. A recap is provided here:

- Standard operating guidelines (SOGs) are written organizational directives that identify a desired goal and describe the general path to accomplish the goal, including critical tasks or cautions. SOGs are formal, permanent documents that are published in a standard format and remain in effect until they are rescinded or amended. They are designed to provide organizational members guidance and direction. SOGs are constructed to assist members in making decisions in both emergency and nonemergency situations, but with the recognition that not all situations are "standard" they allow for situation specific deviation from the SOG if conditions warrant doing so. **FIGURE 15-8** shows an example of a department SOG.
- Standard operating procedures (SOPs) are written organizational directives that establish or prescribe specific operational or administrative methods to be followed routinely for the performance of designated operations or actions. SOPs are formal, permanent documents that are published in a standard format, signed by the fire chief, and widely distributed to

Department Reference Number:

Purpose
In an effort to improve safety and prevent injuries sustained in motor vehicle accidents, the _____ Fire Department provides child car seat inspections to parents and caregivers who are responsible for transporting children under 8 years of age.

Scope
The _____ Fire Department has certified Child Passenger Safety Technicians on staff ready to provide the most updated information and a check of your child seat and, if necessary, perform a complete child seat installation. All technicians are certified through the *Safe Kids Worldwide National Child Passenger Safety Certification Training Program* and the *National Child Passenger Safety Board (National Highway Traffic Safety Administration)*.

The _____ Fire Department team will share:
    The steps of restraint as the child grows:
        Rear to forward-facing car seat
        Forward-facing harnessed seat to a booster seat
        Booster to an adult safety belt
    The benefits of riding properly restrained.
    Safety in and around the vehicle (never leaving children unattended, walking around the vehicle before moving, etc.)
    State laws and best practice recommendations for occupant safety

Process
The _____ Fire Department provides child car seat inspections 7 days a week by appointment only.
Please contact the _____ Fire Department at ( ) XXX-XXXX to schedule an appointment.
Community members are asked to bring the following items to the inspection:
Vehicle owner's manual
Car seat owner's manual
And, if possible, the child who will be properly secured in safety seat
Please allow 1 hour for the installation and the safety check process.

State Law Reference
Child Passenger Protection Act (reference state Vehicle Code here)

FIGURE 15-8 An example of a standard operating guideline (SOG).
Courtesy of Marsha Giesler.

department members. They remain in effect permanently or until they are rescinded or amended. SOPs are developed within the fire department, are approved by the chief of the department, and ensure that all members of the department approach a situation or perform a given task in the same manner. An SOP provides a uniform way to deal with situations. Although many consider SOGs and SOPs to be interchangeable, an SOP can be thought of as more prescriptive with less room for interpretation. Because an SOP is not formally recognized as a *policy*, many organizations have opted to use the term *SOG* for both types of directives to avoid confusion.

- Rules and regulations are directives developed by various government or government-authorized organizations to implement a law that has been passed by a government body. Rules and regulations can also be considered policy. The rules may be established by a local jurisdiction that sets the conditions of employment or internally within a fire department. They do not allow for any latitude or deviation. A fire department rule requiring all department members to wear their seat belts when riding in vehicles is an example of a rule that is clear in intent with no room for excuses or deviation. If potential reasons for necessary deviation are brought to light during the drafting of the order, that exception should be included and clearly outlined in the rules and regulations.
- General orders are short-term directions, procedures, or orders signed by the fire chief as formal documents that address a specific subject, policy, condition, or situation. The orders can remain in effect for various periods of time, ranging from a few days to permanently. Copies of general orders should be available for reference for all members at all stations.

Writing Policy

When writing any of the previously described policies or directives, it is good practice to involve others in the process. Policies should not be written in isolation. Discuss what should be included in the policy, how its implementation will affect members both positively and negatively, and the need for any clarification. Proofread all policies several times to ensure the verbiage is understandable by anyone reading the document. It will be important to keep an open mind and give due consideration to feedback and ideas.

Unless the policy is something that needs to be implemented immediately, create a draft of the document and provide this version to the members. Once consensus is reached, it is time to submit the document for approval and necessary signatures. Review all policies, procedures, and standards periodically to ensure they continue to serve the needs of the organization and its members.

A sample training policy includes some of the following areas (see also Appendix A, *Resources for Fire and Emergency Services Instructors*):

1. **Purpose and scope of the policy.** The purpose for which the policy is created must be stated, as should the scope of whom it covers.
2. **Instructional personnel responsibilities.** Each member involved in the development, delivery, participation, and administration of the training program should be identified, along with that person's level of responsibility.
3. **Training assignments.** The types of training assignments in which members need to participate should be identified as mandatory, regular, company-level, or department-level training sessions.
4. **Documentation of training.** The requirements for documentation of training should be clearly spelled out in the policy. Signatures, record retention, display of progress and completion records, and access to information contained in the policy should be specified. Evaluation of training activities and the ways in which results are compiled and stored should be outlined.
5. **Conduct policies.** Policies should exist for both instructional staff and department members attending training on their conduct during training such as attendance, appropriate behavior, and safety.
6. **Other policy areas.** Requests to attend external training sessions or schools outside the organization might need clarification by members and officers. These policies should be very specific, including any forms that are used to request participation in training. Performance evaluations and periodic progress reports are another area that should be detailed. If individual performance deviations are identified, personal improvement plans and their follow-through should be included.

Adoption and Implementation

When recommending the adoption of a policy, you may be called upon to provide explanations and justifications for the policy. Be prepared to explain why the policy is needed and the purpose it will serve. In addition, some consideration should be given to how the policy will be implemented.

Personnel Management: Selecting Staff

As the Instructor III, you must be able to select staff for the training program as well as evaluate existing staff. In some departments, the training officer may be responsible for all the steps of human resource management (HR) including hiring and firing employees, but more typically the training officer will be responsible for creating and applying the selection criteria for new and existing teaching personnel, with HR completing the remaining steps. In addition, the training officer will often need to create and administer an instructor evaluation system to provide data for program and instructor development. In some jurisdictions, the training officer may use the results of evaluations to make recommendation for promotions, reassignment, and discharge in conjunction with department employee evaluation programs.

In some larger organizations, a formal process for assignment to the training division may exist. For example, during a recruit academy in-service, officers might be assigned to the training division for the duration of the academy to assist in teaching the new recruits. When openings occur in the training division of some larger departments, there is a bid process for in-service persons who seek to be assigned to the training division for a period of several months to several years. In smaller departments, instructors may be members that have instructional duties in addition to their regular duties. In any case, selecting outstanding instructional staff for a training program or division requires careful consideration and a clearly defined process. The selection process is very important because the policy and procedures must not only comply with all applicable legal and ethical issues; they must also ensure that the persons selected to instruct are well-qualified and a good fit for the position.

Departments have individual processes to identify candidates for instructional positions. Candidates may be chosen from within the organization or from outside of the organization. Before staff for instructional positions can be selected, a job description statement and the basic predetermined qualifications should be stated, such as minimum age, years of experience, and education and certification requirements. In many cases, job descriptions for instructional positions already exist within the organization. The Instructor III will need to obtain and review them before beginning the selection process.

If the job descriptions do not exist, then the Instructor III will need to create them before beginning the selection process. Conducting a job analysis prior to creating the description will pay dividends because the job analysis produces a set of knowledge, skills, abilities, and personality characteristics required to perform a specific job. The goal of the job analysis is to collect as much information as possible about the job to write accurate job descriptions that will meet the needs of the organization and the training division.

A job analysis for an instructor position will require that data be collected from several sources. A primary source will be the content of the courses that the instructor is being brought on board to teach. This should be given significant attention. Instructors must be subject-matter experts in the areas that they will teach in. This information will help to develop criteria for the knowledge and skills that the instructor who will be selected must possess. The physical requirements of the courses the instructor will teach must be examined and stated. One way to collect this information may be to observe existing instructors teaching the class and note what they do. Remember that physical requirements must be job related. In addition to the knowledge of the subject matter the courses will cover, the necessary teaching skills and knowledge must be considered and established as part of job analysis. This should include outlining the teaching technologies that the instructor will be required to use.

The job analysis data is then combined with additional information to create a job description. The job description must also include any federal, state, and local requirements for licenses or certification necessary to teach the course. Other tasks associated with delivering the educational lessons should be listed, for example, if the instructor must set up and clean up props or transport trainees. The education and experience the candidate must have in order to be considered should be specified as well.

Candidate Evaluation

The selection process must be systematic and governed by policy and procedure to ensure consistency and fairness. The goal of the selection process is to match the organization's needs with a highly qualified instructor who adds value to the teaching team and works to achieve the organization's goals.

There are a multitude of methods in existence for evaluating and ranking candidates. Any system that is used should clearly specify the criteria the candidates are ranked against and the relative importance of each of those criteria. Using rubrics or matrices based on the criteria is advisable. Ensure candidates are ranked against the evaluation tool and not each other. Be aware of methods of avoiding bias.

The selection process may include several steps including interviews and evaluation of the candidate's performance on actual or simulated teaching tasks. In the case of existing instructors, a review of their performance in current and former teaching assignments would be included.

An interview can take many different forms, but for the purposes of an instructor selection interview, a structured interview is recommended. A structured interview will have a set of predetermined questions that will be asked, in the same order, to all candidates. Providing a scale on which to base a score/rating will help evaluators who are participating in the hiring process to be more objective during this process. The benefit is clear; all candidates are given the same opportunity to share their information.

Selection Policy and Procedures

Instructor selection that is being conducted for an outside person looking to join an agency or an in-service fire fighter who is seeking advancement within the training division needs to follow established policies and process. Assignments of an instructor to a course or a class also should be made following department policies and procedures. In many larger organizations the assignment of instructors to a specific duty is done by a supervising instructor who oversees a specific program. Having an instructor selection and assignment policy in place helps to eliminate potential issues of discrimination, bias, or favoritism. A selection policy also provides aspiring instructors with clear expectations of what skills to develop and certifications to obtain, as well as an understanding of how instructors are selected and assigned. The effort to create a valid policy is well worth the time because it will save time and effort in the future.

Whether selecting an instructor from the outside to fill an opening in a training division or fire department or promoting someone from within an organization, the instructor selection policy and procedures must consider the impact and importance of legal requirements that affect personnel selection.

At the federal level, laws such as the Civil Rights Acts of 1866 and 1871, 1964, and 1991 prohibit all forms of discrimination in the workplace, including the area of promotions. The Equal Employment Opportunity Act of 1972 expands the scope of the 1964 Civil Rights Act, providing its legal coverage to almost all public and private employers of 15 or more people. The Americans with Disabilities Act and the Age Discrimination in Employment Act may also affect the selection policy and procedure. The Federal Equal Employment Opportunity Commission enforces many of the federal laws regarding employment issues and provides oversight and coordination of all federal equal opportunity regulations, practices, and policies, such as the following:

- Title VII of the Civil Rights Act of 1964 prohibits employment discrimination based on race, color, religion, sex, or national origin. Under Title VII, prohibited discrimination can arise from either disparate treatment or disparate impact (42 USCA § 2000e-5(g)(1); 42 USCA § 1981a(a)(1)).

  Disparate treatment discrimination occurs when a member of a protected class is intentionally treated differently from other employees or is evaluated by different standards. An example would be if an African American instructor candidate was not permitted to retake a failed instructor certification test station, but a Caucasian candidate was. Disparate impact or adverse impact is a result when rules that are applied to all employees have a different and more inhibiting effect on a protected class than on the majority. For example, a rule stating employees teaching ladder operations must be at least 6 feet tall may affect more women than men.

- The Age Discrimination in Employment Act of 1967 provides that it shall be unlawful for an employer to (1) fail or refuse to hire or to discharge any individual or otherwise discriminate against any individual with respect to his compensation, terms, conditions, or privileges of employment, because of such individual's age; (2) limit, segregate, or classify his employees in any way that would deprive or tend to deprive any individual of employment opportunities or otherwise adversely affect his status as an employee, because of such individual's age; or (3) reduce the wage rate of any employee in order to comply with this act (29 USCA § 623).

  However, the Age Discrimination in Employment Act permits age discrimination "where age is a bona fide occupational qualification reasonably necessary to the normal operation of the particular business" (29 USCA

§ 623(f)(1) (1976); *U.S. E.E.O.C. v. City of St. Paul*, 671 F.2d 1162, 1164 (8th Cir. 1982). For example, a position that requires a person to operate machinery can include a requirement that they must be 18 years old if state law prohibits those under 18 from operating machinery.

- Under the Americans with Disabilities Act, "[n]o covered entity shall discriminate against a qualified individual on the basis of disability in regard to job application procedures, the hiring, advancement, or discharge of employees, employee compensation, job training, and other terms, conditions, and privileges of employment" (42 U.S.C. § 12112(a)). A qualified individual is "an individual who, with or without reasonable accommodation, can perform the essential functions of the employment position that such individual holds or desires" (42 USCA § 12111(8)).

The statute itself provides that "consideration shall be given to the employer's judgment as to what functions of a job are essential..." (42 USCA § 12111(8)). The statute also expressly provides that "if an employer has prepared a written description before advertising or interviewing applicants for the job, this description shall be considered evidence of the essential functions of the job" (42 USCA § 12111(8)). The Americans with Disabilities Act does not force an organization to make employee accommodations that would cause an undue hardship on the organization.

The basic understanding of these laws serves as a guide to anticipating issues and asking the right questions of the legal professionals and employment law experts. To gain a better understanding of each of these laws, the training officer can review the case law and ask questions.

Whichever processes are used, the overall goal in hiring and recruiting in volunteer and career departments is to ensure that the training department's most important resource, its personnel, are selected in such a way that the new instructor has a good potential of becoming a valuable part of the team.

An important element that also should be taken into consideration when selecting an instructor is the instructor's mindset toward the department's mission purpose and a safety culture. Instructors that create an environment or atmosphere that fosters conduct that does not fit a department's image or reflect a safety mindset are a liability to the program and the department. Training should provide feelings of satisfaction at achieving hard-to-reach goals and team-building that fosters relationships that will last a lifetime—but not at the cost of abuse or endangering students with the attitude "I had to do it as a rookie and so will you." The fire service has a sad history of instructors that have pushed students beyond what a reasonable person should and, in the end, it has cost the life of a student, destroyed the career of an instructor, and brought a department unwanted attention.

Performance-Based Evaluation Plans

A critical element of all training programs is having a plan for evaluating the instructors' performance within a program or division on a regular basis. The information gathered by evaluation of instructors is used not only for instructor development efforts but also for program modification and development.

An evaluation plan details the ongoing system to evaluate instructors. Plans should include the following:

- Timing: Instructor evaluation can take place at different times during a course, such as when a new instructor is giving his or her first presentation or when a seasoned instructor finishes the end of a course. A cycle of when instructor evaluations are completed should be established and followed for consistency.
- Justification: Instructors should be told why the evaluation is being administered.
- Criteria: Set evaluation criteria should be established by policy and known by both the evaluator and the instructor.
- Data sources: The policy should identify the sources of data for evaluations. In addition to direct observation, surveys, questionnaires, and graded scales can be included as indicators of an instructor's ability to deliver a specific presentation, connection with the audience, subject-matter knowledge, and use of materials.
- Data management: The policy should also include how the results are to be used and who should see the results. Confidentiality in the system is critical. Instructor evaluations are personnel records and as such policy must be clear on who has access to them and how they are stored.

A sample method of constructing an instructor evaluation plan might appear as follows:

1. Review the job description duties of the instructor.
2. List the desired criteria to be evaluated.

3. Assemble a form or spreadsheet to collect data.
4. Review the evaluation plan with a supervising officer and gain approval for it.
5. Review the evaluation plan with instructors and inform them of its purpose.
6. Conduct a review of various courses and provide feedback with recommendations for improvement and acknowledgment of positive performance indicators. The instructor evaluation plan is critical to successful management of a training program.

For company-level instruction, it is not a common practice to have on-duty or on-call personnel complete written evaluations of instructor performance. As a program manager or training officer simply participating in the drill or in-service program, you can make personal notes on the instructor's performance and provide constructive feedback in appropriate areas. The training officer is responsible both for providing on-the-spot instructor performance reviews during training he or she witnesses and for supporting and encouraging instructors over the long term. This is similar to what company officers would do for members they are responsible for. This ongoing evaluation of in-service training instructors should not be overlooked and should be included in the overall evaluation of the training program.

There are many ways to construct the evaluation instrument. In some cases, a simple numeric system is used to rate an instructor's performance on a scale of 1–10. Other systems use descriptors such as "needing improvement," "meeting expectations," or exceeding expectations."

Regardless of the system being used, it is important that the training officer provide the evaluators (who may be Instructor IIs) with training on how to complete the evaluation process. One of the most important aspects of any evaluation program is consistency in evaluating instructors. This is especially true when multiple instructors are serving as evaluators. Citing specific examples of both negative and exceptional behavior noted during the evaluation is a good way of validating the evaluation given.

When properly implemented, instructor evaluations can be a useful tool in improving program and instructor performance. When multiple evaluators provide input on evaluations it is still best to have just one individual deliver the evaluation feedback to avoid the impression of ganging up on the instructor. In some cases, instructors are asked to self-evaluate first and then submit their completed forms to the training officer for review and additional input.

Because evaluations are a part of the personnel record, they should be completed in an open and honest manner. More than one department has found itself in a difficult situation when trying to formally discipline an instructor for bad behavior only to have that same instructor refer to multiple positive evaluations on file and ask, "Why have I not been told of these issues before?" With incomplete records, an instructor may challenge a disciplinary action, potentially placing the department in a compromised position.

The evaluation process should be approached from a positive frame of mind and not be used as discipline. The evaluation is a chance to point out both positive instructor behavior as well as areas needing improvement. In cases where instructor performance is not up to department standards, the evaluation feedback should focus on reviewing positive performance and exploring ways to improve the negative or below-standard performance. An emphasis should be placed on correcting any deficiencies noted with a training officer's commitment to assist the instructor in any way possible. It is a good idea to involve the HR staff as early as possible in the case of instructors with poor performance or issues that may lead to disciplinary proceedings.

Instructor evaluations can also be used to mentor the exceptional instructor. In these cases, the evaluation review can focus on setting a career path and discussing opportunities for expanding the instructor's horizons. Additionally, the training officer might be able to suggest how the instructor can further his or her education, training, and experience in preparation for promotional opportunities.

The training officer must standardize the performance-based evaluation plan and process of instructor evaluations in written policy to ensure fairness and the quality of the instructor and program feedback provided by the evaluations. Policies should also set the tone of the evaluation program to ensure that the evaluation and monitoring process is not viewed by the instructors or the training officer as antagonistic or punitive, but rather as a two-way process focused on instructor development to advance the skills and knowledge of the entire cadre of instructors and the program as a whole.

Budgets

As an Instructor III, responsibility expands beyond basic budgeting concepts to preparing and justifying a training budget, as well as preparing purchasing specifications for equipment. Chapter 12, *Program Management and Training Resources*, introduces basic

Job Performance Requirements in ACTION

The management of a training program requires the creation of policies and procedures that direct all members, including the training staff, on the management of instructional resources, staff, facilities, records, and reports that will help determine staff selection and better define instructional responsibilities. Additional requirements of this position include knowledge of methods of constructing instructor evaluation plans and the way in which a training program manager should evaluate equipment and resources used for training delivery.

Instructor I	Instructor II	Instructor III
As part of the instructional team, manage resources, time, and materials efficiently to conserve available funds while providing quality training.	When developing training materials and completing training reports and records, ensure that they are all completed according to department guidelines and practices.	Create policies and manage the training division so that effective training that meets the agency goals is delivered and all applicable compliance issues are met. Instructor selection and performance reviews should be conducted on a regular basis to ensure quality delivery of training.
JPRs at Work	**JPRs at Work**	**JPRs at Work**
There are no JPRs at this level within this chapter.	The Instructor II will plan and budget for training resources. This will include deciding when replaced equipment can be used instead of newer in-service equipment for a course or training session.	The Instructor III will create the policies and procedures for the management of instructional resources, staff, facilities, records, and reports for the training program. Methods of constructing instructor evaluation plans will be a critical function at this level. Knowledge of the methods used to evaluate equipment and resources used for training delivery will allow the Instructor III to present training program evaluation findings to various audiences.

 Bridging the Gap Among Instructor I, Instructor II, and Instructor III

All levels of instructors should be considered members of the training division team. Program management functions will include many recordkeeping and quality assurance reviews and measures. All instructors should complete records and reports thoroughly and correctly.

budgeting concepts and making budget recommendations as an Instructor II.

Recall that a budget is used as a tool for implementing the policies established by the governing body and the goals of the organization. If training has a high priority within the department and within the local government, this will be reflected in the level of funding for training placed in the fire department budget. The opposite is equally true. By laying out the needs of the training department in a systematic manner, a fire department training officer and administrator can show the public, the department leadership, and the members of the governing body what the fire department will require to achieve its training goals. A truly effective budget will also provide the reasons that a specific level of funding is needed to fulfill the fire department's mission, goals, and objectives.

It is also important to stress that a budget can serve as a guide for organizational decision makers, allowing the decision makers to tailor their operations to the available level of funding. In this way, resources can be allocated on a periodic basis. A well-run organization that requests a specific amount of money and then uses it wisely, according to predetermined criteria, will gain great credibility with local government. This credibility will work to the benefit of the fire department. Over time, the department's requests will be accepted as factual and reliable. Development of this type of trust is a critical part of the chief officer's responsibilities, and a training department that supports chief officers in this endeavor will assist not only the chief but entire department as well.

Budget Categories

Two basic categories of **expenditures** are included in typical municipal fire department budgets: the operating budget and the capital budget.

Operating budget items include salaries, supplies, payments for utilities, and other day-to-day operating expenses. The **operating budget** (sometimes called the expense budget or the general fund budget) is a plan for future operations, expressed in financial terms. Basically, formulating a budget involves asking what a fire department intends to accomplish during a given year and how much those activities will cost. A budget allocates financial resources to the different uses for which the money is intended. Because exact amounts cannot be known in advance, a budget reflects a forecast of what will be needed and thus a decision on how much money should be made available for each function. At the same time, the budget forecasts how much will be spent for a particular use.

Capital budget items for a fire department include equipment, buildings, and vehicles. Capital budget items for training may include simulators, props, software and materials, and equipment with a longer life span than 1 year.

When preparing capital budgets, department management must consider long-range need and the fact that, to some extent, capital and expense budgets compete for funds. To purchase all new items in 1 year would be impractical. Their expense would cut into a department's ability to fund other important activities, such as training or perhaps even salaries.

Budget Organization

Budgets may be organized or approached several ways. The most common are line item and program, but there are others. Line item is the most often used approach to budgeting because it is easy to understand and provides significant control. Budget items are divided up by department or cost centers (**TABLE 15-3**).

The line-item budget provides targeted budget figures that allow for better control and management of budgeted funds. Budgets may also be set up by a specific division. This control feature indicates where the dollars in each line-item area are going.

Some departments may use a program budget approach instead of a line-item budget.

In a **program budget** system, expenditures are allocated for specific activities such as training. A department estimates the total expense—including items such as salaries and materials used—for performing a specific function over a given period of time. This estimated amount becomes the basis for measuring actual expenditures for that function. As the year progresses, the department keeps a record of the actual money spent. Officers directly involved with the budget can keep track to see how adequate the budget is for the department's needs and functions, in preparation for next year's budgetary estimates.

Budget Process

An important aspect of leading and managing a training program is the management of the budget. It is a simple fact that to train fire fighters and medical responders to do their jobs it takes equipment and material and that costs money. Proper resource identification, cost analysis, and budget development are all part of the tools necessary for the management of a training program or a training division. A training officer must understand the budget process and how it is undertaken within their AHJ.

TABLE 15-3 A Sample Line-Item Expense Budget

Account Number	Classification	Current Year ($)	Previous Year ($)
001	Maintenance	88,000	75,000
002	Office supplies	12,000	10,500
003	Fire hose	57,000	46,000
004	Radio repairs	15,000	13,500
005	New radios	40,000	15,000
006	Janitorial services	55,000	48,000
007	New computer programs	15,500	—
008	Firefighting equipment	157,000	137,000
009	Breathing apparatus	90,000	84,000
010	Salaries and benefits	10,500,000	9,330,000
Total		11,029,500	9,759,000

The budget process consists of essentially four steps:

1. Formulation
2. Transmittal
3. Approval
4. Management

During the *formulation* phase, fire department managers review past budgets to see how well estimated outlays compared to actual expenditures. They focus on predicting for the future. During this time, suggestions and support information are solicited from each division, bureau, or level of the department so that all ideas can be considered. It is in this phase that the training officer will be called on to prepare a preliminary budget for the training department or training functions.

After the preliminary budget has been drafted, it goes through the *transmittal* process wherein it is reviewed by all interdepartmental stakeholders for final consideration. The fire chief or financial officer must be prepared to answer any questions surrounding the proposed budget.

The next phase is *approval*, which may actually involve several steps before it takes effect for the upcoming fiscal year.

The final step is *management* of the approved budget. Those entrusted with management of a budget section must carefully monitor the spending to prevent excessive overspending or underspending compared with the allocated funds. In either case, a strong deviation from the projected budget can lead to credibility problems for future budgeting. A training officer should work closely with the department administration team in budget development (**FIGURE 15-9**).

Schedules and cycles will differ in accordance with the city or the fire district's adopted calendar. Within each schedule are target dates for submissions, reading, and approvals. The training officer must be aware of these dates and complete all necessary information in accordance with this deadline. Missing the deadlines can result in the loss of necessary line-item requests and future budgeting allocations. Typical to most budget schedules are deadlines for the following:

- Budget, estimates, capital, and goals/highlights
- Draft budget submission to fire chief, village manager, city council, or other governing body
- Budget first reading
- Budget public hearing
- Adoption of budget
- Budget implementation
- Approval

FIGURE 15-9 A training officer should work closely with the department administration team in budget development.
Courtesy of Marsha Giesler.

Remember that the budget cycle and schedule varies from one jurisdiction to another. Learn about the procedures specific to your department and follow those timetables.

Budget Skills Application

Assume the fire chief has asked each division to submit a proposed budget for the upcoming year. He plans to review the draft budgets and will meet with each supervisor to discuss approval or necessary adjustments. Identify and analyze any and all projected budget needs. You will need to collect, organize, and categorize the necessary equipment, materials, personnel, and ancillary costs for all training initiatives for the upcoming year. The use of a budget spreadsheet can be a valuable tool in developing a budget request. A budget spreadsheet contains several columns that show how much money was budgeted in the previous year, how much was actually spent, how much is requested for the current year, how much difference (+/−) there is between the amount requested this year and the previous year's request, and year-to-date spending.

Budgetary Justifications

Training officers often must justify the expense of an item. There are two aspects to such justification:

- Showing need for the item: Showing the need may be obvious, such as replacement of a piece of apparatus that has been destroyed, or the growth of the served area. With nonessential items (e.g., updated computer programs), the need can be shown by demonstrating the benefits in productivity gains.
- Showing why the requested funds are the best way to satisfy the need: Proving the selection of the specific item may be done by providing a comparison with the costs of competing options. If the requested item is the least expensive, that fact could be highlighted. If it is not the least expensive, officers could provide a justification with the reasons for the request (e.g., familiarity with a product or vendor, compatibility).

Department budgets are impacted by competing community needs, local economic trends, emergency disasters, and state, national, and global trends. Defending a training budget can be a very challenging task. This is why it is so important to understand the total budget process, thoroughly document all evaluation results, define cost allocation breakdowns, and research possible options and alternative sources of funding and program delivery. It is imperative that the program impact and outcomes measures be reported with a direct correlation drawn between the program's results and the dollars spent by an organization. Program managers and administrators must consistently seek to build quantifiable and qualifiable evaluation methods into program efforts.

Purchasing Guidelines and Policies

Purchasing rules vary widely among different governments and their agencies. Failure to follow these rules can lead to serious consequences. A purchasing policy is typically part of an organization's overall operations handbook. It will describe the purpose for the policy, restrictions placed on employees, responsibilities of purchasing department employees, and other specific procedures or processes. The policies may regulate the following:

- Qualifications of bidders
- Selection and list of approved vendors
- Approval process
- Approval limits and purchase ranges for various commodities and services
- Limits on purchases made without a formal bidding process
- Procedures for petty cash, quotes, bids, requests for proposal (RFPs), contract extensions, and change orders
- Policies for emergency procurement during a disaster incident
- Procedures for inventory of property or equipment
- Required approval for capital items or travel
- Standard service contracts

Purchasing policies are available for review by department members and city employees. The training officer should be familiar with how their requested purchase for equipment, personnel, and services fit within the established ranges and processes.

It is important to remember that restrictions exist on how public funds are spent. Many states and municipalities have published bid laws that require competitive bidding to ensure that the lowest possible price is received. Care must be exercised in drawing up purchase specifications or RFPs so that they conform to all legal and procedural requirements.

Request for Proposal

An RFP is a solicitation document requesting submittal of proposals in response to a scope of work. RFPs are used as an objective method of contracting for goods or services whereby formal proposals are solicited from qualified vendors. RFPs are generally used to select a vendor for a professional or consulting service, or to implement a system or program. Occasionally, RFPs are used for commodities, such as computers. If an RFP is your preferred method of procurement, this RFP process must be followed.

An RFP should not be used when the service or equipment to be contracted is a standard or common off-the-shelf item. An RFP should not be used if there is an industry standard associated with the service or commodity to be contracted. RFPs are a very time-consuming and potentially costly method of procurement. As such, they should be used only when sealed bidding is not appropriate. An RFP does not need to specify in detail every aspect of how to accomplish or perform the services required.

The bidding process should begin with a pre-bid conference at which the purchasing officer can discuss the department's needs with potential vendors. Getting questions answered this early in the process provides a better understanding of the department's expectations and can eliminate confusion later. Bids are evaluated based on how well they meet the department's needs. Once a bid is awarded, the purchasing officer and approved vendor must agree on the procurement process (whether constructing a new station, building new apparatus, etc.). If the purchase is apparatus or equipment, training on correct usage will also be necessary and should be part of the process.

Writing Specifications

There are times when the training officer will be called upon to prepare specifications for an item or service to be purchased. Purchase specifications are used when goods or services are procured by bid. A purchase specification is a very detailed description of the item and a detailed description of the requirements the vendor selling you the item must meet. In addition, it describes all the details of the transaction, for example warrantee information and how the item is delivered. The description might specify performance or design criteria or both.

Purchase specifications typically contain these essential elements:

- A description of the item to be purchased
- Quantity of the item to be purchased
- Physical specifications
- Performance specifications
- The minimum acceptable quality of the item to be purchased
- Documentation that is to be provided with the purchase
- Services provided at delivery
- Terms of acceptance
- Terms of payment
- Shipping information
- Warrantee information

A good purchase specification is clearly written and provides necessary detail for vendors to determine if they should bid, while at the same time encouraging competitive bidding. The goal is to acquire the best product or service to meet the organization's need at the lowest price. To write good purchasing specifications the training officer needs to start with a detailed understanding of the need that is driving the desired purchase. If the need is clearly understood it is more easily transferred into specifications the product or service must meet. Writing technical documents with legal effect such as purchase specifications is a difficult task and should not be undertaken lightly. It is good practice to seek assistance from department administration and legal advisors.

Purchasing Procedures

Clear and established purchasing rules are the backbone to an easy and efficient program. Whatever systems and programs are put into place, the rules need to be clearly defined and follow all department policies and municipality, state, and federal laws. There are many ways for a department to purchase goods, services, and materials. A training officer should become familiar with mechanisms that have proved to be successful tools (**TABLE 15-4**).

Purchasing Process

During a course development process, a needs assessment is conducted to determine what equipment

TABLE 15-4 Purchasing Systems

System/Mechanism	Description
Purchase of goods	Products, supplies, and equipment are purchased through competitive processes using the bid thresholds as a guide.
Purchase of services	Consultant, professional, and technical services are often obtained through RFPs, which consider skill and experience as well as cost in the evaluation process.
Cooperative purchasing agreements	The city utilizes available state, university, and federal contracts at its discretion.
Sole-source negotiation	Also called noncompetitive negotiation, this method may be used when competition does not exist or would not be in the best interests of the municipality. An example would be scientific equipment manufactured by only one vendor.
Contract methods	There are numerous systems that have been created using contracts for purchasing purposes. Many of these contract orders are established through the bidding process. Once contracts are awarded, city agencies can use limited purchase orders and purchasing cards to purchase miscellaneous items or predetermined items on the contracts directly from vendors.
Request for proposal (RFP)	Professional service contracts over designated amounts are established by means of an RFP process and typically require a board of directors or some level of municipal approval.
Purchase orders	Purchase orders are legally binding documents typically issued by the finance officer of an agency or municipality; they specify the goods or services being ordered or purchased from a vendor for a specific price. Purchase orders are issued for either one-time purchases or for blanket requests when a department will be making purchases periodically against the purchase order throughout the fiscal year. The finance officer typically prepares and processes the purchase orders based on the information provided on the requisitions and other attached documentation submitted by departments through a purchasing system. If a contract or vendor agreement of any type is required by the vendor, it is the responsibility of the requisitioning department to have the agreement reviewed by counsel or its designee. If there is to be a competitive bid process for the purchase, the finance officer will work with counsel in the review.
Blanket purchase orders	A blanket purchase order is issued when it is anticipated that multiple purchases will be made to a vendor for routine goods or services over a specified period of time. A blanket purchase order cannot extend beyond the end of the current fiscal year. After its acceptance by the vendor, purchases may be made against it periodically as required without calling for new purchase orders. For the department to begin using the blanket purchase order, the finance officer transmits the completed blanket purchase order to the auditor/controller for encumbrance against the department's budget. After the blanket purchase order is encumbered, it is transmitted to the vendor. The department can then begin making purchases with that vendor.
Cash reimbursements	At times, a purchase may need to be made from a company that does not accept credit cards or purchase order numbers. An example of this might be during an emergency incident when food supplies are needed to feed crews during a long-term incident. The employee would submit a reimbursement form through his or her supervisor to the finance office for payment.

would be necessary to support and deliver the desired training. This analysis identifies what equipment currently exists and what is needed to support the delivery of the training. An item identified could be the need for a new projector or the need for a training tower; regardless of the size of the item, the process for obtaining the item is the same.

The first step is to evaluate what is needed. For example, if looking to purchase a new projector you would need to research various types and brands of projectors that would meet your needs. This research can be done on the internet, by speaking with other agencies about equipment they use, or by speaking with vendors who can provide in-depth information on their product. Part of this process also includes determining the cost for the projector, service costs, and possible warranties for the projector.

After identifying a specific projector that meets the needs of your training program, you need to write an equipment purchasing request form to purchase the projector. Agencies that are funded and supported with tax dollars are required to follow the purchasing process for the agency. In most states the purchase process is identified and outlined in state or local statues that follow accepted accounting practices. These guidelines are developed to ensure the correct use of public funds and to track all purchases to maintain accountability for the expense.

In writing the purchase request there is the need to be as detailed as possible in what the funds will be used to purchase. In the case of purchasing the projector, a product item number with specific detail would ensure that the correct projector is identified. Depending on the cost of the item, many agencies require that there are two or three proposals of where to purchase the item. In many cases the submitter of the purchase request will do the research to find possible bidders, or this step may be completed by the agency's purchasing office. The purpose for this step is to make sure that the agency is being cost effective in the use of public funds and that there is no question of any ethical wrongdoing by the agency.

Once the purchase request is submitted to the appropriate purchasing agent for the fire department, the projector will be bought and delivered to the agency that submitted the request. Upon receipt of the projector, the person who requested the purchase verifies that the correct item was received and makes notification to the purchasing department that the order was complete and the correct projector has arrived. The key to the purchasing process is that you know how to purchase an item for your agency and what steps are required to obtain equipment and resources to deliver training to your members.

It is important to remember that a training officer has responsibilities in the management of financial resources. Training officers are expected to provide specific input to the planning of each year's budget. As managers, they must also understand what budgets are and how the budgeting process works. They must fully understand their own operational budgetary requests so they are prepared to answer any questions that arise from the members of the local government leadership team.

Understanding all the elements it takes to successfully plan and implement a training budget will help the agency achieve its strategic plans, thus demonstrating fiscal responsibility and the ability to properly manage and operate one of the community's greatest assets.

Training BULLETIN

JONES & BARTLETT FIRE DISTRICT
TRAINING DIVISION
5 Wall Street, Burlington, MA, 01803
Phone 978-443-5000   Fax 978-443-8000

Instant Applications: Program Administration

Drill Assignment

Apply the chapter content to your department's operation, training division, and your personal experiences to complete the following questions and activities.

Objective

Upon completion of the instant applications, fire and emergency services instructor students will exhibit decision making and application of job performance requirements of the instructor using the text, class discussion, and their own personal experiences.

Suggested Drill Applications

1. Determine the types of training resources that will be necessary for a joint fire/EMS training program and write purchasing specifications according to your local purchase policy.

2. Determine the most appropriate type of training record system for a fire/EMS training program, including an implementation plan.

3. Construct performance-based instructor evaluation plans for a training program and determine how they will be incorporated into the program.

4. Review the Incident Report for this chapter and be prepared to discuss your analysis of the incident from a training and policy development perspective and how the prevention of such incidents can be addressed through the development of policies and procedures for training programs involving high temperature/high stress evolutions.

Incident Report: NIOSH Report # 2012-08

Drill Assignment

Review the information in this incident report and prepare a practice presentation for class delivery at the direction of your instructor. Your presentation should include a summary of the incident facts and identify the NFPA standard(s) that could apply to the incident. Use an outline to organize your thoughts. You may be evaluated on your communication skills during your presentation.

Abbotsford, Wisconsin

On March 4, 2012, a 34-year-old male volunteer lieutenant (the victim) lost his life at a theatre fire after the roof collapsed, trapping him within the theatre. At approximately 1215 hours, an on-duty patrol officer (also chief of the victim's fire department) radioed dispatch for a structure fire (flames visible). The 1st due fire department arrived on scene, set up operations on the A-side of the structure, and directed the incoming mutual aid department (victim's department) to the rear of the structure. No fire was visible from the rear. Both departments attacked the theatre fire from opposite sides (A-side and C-side) of the structure establishing their own incident commander/officer in charge, fire-ground operations, and accountability systems. The 1st due fire department initially fought the fire defensively from the A-side, while the victim and two additional fire fighters (FF1 and FF2) entered through the C-side, advancing a hoseline until they met A-side fire fighters near the theatre's lobby (area of origin). The 1st due fire department eventually placed an elevated master stream into operation, directing it into the lobby and then onto the roof while fire fighters were operating inside. Roof conditions deteriorated until the roof collapsed into the structure trapping the victim, FF1, and FF2. FF1 and FF2 recalled speaking with the victim immediately following the collapse, but nothing was heard from the victim following the activation of a personal alert safety system device (PASS). All three were eventually located, removed from the structure, and transported to a local hospital, but the victim had already succumbed to his injuries.

Postincident Analysis Abbotsford, Wisconsin

Key Recommendations

- Recommendation #1: Fire departments should ensure that an effective incident management system is established with a designated incident commander not involved with fire suppression activities.
- Recommendation #2: Fire departments should ensure that a complete situational size-up is conducted on all structure fires.
- Recommendation #3: Fire departments should use risk management principles at all structure fires.
- Recommendation #4: Fire departments should work together to develop mutual aid standard operating procedures for fire-ground operations that support interagency operability and accountability and train on those procedures.
- Recommendation #5: Fire departments should ensure that the incident safety officer (ISO) position, independent from the incident commander, is appointed and effectively utilized at every structure fire meeting the requirements within NFPA 1521: *Standard for Fire Department Safety Officer Professional Qualifications*.
- Recommendation #6: Fire departments should ensure that a rapid intervention crew (RIC) is readily available, on scene, and prepared to respond to fire fighter emergencies.

After-Action REVIEW

IN SUMMARY

- Training record systems are one of the most important management functions assigned to the training officer. A record system must be developed by the Instructor III, who may function as the training officer or training division chief, to document all aspects of the training program.
- Periodically, the training division must compile and submit reports on the activities, inventory, statistics, and successes of the division. These reports can be used for budget justification, equipment purchasing, and many other useful management functions.
- Generally speaking, training reports—like other reports submitted to a supervisor or administrative body—are considered technical reports, and specific guidelines for their creation may exist.
- Training reports should be clear and concise and follow specific organizational steps. A logical sequence of the information contained should be followed throughout each report, and the same sequence should be used whenever the report is used.
- As the Instructor III, you must be able to select staff for the training program as well as evaluate existing staff. In some departments, the training officer may be responsible for all the steps of human resource management including hiring and firing employees, but more typically the training officer will be responsible for creating and applying the selection criteria for new and existing teaching personnel, with HR completing the remaining steps.
- The training officer will often need to create and administer an instructor evaluation system to provide data for program and instructor development.
- The candidate selection process must be systematic and governed by policy and procedure to ensure consistency and fairness. The goal of the selection process is to match the organization's needs with a highly qualified instructor who adds value to the teaching team and works to achieve the organization's goals.
- The selection process is very important because the policy and procedures must not only comply with all applicable legal and ethical issues; they must also ensure that the persons selected to instruct are well-qualified and a good fit for the position.
- As an Instructor III, responsibility expands beyond basic budgeting concepts to preparing and justifying a training budget, as well as preparing purchasing specifications for equipment.
- A budget can serve as a guide for organizational decision makers, allowing the decision makers to tailor their operations to the available level of funding. In this way, resources can be allocated on a periodic basis.
- A well-run organization that requests a specific amount of money and then uses it wisely, according to predetermined criteria, will gain great credibility with local government. This credibility will work to the benefit of the fire department.

KEY TERMS

Access Navigate for flashcards to test your key term knowledge.

Capital budget A budget that is used to purchase an item that would last for several years such as a fire engine.

Expenditures Monies spent for goods or services that are considered appropriations.

Operating budget A budget that is used to pay for the day-to-day expenditures such as fuel, salaries, or fire prevention flyers.

Program budget A system in which expenditures are allocated for specific activities such as training.

REFERENCES

National Fire Protection Association. 2015. *NFPA 1521: Standard for Fire Department Safety Officer Professional Qualifications.* Quincy, MA: National Fire Protection Association.

National Fire Protection Association. 2017. *NFPA 1401: Recommended Practice for Fire Service Training Reports and Records.* Quincy, MA: National Fire Protection Association.

National Fire Protection Association. 2018. *NFPA 1403: Standard on Live Fire Training Evolutions.* Quincy, MA: National Fire Protection Association.

National Fire Protection Association. 2019. *NFPA 1041: Standard for Fire and Emergency Services Instructor Professional Qualifications.* Quincy, MA: National Fire Protection Association.

CHAPTER PRESENTATION EXERCISE

Instructions: Your course instructor will assign individual or group discussions on the key points and teaching tips of this chapter. You or your group will review the chapter teachings and identify the major learning points from the chapter. You should be able to discuss the points, why they are important, and how/where they apply to your responsibilities at your level of instructor training.

REVIEW QUESTIONS

1. An evaluation plan details the ongoing system to evaluate instructors. List and describe some of the aspects that should be included.

2. List the basic information that should be kept in a state certification record file and in all types of training files.

3. For instructor selection interview, what type of interview is recommended and what format should it follow?

SECTION 4

Live Fire Instructor and Live Fire Instructor-in-Charge

CHAPTER **16**   **Live Fire Instructor Introduction**

CHAPTER **17**   **Live Fire Instructor**

CHAPTER **18**   **Preburn Inspection and Planning**

CHAPTER **19**   **Conducting Burn Evolutions**

CHAPTER **20**   **Post Live Fire Evolution**

CHAPTER 16

Live Fire Instructor Introduction

NFPA 1041 JOB PERFORMANCE REQUIREMENTS

7.1.1 For qualification at Live Fire Instructor, the candidate shall meet the requirements of Fire Fighter II as defined in NFPA 1001 or Interior Structural Fire Brigade Member as defined by NFPA 1081, the requirements of Fire and Emergency Services Instructor I as defined in Chapter 4, and the job performance requirements defined in 7.2 through 7.3 of this standard.

7.1.2 A Live Fire Instructor shall demonstrate competency in knowledge and skills in all subjects, methods, and equipment being taught and the objectives contained in NFPA 1403 and identified for the live fire evolution.

KNOWLEDGE OBJECTIVES

After studying this chapter, participating in a structured learning environment, and completing assigned assessments, you will be able to:

- Describe the purpose of NFPA 1403: *Standard on Live Fire Training Evolutions*. (**NFPA 1041: 7.1.2**, p 373)
- Describe the additional NFPA standards that affect live fire training. (**NFPA 1041: 7.1.2**, pp 373–374)
- Define live fire as used in NFPA 1403. (**NFPA 1041: 7.1.2**, pp 372–373)
- Define live fire instructor and live fire instructor in charge as used in NFPA 1403 and NFPA 1401. (**NFPA 1041: 7.1.2**, p 376)

You Are the Live Fire Instructor

Your chief has asked you to set up a live fire training evolution using an acquired structure. The two closest fire departments that run mutual aid with you have been invited to participate. The acquired structure must be burned within 7 days. Your department is relatively small with 20 certified fire fighters and two engines. The acquired structure has been damaged by a previous fire, which your department contained to one room. The first thing to determine is whether there is enough time to remediate any building hazards, such as asbestos, and next, to determine whether there is enough time to develop training objectives and scenarios. When you talk to the mutual aid companies, they are eager to participate; however, some of their personnel are in the middle of training and are not yet certified. They are also not sure what apparatus they can provide.

As you start to plan, decide what the intended learning outcomes are. Also, is live fire training the most appropriate environment for the training based on participant experience, intended learning outcomes, and risk–benefit assessment? For example, teaching hose handling or primary search are more effectively and safely accomplished in a non–live fire environment.

From previous experiences with the department, you have had concerns regarding their personal protective equipment (PPE) not fitting properly and being worn out. Before the day of the drill, you must determine that it meets the current standards.

As you consider the possibility of conducting this live fire training evolution, there are many questions that must be answered (some of these require knowledge of other National Fire Protection Association [NFPA] standards):

1. Will there be enough resources to provide an adequate water source to conduct the live fire training?
2. Do the participants have the training needed to participate in the live fire training evolution?
3. Have the participants been medically screened for response/training?
4. Is there enough time to adequately prepare for this live fire training evolution?
5. Are there enough qualified live fire training instructors and safety officers?

Access Navigate for more practice activities.

Introduction

Live fire training is among the most important training we can do as fire fighters. The training of fire fighters to fight fire, to understand fire dynamics and the conditions they will encounter, and to know how to successfully operate in a hostile environment are among the most important aspects of training. **Live fire** training is also vital for new and existing fire fighters to hone their skills or learn new ones. As an instructor conducting live fire training exercises, you are responsible for the value of the training and the safety of all participants involved. Instructors, in-service fire fighters, and students learning to become fire fighters have died or have been severely injured during training evolutions. Achieving a balance between participant safety and training effectiveness can be a difficult task (**FIGURE 16-1**).

There are thoroughly written standards that relate the requirements of conducting live fire training evolutions. The National Fire Protection Association (NFPA) has published one of the most popular of these standards, NFPA 1403: *Standard on Live Fire Training Evolutions*. The purpose of NFPA 1403 is to provide

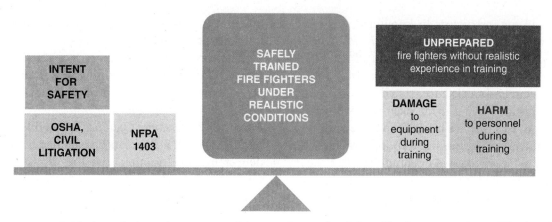

FIGURE 16-1 Achieving a balance between participant safety and training effectiveness can be a difficult task.
Courtesy of Dave Casey.

a process for conducting live fire training evolutions to ensure that they are conducted in a safe, effectively organized manner and that the exposure to health and safety hazards for the fire fighters receiving the live fire training is minimized. NFPA 1403 was designed to set standards on what should be done to mitigate the inherently dangerous conditions of live fire training. A live fire instructor must thoroughly understand the content and application of the 1403 standard.

The Importance of NFPA 1403

Realistically, in-context training is becoming more important because many fire departments are currently experiencing large numbers of retirements, with those replacing the retirees typically having less fire suppression experience. Added to this loss of experience is the significant increase in average home sizes (more than doubled in size since 1973) coupled with significant changes in construction and contents that adversely change fire behavior and increase the danger of internal fire spread and collapse (Perry, 2014). According to NFPA statistics, since the mid-2000s, the number of structure fires has remained relatively unchanged, while the fire fighter death rate at structure fires has increased (Fahy, LeBlanc, and Molis, 2016).

As with so many safety issues like the cost of new, more expensive turnout gear and self-contained breathing apparatus (SCBA), full-size structural and exterior firefighting props can be an expensive burden, especially to smaller, cash-strapped agencies. There are cost-effective means to provide realistic training to better prepare fire fighters. NFPA 1403 provides the structure to conduct live fire training in a manner that as closely as possible provides students with valuable experience to be used when responding to an actual structure fire.

The History of Live Fire Training Evolutions

According to the National Institute for Occupational Safety and Health (NIOSH, 2018), during the period from 2001 to 2014, approximately 11 percent (150 out of 1396) of fire fighter line-of-duty deaths were training related. From 2001 to 2015, 81 training-related fatalities were investigated by NIOSH through the Fire Fighter Fatality Investigation and Prevention Program. Of these fatalities, 55 percent were heart attacks, 28 percent were trauma related, and 17 percent were related to cardiovascular disease and other events. These investigations included 30 deaths due to physical fitness activities, 36 deaths due to apparatus/equipment drills, 14 deaths due to live-burn exercises, 8 deaths due to dive training, and 11 deaths due to other training-associated circumstances including travel to and from the site (**FIGURE 16-2**).

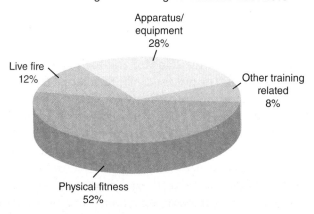

FIGURE 16-2 NIOSH training-related fire fighter fatalities, 2001–2015.
Data from the National Institute for Occupational Safety and Health (NIOSH)

The need for instructors to know and apply the standard is critical. In a 2003 edition of *NFPA Journal*, Texas Fire Chief Bill Peterson, the chair of the NFPA Fire Service Training Technical Committee during much the standard's existence, stated, "We're seeing the same mistakes being made over and over again. The reality of it is this: we haven't found new ways to kill fire fighters during live fire training—we're doing it the same old way" (quoted in Paradise, 2003).

NFPA 1403: *Standard on Live Fire Training Evolutions in Structures*, was released 30 years ago. The vast majority of serving fire fighters started after the standard was adopted. It was revised in 1992, and in 1997, NFPA 1406: *Standard on Outside Live Fire Training Evolutions* was merged with NFPA 1403 to create the *Standard on Live Fire Training Evolutions* that we know today. It was revised again in the years 2002, 2007, 2012, and 2018.

In a 2007 edition of *Firehouse* magazine, Dominic Colletti stated, "Since NFPA 1403 was adopted in 1986, and with more than 30 years of additional experience with the concept of Live Fire Training, no fire fighters or instructors who have participated in Live Fire Training conducted in accordance with the requirements of the standard have ended up in the NFPA or USFA fire fighter death and injury statistics."

The Impact of NFPA 1403 on Live Fire Training

As the live fire training instructor, you need to consider how NFPA 1403 affects your training. NFPA standards are consensus standards that will be used to judge the actions of those conducting live fire training. The application of NFPA 1403 is handled in a variety of ways throughout the nation. Some states require that instructors conducting live fire training be trained or certified as live fire instructors. There are some jurisdictions where this training is required by state statutes. Other states rely on instructors to be knowledgeable of the standard and follow the guidelines provided. The process can be formalized through standard operating procedures (SOPs) or standard operating guidelines (SOGs), or may simply be a directive to instructors. Regardless of how the authority having jurisdiction (AHJ) handles the adoption of the standard, the instructor-in-charge (IIC) must be thoroughly knowledgeable on the standard and assure that it is enforced on the training ground. It is the immediate duty of the IIC to ensure full compliance with NFPA 1403 in all instances. The failure to follow the standard can result in legal actions, possible arrests of those in charge of the training, and needless fire fighter deaths and injuries.

A lot of thought should go into selecting instructors to conduct live fire training. This should be based on meeting the job performance requirements of NFPA 1041: *Standard for Fire and Emergency Services Instructor Professional Qualifications*. NFPA 1403 also requires that all instructors be qualified by the AHJ to deliver live fire training, and the IIC is responsible for full compliance with the standard. Some states have already adopted live fire training programs, and the International Society of Fire Service Instructors (ISFSI) delivers live fire training to instructors throughout the nation. If training is going to involve an acquired structure, the instructors need to have significant experience fighting structural fires. Knowing how to read smoke, knowledge of building construction, understanding fire behavior, and determining fuel quantities are all skills that are learned through experience and cannot be gained simply by reading a textbook. The training props used in live fire training can be as simple as untreated hay inside a container or as complicated as a control panel that works a variety of different devices. Again, knowledge of the training prop and its proper usage is essential.

Following the standard does not guarantee that there will not be injuries during training. However, it will help minimize the risk of the hazards that are encountered. Every fire fighter is aware of the saying "everyone goes home," and safety should be the first priority for all participants. This applies to the training environment as well. At the end of the training, the instructor should be able to say that the participants met specific training objectives and they are better prepared for the real world, and that the live fire training was conducted in the safest possible manner. Remember this standard is a minimum, and the knowledge and competency of the instructors is paramount.

Referenced Standards

As a live fire training instructor, you must be familiar with the information provided in NFPA 1403 and the standards it references. No one is expected to be able to recite the standards, but you should have access to them and be able to look up information as needed. If an answer cannot be found in the standards, it will be up to the AHJ to decide how to handle the concern.

All of the knowledge needed for conducting live fire training evolutions cannot be contained within NFPA 1403. Additional published standards should be consulted. Selected NFPA standards are covered briefly, highlighting areas that may impact live fire training.

NFPA 30: *Flammable and Combustible Liquids Code* addresses flammable and combustible liquids and is referenced in NFPA 1403 in regard to definitions of flammable and combustible liquids by their flash points. When dealing with live fire training, NFPA 1403 addresses not using flammable and combustible liquids in acquired structures or permanent live fire training structures. The AHJ will need to make the decision if flammable or combustible liquids can be used for ignition purposes.

Even where state or provincial law allows less restrictive requirements, fire departments need to consider very carefully the possible impact of deviating from NFPA 1403. State or provincial legislation can protect a department from prosecution, but not from civil actions. Most importantly, while it will be argued vehemently by some that the use of combustible or flammable liquids is safe, the specter of improper use and the dangers of vapors and near-instantaneous fire spread should weigh very heavily in an informed decision. Regardless of the decision, there are other factors to consider such as air quality, amount of smoke produced, and environmental impact.

NFPA 58: *Liquefied Petroleum Gas Code* and NFPA 59: *Utility LP-Gas Plant Code* are also referenced in NFPA 1403. When conducting live fire training using liquid-propane (LP) props, the instructor is required to visually inspect and operate the LP prop prior to the training evolutions. The instructor must be trained on the LP prop by the manufacturers or their designee. However, to properly inspect the LP prop, knowledge of these two standards is also required. There are a variety of props nationwide that include those that are permanently piped, some that have flexible above-ground piping, and some that are actually LP tanks used to train for techniques on how to fight LP-tank fires.

Perhaps the most critical standard for live fire training, NFPA 1001: *Standard for Fire Fighter Professional Qualifications*, directly affects NFPA 1403. Prior to participating in live fire training, participants must be qualified in specific areas of training for Fire Fighter I, as outlined in NFPA 1001. It should be stressed that all participants must be trained in all job performance requirements for Fire Fighter I. This chapter will briefly touch on these job performance requirements; however, it is up to the instructors to verify each participant's qualifications before putting him or her through live fire training.

For acquired structures, NFPA 1142: *Standard on Water Supplies for Suburban and Rural Fire Fighting* is referenced in NFPA 1403. The instructor must know how to calculate the needed water supply to fight the fire based on type of construction, exposures, and water sources. The main goal is to make sure that there is enough water to handle any emergency situation that may occur. Many live fire training evolutions have become full emergency responses due to fire spreading and changing conditions. New in the 2018 edition of NFPA 1403, water supply requirements for permanent props are differentiated from acquired structures. (In NFPA 1403, see Chapter 7, *Acquired Structures*, and Chapter 8, *Gas-Fired and Non-Gas-Fired Structures and Mobile Enclosed Live Fire Training Props*.)

NFPA 1971: *Standard on Protective Ensembles for Structural Fire Fighting and Proximity Fire Fighting* is referenced in NFPA 1403. The live fire training instructor and safety officer are responsible for ensuring that all participants' gear meets NFPA standards, and that they are in a safe, usable condition. This can easily be addressed by conducting a gear check prior to beginning the live fire training event, to check for signs of deterioration and fatigue. All pieces of PPE need to be inspected before being given to the students, including protective coats, trousers, hoods, footwear, helmets, and gloves. It is critical to check the outer lining as well as the inner lining because a damaged inner lining may have been paired with a new outer lining mistakenly, which could leave a fire fighter exposed. After the gear check is completed, the students should dress completely and be checked again for any areas of skin showing.

As the instructor, you need to review NFPA 1975: *Standard on Emergency Services Work Clothing Elements*. This issue needs to be addressed prior to the live fire training evolution as some departments do not meet this standard. Many problems can be solved by simply sending out a notice to the departments/participants about clothing requirements. This will be strongly influenced by the AHJ. You should also review, if station or work uniforms are not worn, what clothing is to be worn under personal protective clothing. Clothing made of all-natural fibers should be considered.

The standard also requires that SCBA be manufactured to meet NFPA 1981: *Standard on Open-Circuit Self-Contained Breathing Apparatus (SCBA) for Emergency Services*. Instructors need to be familiar with all components of the SCBA unit. Depending on the experience levels of the students, you may want to have the group perform an inspection prior to the beginning of the training and before entering the building for live fire training. It is important to remind students to check tank air pressure against the gage reading, check gaskets, assure they have the proper amount of air in their cylinder, and make sure that all safety devices such as low pressure alarms, PASS devices, and bypass or purge valves are working properly. As the instructor, you are responsible for the complete safety of the students including gear checks and SCBA functionality. The extra time that it takes to ensure that the SCBA is in good working condition and safe to use can prevent an emergency from happening.

NFPA 1403 requires that personal alert safety system (PASS) devices be manufactured to meet NFPA 1982: *Standard on Personal Alert Safety Systems (PASS)*. Again, it is your responsibility to ensure that students have been trained on how to operate their PASS device. It is also your responsibility to ensure that every student has a PASS device. Newer SCBA units contain integrated PASS devices. If the unit is not equipped with an integral pass device on the SCBA, then an attachable device must be present. No student should ever enter live fire training without an active PASS device. Check to make sure the PASS device is functioning before entry.

NFPA 1041: *Standard for Fire and Emergency Services Instructor Professional Qualifications* is now a referenced publication. Instructors of live fire training should meet the qualifications of an Instructor I and Live Fire Instructor as a minimum, and the IIC should meet the qualifications of an Instructor II and Live Fire Instructor-in-Charge. Safety and instructional techniques are especially important in live fire training, and the standard outlines the skills and knowledge that an instructor should possess.

NFPA 1407: *Standard for Training Fire Service Rapid Intervention Crews* is another important standard to follow. Rapid intervention teams need to be assigned at live fire training events, and the members on this team need to understand the responsibilities and duties of that assignment and should be trained to meet this standard.

Live fire training provides an opportunity for fire fighters to be educated on cancer prevention steps. The National Volunteer Fire Council has developed a list of 11 steps to reduce fire fighter cancer risks, "*Actions for preventing cancer in the fire service,*" located on their website.

Using NFPA 1403

Strict adherence to NFPA 1403 is strongly advised; it provides a solid framework to help instructor develop an organized training event, and when combined with properly trained and experienced instructors providing the live fire training, it will lead to the safest live fire training environment possible.

Should there be any confusion on what a live fire is, NFPA 1403 defines it as any unconfined open flame or device that can propagate fire to a building, structure, or other combustible materials. An individual participating in a live fire training evolution within the operations zone is defined as a **participant**. By this definition, a participant could be a student, instructor, safety officer, visitor, or other person who is in the operations area.

Although there will be many instructors involved in the live fire training evolution, there are specific differences between a Live Fire Instructor (LFI) and the live fire instructor-in-charge (IIC). An instructor is an individual qualified by the AHJ to deliver fire fighter training who has the training and experience to supervise students during live fire training evolutions. These individuals should meet the criteria outlined in NFPA 1041: Instructor I and Live Fire Instructor as a minimum. Each AHJ will have to determine what qualifications instructors must obtain or possess to be involved in live fire training. The IIC has more responsibilities than an instructor. They are responsible for ensuring that the evolution is conducted safely, that the training structures are prepared to meet NFPA 1403, that there are enough properly trained live fire training instructors involved, and that all training evolutions are conducted following the standard, as well as monitoring all fire ground activities. This role carries a lot of weight and requires someone willing to take on the additional responsibilities of live fire training.

Establishing Clear Objectives

Every live fire training evolution must have a clear purpose other than for the participants to "feel the heat" and simply throw water on flames. The learning objectives and desired outcomes should be clear to all participants, regardless of experience levels. Under the watchful eye of the team of live fire instructors, students should be able to see the effectiveness of fire streams, when applied properly. They should be able to observe the results of the coordinated effort of the attack crew and ventilation crew, and see how ventilation impacts the fire attack effort. They should experience the benefit of proper radio communication, effective crew management, and situational awareness, while operating in and around the fire building. Students should also have the opportunity to observe the instructors as they manage the evolution through the use of the incident command system, personnel accountability system, resource status and management process, and situation status and management process. In short, the students should come away from the exercise with an appreciation for how their knowledge of firefighting methodology and their skills in applying that knowledge led them to a desired outcome. A hot wash conducted after the drill can reinforce meeting stated objectives or where further training needs to occur.

LIVE FIRE TIP

Every instructor involved in live fire training evolutions should have a clear understanding of what this type of training is and what it is not:

- It is *not* a test. There are more appropriate and meaningful ways to test a student's abilities without the danger of live fire. No fire fighter should be placed in dangerous live fire conditions without receiving the necessary background training.
- It is *not* an opportunity for instructors to engage in recreational "training." Instructors must focus on the expected benefit for the student and refrain from the temptation to overload a 10′ × 10′ (3 × 3 m) room with a larger fuel package than is recommended for this size room.
- It is *not* baptism by fire. Students should not fear what instructors may do to them during the evolution. This indicates both a lack of professionalism and a lack of understanding regarding this type of training on the part of the instructor.
- It is *not* an opportunity to discover the limits of human endurance or tolerance of extreme temperature.
- It is *not* an opportunity to test the fire retardant tendencies of our PPE or the integrity of critical structural building components.
- It is *not* an impromptu or improvised activity.

The Importance of the 1001 Prerequisite JPRs

Safety

First and foremost, safety is an *attitude* that must become ingrained into the organizational culture. Everyone must be accountable for compliance with established safety rules, procedures, and policies. On the training ground, it starts with the IIC, who is responsible for both compliance and enforcement. The IIC will set the tone for the training evolution, and it should be expected that everyone follows his or her lead. All of the instructors are role models to recruits, and often to the more experienced participants and even to each other. If their protective clothing is soot stained, their helmets blackened with melted eye protection, what message is sent—that gear represents experience? Or the contaminants in and permeated through the protective clothing won't hurt the instructor? So many vital messages and directives can, without a word, be quickly discounted to the participants by observing instructors. Whether inadvertently, by disdain, habit, or lack of interest, the instructor's use of PPE (and its condition), use of SCBA, and demeanor toward safety procedures that they may in fact preach will all be "read" by the participants, and unintended learning occurs. With the newer information regarding the contamination of protective clothing in turn contaminating fire fighters' bodies, it is imperative that instructors are very strong, positive role models. Many students will want to emulate the instructors and look "seasoned" rather than inexperienced, but wearing PPE with heat damage and that is contaminated with soot and saturated with particulate from smoke can harm the wearer immediately and cause greater harm later on; soot can also obscure reflective strips (**FIGURE 16-3**).

Instructors not wearing the PPE and SCBA correctly, or wearing PPE that should be cleaned or replaced, set a bad example. Our actions should always be consistent with our goal to conduct all training evolutions without injury. Therefore, safety violations must never be dismissed or overlooked but should be corrected at once.

Students must come to understand and appreciate the measures that are in place for their personal safety and protection during training (and response). In today's fire service, rules and policies related to the use of PPE and SCBA are generally understood and accepted by even the newest rookie. The intent of prerequisite training is to develop that same level

FIGURE 16-3 The helmet shown was once yellow. Wearing PPE with heat damage and that is contaminated with soot can harm the wearer immediately and cause greater harm later on.
Courtesy of Dave Casey.

of awareness, acceptance, and willingness to comply with all safety standards. The procedures are somewhat more challenging but equally as important. They generally require repetitive drilling to develop muscle memory and proper technique. This training may focus on fundamental skills such as proper lifting, dragging, and carrying techniques. The training should be comprehensive enough to involve the proper use of all hand and power tools, especially where those tools are used for self-rescue and disentanglement. Whether the training involves a rule, policy, or procedure, it should be made clear that compliance is mandatory and not an option.

Safety is everyone's responsibility. All participants in the live fire exercise, including students, should be encouraged to speak up regarding something that appears to be an unsafe operation or practice. Though the IIC has the authority and responsibility for the safety of the drill, all participants' concerns should be heard and evaluated. A safety briefing conducted before the exercise will establish the importance of safety on participants.

Many changes in the construction industry and the materials used in the furnishings of buildings have further increased the possibility of a **flashover** and **backdraft** in a compartment(s) fire. At the same time, it can be shown that a number of fire fighter fatalities and severe injuries have occurred directly or indirectly as a result of inadequate understanding of fire behavior and of the precautions and tactics that need to be employed. Instructors and students must understand the signs of hostile fire events and the techniques used to prevent them.

In response to these changes, there are new prerequisite training requirements, and *these requirements apply to incumbent fire fighters and instructors alike.* The additional new competencies are contained in NFPA 1403 sections 4.3.2.1 through 4.3.2.5. These prerequisites are just that: they must be met "prior to being permitted to participate in live fire training evolutions." These additional prerequisites encompass five primary topical areas: fire dynamics, health and safety, fundamentals of fire behavior, fire development in a compartment, and nozzle techniques and door control.

When working with fire fighters from another fire department, live fire instructors must require written documentation that the fire fighter's prerequisite knowledge and skill levels match or exceed the objective content for the live fire training evolutions. In addition, the instructor-in-charge may conduct a quick series of skill level assessments by observing as the fire fighters don PPE and perform emergency evacuation procedures, search-and-rescue procedures, and ventilation procedures.

After-Action REVIEW

IN SUMMARY

- Live fire training is among the most important training we can do as fire fighters. The training of fire fighters to fight fire, to understand fire dynamics and the conditions they will encounter, and to know how to successfully operate in a hostile environment are among the most important aspects of training.
- Realistically, in-context training is becoming more important because many fire departments are currently experiencing large numbers of retirements, with those replacing the retirees typically having less fire suppression experience.
- The purpose of NFPA 1403 is to provide a process for conducting live fire training evolutions to ensure that they are conducted in safe facilities and that the exposure to health and safety hazards for fire fighters during live fire training is minimized.
- As the live fire training instructor, consider how NFPA 1403 affects your training. The application of NFPA 1403 is handled in a variety of ways throughout the nation. Some states require that instructors conducting live fire training be trained or certified as live fire instructors. Other states rely on instructors to be knowledgeable of the standard and follow the guidelines provided.

- Regardless of how the AHJ handles the adoption of the standard, the instructor-in-charge must be thoroughly knowledgeable on the standard and assure that it is enforced on the training ground. It is the immediate duty of the IIC to ensure full compliance with NFPA 1403 in all instances.
- As a live fire training instructor, you must be familiar with the information provided in NFPA 1403 and the standards it references.
- All of the knowledge needed for conducting live fire training evolutions cannot be contained within NFPA 1403. Additional published standards should be consulted.
- Safety is an *attitude* that must become ingrained into the organizational culture. Everyone must be accountable for compliance with established safety rules, procedures, and policies. On the training ground, it starts with the IIC, who is responsible for both compliance and enforcement.
- The IIC will set the tone for the training evolution, and it should be expected that everyone follows his or her lead. All of the instructors are role models to recruits, and often to the more experienced participants and even to each other.

KEY TERMS

Access Navigate for flashcards to test your key term knowledge.

Backdraft A deflagration (explosion) resulting from the sudden introduction of air into a confined space containing oxygen-deficient products of incomplete combustion (NFPA 1403).

Flashover A transition phase in the development of a compartment fire in which surfaces exposed to thermal radiation reach ignition temperature more or less simultaneously and fire spreads rapidly throughout the space, resulting in full room involvement or total involvement of the compartment or enclosed space (NFPA 1403).

Live fire Any unconfined open flame or device that can propagate fire to the building, structure, or other combustible materials (NFPA 1403).

Participant Any student, instructor, safety officer, visitor, or other person who is involved in the live fire training evolution within the operations area (NFPA 1403).

REFERENCES

Colletti, Dominic. 2007. "Live Fire Training–Are We Making a Wrong Turn?" *Firehouse*, March 2007. http://www.firehouse.com/article/10505064/live-fire-training-8212-part-1-are-we-making-a-wrong-turn. Accessed December 20, 2018.

Fahy, Rita R., Paul R. LeBlanc, and Joseph L. Molis. 2016. "Firefighter Fatalities in the United States." National Fire Protection Agency.

National Fire Protection Association. 2018. NFPA 1403: *Standard on Live Fire Training Evolutions*. Quincy, MA: National Fire Protection Association.

National Fire Protection Association. 2019. NFPA 1041: *Standard for Fire and Emergency Services Instructor Professional Qualifications*. Quincy, MA: National Fire Protection Association.

National Institute for Occupational Safety and Health (NIOSH). 2018. "Fire Fighter Fatality Investigation and Prevention." Centers for Disease Control and Prevention. Last updated November 27, 2018. https://www.cdc.gov/niosh/fire/default.html.

Paradise, John R. 2003. "Live Burn." *NFPA Journal* (May/June 2003).

Perry, Mark J. 2014. "Today's new homes are 1,000 square feet larger than in 1973, and the living space per person has doubled over last 40 years." *AEIdeas*, February 26, 2014. https://www.aei.org/publication/todays-new-homes-are-1000-square-feet-larger-than-in-1973-and-the-living-space-per-person-has-doubled-over-last-40-years/.

REVIEW QUESTIONS

1. What should the instructor be able to state at the end of a live fire training event?

2. What is the individual participating in a live fire training evolution within the operations zone called?

3. Explain some of the key differences between a live fire instructor (LFI) and the live fire instructor-in-charge (IIC).

CHAPTER 17

Live Fire Instructor

NFPA 1041 JOB PERFORMANCE REQUIREMENTS

7.1.1 For qualification at Live Fire Instructor, the candidate shall meet the requirements of Fire Fighter II as defined in NFPA 1001 or Interior Structural Fire Brigade Member as defined by NFPA 1081, the requirements of Fire and Emergency Services Instructor I as defined in Chapter 4, and the job performance requirements defined in 7.2 through 7.3 of this standard.

7.1.2 A Live Fire Instructor shall demonstrate competency in knowledge and skills in all subjects, methods, and equipment being taught and the objectives contained in NFPA 1403 and identified for the live fire evolution.

7.2.1 Inspect live fire participants' PPE and SCBA, given participants and PPE and SCBA, so that equipment is determined to be serviceable and worn in accordance with manufacturer's instructions.

(A) Requisite Knowledge.
Manufacturers' instructions.

(B) Requisite Skills.
Visual inspection, using an inspection checklist.

7.3.1 Predict stages of fire growth in a compartment, flow path, flashover, rollover, and backdraft, given a live fire evolution, so that a safe environment is maintained.

(A) Requisite Knowledge.
Fire dynamics, including fuel load, fire growth, flow path, flashover, rollover, and backdraft.

(B) Requisite Skills.
Configure fuel loads to meet the objectives of the live fire evolution, recognize changing conditions of the live fire environment.

7.3.2 Supervise a group during a live fire evolution, given a live fire structure or prop and a group of participants, so that instructional objectives are met, crew integrity is maintained, the instructor maintains a position to supervise the crew, fire conditions are monitored, and emergency actions are taken as necessary.

(A) Requisite Knowledge.
Group dynamics, instructor positioning, egress routes, fire dynamics, including fuel load, fire growth, flow path, flashover, rollover, and backdraft.

(B) Requisite Skills.
Supervisory skills, fire suppression operations.

7.3.3 Conduct a personnel accountability report (PAR) upon entering and exiting a live fire structure or prop, given a group of participants in a live fire evolution, so that all participants are accounted for and safety is ensured and maintained.

(A) Requisite Knowledge.
Knowledge of incident management system, authority having jurisdiction (AHJ) personnel accountability procedures.

(B) Requisite Skills.
Use of AHJ's accountability system, ability to recognize inadequacies in the use of the accountability system.

7.3.4 Monitor live fire participants to safeguard participants, given a live fire evolution, so that signs and symptoms of fatigue and distress are recognized and action is taken to prevent injury.

(A) Requisite Knowledge.
Signs and symptoms of fatigue and distress, knowledge of environmental conditions, AHJ safety, rehabilitation, and emergency procedures.

(B) Requisite Skills.
Evaluation of environmental conditions, class management, activation of AHJ emergency procedures.

KNOWLEDGE OBJECTIVES

After studying this chapter, participating in a structured learning environment, and completing assigned assessments, you will be able to:

- Identify all personal protective equipment standards and requirements for live fire training evolutions. (**NFPA 1041: 7.2.1**, pp 383–385)
- Explain the inspection considerations for live fire personal protective equipment. (**NFPA 1041: 7.2.1**, pp 383–385)
- Describe considerations for predicting fire behavior during a live fire training evolution. (**NFPA 1041: 7.3.1**, pp 386–387)
- Describe ways an instructor can use ventilation to control fire growth. (**NFPA 1041: 7.3.1**, pp 386–387)
- Identify the fire behavior knowledge necessary for a live fire training instructor, including fire dynamics and heat release rate. (**NFPA 1041: 7.3.1**, pp 386–388)
- Explain considerations for supervision of recruit fire fighters with little or no live fire experience. (**NFPA 1041: 7.3.2**, pp 388–390)
- Recognize common burn evolution events with the potential to develop into emergencies. (**NFPA 1041: 7.3.2**, pp 389–390)
- Explain the purpose of a personnel accountability report. (**NFPA 1041: 7.3.3**, p 385)
- Identify actions to be taken if a participant is unaccounted for. (**NFPA 1041: 7.3.3**, pp 385–386)
- Describe the cardiovascular and thermal responses to firefighting. (**NFPA 1041: 7.3.4**, pp 393–399)
- Describe how firefighting activity and turnout gear affect cardiovascular and thermal strain. (**NFPA 1041: 7.3.4**, pp 394–395)
- Describe how to prevent injuries, such as heat illness, during firefighting activity and training. (**NFPA 1041: 7.3.4**, pp 393–399)
- Describe the warning signs for heat illnesses that may occur in firefighting activity and training. (**NFPA 1041: 7.3.4**, p 398)
- Explain the function and importance of rehabilitation for all participants in live fire training. (**NFPA 1041: 7.3.4**, p 400)

SKILLS OBJECTIVES

After studying this chapter, participating in a structured learning environment, and completing assigned assessments, you will be able to:

- Inspect participant PPE to verify its condition and use are acceptable for a planned evolution. (**NFPA 1041: 7.2.1**, pp 383–385)
- Adjust fuel loads and ventilation to maintain safe environment during a live fire evolution. (**NFPA 1041: 7.3.1**, pp 386–387)
- Maintain crew integrity and safety while operating as a live fire training instructor working with interior crews. (**NFPA 1041: 7.3.2**, pp 385–390)
- Conduct a personnel accountability report. (**NFPA 1041: 7.3.3**, p 385)
- Call a mayday for a missing participant. (**NFPA 1041: 7.3.3**, pp 385–386; 390)
- Monitor live fire evolution participants for signs of distress and take action to prevent injury. (**NFPA 1041: 7.3.4**, pp 387; 398–399)

You Are the Live Fire Instructor

You have just been given an assignment to be the instructor-in-charge (IIC) of a live fire training evolution. Three local fire departments will be present to take part in the exercises. The training is scheduled for early July and will be conducted at a regional training facility. There are many aspects that need to be planned before this training can be a success. As the training date is approaching, you ask yourself the following questions:

1. What are the primary physiological threats that I must consider to ensure the safety of the instructors and students?

2. How will a well-run rehabilitation sector help me address the safety of my students?

 Access Navigate for more practice activities.

Introduction

Live fire training is the pinnacle of a recruit's training experience in an academy. NFPA 1403 requires that prior to exposing a recruit to live fire training, they must meet the job performance requirements in NFPA 1001: *Standard for Fire Fighter Professional Qualifications*. Live fire training should *not* be the testing ground for these skills. Instead, it should bring together all of the skills they have successfully demonstrated up to this point in their training.

The realism that can be achieved in the live fire environment is its greatest advantage. Conversely, that same realism if not properly managed can be its greatest disadvantage as the fire dynamic can suddenly spiral out of control. It has the potential to overwhelm the command and control systems and their available resources, with potentially disastrous results. Some of the inherent dangers of fighting fire in real structures are present during training burns. A thorough understanding by the instructors of fire dynamics under the conditions presented is necessary, along with an understanding of physiological demands of live fire firefighting on the students and instructors. Failure to recognize these dangers has proven to be very costly. Accordingly, it is imperative that these training exercises be conducted with strict adherence to NFPA 1403: *Standard on Live Fire Training Evolutions*.

Preparing for a Live Fire Evolution

An instructor supervising a group of students during a live fire training truly has the students' lives in his or her hands. There is no room for error or lax adherence to safety rules. Live fire instructors must ensure students are prepared for the evolution, including wearing appropriate gear that is being correctly used to participate in the evolution.

Protective Clothing and Self-Contained Breathing Apparatus

Live fire training instructors are responsible to ensure that all required personal protective equipment (PPE) is in serviceable condition and properly worn. The live fire instructor (LFI) assigned to be the safety officer (SO) for the evolution needs to ensure that all students, instructors, safety personnel, and any other participants wear their equipment appropriately. NFPA 1403 expands on this to say that all participants must wear all protective clothing and equipment "according to manufacturer's instructions." The LFI must ensure that all PPE meets this requirement prior to entry into the structure. LFIs need to be familiar with NFPA 1851: *Standard on Selection, Care, and Maintenance of Protective Ensembles for Structural Fire Fighting and Proximity Fire Fighting* and NFPA 1852: *Standard on Selection, Care, and Maintenance of Open-Circuit Self-Contained Breath Apparatus (SCBA)*. LFIs must be familiar with inspection and the requirements of protective clothing, SCBA, personal alert safety system (PASS) device, and uniform apparel. This portion of NFPA 1403 becomes problematic and can, if followed verbatim, require the LFI and his or her staff to read the labels on each piece of participant gear to determine if the gear is compliant to the standard.

This would include turnout coat, pants, hood, boots, helmet, uniform pants and shirt, and SCBA with an integrated or separate PASS device. When conducting training with only your own fire department, the instructors must have knowledge of compliance with NFPA 1971: *Standard on Protective Ensembles for Structural Fire Fighting and Proximity Fire Fighting*; NFPA 1981: *Standard on Open-Circuit Self-Contained Breathing Apparatus (SCBA) for Emergency Services*; NFPA 1982: *Standard on Personal Alert Safety Systems (PASS)*; and NFPA 1975: *Standard on Station/Work Uniforms for Fire and Emergency Services*. When training involves other fire departments, documentation of compliance must be obtained and onsite inspections conducted. All of the aforementioned gear should be compliant in manufacture, but its condition still needs to be checked, and the pants and coats should be of the same brand and ensemble type. With high-backed pants versus bib-cut pants and standard cut, and coats with similar variations, it is important to make sure that the pants match the coat's style and manufacture.

The inspection of the PPE and SCBA needs to be documented, and the LFI/SO must also ensure that SCBA is worn anytime a participant could be exposed to an **immediately dangerous to life or health (IDLH)** or potential IDLH atmosphere. Participants should be aware of the dangers of clothing made from certain synthetic materials that can melt and burn the wearers even if they are using clothing that meets National Fire Protection Association (NFPA) standards. Whenever possible, all clothing worn should comply with NFPA 1975.

When multiple agencies or individuals from different agencies are allowed to participate in training, a more thorough inspection is necessary. In this case, it would be wise to have the outside agencies provide documentation to the host agency that all of the protective clothing and SCBA have been checked and are compliant. The SO and staff should understand that each participant is not going to bring the written manufacturer's instructions for each item. Still, due diligence is necessary to verify that the protective clothing appears to be in good condition. The LFI/SO should check the following:

- Turnout coat and pants
  - They should be of the same ensemble.
  - Outer shell: There should be no obvious signs of contamination, heat discoloration, burns or charring, tears, holes, fraying, or weakening of material.
  - Check the NFPA compliance tag.
  - Ensure proper fit on participants.
  - Liner: There should be no thermal damage, delaminating of the moisture barrier, tears, heat discoloration, holes, or fraying.
  - Hardware: Ensure that snaps, zippers, Velcro, or other closures are functional.
- Hood
  - There should be no tears, holes, or fraying.
  - Ensure proper fit on participants. The hood should not be stretched out.
  - The National Fire Protection Association (NFPA) compliance tag should be checked.
- Helmet
  - Shell: There should be no obvious contamination, cracks, holes, weakened material, burns, or charring.
  - Liner: Check for thermal damage and damage to the impact shell.
  - Check the earflaps for functionality.
  - The hardware should be functional and properly adjusted for good fit.
  - The NFPA compliance sticker should be checked.
  - Check the strap to make sure it is in good condition and functions properly.
- Gloves
  - Outer shell: There should be no obvious contamination, burns, or charring, and no tears, holes, or fraying.
  - Liner: There should be no moisture-barrier delaminating, tears, holes, or fraying.
  - Ensure proper glove-to-coat interface.
  - The NFPA compliance sticker should be checked.
  - Ensure proper fit on participants.
- Boots
  - There should be no contamination, tears, holes, fraying, weakened material, burns, or charring.
  - The liner should show no thermal damage.
  - There should be no moisture-barrier delaminating or tears, holes, or fraying.
  - Hardware: Ensure that snaps, zippers, Velcro, and other closures are functional.
  - The NFPA compliance tag should be checked.
  - Ensure proper fit on participants.
- SCBA
  - Cylinder(s): There should be no physical damage, contamination, or thermal damage. The hydrostatic test should be current.
  - Harness: There should be no physical damage, fraying of straps, contamination, or thermal damage. The hardware should be complete and functioning.

- Regulator and hoses should be intact and functional, and there should be no physical damage affecting use or contamination.
- Face piece should be intact, lens visibility should be good, and the straps and headpiece should be intact and not frayed, and should seal properly.
- PASS device should be checked.
- SCBA should be NFPA compliant.

Live Fire Evolutions

As an instructor of live fire training, you will have the responsibility of creating realistic and challenging evolutions for live fire training. According to NFPA 1403: *Standard on Live Fire Training Evolutions*, a live fire training evolution is a set of prescribed actions that results in an effective fire-ground activity.

Personal Protective Equipment Use

Checking PPE is not a "once and done" action. Each participant, including instructors, shall be checked before entering a live fire evolution, to assure that PPE is being worn correctly and that there are no defective pieces of equipment. There should be no exposed skin prior to personnel entering into the IDLH environment. Also, prior to beginning the evolution, the safety officer should inspect everyone's SCBA to confirm that they are beginning the training evolution with an adequate air supply. Inspection of PPE and SCBA must be done prior to every training evolution. It is not sufficient or acceptable to check all of these items at the beginning of the training day and never reevaluate them.

Care must be exercised with wet PPE being used again in a high-heat area. Wet gear transmits heat more quickly and there is the concern for steam burns. At this time, it may be necessary to review the requirement to wear and use the equipment while within the operations area. The use of scene tape or barrier tape reinforces this safety practice, especially for recruit fire fighters (**FIGURE 17-1**). At permanent live fire training structures, a painted line with signs can serve the same purpose.

Individual fire departments should already have policies on the use of PPE and SCBA, and related safety concerns. This needs to include decontamination of personnel after exiting the acquired structure, or personnel participating with liquid fuel fires, as well as the removal of contaminated hoods, coats, and so on.

FIGURE 17-1 The operations area where PPE and SCBA are required should be clearly defined.
Courtesy of Dave Casey.

In the last few years, it has become known that there are a significant number of toxins being absorbed into the PPE and then into the wearer either by breathing them in from PPE that is off-gassing, or even dermally through open, sweating skin pores.

Instructors must examine their behavior and habits to make certain they are conforming with current practice. Modeling these good behaviors helps to instill this as "important" to the recruits. Sometimes this means that instructors must break longtime bad habits such as leaving thermal hoods around their neck after removing their SCBA mask. Some of the "new" skills that must be demonstrated and modeled include decontamination, such as how to effectively use sanitation wipes, gross decontamination of the PPE, and proper instruction on more complete cleaning.

Accountability

The live fire instructor must be able to maintain the integrity of the crew and have accountability for them throughout the entire evolution. All participants must understand that crew accountability is in place for one reason: safety. It must not be compromised for any reason.

A live fire instructor must keep track of his or her students, just like a company officer keeps track of his or her crew. To prepare, a personnel accountability report (PAR) is conducted outside of the structure or before the evolution begins, ensuring all participants are present, properly geared up, and ready to go. A head count must be done prior to entry, once inside, and again upon exiting. Any time a student cannot be accounted for, it must be reported immediately. If this happens, all participants must exit the structure as the evolution has now gone from a live fire training evolution to a mayday situation, or one where a fire fighter is missing, lost, or in need of immediate assistance. In any case, the live fire training evolution needs to be stopped and the unaccounted-for fire fighter must be found. It is important to note that calling a PAR has become lax at times. When exiting a structure, the team leader will ultimately radio something to this extent: "Engine 10 exiting building, PAR." It would be a good practice to get students in the habit of calling, "Engine 10 exiting building, PAR 4."

Fire Behavior and Structural Fire Dynamics

When looking at the fire dynamics in a structure, the instructor must understand fire behavior and all of the elements that go along with it. Instructors must be able to effectively teach fire behavior and recognize and disclose adverse and changing conditions that could endanger participants. Reading smoke or fire behavior, understanding room-and-contents versus deep-seated structure fires, and recognizing normal fire spread and extreme or hostile fire behavior are all skills that you need to be able to observe and correctly identify during suppression operations. It is very important to understand terminology and fire behavior. Under the definitions section of the standard, there are many terms relating to fire behavior. It is also very important to be knowledgeable on how **fuel loads** will affect fire behavior. In addition, you must understand the breakdown of fuels and the consequences if too much fuel is used, which could result in **flameover** (or rollover), **flashover**, or backdraft if not properly ventilated.

Understanding the **flow path** in all structures is paramount when determining where a fire will go, controlling the amount of oxygen available to the fire, and the extension of the fire if using an acquired structure. Furthermore, the importance of fire behavior is critical, and both the IIC and the safety officer have the responsibility to assess rooms for fire growth potential.

Instructors and the safety officer must also understand how to develop a **ventilation-controlled fire** evolution. As you ventilate a structure, the flames and heat will be pulled to any new opening to seek more oxygen. It is critical to be able to determine where the fire should go, and what impact ventilation has on the fire. Not all instructors will be familiar with these concepts, so demonstrating them to the instructor s is important so that they have an understanding of a ventilation-controlled evolution. If fire growth or fire behavior presents a hazard to participants, the LFI should speak up because the IIC or safety officer can terminate the incident. The IIC and safety officer are also responsible for documenting the fuel load.

Instructors need to understand that most fuels in structural room-and-contents fires start out as solid materials. As these materials are heated, they go through a process called **pyrolysis**, which is the decomposition or off-gassing of a material that, once it reaches its ignition temperature, will burn openly. During this process, smoke, or the incomplete combustion of materials, is produced. This smoke contains many toxic gases, such as carbon monoxide, hydrogen chloride, hydrogen cyanide, phosgene, and chlorine. The most abundant of these gases is carbon monoxide, with a flammable range of 12.4–74 percent, and an ignition temperature of 1128°F (609°C).

Flow paths, flame spread, and smoke can be affected by wind and weather conditions. These conditions also affect students. Remember to be watchful of weather conditions including wind velocity and wind direction, and know how they affect live fire evolutions. The effects of weather are especially critical when using outdoor props.

Checking the weather ahead of time only makes sense. There are many weather apps for smartphones, including those from local television stations that often include radar. Conduct a final check for possible weather changes immediately before the actual ignition.

The use of outdoor props or any specialty prop requires that instructors and the safety officer have complete training on those props. This training must be performed by an individual authorized by the gas-fueled training system and specialty prop manufacturer.

> **LIVE FIRE TIP**
>
> Remember from Fire Fighter I and II training that 1 gallon (3.785 liters) of water will absorb over 9000 BTU. This number is found by using the following formula:
>
> 212°F − 60°F = 152 BTU per pound of water
>
> 152 BTU × 8.35 (weight of 1 gallon of water) = 1269.20 BTU
>
> Due to the law of latent heat of vaporization, water will absorb an additional 970.3 BTU per gallon.
>
> 970.3 BTU/gallon ×
> 8.35 (weight of 1 gallon of water in pounds) = 8102.01
>
> Add 1269.20 + 8102.01 = 9371.21 BTU per gallon. One gallon of water will absorb a little over 9000 BTU.

Heat Release Rate

When looking at materials in today's homes, most are not made of wood but of many different types of fuel materials, each having a different heat of combustion and **heat release rate**. Heat release rate is the rate at which heat energy is generated by burning. The scientific community uses Celsius, joules, calories, and kilowatts to measure heat release rate. Remember that 1 BTU (British Thermal Unit) per second is roughly equal to 1 kilowatt (1.055), and that 1 BTU is equal to 1055 joules or 252 calories. BTU is the amount not the intensity of heat. In the fire service, BTUs are the standard unit of measure.

TABLE 17-1 lists some common materials and the approximate heat release rate in BTUs per minute. Looking at the table, it is clear why only Class A materials with known burning characteristics should be used for the live fire training evolution and why other materials must be removed. A pound of wood will generate between 7000 and 8000 BTU per pound. On the other hand, the same volume of polyurethane will generate 16,000 to 20,000 BTU per pound. A pound of polyurethane will give off up to three times more heat than a pound of wood. Polyurethane's rate of burning is faster than that of wood. Further, the smoke from the polyurethane or other plastics contains greater levels of toxins and generally more unburned fuels, increasing the dangers considerably. Knowing the heat release rates of different materials is extremely important for the safety of the participants.

TABLE 17-1 Heat Release Rate of Common Materials

Material	BTU per min
1 lb (0.45 kg) of wood	7000–8000
Cotton mattress	7969–19,922
Styrofoam	18,400
Gasoline (2 sf)	18,400–22,768
Gasoline (gallon)	115,000–125,000
Dry Christmas tree	28,460–36,998
Propane per pound	21,857
Propane per gallon	91,800
Polyurethane mattress	46,105–149,699
Polyurethane sofa	177,590

Based on data from Kales S. N., E. S. Soteriades, S. G. Christoudias, and D. C. Christiani. 2003. "Firefighters and on-duty deaths from coronary heart disease: a case control study." *Environmental Health* 2(1):14.

Smoke

Usually, the higher the BTU, the more smoke that is produced in a fire. It is critical for instructors to understand fire behavior and to learn how to read smoke. Remember, smoke is fuel and fuels behave in relatively predictable ways. Smoke volume, smoke velocity, smoke density, and smoke color are key factors in reading smoke.

Smoke volume and color help determine how much fuel and what type of fuel is burning. Smoke velocity will help determine the pressure that is being built up, whether it is a lazy smoke (laminar) or it is agitated (turbulent) smoke (**FIGURE 17-2**). Smoke density can help determine how much fuel is in the smoke. The denser the smoke, the more fuel there is to burn. Smoke color, from white to gray to black, is also an indicator of fire intensity, with black smoke being the hottest of the three. Lastly, the rate of change of smoke in a structure may help determine if a flashover is about to occur (**FIGURE 17-3**). This can be observed from inside or outside, but it is more easily read from outside the structure where the instructors can see the whole picture.

Inside the structure, the smoke may become darker, more aggressive, and much hotter, and it may bank to the floor. Fingers of flame or rollover may not be seen, due to the density of the smoke and the height of the ceiling. It is the instructor's duty to recognize these signs as potentially dangerous conditions, and cool the upper atmosphere or get out.

On the outside, the smoke may go from lazy to aggressive, become darker, and **vent point ignition** could occur (**FIGURE 17-4**). Vent point ignition occurs when the smoke is at its ignition temperature but is lacking oxygen and is too rich to burn while in the structure. The smoke receives oxygen and falls within

FIGURE 17-2 Agitated, turbulent smoke is indicative of an imminent flashover.

FIGURE 17-3 A flashover occurs when surfaces exposed to thermal radiation reach ignition temperature, then fire spreads rapidly throughout the structure.

FIGURE 17-4 Vent point ignition occurs when smoke and unburned gases escape to the outside and then ignite.

its flammable range as it exits an opening and bursts into flame. This is a sign of impending flashover. Remember that carbon monoxide is the most prevalent gas in the smoke, with an ignition temperature of 1128°F. How rapid is the rate of change of these factors and what may be influencing them? These could be signs of flashover and must be carefully but quickly analyzed by the instructor. If a flashover occurs, it only takes seconds to injure or kill.

LIVE FIRE TIP

A thermal imager (TI) can help the instructor to see through the smoke to get a better understanding of the temperatures at different levels in the room. A TI can also be used by an instructor to monitor students and their locations and reactions to the environment in which they are working. To become proficient, an instructor must be trained on the use of TIs to the point where using one becomes second nature.

Also, remember that the more times you burn within a structure, the more the contents and the structure are absorbing heat. This will reduce the amount of radiant heat it can absorb and lessen the time required for the material to off-gas, and more importantly, lessen the time for flashover to occur.

Students in the Live Fire Training Environment

Before conducting any type of class, know who your audience is going to be. This is especially true when conducting the high-risk training evolutions associated with live fire training. Of course, instructors would never place a student who is not adequately prepared to perform basic firefighting skills in a live fire environment. Consider the following questions:

- Should a new recruit be faced with the same level of difficulty and skill requirements as an experienced fire fighter?
- Would a group of experienced fire fighters in a simple search-and-rescue evolution with a small fire in a steel barrel question the relevance and quality of the evolution?
- How can an instructor ensure that all training needs are met, without putting anyone in danger or wasting anyone's time?

Recruit Students

In live fire training, we are posed with a challenge that does not exist in the other types of trainings. Fire service recruits are the newest and least experienced members of our organization. Introducing recruits into the most hazardous types of training requires extreme vigilance and careful preparation by all instructors involved in the live fire training. New recruits lack the experience of working with a crew at structural fires. Remember that the natural instincts that many fire fighters have acquired over time are not developed in the new recruit.

As recruits, these students have not mastered all of the skills necessary to be proficient fire fighters. They are still developing and comprehending their fire-ground skills. When introducing these newest of fire fighters to live fire training evolutions, instructors should be prepared for their students to sometimes react incorrectly (freeze, panic, or try to stand). For example, a smoke condition that would cause an experienced fire fighter to step back, slow down, and assess the situation may not get a second glance from new fire fighters because they do not fully understand smoke and fire behavior yet. On the contrary, a situation where a window breaks suddenly for ventilation, which may be a sign of relief to experienced fire fighters, may create a state of panic in recruits, or it could influence the flow path in the structure, possibly placing members in danger. Instructors must constantly keep their focus on the behavior of recruits during evolutions and evaluate their reactions and performance (**FIGURE 17-5**).

New recruits are still learning how to follow orders and maintain awareness of their surroundings. They may not recognize a dangerous situation or condition. For this reason, an instructor must be close at hand to keep the recruit safe and to instruct them on what to be looking for when fighting a fire.

Live fire training instructors should rehearse potential scenarios that may occur when introducing new recruits to live fire training. Some common untoward reactions seen in new recruits during live fire training include the following:

- Becoming lost or disoriented
- Not following instructions (e.g., nozzle operation)
- Standing up in high-heat and low-visibility conditions
- SCBA malfunctions during live fire conditions
- Apprehension, hyperventilation, and panic
- Moving in too quickly and getting too close to the fire
- Not handling tools properly

Each of these reactions poses a different challenge for the instructor. The safety of the participants is dependent on how quickly the situation is identified and rectified. Instructors should practice how they will correct these problems prior to conducting live fire training, especially with new recruits. Instructors need to react quickly to situations, preventing injury to students and participants.

Student Psychology: Fire Fighter Style

The live fire instructor must have a solid understanding of student psychology and know how to modify his or her own training techniques to maintain student motivation. Whether a live fire training instructor (LFTI) is leading rookies or certified fire fighters into a live fire, he or she must consider both the same. One may have no live fire experience while the other may have fought many fires in his or her career. In both cases, they are students in a learning environment, and instructors will be responsible for both. Just because someone has been on the job for a long time, it cannot be assumed that the individual has the experience to go with it. Remember the saying, "Do they have 20 years of experience or do they have 20 1-year experiences?" There is a great difference, and the LFTI candidate must understand this difference. How an individual reacts will also change daily. Just as a fire fighter has good and bad days, physical and mental health issues can cause a student to become distracted and a seasoned fire fighter can make a recruit-level mistake. As the instructor, you must constantly monitor the students and environment to ensure everyone's safety.

Brand new fire fighters are more likely to listen and do exactly as they are told, whereas a seasoned fire fighter "knows all they need to know." Do not be afraid to "untrain" a trained fire fighter of his or her poor practices. Often the experienced individual will be much more difficult to accept new or changing practices or concepts, whereas the newer member is often eager to absorb knowledge.

To a degree, an instructor's teaching style and demeanor will depend on the students. The instructor must reassure the students while keeping a close watch on their behavior and body language. As an example, a student who is continually readjusting a SCBA mask may not be comfortable wearing it. Therefore, the instructor should be close to that student when entering the burn building. An instructor candidate must learn the subtle signals and clues of body language to discern which students need to be closely monitored.

Live fire instructors must recognize and respond to events with the potential to develop into emergencies including:

- The crew is having difficulty advancing the hose during line advancement. Instructors must determine if the hose is stuck on a corner and if somebody needs to be positioned at the corner to feed the hose.

FIGURE 17-5 An exterior prop with clear views into the interior will allow instructors to closely monitor the actions of recruits during training evolutions.

- The fire stream is insufficient due to the wrong nozzle selection on variable gallon nozzles. Instructors need to recognize the problem and correct it.
- Kinks in the hose result in poor quality of fire stream. Instructors need to recognize the problem and correct it.
- A fire fighter or instructor gets pinned against the wall. Instructors need to determine what actions need to be taken.
- Contact is lost with one of the students. Instructors need to locate the student rapidly and safely.
- Students become bunched up and cannot move. Instructors need to separate them.
- A PASS device goes off. Instructors need to check for a problem, and if there is a problem, make a mayday call.
- A fire ensues above and behind an attack line crew. Instructors need to advise the attack crew and get it under control.
- The room-and-contents fire has grown out of control. Instructors need to determine what actions need to be taken.
- The fire has spread into the cockloft, attic, or some other concealed space (**FIGURE 17-6**). Instructors need to determine what actions need to be taken.
- An emergency requires quick action. Instructors must demonstrate an understanding of the difference between emergency traffic, like a power line down on the Charlie side, and a mayday call, like a fire fighter lost, fallen, trapped, or low on air.

FIGURE 17-6 Once a fire has gotten into the attic or other concealed spaces in a structure, it is no longer a training session. If this is a training event for new recruits, they should be replaced with seasoned, trained fire fighters who are certified to handle structure fires.

LIVE FIRE TIP

1. Vent point ignition is being communicated from outside: This is an indicator that flashover is possible and the instructor must be aware of this and cool the atmosphere.
2. Heavy black smoke with intense heat is suddenly banking down: This could be an indicator of an impending flashover.
3. The fire has gotten into the cockloft, attic, or some other concealed space: This is not a live burn evolution anymore.
4. Sudden changes in heat, flames, or visibility could also be an indicator of flashover.

Physiological Aspect of Fire Training

Firefighting is an inherently dangerous and physically demanding occupation. Every day, fire fighters are faced with potentially life-threatening challenges, including burn injury, asphyxiation, collapse, and entrapment. Less appreciated, however, are the physiological consequences that threaten fire fighters. The combination of strenuous work, heavy and encapsulating PPE, hot and hostile fire conditions, and high adrenaline levels leads to significant levels of cardiovascular and thermal strain during firefighting.

Live fire training is necessary to prepare fire fighters for the dangerous and challenging environment in which they are expected to perform. It places fire fighters in high heat environments with live fire conditions. These intense training sessions can create high levels of cardiovascular and thermal strain, and thus increase the risk for heat-related injuries and sudden cardiac events. Live fire training instructors, in particular, are often exposed to severe heat conditions for prolonged periods of time. The exposure to such severe conditions creates a challenging and potentially dangerous situation for both instructors and students.

Cardiovascular and Thermal Strain of Firefighting

As a result of the combination of heavy work, heavy and encapsulating PPE, and hot and hostile environmental conditions, firefighting creates significant physiological strain, affecting nearly every system of the body. The greatest risks to the fire fighter come from the ensuing cardiovascular and thermal strain. NFPA 1403: *Standard on Live Fire Training Evolutions* mandates that training evolutions be done in such a way as to minimize the exposure to health and safety hazards for the fire fighters involved. In terms of human physiology, this means recognizing the thermal and cardiovascular strain associated with firefighting and the resulting deleterious impact on cognitive function, and pursuing measures to minimize the associated risks. The length and number of evolutions must be based on the nature of work to be performed, the physical stress of the work, and the exposure time in a high temperature environment, among other factors such as weather.

Thermal Strain

Core temperature, or the temperature of the central part of the body, rapidly increases during firefighting activity. Several studies have reported an increase in core temperature of approximately 1.3°F (0.7°C) after short bouts (20 minutes) of live fire drills (Colburn et al., 2011; Horn et al., 2011; Smith et al., 2011). A more recent study that focused on repeated bouts of firefighting activity over approximately 3 hours, as typically occurs during training, showed that core temperature rose more quickly and reached higher levels with successive bouts of work, with core temperature increasing by approximately 3°F (1.8°C) and exceeding 102°F (38.9°C) for some fire fighters by the end of the last work cycle (Horn et al., 2013). Thermal stress is also a primary concern for fire instructors during training drills. A British study that measured the core temperature of fire instructors reported an average increase in core temperature of 1.8°F (1°C) over the 40 minutes of live fire training evolutions (Eglin, Coles, and Tipton, 2004). Average core temperature increased to 101.3°F (38.5°C), and 8 of the 26 instructors had core temperatures over 102°F (38.8°C) after just 40 minutes of data collection. In a companion study, researchers investigated the ability of live fire training instructors to perform a simulated rescue after 40 minutes of a live fire training evolution (Eglin and Tipton, 2005). Ten minutes after the live fire training evolution, fire instructors were required to drag a 187-lb (84.8-kg) dummy 98 feet (29.9 meters). In six out of seven trials, they were able to do so. The authors concluded that *most* of the fire instructors were able to perform a rescue task after live fire training evolutions, but they were approaching their physical limit. Importantly, this study only investigated instructor's abilities to perform a rescue following a single 40-minute evolution and a 10-minute recovery period. It is common for instructors to perform multiple training evolutions in a single day and an emergency may happen without the benefit of a rest period. The ability to be able to perform a rescue is an important safety consideration for all instructors.

In addition to elevated body temperature, and in part because of elevated body temperature, firefighting also causes profuse sweating. The body sweats in an attempt to cool itself through a process called evaporative cooling. Unfortunately for fire fighters, PPE creates a warm, moist, and stagnant air layer next to the skin, which severely limits the evaporation of sweat. Without the evaporation of the sweat, the body becomes unable to use evaporative cooling as a method for temperature control. Profuse sweating can also decrease plasma volume, placing additional strain on the cardiovascular system and further impairing thermoregulation.

Researchers found a 15 percent reduction in plasma volume following approximately 18 minutes of strenuous simulated firefighting activity (Smith et al., 2001b). Sweat loss of 2.8 pounds (1.3 kg) per hour has been reported during exercise in a hot environment while wearing PPE. Sweat loss is a major concern during live fire training because individuals may be engaged in training over a relatively long time period. Furthermore, training on consecutive days may present an additional challenge if individuals do not fully rehydrate in between training days.

> **LIVE FIRE TIP**
>
> Firefighting training causes significant cardiovascular and thermal disruption, including the following:
> - Increased core temperature
> - Profuse sweat loss
> - Near maximal heart rate
> - Decreased stroke volume
> - Decreased plasma volume
>
> These physiological changes can lead to life-threatening pathological conditions including heatstroke and sudden cardiac events in extreme cases.

Cardiovascular Strain

Research studies have found that strenuous firefighting activities lead to near maximal heart rates that remain high for extended periods of time during fire suppression activities (Smith, Manning, and Petruzzello, 2001a; Horn et al., 2013). As early as the mid-1970s, there were studies reporting the heart rates of on-duty fire fighters. The researchers (Barnard and Duncan, 1975) documented one individual with a heart rate over 188 beats per minute for a 15-minute period compared to the average adult's heart that beats between 60 and 100 times per minute. While most fire fighters realize that firefighting results in an increase in heart rate and core temperature, relatively few fighters are aware of other cardiovascular changes that may have important implications for their health and safety. Firefighting affects all components of the cardiovascular system: heart, vessels, and blood (**FIGURE 17-7**). A summary of some of the major cardiovascular changes by component is provided here.

Several studies have found that heart rate reaches maximal or near maximal levels during strenuous firefighting. Furthermore, similar to core temperature, heart rate was found to increase with successive work cycles during prolonged firefighting (around 3 hours) (Horn et al., 2013). Blood pressure also increases during firefighting. The increases in heart rate and blood pressure increase the workload of the heart (myocardial work demand).

Evaluations done before and after firefighting drills have shown that **stroke volume** (the amount of blood ejected with each beat of the heart) decreases by 13–30 percent following strenuous firefighting (Smith et al., 2001a; Fernhall et al., 2012). This finding is especially troubling because the amount of blood being pumped out of the heart is decreased at the very time it is most needed. During firefighting, blood needs to be delivered to many areas of the body, such as the following:

- The working muscle, to support contraction
- The skin, to cool the body
- The heart, to support increased work associated with elevated heart rate
- Other vital organs including the brain

There is an increase in muscle and skin blood flow following firefighting. This change is consistent with the need to deliver more blood to the working muscle and to the skin (to dissipate heat). An increase in arterial stiffness has also been shown following extended firefighting training drills (Fahs et al., 2011). This finding could have both positive and negative effects with firefighting, and additional studies are required to recognize its implications.

Firefighting leads to an increase in sweating, which can lead to dehydration and decreased plasma volume. The decrease in plasma volume increases the viscosity (thickness of the blood). There is an increase in platelet number and platelet activity after firefighting that persists for at least 2 hours after firefighting, even when rehab is used. In addition to activating platelets, firefighting affects several clotting proteins and disrupts the balance between blood clot formation and breakdown. For several hours after firefighting activity, the enhanced clotting potential increases the risk of clot formation, which may lead to arterial occlusion and ischemia or a fatal arrhythmia (Smith et al., 2016).

FIGURE 17-8 summarizes the changes in the components of the cardiovascular system—cardiac, vascular, and blood/coagulatory—that have been documented during firefighting activity.

Cognitive Impairment

Firefighting stresses all systems of the body and leads to fatigue. The familiar feeling of fatigue experienced by instructors and trainees is due in part to the thermal and cardiovascular strain associated with firefighting. These physiological changes also affect cognitive function. It is well known that cognitive changes occur with heat illness, but subtle changes in cognitive function that are relevant to safety in an emergency

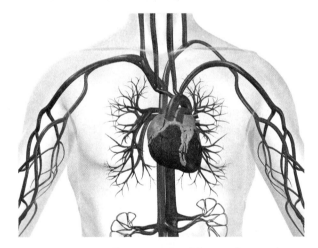

FIGURE 17-7 Components of the cardiovascular system: heart, vasculature, and blood.
© Sebastian Kaulitzki/ShutterStock, Inc.

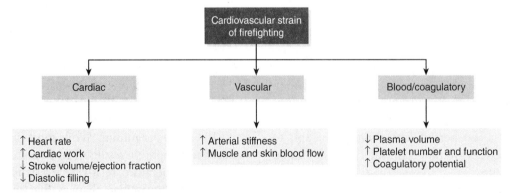

FIGURE 17-8 The cardiac, vascular, and blood/coagulatory responses to firefighting activity.
© Jones & Bartlett Learning.

environment may occur before heat illness is present. A study that investigated three short periods of firefighting (about 7 minutes each) with a 10-minute break between the second and third trial found a tendency for the error rate to increase on a computerized performance test (Smith et al., 2001a).

Factors Affecting Cardiovascular and Thermal Strain

The magnitude of the thermal and cardiovascular strain experienced by a fire fighter depends on several interrelated environmental and personal factors (**FIGURE 17-9**).

Environmental Conditions

Environmental conditions are a major contributing factor to the physiologic stresses of firefighting (Selkirk and McLellan, 2004). In most cases, this means that the heat of the fire contributes to the total heat stress of the individual. Performing any task in a hot, oppressive environment creates greater physiological strain than performing the same task in moderate conditions. The environment in which fire fighters and live fire training instructors conduct live fire training varies greatly from a moderate level of heat to conditions so severe that they can only be tolerated for a brief period of time before damage to PPE can occur and physical injury can result.

An instructor must consider the potential for burn- and heat-related injuries when working in different conditions. Heat exposure can be in the form of the ambient environment or direct exposure to flames or another heat source. Heat exposure poses different challenges to the human body depending on the absolute temperature, or heat flux, and the duration of exposure. High ambient conditions combined with direct radiant heating will increase the effects on the human body. The effects of heat and the length of time the human body can sustain such conditions will depend on the intensity of the heat source and the duration of the exposure.

A chart that identifies routine, hazardous, extreme, and critical exposures that fire fighters may encounter is provided here (**FIGURE 17-10**). Individuals should not be exposed to critical conditions in training settings. Fire instructors must be keenly aware of fire dynamics during training and not expose themselves or students to untenable conditions in a flow path. They must consider the cumulative nature of heat stress even when students are exposed to "routine" fire conditions for prolonged or repeated bouts of firefighting. It is important to keep in mind that even exposure to "routine" conditions presents a considerable challenge to the human cardiovascular and thermoregulatory systems.

The fire environment also contains smoke, which contains numerous compounds, many of which are known or suspected carcinogens, and particulate matter. The toxins in smoke can cause multiple acute and long-term health concerns, including respiratory issues, triggering of a cardiac event, and increased risk of cancer.

Work Performed

On the fire ground and in training scenarios, there are a number of different tasks performed, including climbing stairs with heavy loads, performing a search, advancing a line, forcing open a door, hauling pallets and hay, and overhauling a room. The individual tasks and the intensity at which they are performed have a strong influence on physiological responses to firefighting. Clearly, the work being performed has a major influence on the physiological response of the body—the more strenuous the work and the longer the duration, the greater the thermal and cardiovascular responses. Physical work creates metabolic heat, which leads to an increase in body temperature, and thus adds additional cardiovascular strain. It is important to recognize the cumulative effect of performing multiple evolutions or prolonged work throughout the day. In a training environment, it is critical to account for additional work the instructors and trainees do in setting up for evolutions and cleaning up after drills. This work adds to the cumulative work of the day, as does physical fitness training that may be done as part of a training academy.

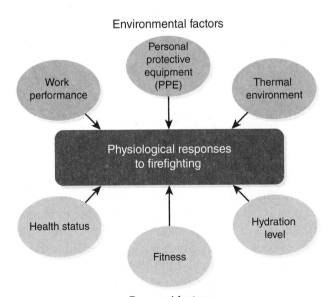

FIGURE 17-9 Factors affecting physiological responses to firefighting.
© Jones & Bartlett Learning.

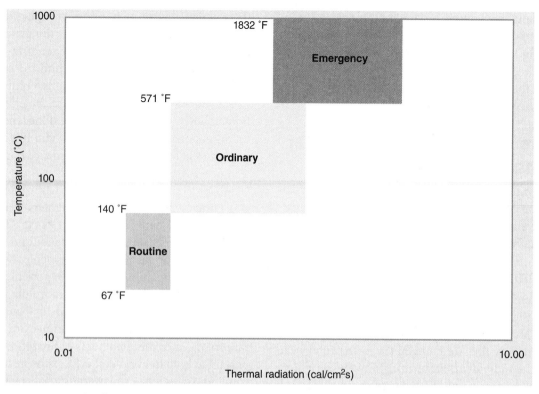

FIGURE 17-10 Levels of exposure encountered during firefighting.
Courtesy of the USFA.

Personal Protective Equipment

PPE is necessary to protect fire fighters from burn and inhalation injuries. Because of its weight and restrictive properties, PPE adds to a fire fighter's work. And improperly fitting PPE can add even more physiological strain on an individual. PPE also interferes with heat loss because of encapsulation. Thus, PPE adds to the cardiovascular and thermal strain associated with firefighting. In one laboratory study (Smith et al., 1995), fire fighters were asked to walk in full PPE with a SCBA. The fire fighters reached heart rates around 178 beats per minute, which was, on average, 55 beats per minute higher than when they walked in a station uniform with a SCBA.

Wearing PPE is absolutely essential to safety in training evolutions; however, it does add to the physiological stresses that the live fire training fire fighter encounters. Thus, a live fire training instructor should be aware of the amount of time that trainees are in full gear. *During debriefing periods, breaks, and rehabilitation, gear should be removed to allow for recovery.*

The air outside is much cooler than the air inside of a fire fighter's gear, which will cause sweat to evaporate as gear is removed, giving the fire fighter a feeling of being cool. However, body temperature does not decrease as quickly as the mind perceives it to. In fact, body temperature often continues to rise even after firefighting activity has ended, and it remains elevated for up to an hour after firefighting. Therefore, to decrease body temperature, it is necessary to doff the bunker coat and pants and to intentionally cool the body.

> **LIVE FIRE TIP**
>
> PPE is essential to protect fire instructors and students from burn injuries and smoke inhalation, and it should be worn in compliance with manufacturer's recommendations and local policies during live fire training. However, the live fire training instructor must also understand how the weight and insulative properties of PPE add to heat stress and cardiovascular strain. PPE should be doffed, when appropriate, to allow body temperature to decrease.

Individual Characteristics

A fire fighter's age, sex, and body size all affect the physiological responses to firefighting activity. In general, the risk of heart attack while performing firefighting work increases as the age of the fire fighter increases (Kales et al., 2007). The number of fatalities due to sudden cardiac events versus other causes during training over a 10-year period is shown in **FIGURE 17-11**. Note that as age increases to 51–55 years, so does the percentage of fatalities due to sudden cardiac events (Fahy, 2017). What may be surprising is the number of cardiac events that occur in young fire fighters. Instructors must not become complacent when working with young fire fighters.

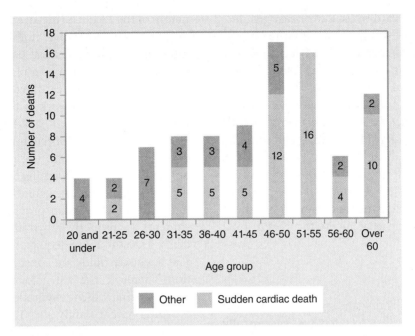

FIGURE 17-11 Training fatalities by age and cause of death (2007–2016).
Rita F. Fahy, U.S. Firefighter Deaths Related to Training, 2007–2016. National Fire Protection Association, 2017.

Excess body fat creates additional cardiovascular strain on a fire fighter by adding to the metabolic work that must be done to move his or her body mass. Excess body fat also increases the thermal strain by providing insulation, impeding the range of motion and mobility, and interfering with heat dissipation. Additionally, and of graver consequence, obesity increases the risk of coronary heart disease (CHD), duty-related death, and disability and the risk of injury (Kales et al., 2003).

Medical Conditions

Firefighting activities can be extremely strenuous. Fire fighters should be in good health to be able to operate safely on the fire ground or in a training situation because of the combination of heavy work, severe heat, and mental stress. *A live fire training instructor who is not medically cleared for firefighting presents a risk not only to himself but also to the students under his or her command.* A fire fighter with a preexisting cardiovascular disease is more susceptible to cardiac events on the fire ground because of the additional cardiovascular strain associated with firefighting. In fact, a retrospective analysis of sudden cardiac events over a 10-year period found that 75 percent of fire fighters who suffered fatal cardiac events had preexisting cardiovascular conditions.

Not only a diagnosis of cardiovascular disease but simply having cardiovascular disease risk factors is associated with a severely increased risk of CHD-related death. As shown in **TABLE 17-2**, the increased odds of CHD-related death range from 3- to 12-fold for obesity, high cholesterol, diabetes mellitus, hypertension,

TABLE 17-2 Risk of Duty-Related CHD by Risk Factor

	Odds Ratio
Obesity (BMI ≥30 kg/m^2)	3.1
Cholesterol (≥200 mg/dL)	4.4
Diabetes	10.2
Hypertension	12.0
Current smoking	8.6
Prior diagnosis of CHD	35.0

BMI, body mass index; CHD, coronary heart disease. Odds ratio represents the likelihood that an outcome will occur (here, cardiac fatality) given a particular exposure (here, a cardiovascular disease risk factor).

Based on data from Kales, Stefanos N., Elpidoforos S. Soteriades, Stavros G. Christoudias, and David C. Christiani. "Firefighters and on-duty deaths from coronary heart disease: a case control study." *Environmental Health* 2, no. 2 (2003): 14.

and current smoking (Kales et al., 2003). The greatest increased risk of duty-related cardiac death is among those who have a prior diagnosis of CHD. In terms of traditional risk factors that are assessed during routine medical evaluations, hypertension was associated with the greatest risk of CHD, with hypertensive fire fighters having a 12-fold increased risk compared with fire fighters with normal blood pressure.

Importantly, these cardiovascular disease risk factors are modifiable—if you have one or more of these risk factors, you can change your behaviors to reduce your risk of developing cardiovascular disease and/or suffering a duty-related cardiac death.

NFPA 1582: *Standard on Comprehensive Occupational Medical Program for Fire Departments*, identifies the 13 essential tasks of firefighting and provides guidelines for the medical clearance of fire fighters. In order to ensure that fire instructors can safely engage in live fire training and meet the responsibilities they have to their students, they should have a medical examination that meets the NFPA 1582 standard.

Fitness Level

Firefighting is physically demanding work, and a high level of fitness is necessary to safely and successfully perform the job duties. A fit fire fighter can perform the same amount of work with less cardiovascular and thermal strain than a less fit fire fighter. A fit fire fighter also has a greater energy reserve needed to perform more work. Fire fighters need to possess a high degree of muscular fitness (muscle strength and endurance) and cardiovascular fitness (aerobic fitness). Cardiovascular fitness is especially important because it does the following:

- Increases the efficiency of the heart
- Improves work capacity
- Increases plasma volume
- Improves thermal tolerance
- Decreases tendency of blood to clot
- Enhances the ability of blood vessels to dilate to allow more blood to be supplied to muscles

The components of a health-related fitness program are outlined in NFPA 1583: *Standard on Health-Related Fitness Programs for Fire Department Members*. As an instructor, you have an obligation to pay careful attention to your own fitness level because of the following:

- Your fitness level directly affects your risk of heat-related illness or sudden cardiac events during training.
- Your fitness level affects your ability to perform strenuous firefighting activity, including the possibility of needing to rescue a participant.
- You are a role model for your students, and your approach to fitness will affect your students' attitudes toward fitness and their respect for you.

Hydration Status

Firefighting leads to a large amount of sweat loss due to the heavy work, hot conditions, and the impermeable nature of the PPE. During strenuous work in hot conditions, or in protective clothing, humans can lose more than 2 quarts of sweat per hour. This sweat loss contributes to a decrease in plasma volume, adds additional strain on the cardiovascular system, and decreases the ability to work in the heat, also known as **thermal tolerance**.

Live fire training instructors need to be well hydrated before they begin training drills and need to be vigilant about consuming water or sports drinks throughout the training evolutions. Likewise, instructors must be diligent in providing opportunities for students to properly rehydrate during the training drills. Thirst is an inadequate mechanism of ensuring that adequate fluids are consumed to avoid dehydration. Therefore, the instructor should have structured breaks during which rehydration is emphasized.

Water is constantly lost from the body by urinating, sweating, and breathing, so it is critical to replenish the supply. A sedentary person requires approximately 3 liters (1 liter = 1.06 quarts) of fluid per day (Institute of Medicine, 2005). Exercise or work under hot conditions, such as those routinely encountered during firefighting, can dramatically increase the need for fluid. Humans routinely lose 1–2 liters of sweat per hour when working in hot and humid environments. Given that training often occurs over a period of several hours, this high rate at which sweat is lost can lead to severe dehydration. Fire fighters should be mindful to consume adequate fluids to replace what is lost during normal fluid turnover plus what is lost with sweating. Furthermore, fluid ingestion should consist primarily of pure water, with electrolyte-replenishing liquids used only as supplements when sweating profusely. Caffeinated beverages such as coffee, tea, and soda can act as diuretics and increase water loss, exacerbating rather than helping dehydration. An additional concern is that many beverages, including commonly available energy drinks, soda, and sweetened tea, include a lot of sugar, which is often associated with excess body weight, another factor that increases heat stress. Common energy drinks often include caffeine and other ingredients that can increase metabolic rate and add to heat stress. These beverages should be avoided in a training or operational setting.

Given the detrimental effects of dehydration, and evidence that many fire fighters are dehydrated even before beginning work, it is a good idea to monitor your hydration status. The easiest way to do this is with the use of a simple urine chart (**FIGURE 17-12**). These charts can be placed in restrooms to serve as a reminder to instructors and students that urine should be a light color and should not have a strong odor. Urine color

1	PROPERLY HYDRATED
2	If your urine matches the colors 1–3 (above the red line), you are properly hydrated and should continue to consume fluids at the recommended amounts
3	
4	DEHYDRATED
5	If your urine matches the colors 4–8 (below the red line), you are dehydrated and at risk for cramping and heat illness!
6	
7	YOU NEED TO DRINK MORE WATER OR SPORTS DRINK
8	

FIGURE 17-12 Urine color and hydration.
© Jones & Bartlett Learning.

is affected by many factors, including medication and diet, so it is wise to have individuals follow a hydration program that ensures that they are well hydrated. From there, one should be able to determine the proper color of his or her urine. Students and instructors should be reminded regularly to consume fluid that ensures light urine or urine that is consistent with their urine color in a well-hydrated state.

Heat Emergencies

Heat illnesses can be fatal, and there are far too many instances of fire fighters suffering fatal heatstroke during training. Instructors have an obligation to be alert to the dangers of heat illness and to help decrease the number of fatal heat illnesses that occur in training. The United States Fire Administration's report, *Emergency Incident Rehabilitation* (2008), documents seven case studies in which fire fighters have died due to heat illness following work or training. These tragic case studies should be read by every fire instructor because of the lessons that can be learned from them, the first of which is that fatal heatstroke can occur during training. Another important and related lesson is that heatstroke can affect young and old fire fighters alike. In fact, young recruits are especially vulnerable because of their extreme motivation to succeed in the training environment. Under these conditions, many young fire fighters push themselves far beyond what is physiologically safe.

Working in protective clothing with high ambient temperatures presents a serious challenge to the thermoregulatory and cardiovascular systems. If the cardiovascular system cannot meet the simultaneous demands of supplying adequate blood to the muscles and maintaining thermal balance, heat illness may ensue. Heat illness covers a spectrum of disorders from heat rash to life-threatening heatstroke. Heat-related illnesses occur because the body is unable to maintain thermal balance.

Thermal Balance

Normally, the body regulates its internal body temperature within a narrow range, despite wide variations in environmental temperatures. The process by which the body regulates body temperature is called **thermoregulation**. The human body typically regulates its temperature around 98.6°F (37°C). It is important to maintain a temperature within approximately 1.5°F (0.7°C) of this temperature because changes in body temperature dramatically affect biological function. Greater temperature changes can alter chemical reactions and, ultimately, directly damage body tissue.

Body temperature results from a balance between heat gain and heat loss. Heat can be gained from the environment, when the ambient temperature is higher than the body temperature, and from metabolic heat produced by the body as a result of muscular work. Heat can be exchanged from the body through four processes: **radiation**, **conduction**, **convection**, and evaporation. The extent of heat gain or loss through these processes depends on environmental conditions such as ambient temperature, relative humidity, and wind speed.

The effectiveness of heat exchange between an individual and the environment is affected by five factors:

1. Thermal gradient
2. Relative humidity
3. Air movement
4. Degree of direct sunlight
5. Clothing worn

The greater the difference between two temperatures, known as the **thermal gradient**, the greater the heat loss is from the warmer of the two. Typically, the body is warmer than the environment, so heat moves down its thermal gradient to the environment. More heat is lost in cooler environments because the thermal gradient is greater. In firefighting situations, heat may be gained by the body.

The body's use of evaporative cooling techniques is decreased in high-humidity conditions because the air already contains an abundance of water vapor. Evaporative cooling is largely determined by the relative humidity; on humid days, evaporative cooling is limited. Although a fire fighter may sweat profusely when humidity is high, the body doesn't properly cool down because the sweat does not evaporate as effectively.

Air movement increases convective heat loss from the skin to the environment. Thus, on windy days more heat is lost from the body. Conversely, when a fire fighter's skin is covered by heavy PPE, heat loss is minimized.

Direct sunlight can add considerably to the radiant heat load of an individual, just as shade or cloud cover can often provide significant relief from heat. Measures should be taken to avoid having students spend prolonged periods of time in PPE in direct sunlight on hot days, as it can cause dangerous increases in body temperature.

When the body is in thermal balance, the amount of heat lost equals the amount of heat produced, and body temperature remains constant. However, when heat produced (and absorbed) exceeds heat loss, an increase in body temperature occurs. This increase in body temperature places considerable strain on the cardiovascular system, hastens fatigue, promotes sweating and loss of plasma volume, and may lead to serious heat illnesses.

Heat Illness

Heat illness includes a spectrum of disorders, resulting specifically from the combined stresses of exertion and high-heat situations. Heat illness affects many systems of the body and can cause elevated core body temperature and impaired thermoregulation. It varies greatly in severity from heat rash, sunburn, and heat cramps, to heat exhaustion and heatstroke. One of the greatest challenges in dealing with heat illnesses is distinguishing among disorders because they frequently overlap and can evolve into different forms over time. A summary of the causes, signs, symptoms, and treatments for heat illnesses is provided in **TABLE 17-3**. The most common and serious heat illnesses encountered during live fire training are heat cramps, heat exhaustion, and heatstroke. Symptoms of heat illness may be nonspecific, particularly in the early stages. As heat illnesses progress, so does the severity of signs and symptoms.

Cardiac Emergencies

Sudden cardiac events are the leading cause of line-of-duty deaths among fire fighters. Each year, approximately 45–50 percent of fire fighter deaths are caused by sudden cardiac events (Fahy, LeBlanc, and Molis, 2016). In addition to the fatalities, approximately 800–1000 fire fighters suffer nonfatal cardiovascular disease events while on duty (Haynes and Molis, 2016).

TABLE 17-3 Heat Illness Classifications

Classification	Cause	Signs and Symptoms	Treatment	Prevention
Heat rash (also called prickly heat or miliaria)	Hot, humid environment; clogged sweat glands.	Red, bumpy rash with severe itching.	Change into dry clothes and avoid hot environments. Rinse skin with cool water.	Wash regularly to keep skin clean and dry.
Sunburn	Too much exposure to the sun.	Red, painful, or blistering and peeling skin.	If the skin blisters, seek medical aid. Use skin lotions (avoid topical anesthetics) and work in the shade.	Work in the shade; cover skin with clothing; apply skin lotions with a sun protection factor of at least 15. Fair people have greater risk.
Heat cramps	Heavy sweating depletes the body of salt, which cannot be replaced just by drinking water.	Painful cramps in arms, legs, or stomach that occur suddenly at work or following work. Heat cramps are serious because they can be a warning sign of other more dangerous heat-induced illnesses.	Move to a cool area; loosen clothing and drink a commercial fluid replacement beverage (sports drink). If the cramps are severe or don't go away, seek medical aid.	Reduce activity level and/or heat exposure. Drink fluids regularly. Workers should check each other to help spot the symptoms that often precede heatstroke.

Classification	Cause	Signs and Symptoms	Treatment	Prevention
Heat exhaustion	Fluid loss, inadequate salt, and the cooling system begins to break down.	Heavy sweating; elevated body temperature; weak pulse; normal or low blood pressure; person is tired and weak or faint, has nausea and vomiting, is very thirsty, or is panting or breathing rapidly; vision can be blurred.	**GET MEDICAL AID.** This condition can lead to heatstroke, which can kill. Remove gear, move the person to a cool shaded area. Loosen or remove excess clothing; provide sports drink. Use active cooling (forearm immersion, misting fans, or cold towels) to lower core body temperature.	Reduce activity level and/or heat exposure. Drink fluids (water and sports drink) regularly to compensate for sweat loss. Monitor participants frequently to assess symptoms.
Heatstroke	Elevation of body temperature due to a breakdown of thermoregulatory mechanisms. Caused by depletion of salt and water reserves. Heatstroke can develop suddenly or can follow from heat exhaustion.	Body temperature is very high and any of the following: the person is weak, confused, upset, or acting strangely; has hot, dry, red skin; a fast pulse; headache or dizziness. In later stages, a person can pass out and have convulsions.	**IMMEDIATELY TRANSPORT TO A MEDICAL FACILITY.** This is a life-threatening emergency. If transport is delayed, immediately immerse body in cold water.	Reduce activity level and/or heat exposure. Drink fluids (water and sports drink) regularly to compensate for sweat loss. Monitor participants frequently to assess symptoms.

Reproduced with permission from NFPA 1584: *Standard on the Rehabilitation Process for Members During Emergency Operations and Training Exercises*, Copyright © 2008, National Fire Protection Association. This reprinted material is not the complete and official position of the NFPA on the referenced subject, which is represented only by the standard in its entirety.

Sudden cardiac events also occur during fire fighter training. In fire fighters older than 35 years of age, approximately 68 percent of fire fighter fatalities during training were the result of sudden cardiac events (Fahy, 2017). A live fire training instructor must understand the factors that increase the risk of sudden cardiac events and do everything in his or her control to lessen these risks.

Prevention of Cardiac Emergencies

Live fire training often involves performing strenuous muscular work while wearing heavy personal protective clothing under hot and hostile conditions. This level of exertion can trigger a sudden cardiac event in individuals with underlying cardiovascular disorders. In order to minimize the risk of cardiac emergencies, fire instructors should do the following:

- Be aware of the risk factors for cardiovascular disease.
- Work to ensure that students have medical clearance to engage in structural firefighting.
- Be aware of signs and symptoms of a heart attack.
- Establish incident scene rehabilitation that provides medical monitoring of personnel.

Heart attacks can occur suddenly and be associated with intense symptoms, or they may have a gradual onset. A fire fighter complaining of severe chest pain and radiating pain in the arm is showing signs of a possible heart attack. In less obvious cases, a fire fighter may convey vague information such as needing to rest or not feeling well. In some cases, the fire fighter may believe he is suffering from indigestion. It is very important that information of this sort be taken seriously, and that a fire fighter who makes such complaints receives medical attention. In the case of a heart attack, the sooner treatment is initiated, the greater the chance of survival. Perhaps the most important initiative that can be taken to prevent cardiac emergencies is to have an effective incident scene rehabilitation area staffed with dedicated, trained emergency medical personnel responsible for monitoring the medical conditions of participants.

Incident Scene Rehabilitation

Incident scene rehabilitation for live fire training is defined as an intervention designed to lessen the physical, physiological, and emotional stresses of firefighting with a goal of improving performance and decreasing the likelihood of on-scene injury or death. Effective incident scene rehabilitation can mitigate the effects of some of the detrimental physiological problems of firefighting by providing rest, rehydration, and cooling of fire fighters. Incident scene rehabilitation can also help identify medical problems early, and may prevent potentially serious consequences by providing appropriate medical monitoring.

Crew Resource Management

Sam Spatzer is an attorney who has been involved in civil litigation concerning the failure to follow NFPA 1403. He suggests putting some of the basics of the standard into the recruit curriculum, so recruits know if they are not given a walkthrough or an emergency briefing, or there are not PPE inspections, that they need to ask. Spatzer went on to suggest that the "buck stops at the top," and, "if you want accountability, have the fire chief and training chief sign off on compliance to the NFPA 1403 standard and legal requirements. Take away the excuse of plausible deniability. The fire service has had the standards for some time now, but what good is it if it isn't followed? Enforcement must come internally from the top down, and not depend on the state fire marshal or other outside agencies for enforcement." He also suggests that avoidance may come from even the newest recruit if they know what is supposed to be done, and if they have a way without fear of repercussion to let the leadership know of problems. Utilize a crew resource management approach to safety. Everyone on the evolution should be empowered to draw attention to a potential problem without fear of repercussions.

After-Action REVIEW

IN SUMMARY

- Live fire training is the pinnacle of a recruit's training experience in an academy. NFPA 1403 requires that prior to exposing a recruit to live fire training, they must meet the job performance requirements in NFPA 1001: *Standard for Fire Fighter Professional Qualifications*.
- Live fire training should *not* be the testing ground for these skills. Instead, it should bring together all of the skills students have successfully demonstrated up to this point in their training.
- Live fire instructors must ensure students are prepared for the evolution, including wearing appropriate gear that is being correctly used to participate in the evolution.
- Live fire training instructors are responsible to ensure that all required PPE is in serviceable condition and properly worn. The instructor assigned to be the safety officer for the evolution needs to ensure that all students, instructors, safety personnel, and any other participants wear their equipment appropriately.
- You must be able to maintain the integrity of the crew and have accountability for them throughout the entire evolution. All participants must understand that crew accountability is in place for one reason: safety.
- Reading smoke or fire behavior, understanding room-and-contents versus deep-seated structure fires, and recognizing normal fire spread and extreme or hostile fire behavior are all skills that you need to be able to observe and correctly identify during suppression operations.
- Understanding the flow path in all structures is paramount when determining where a fire will go, controlling the amount of oxygen available to the fire, and the extension of the fire if using an acquired structure.
- Before conducting any type of class, know who your audience is going to be. This is especially true when conducting the high-risk training evolutions associated with live fire training.

KEY TERMS

Access Navigate for flashcards to test your key term knowledge.

Conduction Heat transfer to another body or within a body by direct contact (NFPA 1403).

Convection Heat transfer by circulation within a medium such as a gas or liquid (NFPA 1403).

Core temperature The body's internal temperature.

Flameover (or rollover) The condition in which unburned fuel (pyrolysate) from the originating fire has accumulated in the ceiling layer to a sufficient concentration (i.e., at or above the lower flammable limit) that it ignites and burns. Flameover can occur without ignition of or prior to the ignition of other fuels separate from the origin (NFPA 1403).

Flashover A transition phase in the development of a compartment fire in which surfaces exposed to thermal radiation reach ignition temperature more or less simultaneously and fire spreads rapidly throughout the space, resulting in full room involvement or total involvement of the compartment or enclosed space (NFPA 1403).

Flow path The movement of heat and smoke from the higher pressure within the fire area towards the lower pressure areas accessible via doors, window openings, and roof structures (NFPA 1410).

Fuel load The total quantity of combustible contents of a building, space, or fire area, including interior finish and trim, expressed in heat units or the equivalent weight in wood (NFPA 1403).

Heat release rate The rate at which heat energy is generated by burning (NFPA 921).

Immediately dangerous to life or health (IDLH) Any condition that would pose an immediate or delayed threat to life, cause adverse health effects, or interfere with an individual's ability to escape unaided from a hazardous environment.

Incident scene rehabilitation A function on the fireground that cares for the well-being of the fire fighters. It includes physical assessment, revitalization, medical evaluation and treatment, and regular monitoring of vital signs.

Pyrolysis A process in which material is decomposed, or broken down, into simpler molecular compounds by the effects of heat alone; pyrolysis often precedes combustion (NFPA 921).

Radiation Heat transfer by way of electromagnetic energy (NFPA 1403).

Stroke volume The volume of blood pumped by the left ventricle with each contraction of the heart.

Thermal gradient The rate of temperature change with distance.

Thermal tolerance The body's ability to cope with high heat conditions.

Thermoregulation The process by which the body regulates temperature.

Ventilation-controlled fire A fire in which the heat release rate or growth is controlled by the amount of air available to the fire (NFPA 1403).

Vent point ignition Smoke that is at or above its ignition temperature and is lacking oxygen. The smoke will ignite as it exits the opening and falls within the flammable range.

REFERENCES

Barnard, R. James, and Henry W. Duncan. "Heart rate and ECG responses of fire fighters." *Journal of Occupational Medicine* 17 (1975): 247–250.

Colburn, Deanna, Joe Suyama, Steven E. Reis, Julia L. Morley, Fredric L. Goss, Yi-Fan Chen, Charity G. Moore, and David Hostler. "A comparison of cooling techniques in firefighters after a live burn evolution." *Prehospital Emergency Care* 15, no. 2 (2011): 226–232.

Kales, Stefanos N., Elpidoforos S. Soteriades, Costas A. Christophi, and David C. Christiani. "Emergency duties and deaths from heart disease among fire fighters in the United States." *New England Journal of Medicine* 356, 12 (2007): 1207–1215.

Kales, Stefanos N., Elpidoforos S. Soteriades, Stavros G. Christoudias, and David C. Christiani. "Firefighters and on-duty deaths from coronary heart disease: A case control study." *Environmental Health* 2, no. 1 (2003): 14.

Eglin, Clare M., Sue Coles, and Michael J. Tipton. "Physiological responses of fire-fighter instructors during training exercises." *Ergonomics* 47 (2004): 483–494.

Eglin, Clare M., and Michael J. Tipton. "Can fire fighter instructors perform a simulated rescue after a live fire training exercise?" *European Journal of Applied Physiology* 95 (2005): 327–334.

Fahs, Christopher A., Huimin Yan, Sushant Ranadive, Lindy M. Rossow, Stamatis Agiovlasitis, George Echols, Denise Smith,

Gavin P. Horn, Thomas Rowland, Abbi Lane, and Bo Fernhall. "Acute effects of firefighting on arterial stiffness and blood flow." *Vascular Medicine* 16, no. 2 (2011): 113–118.

Fahy, Rita F. 2017. *U.S. Firefighter Deaths Related to Training, 2007–2016*. Quincy, MA: National Fire Protection Association.

Fahy, Rita R., Paul R. LeBlanc, and Joseph L. Molis. 2016. "Firefighter Fatalities in the United States." Quincy, MA: National Fire Protection Agency.

Fernhall, Bo, Christopher A. Fahs, Gavin Horn, Thomas Rowland, and Denise Smith. "Acute effects of firefighting on cardiac performance." *European Journal of Applied Physiology* 112, no. 2 (2012): 735–741.

Haynes, Hylton G., and Joseph L. Molis. *U.S. Firefighter Injuries – 2015*. Quincy, MA: National Fire Protection Association; 2016.

Horn, Gavin P., Sue Blevins, Bo Fernhall, and Denise L. Smith. "Core temperature and heart rate response to repeated bouts of firefighting activities." *Ergonomics* 56, no. 9 (2013): 1465–1473.

Horn, Gavin P., Steve Gutzmer, Christopher A. Fahs, Steve J. Petruzzello, Eric Goldstein, George C. Fahey, Bo Fernhall, and Denise L. Smith. "Physiological recovery from firefighting activities in rehabilitation and beyond." *Prehospital Emergency Care* 15, no. 2 (2011): 214–215.

Institute of Medicine. *Dietary Reference Intakes for Water, Potassium, Sodium, Chloride, and Sulfate*. Washington, DC: The National Academies Press; 2005.

Selkirk, Glen. A., and Thomas M. McLellan. "Physical work limits for Toronto fire fighters in warm environments." *Journal of Occupational and Environmental Hygiene* 1, no. 4 (2004): 199–212.

Smith, Denise L., Jacob P. DeBlois, Stefanos N. Kales, and Gavin P. Horn. "Cardiovascular strain of firefighting and the risk of sudden cardiac events." *Exercise and Sport Sciences Reviews* 44, no. 3 (2016): 90–97.

Smith, Denise L., Steven J. Petruzzello, Eric Goldstein, Uzma Ahmed, Krishnarao Tangella, Gregory G. Freund, and Gavin P. Horn. "Effect of live-fire training drills on firefighters' platelet number and function." *Prehospital Emergency Care* 15, no. 2 (2011): 233–239.

Smith, Denise L., Timothy S. Manning, and Steven J. Petruzzello. "Effect of strenuous live-fire drills on cardiovascular and psychological responses of recruit fire fighters." *Ergonomics* 44, no. 3 (2001a): 244–254.

Smith, Denise L., Steven J. Petruzzello, Mike A. Chludzinski, John J. Reed, and Jeffrey A. Woods. "Effect of strenuous live-fire firefighting drills on hematological, blood chemistry and psychological measures." *Journal of Thermal Biology* 26 (2001b): 375–379.

Smith, Denise L., Steven J. Petruzzello, Jeff M. Kramer, Sarah E. Warner, Bradley G. Bones, and James E. Misner. "Selected physiological and psychobiological responses to physical activity in different configurations of firefighting gear." *Ergonomics* 38, no. 10 (1995): 2065–2077.

National Fire Protection Association. 2018. NFPA 1403: *Standard on Live Fire Training Evolutions*. Quincy, MA: National Fire Protection Association.

National Fire Protection Association. 2019. NFPA 1041: *Standard for Fire and Emergency Services Instructor Professional Qualifications*. Quincy, MA: National Fire Protection Association.

United States Fire Administration. *Emergency Incident Rehabilitation*. Washington, DC: United States Fire Administration; 2008.

REVIEW QUESTIONS

1. You have not worked with this particular department in a live fire training burn before, but you have responded to fires with their company and have noticed significant problems with their PPE and SCBA. How will you handle this?

2. Explain reasons that a mayday call would be appropriate.

3. Why is it important to follow NFPA 1403: *Standard on Live Fire Training Evolutions* for any live burn evolution?

CHAPTER 18

Preburn Inspection and Planning

NFPA 1041 JOB PERFORMANCE REQUIREMENTS

8.2.1 Prepare a preburn plan in compliance with NFPA 1403, given the authority having jurisdiction (AHJ) policy and procedures for live fire training evolutions, the facility policies applicable to evolutions, learning objectives, and all conditions affecting the evolution, so that learning objectives are developed, the plan meets all AHJ requirements, existing conditions are identified, and the plan meets the developed learning objectives.

(A) Requisite Knowledge.
NFPA 1403, components of learning objectives, AHJ and facility policies and procedures, hazards associated with live fire training, fuel packages, burn room size, ventilation strategies, time between sequential burn evolutions, evidence-based practices for fire control, and training procedures.

(B) Requisite Skills.
Learning objective development, technical writing, preburn plan development.

8.2.2 Conduct a preburn inspection of the structure or prop, given a structure or prop for live fire training, so that structural damage is identified, structural preparation is determined, and safety concerns are identified and addressed prior to the live fire evolution.

(A) Requisite Knowledge.
Facility requirements, structure or prop considerations.

(B) Requisite Skills.
Observation techniques, inspection and evaluation skills.

8.2.3 Calculate the minimum water supply required for a live fire evolution in compliance with NFPA 1403, Section 4.12, given a structure or prop so that the required minimum water supply is determined.

(A) Requisite Knowledge.
NFPA 1403, fire flow calculations.

(B) Requisite Skills.
Calculation of water supply requirements, development of water supply documentation

8.2.4 Calculate the minimum water flow application rate for a live fire evolution in compliance with NFPA 1403, Section 4.12, given a structure or prop so that the required minimum water flow application rate is determined.

(A) Requisite Knowledge.
NFPA 1403, fire flow calculations, capacity of hose lines, fire-ground hydraulics.

(B) Requisite Skills.
Calculation of minimum water flow application rate.

KNOWLEDGE OBJECTIVES

After studying this chapter, participating in a structured learning environment, and completing assigned assessments, you will be able to:

- List the minimum content areas of a preburn plan for live fire training needed to comply with NFPA 1403. (**NFPA 1041: 8.2.1**, p 405)
- Explain why learning objectives and a list of training evolutions must be generated for a live burn exercise. (**NFPA 1041: 8.2.1**, p 405)

- Identify participants and their roles in live fire training. (**NFPA 1041: 8.2.1**, p 406)
- Identify water supply needs for live fire training. (**NFPA 1041: 8.2.1**, pp 406–409)
- Identify the appropriate fuel materials for live fire training. (**NFPA 1041: 8.2.1**, p 411)
- Identify apparatus needs for live fire training. (**NFPA 1041: 8.2.1**, p 409)
- Identify parking and areas of operation for live fire training. (**NFPA 1041: 8.2.1**, pp 411–412)
- Establish appropriate control measures for visitors and spectators. (**NFPA 1041: 8.2.1**, p 412)
- Explain why building and site plans are necessary for a burn plan. (**NFPA 1041: 8.2.1**, pp 411–412)
- Explain the need for an emergency plan to be included in a live fire plan. (**NFPA 1041: 8.2.1**, pp 412–413)
- Describe the general order of operations for live fire training. (**NFPA 1041: 8.2.1**, pp 414–415)
- Identify safety hazards in live fire training and describe the steps to mitigate risk during the preparation and operation phases of live fire training. (**NFPA 1041: 8.2.1**, pp 416–427)
- Specify personal protective equipment (PPE) requirements for live fire training evolutions. (**NFPA 1041: 8.2.1**, pp 414–416)
- Describe how to perform a preburn inspection on an acquired structure. (**NFPA 1041: 8.2.2**, pp 416–423)
- Describe how to conduct a preburn inspection on props and facilities for live fire training. (**NFPA 1041: 8.2.2**, pp 425–426)
- Identify accepted methods for calculating water supply requirements. (**NFPA 1041: 8.2.3, 8.2.4**, pp 406–409)

SKILLS OBJECTIVES

After studying this chapter, participating in a structured learning environment, and completing assigned assessments, you will be able to:

- Calculate and document the minimum water supply required for a live fire evolution in accordance with NFPA 1403, Section 4.12. (**NFPA 1041: 8.2.3 , 8.2.4**, pp 406–409)
- Prepare a written preburn plan for an acquired structure to use for live fire training exercises for compliance with NFPA 1403. (**NFPA 1041: 8.2.1**, pp 406–416)
- Conduct a preburn inspection on an acquired structure for live fire training. (**NFPA 1041: 8.2.2**, pp 416–423)
- Conduct a preburn inspection on a facility for live fire training. (**NFPA 1041: 8.2.2**, pp 423–425)
- Conduct a preburn inspection on a prop for live fire training. (**NFPA 1041: 8.2.2**, pp 425–426)

You Are the Live Fire Training Instructor

You are the live fire training officer in your fire department. The department owns a Class A prop. You are instructed to conduct an in-service training for current fire fighters. Several of the department's chiefs want to run the training like it's a "regular fire," letting fire fighters pull up, advance lines, search for victims, and finally, put out the fire.

1. What immediate problems do you see with this method?

2. While remaining compliant with NFPA 1403, what training objectives will be as close as you can get to the chiefs' directives?

Access Navigate for more practice activities.

Introduction

Organizing a live fire training evolution requires time and work. If the training is to be valuable and safe, the instructor(s) must take the time and effort to plan and organize the event. Cost is also a reality. In order to make sure the evolution is safe for everyone, there must be a dedication of resources to prepare the structure, conduct the evolutions, and conduct a postincident analysis. Well-developed training session objectives will support safety and effectiveness as well as give the most benefit to the crews that are conducting the training. Agreed-upon training objectives are valuable in evaluating performance after the drill.

NFPA 1403: *Standard on Live Fire Training Evolutions*, uses the term **preburn plan** to refer to an overall plan for how the live fire training evolution will be conducted. The preburn plan should include approximately 20 items. This number may vary depending on what type of structure or prop will be used. This chapter will cover planning issues and preburn tasks common to live fire training evolutions.

Initial Evaluation of the Site

Regardless of the type of structure being used, before any planning is done, the first question asked should be, "Will the acquired structure or permanent live fire training structure allow for what the fire department wants to accomplish with this training?" It is important to specifically define the goals and objectives of the training evolution(s) to ensure that the structure will meet those needs. Note that NFPA 1403 must be applied with the understanding that every structure is unique. A few of the factors that would affect how 1403 is applied to a specific structure include the number of stories; construction type; size of burn room; the fuel type used; and type and number of means of egress.

Along with the department's training needs, access to the training site will play a determining role in the initial evaluation. A logical initial consideration should be the location of the structure relative to the location of the fire department. There may be a wonderful, state-of-the-art live fire training structure 3 hours away from the department, but is it feasible to get personnel and apparatus there while ensuring proper coverage back home? Especially with volunteer departments, distance can quickly rule out training traditionally scheduled for evenings. In addition, career departments can incur overtime costs, either for covering on-duty training participants, or when sending off-duty fire fighters to training.

It is also important to consider the site's physical location. If the department has been given access to an acquired structure, the location can be an impedance to its use. Consider a structure that is in a rural area with only a narrow road accessing it, and trees completely surrounding the building. During the planning phase, staging of apparatus, water shuttle operations, and accessibility would need to be evaluated. Although the structure may be ideal, there are also the logistics of clearing property to ensure enough apparatus on scene, and you must consider access for shuttle operations. These factors may rule out the use of an isolated structure for live fire training.

Adjacent properties and infrastructure are of vital concern when training with live fire. The obvious concerns with adjacent structures are instances where fire can spread or where smoke can adversely affect such structures as schools, patient care facilities, child care facilities, commercial businesses, or residential dwellings. Infrastructures, such as roadways, airports, and railroads, are also concerns for live fire training. Some examples could include fire schools located on community college properties that cannot accommodate smoke during the week when classes are in session, or a fire academy located on county fair property being used for fairs or events. These are the types of planning issues to consider before setting a date for the training evolutions. Once the date and time is set, the persons in charge of these properties must be notified. Surrounding streets and highways need to be identified and surveyed for potential effects of the smoke, and safeguards taken to eliminate hazards to motorists. This could include placement of signs, warning devices, or coordination with law enforcement as needed. Notifying area stakeholders of live burn training may minimize false alarms due to training activities. Notification also provides an opportunity to develop relationships within the community.

Permits

As part of the evaluation of the site, consider the required permits. These can include municipal or state (provincial) environmental permits, municipal demolition permits, and in limited cases, approval by the state fire marshal or other authority to conduct the training. The building must be checked/tested for asbestos. Because of the health risk, there are legal requirements for the inspection and testing, as well as for the mitigation, removal, and disposal of asbestos.

The state (or provincial) environmental authority may need to be contacted as there will be concerns regarding water supply and how close a burn evolution is to a water source. Regardless of whether environmental permitting is required by the authority having jurisdiction (AHJ), runoff from any live fire training must comply with AHJ requirements. It is very important to understand that runoff entering the ground or local water streams can contaminate local water resources and potentially drinking water.

The AHJ and the appropriate environmental authority will also be concerned with fire by-products introduced into the atmosphere. As such, they will want to see the reports to ensure any asbestos has been removed. Your AHJ may require that you obtain a burning permit or have filed documentation with the local dispatch center. As an instructor, it is your responsibility to know the AHJ's permit requirements.

The permits that you obtain should be kept on file with your incident action plan (IAP) and other documentation. You may receive a request for these permits from contractors who are cleaning up the site; state, federal, and local agencies; and other affected agencies. These documents need to be maintained so that the AHJ can provide them to these outside agencies.

Once the location of the live fire training evolutions is finalized, it is then time to develop the preburn plan.

Developing the Preburn Plan

Learning Objectives

Those involved in the planning need to determine what they wish to accomplish to develop the specific learning objectives of the training evolution(s). This can be more difficult than it sounds. The required learning objectives need to be defined so that the individual evolutions can be specifically designed to meet those objectives. The instructors should be fully briefed and prepared to meet the learning objectives. This will ensure effectiveness of the post-training review in assessing performance against training objectives.

The participants and their experience must be analyzed to match their skill sets to the evolutions. Clearly, personnel fresh out of Fire Fighter I training have fewer skills and less experience than seasoned personnel do. But even with fully trained personnel, live fire training is rarely the time for learning new skills, except under the most controlled circumstances. The average engine company will have considerable differences in the experience and training levels of their members, and this differential must be considered.

Participants

All personnel participating in live fire training evolutions must be properly trained before the evolution. Prerequisite training for students is covered in NFPA 1001: *Standard for Fire Fighter Professional Qualifications*.

The participating student-to-instructor ratio, according to NFPA 1403, should not be greater than 5 to 1. This will ensure that there are enough knowledgeable eyes watching over the students' actions. If there are large groups present, the classes should be planned for a long period of time. When there are situations involving extreme temperatures, additional instructors may be needed. It is the duty of all instructors involved to monitor and supervise all students throughout the course of the live fire training evolutions. Without proper monitoring and supervision, the results can be catastrophic. The well-being of instructors should be monitored by the instructor-in-charge (IIC) and safety officer (SO). This may be difficult but is essential in ensuring instructor safety and effectiveness in conducting training. AHJ guidelines should be followed in conducting training in severe weather conditions.

LIVE FIRE TIP

Because the training may include fire fighters from various skill levels, the objectives will most likely be different for each level of experience. The recruits in training have very limited fire experience, so their objectives will be quite different from those of veteran fire fighters. The experience levels of the instructors must also be considered when planning the objectives.

Water Supply

When it comes to water supply for live fire training, instructors have varying opinions on how much water should be available, what the water source(s) need to be, and how to maintain the water. On top of this, there are many times when instructors get nervous over the idea that they must do calculations to determine the amount of water needed at an evolution. The intent of NFPA 1403 is to ensure that training is conducted in a safe environment. This means that if the attack line should lose its ability to fight fire due to a mechanical or other emergency, the backup line must still have pressure and enough water to put out the fire and allow the attack team to escape safely.

To this end, the instructor-in-charge, along with the safety officer, is required to determine the amount of water needed for the live fire evolution as well as the amount needed for any unforeseen emergency. This requires that they determine the amount of water needed, identify the source(s) of the water, and assure maintenance of the water supply for both attack and backup lines.

The instructor-in-charge and the safety officer will determine the rate and duration of waterflow necessary for each individual live fire training evolution, including the water necessary for control and extinguishment of the training fire, the supply necessary for backup lines to protect personnel, and any water needed to protect exposed property. To determine the amount of water, NFPA 1403 refers the instructor to NFPA 1142: *Standard on Water Supplies for Suburban and Rural Fire Fighting*, which determines the minimum requirements for water supplies for structural firefighting. In areas where there is not a reliable municipal water resource, the instructor will need to be familiar with the water calculations in NFPA 1142. NFPA 1403 also requires a minimum reserve of additional water in the amount of 50 percent of the fire flow demand determined to handle exposure protection or unforeseen situations.

Determining the Required Water Supply

There are two water requirements to determine. The **fire flow rate** is the amount of water pumped per minute in gallons (gpm) or liters (lpm) needed to extinguish a fire. Then NFPA 1142 sets the **minimum water supply**, which is the total quantity of water needed for fire control and extinguishment. To begin determining how much water will be needed, the instructor must know the occupancy hazard, type of construction, structure dimensions (length, width, and height), and any exposures that exist.

Remember that the fire flow rate (FFR) and the minimum water supply (MWS) needed for the evolution are two different calculations. For a nozzle that flows a minimum of 95 gpm (360 lpm), which is the minimum allowed for live fire training, use the National Fire Academy's Fire Flow Rate formula, as follows:

$$FFR = (l \times w) \div 3$$

Where:
l = length of room/structure
w = width of room/structure

Per NFPA 1142, the MWS requirement formula is equal to the total volume (TV) of the structure divided by the occupancy hazard classification (OHC) number multiplied by the construction classification (CC) number. If there is an exposure hazard (EH), the result is then multiplied by 1.5. This formula is expressed as follows:

$$MWS = (TV \div OHC) \times CC \times EH$$

The occupancy hazard classification numbers and the construction classification numbers also come from NFPA 1142.

The occupancy hazard classification numbers are a series of numbers between 3 and 7 that represent the hazard levels of certain combustible materials.

Occupancy hazard classification:

3. **Severe hazard occupancies:** Explosives and pyrotechnics manufacturing and storage, flammable liquid spraying
4. **High hazard occupancies:** Warehouses, building materials storage, department stores, exhibition halls, auditoriums, theaters, upholstering with plastic foams
5. **Moderate:** Quantity or combustibility of contents is expected to develop moderate rates of spread and heat release
6. **Low:** Quantity or combustibility of contents is expected to develop relatively low rates of spread and heat release
7. **Light:** (Dwellings) Quantity or combustibility of contents is expected to develop relatively light rates of spread and heat release

The construction classification numbers are a series of numbers between 0.5 and 1.5 that relate to the type of building construction of an acquired structure:

0.5	Type I construction (fire resistant)
0.75	Type II construction not qualifying for Type I (noncombustible)
1.0	Type III construction ([ordinary] exterior noncombustible or interior of wood)
0.75	Type IV construction (heavy timber exterior and interior)
1.5	Type V construction (wood frame)

As a rule of thumb and according to NFPA 1403, attack lines, backup lines, and RIC hose lines should each be capable of flowing a minimum of 95 gpm (360 lpm), and two exposure lines should be capable of flowing 200 gpm (758 lpm) each. If all lines were operating at once, this would require a minimum of 700 gpm (2653 lpm). When calculating fire flow, all possible lines must be considered.

Water Supply Source

Now that the MWS needs for the live fire evolution have been determined, the instructor needs to determine the source of the water. The instructor needs to consider water sources based on the location of the training. Rural water supply and urban water supply sources have different concerns; however, the intent is to have reliable and valid separate water sources. When deciding the reliability of the water source, the instructor-in-charge needs to ask, "If my primary attack line water source is lost, or the pumper drafting should malfunction, will the backup line continue to have an adequate water source?"

If two pumpers are drafting from the same water source such as a river or pond, the instructor-in-charge must ensure that there is enough water to supply both pumpers and both backup lines. If possible, an added safety would be to have two dump tanks to supply the two pumpers. Regardless, the water source has to be enough, and two separate apparatus should be used to supply the lines.

EXAMPLE 1

A one-story, wood-frame dwelling, measuring 20′ × 20′ × 10′ (6.1 m × 6.1 m × 3.05 m), with a standard 4′ (1.2 m) from attic floor to ridgepole and one exposure, requires what MWS?

$$MWS = (TV \div OHC) \times CC \times EH$$

First calculate the TV of the structure. This is done by multiplying length by width by height. For structures with a pitched roof, the height is equal to the wall height + half the height of the pitch. In this example, the height is $10' + (1/2 \times 4') = 12'$

$$TV = (20 \times 20 \times 12)$$
$$= 4800 \text{ ft}^3$$
$$OHC = 7 \text{ (light)}$$
$$CC = 1.5 \text{ (wood-frame)}$$
$$EH = 1.5$$

Therefore:

$$MWS = (4800 \div 7) \times 1.5 \times 1.5$$
$$= 685.7 \times 1.5 \times 1.5$$
$$= 1542.8 \text{ or } 1543 \text{ gallons}$$

Using these calculations, the MWS needed is 1543 gallons, but in this case there is an exception. According to NFPA 1142, the MWS required for any structure without EHs is 2000 gallons (7570 L) and for structures with EHs, it is 3000 gallons (11,355 L). According to NFPA 1142, an EH is any structure within 50 ft (15.2 m) of another building and 100 ft² (9.3 m²) or larger in area. So in this case, the MWS is 3000 gallons (11,355 L).

Now that the required MWS has been calculated, what is the minimum FFR required for this structure?

$$FFR = (l \times w) \div 3$$
$$= (20 \times 20) \div 3$$
$$= 400 \div 3$$
$$= 133 \text{ gpm}$$

133 gpm (504 lpm) would be the minimum FFR required for 100 percent involvement of the structure. Round down from 133 to 125 gpm (504 to 474 lpm) because of nozzle settings. If this example were of a two-story house, we would have to calculate an additional FFR at 25 percent per floor, up to five floors (see Example 2). You also need to make sure the required FFR takes into consideration all of the lines that may have to flow. All hose lines, attack lines, backup lines, and any exposure lines must be added together to make sure you have adequate water supply.

EXAMPLE 2

A two-story, wood-frame dwelling is 50′ × 24′ (15.2 m × 7.3 m). Each story is 8′ (2.4 m) high with a pitched roof that is 8′ (2.4 m) from attic floor to ridgepole. What is the approximate MWS needed?

$$MWS = (TV \div OHC) \times CC$$

Calculate the TV of the structure, which is equal to length × width × height. In this case, the height is equal to:

$$\text{Height} = 8 + 8 + (1/2 \times 8)$$
$$= 8 + 8 + 4$$
$$= 20$$
$$TV = (50 \times 24 \times 20)$$
$$= 24,000 \text{ ft}^3$$
$$MWS = (24,000 \div 7) \times 1.5$$
$$= 3428.6 \times 1.5$$
$$= 5142.9 \text{ or } 5143 \text{ gallons}$$

The MWS required for this structure is 5143 gallons (19,468 L). What is the minimum FFR required for this structure?

$$FFR = (l \times w) \div 3$$
$$= (50 \times 24) \div 3$$
$$= 1200 \div 3$$
$$= 400 \text{ gpm}$$

400 gallons per minute (1514 lpm) would be the minimum FFR required for 100 percent involvement of the structure; however, since this is a two-story house, we must add an additional 25 percent of the calculated FFR for each floor, up to five floors. Therefore:

$$0.25 \times 400 \text{ gpm} = 100 \text{ gpm (25 percent for additional floor)}$$
$$100 \text{ gpm} + 400 \text{ gpm} = 500 \text{ gpm total}$$

The total FFR would be 500 gpm (1893 lpm) and the MWS would be 5143 gallons (19,466 L).

If the training area is in an area with a municipal water source, the instructor needs to ask other questions such as, "If the main pipeline should rupture, are both hydrants affected?" and "Do municipal water sources have backup generators in case of loss of power?" If a municipal water supply system is used, two pumpers on two different hydrants should be used. This assures that if the attack line loses its water, the backup line should still have enough to protect the attack line and allow for escape. If the instructor-in-charge feels that there needs to be additional safety precautions, portable water tanks could be used at one of the hydrants or a pumper could be used to assure there is an additional water source in an emergency. When using a fire hydrant, you need to make sure pitot readings are done and adequate water supply with residual pressure are calculated.

In addition, make sure that exposures are protected from radiant heat (**FIGURE 18-1**).

CHAPTER 18 Preburn Inspection and Planning

FIGURE 18-1 Exposures must be protected from radiant heat damage.
Courtesy of John Buckman.

LIVE FIRE TIP

A technique that will assist with water supply and live fire training is to use different colored hoses to ensure that instructors and students know which lines support which function. For example, the attack line could be red, the backup line could be yellow, and the RIC line could be green. If there is not enough colored hose to run from the engine, a different colored hose could be used at the nozzle only. There should be no confusion as to which line is being used for what function. The exposure lines should be clearly identified by divisions, as in exposure A/B, which would cover two sides of the structure, or exposure A and exposure C, where each line is covering only one side of the structure.

Apparatus Needs

Fire apparatus includes engines (pumpers), aerial devices, and possibly water tenders. An engine will be needed as the primary supply for attack lines, with a second engine being used for the backup line(s). Depending on the site arrangement, it may be necessary to have one engine at the hydrant to pump the supply line(s) and another at a static water source. This will be dependent on hydrant locations, available fire flow, and the need for alternate water sources.

Building Plan

A building plan is a necessary piece of the planning stage. It will serve as a blueprint to be used by instructors and students before and during the live fire training evolutions. The diagram of the building plan needs to include the dimensions of the building and the dimensions of all of the rooms, windows, and doors. This diagram does not need to be drawn to scale. All features of the training areas, including primary and emergency ingress and egress routes, and areas that are off limits to participants should be clearly labeled in the diagram. All rooms that will have burn evolutions in them should be numbered, on the diagram and in the structure itself, in the sequence in which they will be burned. Fuel loading should be indicated and then confirmed and finalized on a hard copy of the building plan on the day of the evolutions. The building plan should include the exact locations of each burn set, and how much and what type of material will be used, for example, "four pallets horizontally stacked with a bale of excelsior."

Follow local protocol when marking the sides of the actual building for clarity and for the purpose of reinforcing the procedure to the participants. In general, the accepted protocol is to label the front of the building as side A, and moving clockwise around the building, label sides B, C, and D (**FIGURE 18-2**). There are local and regional variants, so it is best to check first before assigning designators. Each door and window also needs to be numbered and marked on the plans *and* the actual building. Number each window and door sequentially on *each* side, so that each is uniquely identifiable. The front door is typically labeled as Door A1 and the first window on the left side is typically Window B1. Again, local protocol should always be followed. Be sure to also mark north on the plan.

Fuel Needs

Once burn locations have been identified, the fuels to be used must be determined. Only wood and wood-based products or untreated hay/straw are to be used in live fire training unless a fuel-fired building or prop is being used. Instructors and SOs should be familiar with the type of fuel that is used in buildings and props. Liquefied versions of these gases are not permitted inside the live fire training structure (NFPA 1403: 4.13.3).

FIGURE 18-2 Physically mark the building's sides, doors, and windows.
Courtesy of Shawn Morgan, Corvallis Fire Department.

NFPA 1403 does allow a combustible liquid with a flashpoint above 100°F (38°C) in a live fire training structure or prop that has been specifically engineered to accommodate a defined quantity of the fuel (NFPA 1403: 4.13.3.1). This would be a very unusual feature. There are areas that allow the limited use of a combustible liquid to start a fire under certain conditions and restrictions, and while legal in those areas, the associated liability would be high. Only certified maintenance personnel should perform the maintenance of these fuel-fired props. When untrained personnel try to adjust the amount of fuel, it can result in unexpected fire behavior and serious injuries.

The amount of fuel must be based on factors such as the size of the room, ventilation capabilities, weather, objectives, and the level of training being conducted. *Only the amount of fuel needed to create the needed fire size should be used.* The idea of live fire training is not to make the fire "hot" to see who is tough enough to handle it; it is to ensure personnel know the proper techniques of extinguishment and ventilation. An excessive fuel load can contribute to unusually dangerous fire behavior.

The IIC and SO must assess the fire room for factors affecting the growth, development, and spread of the fire.

The IIC is concerned with the safety of participants and the conditions that can lead to flashover:

1. The heat release characteristics of materials used as primary fuels
2. The preheating of combustibles
3. The combustibility of wall and ceiling materials
4. The room geometry

Preparation of the rooms to ensure proper fire size will take time and effort. Tarpaper, upholstered furniture, and carpeting must be removed from the structure. Holes need to be repaired and the structure made safe. Other fuels that are not permitted include pressure-treated wood, rubber, plastic, and polyurethane foam. These items must be removed so as not to add to the fuel load, as must any debris inside or outside the structure that could burn in an unanticipated way. All possible sources of ignition, other than those used to ignite the fire by the ignition officer, must be removed from the operations area.

Fuel loading must be determined to avoid flashover and backdraft or any other hostile fire event. Flashover training can be valuable if performed correctly. Flashover props are designed to allow fire fighters to see the conditions develop from a safe observation area. If a controlled flashover is planned, additional safety measures for providing a safe observation area must be in place.

Ignition of the fuel load must also be considered. Ignition must be accomplished in a safe and efficient manner. A variety of ignition sources ranging from lighters, kitchen matches, and flares to a propane tank with an extended wand have commonly been used. Whatever is chosen to ignite the fire, the source needs to be removed immediately after igniting the fire. Lighters should not be placed in pockets of turnout gear because they could be exposed to high temperatures and ignite. If fuels are being used in external props, caution needs to be used to ensure the liquid is not spilled on protective clothing.

The use of flammable gas shall be permitted only in training structures designed for their use. Liquefied

---

The following is a flashover addendum to the operational plan, as noted in NFPA 1403: A.4.13.7:

1. The lead instructor should identify the fire growth observation area prior to ignition of a live fire. The area should be out of the exhaust portion of the flow path. Students and instructors should have a charged hose line in the observation area that has a fire stream capable of reaching the ignition room and suppressing the fire. Participants should be in the observation area prior to ignition.
2. Charged hose lines should be placed in position prior to ignition of fire. The hose line should be used to control temperature and fire growth from the observation area.
3. Observation areas should be on the same level as or below the level of the fire with direct and unimpeded access to an exit.
4. No students or instructors should be in the fire room after ignition.
5. The identification of the potential flow path should be communicated to all students and instructors prior to ignition. The lead instructor should designate the flow path.
6. No unidirectional flow paths should be created that exhaust over the fire fighters. If weather or the fire creates a potentially hazardous change to the flow path, the interior instructor should be notified immediately and personnel should exit the structure or take other action to maintain the safety of participants.
7. The interior instructor should coordinate ventilation with exterior personnel to complete the ventilation to achieve the desired fire affect. After charged hose lines are placed, and instructors and students are located in the observation area, ventilation should be coordinated.
8. The IIC should use an assessment, such as the following equation, to estimate the minimum heat release rate needed to flash over the ignition room based on available ventilation:

$$Q = 750 A_0 \sqrt{H_0}$$

where

Q = Minimum heat release rate (kW) needed for flashover
A_0 = Area of opening (m²)
H_0 = Height of opening (m)

natural gas and propane are not permitted inside the training structure. They remain in the liquid state only where they are stored under pressure; a leaking liquid propane pipe has the potential to cause the area inside the space to reach an explosive level.

Once rooms have been identified, fuel load determined, and conditions identified that will affect fire behavior, documentation must be completed. This documentation includes the fuel materials, wall coverings and ceiling materials, type of construction including roof and combustible void spaces, and the dimensions of the room. Characteristics of the structure that affect flow path of the fire must also be identified and contained in documentation. *No flow paths that exhaust over fire fighters should be created.*

Plotting the expected avenues of fire spread and buildup of the fire provides an extra level of safety. The IIC should determine primary and secondary exit paths for the exercises in the case of unexpected situations that may cause blocked exit routes.

If a live fire training structure or prop using gas is used during training, instructors must ensure pressurized props are equipped with remote fuel shut-offs outside of the safety perimeter. These shut-offs should be clearly marked on building or site plans so that instructors can shut down the fuel supply should an emergency occur. These shut-offs should be within site of the prop and the entire field of attack for the prop. A safety person who is trained in the operation of the shut-off valve and prop operation should be assigned to stay at the valve at all times during the evolution and should have the authority to shut off the fuel supply to the prop when necessary. This individual should have communications with the SO and instructors on scene. Liquefied petroleum gas props must have the safety features identified in NFPA 58 and 59. Regardless of the type of prop, runoff considerations must be in place and an oil–water separator must be used to clean the runoff.

If a gas-fired prop is used to train personnel on dealing with a failure of a safety device, the failed part should be located downstream from the correctly functioning safety feature. Should an emergency occur, the correctly operating safety valve would be shut down, stopping the flow of fuel to the failed training section and reducing the chance of injuries.

Acquired vehicles must be prepared before the evolutions to make them safe. These preparations include removing, venting, or draining all fluid reservoirs, tanks, shock absorbers, drive shafts, and other gas-filled closed containers. If not made safe, these items can become projectiles when heated. Air bag cylinders in the vehicle's upright posts (A, B, C, D posts) as well as the side curtain areas need to be made safe.

Also, many newer cars have more than one battery, with the second located in various areas such as under the back seat or the trunk. Vehicle props can use wood products or untreated hay/straw as ignition sources, or they can be gas-fueled. Runoff will be a consideration as well as placement. There have been occasions where a vehicle was placed in a field, and once ignited, the fire spread to the field and became a wildland fire. Planning will prevent this type of situation from arising. Another consideration on vehicle fires is their construction. Some vehicles may contain flammable metals. Anytime an evolution may involve flammable materials, there must be planning for extinguishing agents. Flammable metals react violently to water and because of their burn characteristics, there must be enough extinguishing agent to put out the fire as well as a 150 percent reserve for use by backup crews.

This information will be used throughout the evolution and must be monitored by the IIC and the SO. If, during an evolution, the fire behavior reacts in an unanticipated way or the environment presents a potential hazard, the IIC or SO should stop the evolution immediately and not resume until the hazard has been reduced.

Site Plan

A site plan must be developed that shows how the building is oriented on the property and all other aspects of the layout, with specific measurements and north labeled. Whenever possible, the building plan, site plan, and any other plans should all be on the same axis—drawn in the same direction in relation to the paper. The structure, with the sides identified, command post, rapid intervention crew (RIC), rehabilitation/medical area, operations area, staging area, placement of apparatus, primary and secondary water sources, hot zones, self-contained breathing apparatus (SCBA) refill, emergency meeting place, parking area, exposures or outbuildings, driveway, hazards in terrain, underground and overhead, significant changes in elevation, fencing, and gates all need to be clearly identified on the site plan. Any hazardous areas such as the locations of overhead power lines, septic tanks, and natural or man-made barriers should be identified as well. On the day of the evolutions, bring extra copies that include only the permanent features in case the weather causes changes and functional assignments have to be relocated.

The site plan must be made available to the various involved communication (dispatch) centers that may be needed for dispatch assistance in case of problems. For live fire training in acquired structures, emergency units should be alerted to the date and time so that they are fully aware of the training event. In many areas, police, fire, and EMS are provided by different agencies and could be in different centers. Sometimes, there are even multiple providers, such as local and county law enforcement. Regardless of how they are structured, all of the aforementioned need to be informed of the evolutions. This should be done prior to the day of the evolutions so that routes can be planned for units that may be dispatched to the training site. Participating personnel may not be familiar with the neighborhood. The width of the roads, tree canopy, traffic congestion, and other factors are all important. The entrance and exit routes for emergency and regular use and the various parking areas should all be provided to participants ahead of time.

Parking, Staging, and Areas of Operations

Areas for the staging, operating, and parking of all involved fire apparatus, as well as for police vehicles or press, need to be planned for and noted on the site plan.

Signs, traffic cones, caution tape, and other markings can be a great help on the day of the evolutions (**FIGURE 18-3**). Critical areas that need to be clear of civilian vehicles should be marked "Reserved" or barricaded the day before the evolutions.

Emergency medical services (EMS) with transportation capabilities must be present on the scene of acquired structure live fire training to handle emergencies of participants; ambulances need to be parked where they can quickly access participants in the event of an injury. Parking for emergency vehicles that are *not* involved in the evolutions needs to be set aside so as not to interfere with fire-ground operations. Ambulances and EMS vehicles need to have clear, easy access to the exit, so these areas should also be marked with "No Parking" signs.

Visitors and Spectators

The vehicles of participants and spectators can quickly congest the entrance and exit routes to the training site, which may be needed for emergency response vehicle access. This congestion can spread to the surrounding streets. You must establish control measures to keep pedestrian traffic clear of the operations area; therefore, the entrance and exit routes must be designated and monitored during the training evolutions to ensure their availability in the event of an emergency.

Visitors and spectators can not only cause traffic congestion but also provide safety concerns during the operation. A restricted area outside the area of operations for spectators should be established and clearly marked. Barricade tape, cones, or other devices should be placed. Areas set aside for spectators are more of a concern for acquired structures than they are for permanent props. This needs to be addressed in the planning phase, and spectators must be sequestered in an area away from harm—mainly smoke, but also vehicle traffic, heat, and other hazards. A training evolution can be used as a positive community or media event if structured properly. Assigning a public information officer (PIO) to explain the purpose of the training and allowing the community/media to witness it is an excellent public affairs event; however, these observers must not impact or adversely affect the training evolution. If visitors are allowed in the operations area, their safety must be assured. A visitor inside the operations area must be escorted at all times, preferably by fire department personnel, and must be equipped and wearing appropriate protective clothing. This area should be clearly marked on the site plan.

Emergency Plans

Unlike an emergency response, a live fire training session allows all necessary personnel to meet before the fire is ignited. Hose lines, tools, RIC, safety personnel,

FIGURE 18-3 Signs can be a great help on the day of the live fire evolutions.
Courtesy of Dave Casey.

and a walkthrough of the structure are prepared ahead of time.

The emergency plan needs to cover foreseeable issues such as medical problems, fire control, the need for RIC, and unrelated emergencies that affect the training site or participants. Elements include the following:

- RICs should be fully staffed with experienced personnel who are trained to NFPA 1407. This team of members is also referred to as the rapid intervention team (RIT) and the rapid intervention group (RIG). A federal two-in/two-out rule is standard, but outside of that, there must be enough staff on the RIC to enter, locate, rescue, and remove trapped or lost fire fighters. In permanent props, the danger of collapse or failure is almost negligible, so staffing can be less than it would be for an acquired structure. Regardless, the RIC must be properly prepared for immediate deployment.
- The building evacuation plan must be created and known by all personnel. The building evacuation plan needs to include an evacuation signal that can be heard by all participants during interior evolutions. The signal must be tested and demonstrated. Aerosol air horns or a megaphone that feeds directly into the interior may be necessary. As part of the building evacuation plan, participants should be instructed to report to a predetermined location (rally point) for roll call or **personnel accountability report (PAR)** if evacuation of the structure is signaled. Instructors should immediately report any personnel unaccounted for to the IIC, who will deploy the RIC. Keep in mind that recruit fire fighters may not have participated in PAR drills before.
- The possibility of losing one of the two water supplies needs to be addressed in the plan. Training evolutions need to stop, and personnel need to be relocated to the exterior while the supply is reestablished or replaced.

LIVE FIRE TIP

Although NFPA 1403 requires participants to be made aware of the exits from an acquired structure, every fire fighter, regardless of tenure, should be trained to constantly identify hazards and alternative escape routes during interior fire suppression operations, including training exercises. Fire fighters in Seattle, Washington, are taught from their first day at the academy to know at least two means of egress from whatever room they enter.

As part of the emergency plan, any time the IIC determines that fuel, fire, or any other condition poses a potential hazard, the training evolution should be stopped immediately. If an exercise is stopped, it should only be restarted once the hazard identified has been resolved and after the **Go/No Go sequence**. The Go/No Go sequence (**TABLE 18-1**) is a verification method by radio that ensures that all positions are ready to initiate operations. By using this defined sequence of actions, everyone is separately accountable for all teams and participants they can see, not just their own. This gives the IIC and the SO more eyes around the site, checking for preparation and ensuring that each team is ready for ignition. Fire growth dynamics must be carefully assessed. Consider the presence of combustible void spaces and take steps to ensure that the fire is not able to gain unexpected growth in such areas.

Nonexercise Emergencies

There may be times when other emergencies occur. Anything from on-scene problems to a neighborhood emergency can require the attention of emergency personnel. If there is ever a question of divided attention, the evolution needs to stop. Often, there are enough unengaged personnel to handle such emergencies. However, the IIC must make the determination of whether or not to proceed. Emergency disengagement of apparatus and participants to unrelated emergencies must

TABLE 18-1 The Go/No Go Sequence

From the IIC/Command	Reply by Team or Function
Staging: Go/No Go?	Staging: Go
Rehab/Medical: Go/No Go?	Rehab/Medical: Go
Engine 1: Go/No Go?	Engine 1: Go
Engine 2: Go/No Go?	Engine 2: Go
Entry: Go/No Go?	Entry: Go
RIC: Go/No Go?	RIC: Go
Backup: Go/No Go?	Backup: Go
Attack: Go/No Go?	Attack: Go
Safety: Go/No Go?	Safety: Go

be included in the plan. The SO can play a critical role in that decision-making process.

Weather

The weather can be a determining factor in the decision to cancel or continue with training. The preburn plan should include measures to shelter from the sun, for rehydration, and for cooling.

Storms and other weather systems often come with winds that can drastically change fire dynamics, heavy rain that can hamper operations, and lightning that will curtail training. Acceptable weather parameters of the plan should include wind direction and/or speed that would preclude safe operations, and "red flag" conditions that would preclude open burning, such as high winds and low humidity. The weather forecast, preferably the National Weather Service Point forecast, needs to be checked for expected untoward weather conditions that could cause problems. This should also include a final check for possible changes in weather conditions prior to ignition.

Lightning is a training ground weather hazard that exists everywhere, and it can occur with or without rain. Many training centers and agencies use lightning detection equipment. The Occupational Safety and Health Administration (OSHA) and the National Oceanic and Atmospheric Administration (NOAA) advise that "lightning is unpredictable and can strike up to 10 miles from any rainfall." They recommend that personnel take shelter early and remain in the shelter for at least 30 minutes after hearing the last sound of thunder (OSHA/NOAA, 2016).

List of Training Evolutions

As part of the preburn plan, you will develop a list of the training evolutions to be conducted. First, learning objectives covering all live fire evolutions are written and included in the plan. The list of training evolutions should be developed following the development of the objectives. The procedures for these evolutions should include the instructors and support personnel that are needed and the number of students for each evolution. This list will vary depending on the site being used, the availability of instructors, and the number of students.

Order of Operations

The **order of operations** is a list of the steps, in sequential order, on how to conduct each live fire training evolution. The order of operations is determined after the list of the training evolutions is developed and should be noted in the preburn plan.

On the day of the training evolution, such an order includes the following:

1. Set up according to the site plan, unless conditions dictate a change. It is a good idea to have several copies of the site plan that show only the building and not the locations where the functional areas are to be placed. Revisions can then be made to the functional areas when needed. Any changes *must* be reviewed during the briefing of the live fire training instructors and participants.

2. Conduct a briefing of all instructors with clear objectives and a consensual verbal understanding. It is important to brief the instructors first, before the preburn briefing, because if there is anything overlooked or unclear, they will usually catch it before the next step, the preburn briefing.

3. A preburn briefing is required, along with a walk-through of the entire structure, prior to any ignition of the burn set. The IIC and SO conduct a walk-through with all of the instructors first and then with the students. It may be necessary to demonstrate an evolution if there is any question or clarification needed. All instructors must have a clear understanding of the learning objectives for the training evolution. Part of the preburn briefing is to point out the locations of burn sets as well as review the sequence of burns. The rooms should be numbered and labeled, including the exits, ventilation points, and primary and secondary means of egress. All participants need to be aware of the operations area where personal protective equipment (PPE) is to be worn. All participants must also be aware of location of the fire ground (warm zone) where a helmet may be the only piece required.

   If a prop is being used, the preburn briefing ensures students' knowledge of the prop, how the prop functions, and familiarity with the prop.

4. Assignments are given and personnel must report to their positions. At this time, all lines are flowed to the proper pressure. Once all lines have been flowed and the pressure is verified, the instructors and students are in place for a Go/No Go sequence. The Go/No Go sequence is the same process that NASA follows during the launch of a spacecraft. The Go/No Go sequence is stopped if even one thing is not right.

5. The Go/No Go sequence is the last step of the order of operations before ignition (Table 18-1). The verbal confirmation by radio communication ensures that all participants are ready for action. By radio, the IIC says, "All personnel stand by for a Go/No Go." Start with the support crew, followed by the attack crew, with pump operators

checked before the entry crews. Once the pump operators confirm that they are ready, the IIC of each crew is checked, and with each confirmation, it progresses to the next crew. The order is typically as follows:

- All positions share the responsibility and can stop the process. Any position seeing less than 100 percent preparedness needs to give a No Go report. An example of a No Go would be the SO observing students still getting their equipment on. Even if all of the positions had advised Go, the SO must stop the process.
- Prior to the SO giving a Go signal, he or she must inspect the structure to make sure it is clear of any occupants and that personnel are ready. This includes protective clothing and breathing apparatus inspection.
- After the SO advises a Go, the IIC would declare, "We have a Go for ignition."
- Ignition is the last step in the Go/No Go sequence. As noted earlier, the utilization of specific terminology is vital. One of the most important is the terminology to identify the instructor responsible for lighting the fire and the directive to light the fire. The order for ignition should be along the lines of "Ignition, you have a Go," or another very clear directive that everybody knows ahead of time. Ignition will then advise command when there is "fire in the hole," or a similar, predetermined phrase. For Class A fires in permanent live fire training structures or acquired structures, the SO will need to make sure the torch is returned to the staging area.
- Any time operations are shut down, a Go/No Go sequence should take place before continuing the live fire training evolution. An example would be after a burn room that is being used is no longer usable, and it is necessary to move to another burn room. The fire in the room being moved away from must be fully extinguished. Remember, only one fire at a time is allowable in an acquired structure. Repeating the sequence would also be done after an extended break, like after lunch, or after putting the fire out and before moving to checking for structural integrity.

Emergency Medical Plan

An emergency involving an instructor or student can be emotional and cause confusion. Having a plan in place is necessary to ensure personnel receive treatment rapidly.

NFPA 1403 requires that basic life support EMS be available onsite to handle injuries. For acquired structures, basic life support EMS with transport capabilities is required onsite. Advanced life support should be provided onsite when it is available. Planning for communications with EMS, where EMS will be located, where RIC will bring the injured participant, and locating an appropriate medevac landing zone can save valuable time in case of an incident.

The ability for a local fire department to have an advanced life support transport unit on scene varies tremendously. Costs can be involved, especially when the fire department is not the local EMS provider. The local EMS provider needs to be apprised of the location and time of the evolution.

Planning needs to include identification of a landing zone with global positioning satellite (GPS) coordinates for helicopter transport (Medevac). Coordination with the provider is necessary to meet their landing zone and planning requirements, and to reduce the possibility of confusion or error in the time of need. The property owner of the intended landing zone must be made aware of the plans. Some of the more appealing sites, such as large school fields, can create havoc with the facility and may not actually be available at the time of need.

Written reports, including patient care records, shall be filled out and submitted on all injuries and on all medical aid rendered.

Communications Plan

The communications plan will provide for coordination among the IIC, the interior and exterior divisions, the SO, and any external requests for assistance. Prior to beginning any evolution, all interior crews and command staff throughout the structure and fire ground need to be equipped with working two-way radio communications. Operations need to be conducted on a dedicated radio channel that will not be used for dispatching, or for any other use, during live fire evolutions.

During a training fire that took the lives of two fire fighters, communications became a critical issue when units responding to the IIC's request for assistance interfered with vital on-scene communications. The communications plan needs to include the ability for the IIC to communicate with the dispatch center and incoming units on a different radio channel, and a physically different radio, than on-scene operations. In case of an emergency, the plan should include an aide for the IIC to operate on two radio channels as needed. Radio communication capabilities need to be confirmed during the planning stage for on-scene radio-to-radio and from on-scene to dispatch communications.

Staffing and Organization

The preburn plan should include a list instructors and support personnel needed for each training evolution to be conducted, with assignments and rotation of instructors and students. Safety officer and fire control teams must be identified. Review this topic in Chapter 17.

Protective Clothing and Self-Contained Breathing Apparatus

The SO needs to ensure that all students, instructors, safety personnel, and any other participants wear their equipment appropriately. Review this topic in Chapter 17.

Rehabilitation Plan

There is a wealth of information available regarding fire fighter rehabilitation at emergency scenes and training sites. The rehabilitation plan needs to clearly define how the specific drill will address rehabilitation. The following aspects should be noted:

1. Mandatory use for all participants
2. Number of consecutive evolutions for instructors and other participants
3. Specific cooling measures being used (e.g., shade, cooling fans, forearm immersion, hydration fluids, seating), and warming methods for cold-weather operations.
4. Site (onsite plan) with contingencies for wind changes, etc.
5. Supplies and equipment needed
6. Medical emergency plans
7. Staffing needs

Agency Notification Checklist

It is a good idea to keep a checklist to confirm the training with *any* agency that was previously notified of the evolutions. Notification is required prior to starting any evolution for any agency that plays a critical role in the training. These agencies may include law enforcement, EMS, environmental organizations, local utilities, and any other locally required notifications. In many areas, it also includes the local air traffic control tower. It is strongly suggested to keep contact names and numbers of all points of contact on the checklist.

Demobilization Plan

Through the use of a demobilization plan, personnel, equipment, and fire apparatus can be put back into service in an organized way. The department's policies should generally assist in the development of the plan, but in the absence of such policies, it should be determined ahead of time if apparatus and equipment can be returned at less than 100 percent. Any apparatus that may be needed to respond to an emergency needs to have the SCBA serviced, the booster tank filled, and preconnected hose lines repacked and ready. This includes apparatus that may be used directly following the training evolutions and those that may need to be called for emergencies during the evolutions. Part of the demobilization plan can include bringing fresh hose to repack, personnel assigned to fill SCBA cylinders, refueling the apparatus, and so on.

Any participant who is known or suspected of sustaining an injury during the training should have the appropriate documentation completed before they depart the drill site.

Preburn Inspection

Inspection and Preparation of Acquired Structures

After determining that a structure is usable and the owner and fire department agree to terms, a more detailed inspection of the structure and property should be performed followed by preparation. The instructor-in-charge is responsible for ensuring the acquired structure is prepped to meet NFPA 1403. Although there is no rule or requirement that states this inspection be separate from the initial evaluation, it is often better to do so for organizational purposes.

Access

Access to the training site can be a problem, just as it can be in a hostile fire situation. However, unlike emergency responses, access to the training site can be planned.

- **Access to and from the training area:** Locate a staging area for apparatus and parking for other vehicles that will not be used on the immediate scene. Remember to locate an area for in-service apparatus that will be considered available for response so that they do not get blocked by parked vehicles, hose lines, or spectators.
- **Access to the property:** Fences or walls that limit access to the site may have to be removed. Make sure to keep areas wide open for unexpected emergencies.
- **Access around the property:** Trees, brush, and surrounding vegetation that create a hazard to

CHAPTER 18 Preburn Inspection and Planning

participants or limit access to the scene must cut back or removed. High grass and weeds need to be cut clear in the operations area so any hidden hazards can be exposed. These include toxic weeds, insect hives, or vermin that could present a potential hazard. This removal may require professional assistance, especially with large bee colonies and similar hazards.

- **Underground dangers:** Septic tanks or other buried tanks must be identified and clearly marked and/or barricaded to ensure apparatus do not park above underground tanks. Drain fields that the owner expects to reuse need to be avoided. Both fire fighters and fire apparatus have been known to fall into septic tanks. Mark areas to be avoided (**FIGURE 18-4**).
- **Nonparticipant access:** The press, spectators, law enforcement, and EMS will all be present on the day of training. Be prepared for this influx of spectators and plan ahead of time for parking and the area from which they will observe.

FIGURE 18-4 Septic tanks and other hazards should be marked for safety.
Courtesy of Dave Casey.

Entry and Egress for the Structure

All entry and egress routes must be planned for ahead of time and must be known to all participants. These routes must also be monitored during the evolutions to ensure they are clear in the event of an emergency. In addition, the primary travel distance to the outside should be considered. Likewise, instructors and RIC should be able to rapidly access the interior crews' status in a moment's notice. In selecting what rooms are to be used for the live burns, any room with limited access should not be used for live fire training purposes unless a door can be cut out. Fires cannot be located in any exit paths so that participants always have a direct path to the outside.

The following are precautions to take when selecting rooms and preparing the structure for live fire training:

- Trees, brush, and surrounding vegetation that create a hazard to participants or limit entry and egress should be removed.
- Interior and exterior doorways need to be easily accessible, with no obstructions, and no drop-offs that could cause injury. Disable and remove locking mechanisms. Remove pneumatic or other types of door closers, along with storm doors, screen doors, and secondary security doors.
- Windows need to be easily accessible for emergency egress, and when windows are covered for the training evolutions, these coverings must be removable from inside or out by hand, without tools. Consider using hose hinges or other methods (**FIGURE 18-5**). Any window locks or security blocks, glass, hardware, horizontal or vertical cross pieces, window air conditioners and fans, or anything else that could hinder a fire fighter from exiting should be removed. There must also be an exterior escape path that is not blocked by plants or other obstructions. If the window sill is high or too small, cut the wall below it to make it accessible as an exit, if there are not sufficient means of egress for the evolution planned (**FIGURE 18-6**).
- Even if there is not intent to use a cellar/basement (which should be extremely limited if at all), windows and doors to that area need to be readily available for emergency egress.
- Make doors where needed. Sometimes long hallways or large rooms make for long escape routes. Keep escape routes short and clear, and cut through walls if needed.
- Stairs, porches, and railings need to be in serviceable condition and not hazardous to exiting fire fighters.

FIGURE 18-5 Hose hinges keep window coverings in place but are easily knocked down.

FIGURE 18-6 A window can be made into a doorway by cutting and removing a section of the wall below the window.

Exterior Preparation
Asphalt and Asbestos

In an initial assessment, the exterior was assessed for sagging or other visual signs of potential structural damage. Now the actual exterior surface covering must be evaluated. Although there are a number of environmental concerns, the most common concerns are asphalt and asbestos. Be sure to know the applicable requirements in your jurisdiction for permitting, testing, removing, and disposing of such substances. Most areas have requirements for asbestos testing. Many homes and buildings built before 1985 have asbestos either in the roofing, flooring, insulation around the heater and plumbing, or possibly in exterior or interior wall surfaces. Starting in the 1920s, the National Board of Fire Underwriters recommended that homeowners use asbestos siding and roofing instead of the less fire-resistant wood materials. By 1979, the U.S. government outlawed the use of asbestos in many building products because of its physiological effects on humans. This ban included asbestos cement siding shingles; however, they are still found in structures today.

Siding is generally not an issue, unless it is asphalt or asbestos. Composite siding made of asphalt-impregnated fiberboard, with the surface granules similar to asphalt roof shingles, was popular on less expensive wood-frame houses built before 1950. The asphalt siding often has a pattern that makes it look like stone or brick. Also, cement asbestos board was used for lap siding and wall shingles. Asphalt shingles, paper shingles, or tarpaper roofing may have to be removed due to environmental requirements. It is important to note that products containing asbestos cannot be identified by sight alone, and further testing may be needed.

Houses built before 1980 should be suspected of having asbestos. Asbestos can be found in many places in a building, such as on furnaces, boilers, hot water pipes, ceiling tiles, drywall, flooring, roofing felts and shingles, and exterior siding shingles. Any asbestos found must be addressed, typically by a licensed contractor.

Asbestos siding was manufactured in a wide range of colors and patterns, but it does have some characteristics that may help identify it. Asbestos siding shingles are usually 12″ × 24″ (304.8 mm × 609.6 mm). They may have grooves or a woodgrain pattern pressed into the cement, or they may be smooth. Each tile usually has two or three nail holes at the bottom of each shingle. Another popular type of asbestos siding came in corrugated sheets of various lengths. These sheets were used in the same way that corrugated metal sheeting was used. It can be recognized by the corrugation, but these sheets were seldom used in home construction.

Utilities

Once the building construction is secure and all measures have been taken to mitigate any dangers, the utilities must be secured. This requires going beyond the household or building service disconnects and shut-offs. There have been unfortunate experiences during live fire exercises where the electrical service lines going to the weatherhead were found to still be charged. Electrical service lines must be disconnected from the pole to the weatherhead (**FIGURE 18-7**). Natural gas must also be shut off at the distribution line. In both cases, this disconnect must be done by the service provider, and that provider should be the point of contact if there are any questions. Liquid propane gas (LPG), fuel oil, and other such tanks must be moved away from the building. Any tank that cannot be removed must be rendered inert, most often by filling it with sand. In addition, tanks that cannot be removed shall be vented to prevent an explosion or rupture. Check with the service provider if any doubt exists.

Marking the Building

Mark the sides of the building according to local protocol. This procedure is done to eliminate confusion and to reinforce the procedure with all of the participants. The accepted protocol is generally for the front of the building to be marked as side A, with sides B, C, and D designated in a clockwise fashion. Each floor needs to be designated as Division 1 for the first floor, Division 2 for the second, and so on. However, there are local and regional variants, so you should check with the AHJ (**FIGURE 18-8**).

Ventilation

Roof ventilation is another aspect of exterior preparation that needs to be readied. This process will have to wait until the interior fire locations are finalized. Emergency ventilation must be planned for in order to limit fire spread and improve interior conditions. Neither the primary nor secondary egress points should be used for normal room ventilation. The ventilation opening must be placed in such a way as to draw the products of combustion *from the means of egress*, not toward the means of egress. This gives the fire fighters a way out while the products of combustion are drawn away from them. When deciding where to put the ventilation openings, keep in mind fire behavior, the layout of the building, the burn room(s), hallways, and the primary and secondary means of egress. You can use existing roof openings that are normally closed but could be opened in the event of an emergency. Openings can be precut panels or hinged covers. The ventilation opening must be cut during the preparation stage and not while any live fire evolutions are taking place (**FIGURE 18-9**). Do not place fire fighters on the roof during a live fire training evolution. A pivot point and cable can be used to secure the covering so that it can be opened and closed from the ground as needed.

Chimneys

Chimneys need to be checked for stability to make sure they will not fail during interior operations. Once the structure is compromised, it becomes more likely

FIGURE 18-7 Ensure that the electrical service lines are disconnected from the pole to the weatherhead.

FIGURE 18-8 The sides of the building should be physically marked according to local protocol.

FIGURE 18-9 Roof ventilation openings are made during preparation, not during the actual fire. Hinges, pivot, and cable allow remote release from the ground.

that the chimney will collapse, endangering personnel operating inside and outside. Apart from the collapse concern, chimneys can allow fire spread into attics and upper floors during house fires. Although this is an infrequent problem, it should be considered. Outside chimneys should be removed before training takes place.

Additional Hazards

Any toxic weeds, insect hives, or vermin that could present a hazard to fire fighters must also be removed from the area. The last part of exterior preparation should be to remove any exposures or combustible materials outside, including storage sheds, detached garages, and materials from demolition and such. These items must either be removed or protected from unintended ignition or fire spread.

Interior Preparation
Common Hazards

Most of your preparation efforts will be spent inside, as the interior will be where the majority of operations take place. Most often, buildings donated for live fire training purposes have not been occupied for some time. Some buildings may have been used by trespassers for shelter or for illicit use. Hazards from these nontraditional uses can be present in the form of drug paraphernalia, broken glass, weapons, and even infectious clothing or bedding. Caution should be exercised during the preparation stage to protect personnel from these hazards, as well as from unseen structural dangers. It is important to use PPE, especially safety shoes, helmets, gloves, eye protection, and, if using power equipment, hearing protection, etc.

A systematic manner of preparation is necessary to accomplish the desired level of security. First and foremost, determine the building's utility status. Next, before interior preparation begins, check for environmental hazards, such as insects in or around the structure that could have an effect on the building's preparation or the evolution itself. Environmental hazards include contamination from past storage in homes and in businesses. Contamination may include pesticides and other chemicals, depending on the building's previous use. Illegal drug manufacturing has become an issue in many areas, and the conditions associated with these locations can dangerously contaminate a structure. Unfortunately, with abandoned structures, vagrants or drug users may have occupied them and left behind biological issues in addition to dangerous drug paraphernalia, broken glass, and trash. It may be wise to include biohazard protection such as latex gloves, masks, bags, and needle containers. Even storage found in normal homes can provide dangerous contamination from household pesticides, fertilizers, paints, and other materials. It may simply be safer to remove contaminated wood shelving or flooring. Businesses can be more difficult when it comes to environmental concerns, and it is important to determine what the building was previously used for. Consider checking past inspection reports and preburn plans.

One practice currently used is to spray the inside of the building with a bleach solution using a pump sprayer, starting from the farthest point from the exit and working toward the exit. This will kill off any insects and disinfect the area prior to removing any items. The suggested concentration is 2 cups of bleach per gallon of water (0.48 L of bleach to 3.785 L of water) in a sprayer. Be sure to use a mask when spraying the structure. Spray the structure 1 or 2 days prior to building preparation to allow for drying. Some other suggested solutions are as follows:

- 1 tablespoon of regular bleach per gallon of water, for staph and *E. coli*.
- ¾ cup of bleach per gallon will kill feline parvovirus and canine parvovirus.
- 1¾ cup of bleach per gallon will kill *Mycobacterium bovis* (tuberculosis).

Furniture

Furniture needs to be removed during the interior preparation. It may be tempting to leave furniture in the structure that appears to be wood, but do so with great caution. Most furniture today is not made of solid wood, but rather pressboard with an exterior laminate to look like real wood. Only Class A materials with known burning characteristics may be used for the burn sets (materials to fuel the fire). Until recently, NFPA 1403 did not specifically prohibit the use of burning furniture, as it did the use of flammable or combustible liquids. Previous editions of NFPA 1403 required that the fuels utilized have known burning characteristics and be as controllable as possible. With furniture, it is not always feasible to determine the construction materials used. Many of the commonly used products give off considerably more heat, smoke, and toxins than would be expected.

Decontamination of Props

A bleach solution will kill the following:

- Bacteria: *Staphylococcus aureus* (staph), *Salmonella enterica*, *Pseudomonas aeruginosa*
- *Streptococcus pyogenes* (strep), *E. coli*, *Shigella dysenteriae*
- Fungi: *Trichophyton mentagrophytes* (causes athlete's foot), *Candida albicans* (a yeast)
- Viruses: Rhinovirus type 37 (a type of virus that can cause colds), Influenza A, Hepatitis A virus
- Rotavirus, Respiratory syncytial virus (RSV), HIV-1, Herpes simplex type 2, Rubella virus
- Adenovirus Type 2, Cytomegalovirus

NFPA 1403 specifically prohibits the use of materials found onsite where the fire department cannot verify the environmental or health hazards associated with the materials, such as exposure to chemicals not readily apparent. Unknown chemicals pose dangers, not just in the obvious way of inhaled chemicals but also through contact. An unknown contaminant can get on protective clothing and could be later inhaled or contacted. For this reason, unknown materials are prohibited, along with known materials such as pressure-treated wood, rubber, and plastics. Further, straw or hay that is known to be treated with pesticides or harmful chemicals is not allowed for the same concerns.

LIVE FIRE TIP

Per NFPA 1403, ordinary combustibles such as clean wooden pallets, pine excelsior, and hay and straw (not chemically treated) are allowable fuels. *Clean wooden pallets* means that they are free from any noticeable spilled material that may have soaked into the wood, such as oils, pesticides, or other material that may cause an unforeseen condition or create a hazard.

Flooring

After the furniture has been removed, the rugs, carpeting, padding, and tack strips can be removed. This will expose the structure's actual floor. All combustible

FIGURE 18-10 For safety, expose the structure's actual floor by removing all additional layers of flooring.

material must be removed, especially linoleum, which is a solid petroleum product (**FIGURE 18-10**). Once linoleum is heated, it can act like flammable liquid pouring on the floor. Sheet vinyl (including the backing or underlayment), vinyl tile, and vinyl adhesive may all contain asbestos and must be removed. Remember, personnel will be crawling inside of the structure, so broken glass, debris, carpet tacking strips, vinyl flooring thresholds, nails, and other sharp objects are all crawling hazards and need to be removed. Holes in the flooring can lead to unexpected fire spread. Any holes in the floors need to be covered in a manner that will not cause harm, and must be able to bear the weight of fire fighters. Openings in the floor or ceiling created because of equipment being removed or other renovations should be evaluated for reducing structural stability. Fires should never be set under exposed structural members.

Walls and Ceilings

Walls and ceilings need to be checked for low-density combustible fiberboard or other combustible interior finishes. Low-density combustible fiberboard has contributed to the deaths of many fire fighters, and was a major factor in fire spread in the fires at Our Lady of Angels School (Chicago, Illinois), Hartford Hospital (Hartford, Connecticut), and Opemiska Social Club (Chapais, Quebec). Be very careful with unconventional interior finishes such as burlap, carpeting, artificial turf, and other treatments as they may cause rapid fire spread and unexpected smoke production, along with greater toxicity. Ceiling fans and large light fixtures also need to be removed so that they do not fall on participants during live fire exercises. Any holes that may allow fire to travel into the concealed void spaces must be covered.

Windows and Doors

Windows, as previously mentioned, need to be available for emergency egress. Window openings can be covered to keep smoke in and control flame spread. Be careful not to seal the windows airtight because this can inadvertently create flashover conditions. Depending on the size of the windows, a small space at the bottom of the window can be left open to allow for ventilation to reduce flashover concerns. A small opening can also be made at floor level to allow for the introduction of air. If using a chainsaw, the width of the chainsaw bar will work, and either a three- or four-sided opening should suffice.

Interior doors must also be made safe. Remove any hardware that may snag or catch on PPE (**FIGURE 18-11**). Either remove or secure doors that need to remain open. It is recommended to clearly mark doors that are not exits, such as closets or bathrooms without usable windows, and consider covering them so that a fire fighter under adverse conditions is not confused by the door frame and door (**FIGURE 18-12**).

FIGURE 18-12 Clearly mark doors that cannot be used as exits.

Kitchens

Cabinets and kitchen appliances are considered fixed contents and their removal will vary depending on what they are made of. Some live fire training instructors may want to leave in kitchen cabinets or bathroom cabinets for the live fire evolution. Depending on whether they are solid wood or composite-board, a burn test may be needed to determine the burn characteristics. If in doubt of the type of material, it is always best to remove it.

Commercial fixtures, such as large coolers or freezers, warrant careful review. Such commercial fixtures also frequently have large voids behind or around them to allow a service person to access them for repair. Such voids need to be considered. Closed containers, including water heaters and air conditioner compressors need to either be vented or removed (**FIGURE 18-13**). Cans of products, especially aerosol cans and smaller items, need to be removed.

Oil Tanks

Oil tanks and similar closed vessels that cannot be removed from the structure must be vented to prevent an explosion or overpressure rupture. Enforce strict safety practices when ventilating these tanks. Also, any hazardous or combustible atmosphere within the tank or vessel shall be rendered inert to prevent an unexpected explosion. All hazardous structural conditions shall be removed or repaired so as to not present a safety problem during use of the structure for live fire training evolutions. The area inside the tank or vessel should be filled with dry sand to render the internal atmosphere inert. Water or other liquids should never be used for this purpose.

Attics

The attic space needs to be inspected for hazardous contents. Air handlers, hot water heaters, gas heaters,

FIGURE 18-11 Any hardware that may snag or catch on PPE must be removed.

FIGURE 18-13 Vent or remove closed containers that could rupture when heated.

FIGURE 18-14 Fluorescent spray paint can be used to mark the way out.

storage, and other items could be a collapse hazard and some items could contribute to fire load or adverse conditions. The attic space should be accessible so that a fire could be controlled with ventilation during evolutions, if necessary. Place a piece of plywood over the opening on the ceiling and on the roof, securing them both just enough to hold them in place so that they may be removed when needed. Determination of when to open the space will depend on the layout of the building, the burn room, hallways, and the primary means of egress.

Never omit thorough checks of the attic, and look closely because water tanks, HVAC units, and other items may be hidden from view behind walls or hatches. Failing to remove these could cause a collapse under fire conditions.

Exits

Primary exits and exit routes need to be clearly marked and evaluated before each live burn. There are various ways to show exit routes, including the use of a light rope (illuminated ropelike cord), a strobe light, or other light at the primary exit point. Bright fluorescent paint on the floor can also help indicate the way out (**FIGURE 18-14**). Some live fire training instructors suggest marking the exit pathways at baseboard level or on the floor next to the wall area to prevent the paint from being worn off by crews crawling with hose lines. Participants in the training evolution need to be made aware of the exits and any markings used prior to each evolution.

Stairs and railings must be made safe. Any broken or missing stair treads need to be replaced, the stairs must be clear of trip hazards, and the weight-bearing capability must be checked. The handrails and balusters also need to be secure.

Preparation and Inspection of Props and Facilities

The preparation of burn plans and inspection of props and facilities is where the majority of your time will be spent. For both gas-fired and non-gas-fired live fire training structures, NFPA 1403 states, "Strict safety practices shall be applied to all structures selected for live fire training evolutions." This point cannot be stressed enough. During the planning and preparation phase, all live fire training structures must be visually inspected for damage *before* any training evolutions begin. If any damage is found, it must be documented and reported back to the owners of the structure. If damage is severe enough to endanger participants, the training shall be canceled.

The preparation and inspection for all live fire training structures includes the following:

- It is the responsibility of the instructor-in-charge to coordinate overall live fire training structure fire-ground activities, so as to ensure proper levels of safety. This includes the preparation stage.
- Look for signs of an obvious lack of structural integrity like a sagging roof or floors, and cracks in brick or masonry walls. Do the walls look straight or are they leaning or bulging? Check the doors and windows for easy operation, no obstructions, and no rubbing that could cause the door or window to stick. Hinges should be firmly attached. All doorways and windows need to be clear for unimpeded travel.
- Housekeeping—not only must emergency entry and egress be clear and operable, any area within the prop must be clear from obstructions, debris on the floor, and so on. Small items like staples and nails are painful to knees and palms of hands. Such props tend to "collect" materials that could impede emergency operations or interfere with training evolutions (**FIGURE 18-15**).
- Search the structure for unauthorized persons, animals, or objects.
- Check the stairs and railings to make sure they are stable and intact.
- Roof vents (if present) need to be checked for operation and soundness.
- Carefully walk the floors and look for any debris that could get in the way of the training evolution, or that could potentially harm fire fighters.
- Any damage noted needs to be documented and the appropriate steps need to be taken for repair.
- If there is concern that the damage is significant enough to affect the safety of the students, training cannot be permitted.
- Devices such as automatic ventilators, mechanical equipment, lighting, manual or automatic sprinklers, and standpipes, which are necessary for the live fire training evolution, shall be checked and operated prior to any training evolution to ensure proper operation.
- All safety devices, such as thermometers, oxygen monitors, toxic and combustible gas monitors, evacuation alarms, and emergency shutdown switches need to be checked to ensure proper operation.
- Any unidentified materials, such as debris found in or around the structure, which could

A

B

FIGURE 18-15 A. Obstructions can interfere with training evolutions and emergency procedures. **B.** Fuel should not be stored so that it is an obstruction or could cause unexpected fire spread.
Courtesy of Dave Casey.

burn in unanticipated ways, react violently, or create environmental or health hazards must be removed from where they may unintentionally ignite or be used as fuel.
- Awareness of weather conditions, wind velocity, and wind direction is required by NFPA 1403 and is always a good idea. Ambient heat and humidity can certainly affect the participants. A final check for possible changes in weather conditions should be done immediately before actual ignition.
- Make sure all possible sources of ignition, other than those that are under the direct supervision of the person responsible for the start of the training fire, are removed from the operations area.
- Check the burn set locations or gas-fired systems to ensure there will be ample room around them for attack hose lines, as well as backup lines, to operate freely.

When training on either gas-fired or non-gas-fired live fire training structures, there are specific guidelines given by NFPA 1403 for each respective structure.

Preparation and Inspection of Gas-Fired Live Fire Training Structures and Mobile Props

When training in gas-fired live fire training structures and with mobile props, NFPA 1403 provides specific guidelines for the inspection of the structure. In addition to the general guidelines that pertain to both gas-fired and non-gas-fired live fire training structures, the following must also be met to ensure the complete safety of all participants.

After the visual check of the rooms, the instructors shall run all of the gas-fired props and systems with students to ensure the correct operation of devices such as the gas valves, flame safeguard units, agent sensors, combustion fans, and ventilation fans. It is critical for all participants, instructors as well as students, to be comfortable with the controls in case of an emergency. All safety devices must be operational.

NFPA 1403 contains several items that the facility manager, not the instructor, has control over. The instructor, however, still needs to be aware of the following:
- The selection of LPG, compressed natural gas, or butane is not the instructor's decision, as the burners for the system are set for a specific fuel type when designed and built. These flammable gases shall only be permitted in structures designed for their use. The liquefied versions of these gases shall not be permitted inside the live fire training structure.
- On permanent gas-fired props, engineers have assessed the fire room environment for factors that can affect the growth, development, and spread of fire. It is important that the instructor-in-charge also be very familiar with the fire room environment for those factors. After operating at its most severe settings for several evolutions, how is the room's tenability? If a window or an exterior door is opened and wind is introduced, how will that affect interior conditions?
- Do not allow burn barrels with Class A or other materials, or other additional fire- or smoke-producing devices, to be used in a gas-fired live fire training structure without the manufacturer's and the AHJ's approval. Damage to the atmospheric monitoring equipment can occur, and if located in unprotected areas, damage to the structure can occur, depending on the intensity.
- Ensure that nothing is placed on the gas-fired props, such as Class A products.
- Watch for debris that could hinder the access or egress of fire fighters, and be sure to remove it prior to the beginning of the next training evolutions.
- Any time the instructor-in-charge determines that fuel, fire, or any other condition represents a potential hazard, the training exercise shall be stopped immediately. If an exercise is stopped, it should only be restarted once the hazard identified has been resolved. The Go/No Go sequence will be restarted only after the instructor-in-charge and safety officer determine that it is safe to do so.

Preparation and Inspection of Non-Gas-Fired Live Fire Training Structures and Mobile Props

NFPA 1403 gives specific details about the inspection of non-gas-fired live fire training structures. In preparation for live fire training, a careful visual inspection of the structure shall be made by the instructors and the safety officer to determine signs of damage or stress to the floors, walls, stairs, and other structural components. NFPA 1403 requires several of the same steps for non-gas-fired live fire training structures as for acquired structures. Some of these include the following:

- Property adjacent to the training site that could be affected by the smoke from the live fire training evolution, such as railroads, airports, or heliports, and nursing homes, hospitals, or other similar facilities shall be identified.

- The persons in charge of these properties shall be informed of the date and time of the training evolution.
- Streets or highways in the vicinity of the training site shall be surveyed for potential effects from live fire training evolutions, and safeguards shall be taken to eliminate any possible hazard to motorists.
- The instructor-in-charge shall document fuel loading, including all of the following:
  1. Furnishings
  2. Wall and floor coverings and ceiling materials
  3. Type of construction of the structure, including type of roof and combustible void spaces
  4. Dimensions of room

As with acquired structures, the importance of identifying the exact burn locations is critical, and the final decision is made by the instructor-in-charge and safety officer, ideally with considerable input from the participating senior instructors. With their input, the instructor-in-charge needs to carefully evaluate each fire room and burn location for factors that can affect the growth, development, and spread of fire. Fire spread beyond the room of origin is generally not an issue in non-gas-fired live fire training structures, except when fueling takes place too close by. Flashover also generally has not been an issue unless too much fuel is used. However, if materials other than pallets and hay or excelsior are allowed, the fire conditions can become more intense.

Fuel Load

An excessive fuel load can contribute to conditions that create unusually dangerous fire behavior. Only use the amount of fuel necessary to meet the learning objectives. Avoid conditions that could cause flashover or backdraft, which increase the risk to personnel to an unacceptable level and could also cause damage to the non-gas-fired live fire training structure. Local policy needs to be very specific regarding ignition and restocking of the fires. For example, a policy may state that no more than three pallets of a specific size and half a bale of excelsior should be used in a particular room. Be on the lookout for any debris that may hinder the access or egress of fire fighters. Remove any obstacles prior to the beginning of the next training evolution. Even small objects that seem harmless can get caught on turnout gear and potentially cause a major catastrophe.

Safety

Injuries and fatalities have occurred during past live fire training because of a deviation from the preburn plan, especially regarding improper fuel loading. Several reminders are applicable to help ensure a high level of safety:

- The instructor-in-charge and the safety officer need to ensure that the primary and secondary exit paths do not conflict with the expected avenues of fire spread and burn set locations.
- The burn set should generally be placed in a corner of the room, but permanent interior burn sites and protective lining may dictate burn set locations.
- Keep in mind where the fire stream will be directed into the room. Try not to place the burn set directly across from the door where the nozzle will be operating from.
- The burn set cannot be located in any designated exit path.
- No burn room should be used that does not have at least two separate means of egress.
- Hearths or fire boxes can be used to protect the ceiling and walls from direct flame impingement. One type of vertical burn rack holds several pallets with excelsior and has a top on it to direct flames outward, so they do not damage the ceiling and the floor (**FIGURE 18-16**). Some burn racks can be relocated between drill sessions.

FIGURE 18-16 Vertical burn racks hold several pallets, excelsior, or hay, with a lateral flame deflector.
Courtesy of American Fire Training Systems.

- The use of a steel "drawer" or large pan where pallets and excelsior are placed is becoming more common (**FIGURE 18-17**). These pans can help enforce restrictions on fire loading, and also help to keep staples or nails and debris contained for safety and easier removal. They can be permanently installed to slide out to the outside to be dumped. They can also be fitted with a hearth cover to protect the walls and ceiling. These are most often found in modified shipping containers.
- Each room that is to be used for live fire training evolutions must be prepared using Class A materials only.

FIGURE 18-17 Large pans or drawers hold pallets during training to keep nails, staples, and debris together and off the floor. Drawers slide to the exterior for easier cleaning.

After-Action REVIEW

IN SUMMARY

- Organizing a live fire training evolution requires time and work. If the training is to be valuable and safe, the instructor(s) must take the time and effort to plan and organize the event.
- In order to make sure the evolution is safe for everyone, there must be a dedication of resources to prepare the structure, conduct the evolutions, and conduct a post-incident analysis. Well-developed training session objectives will support safety and effectiveness as well as give the most benefit to the crews that are conducting the training.
- Regardless of the type of structure being used, before any planning is done, the first question asked should be, "Will the acquired structure or permanent live fire training structure allow for what the fire department wants to accomplish with this training?" It is important to specifically define the goals and objectives of the training evolution(s) to ensure that the structure will meet those needs.
- Those involved in the planning need to determine what they wish to accomplish to develop the specific learning objectives of the training evolution. The required learning objectives need to be defined so that the individual evolutions can be specifically designed to meet those objectives. The instructors should be fully briefed and prepared to meet the learning objectives. This will ensure effectiveness of the post-training review in assessing performance against training objectives.
- All personnel participating in live fire training evolutions must be properly trained before the evolution. Prerequisite training for students is covered in NFPA 1001: *Standard for Fire Fighter Professional Qualifications*.
- The IIC and the SO will determine the rate and duration of waterflow necessary for each individual live fire training evolution, including the water necessary for control and extinguishment of the training fire, the supply necessary for backup lines to protect personnel, and any water needed to protect exposed property.
- To determine the amount of water, NFPA 1403 refers the instructor to NFPA 1142: *Standard on Water Supplies for Suburban and Rural Fire Fighting*. In areas where there is not a reliable municipal water resource, the instructor will need to be familiar with the water calculations in NFPA 1142.
- NFPA 1403 also requires a minimum reserve of additional water in the amount of 50 percent of the fire flow demand determined to handle exposure protection or unforeseen situations.
- After determining that a structure is usable and the owner and fire department have agreed to terms, a more detailed inspection of the structure and property should be performed followed by preparation.
- The IIC is responsible for ensuring the acquired structure is prepped to meet NFPA 1403. Although there is no rule or requirement that states this inspection be separate from the initial evaluation, it is often better to do so for organizational purposes.

- The preparation of burn plans and inspection of props and facilities is where the majority of your time will be spent. For both gas-fired and non-gas-fired live fire training structures, NFPA 1403 states, "Strict safety practices shall be applied to all structures selected for live fire training evolutions." This point cannot be stressed enough.
- During the planning and preparation phase, all training structures must be visually inspected for damage *before* any training evolutions begin. If any damage is found, it must be documented and reported back to the owners of the structure. If damage is severe enough to endanger participants, the training shall be canceled.

KEY TERMS

Access Navigate for flashcards to test your key term knowledge.

Acquired structure A building or structure acquired by the authority having jurisdiction from a property owner for the purpose of conducting live fire training evolutions.

Fire flow rate The amount of water pumped per minute (gallons per minute or liters per minute) for a fire. There are several different formulas that are commonly used to calculate this.

Go/No Go sequence A verbal confirmation via radio communication that each and every participant is ready for action in the live fire environment.

Minimum water supply The quantity of water required for fire control and extinguishment. (NFPA 1142)

Order of operations The sequence of steps to conduct a procedure. In this context, the steps are in proper sequence to conduct the live fire evolution, but order of operations could also refer to any emergency scene operation. Most commonly refers to the sequence a mathematical equation is solved.

Personnel accountability report (PAR) A verification by the person in charge of each crew or team that all of their assigned personnel are accounted for.

Preburn plan A briefing session conducted for all participants of live fire training in which all facets of each evolution to be conducted are discussed and assignments for all crews participating in the training sessions are given.

REFERENCES

National Fire Protection Association. 2018. NFPA 1403: *Standard on Live Fire Training Evolutions.* Quincy, MA: National Fire Protection Association.

National Fire Protection Association. 2019. NFPA 1041: *Standard for Fire and Emergency Services Instructor Professional Qualifications.* Quincy, MA: National Fire Protection Association.

OSHA and NOAA. 2016. "Fact Sheet: Lightning Safety When Working Outdoors." Accessed December 5, 2017. https://www.weather.gov/media/owlie/OSHA_FS-3863_Lightning_Safety_05-2016.pdf.

REVIEW QUESTIONS

1. What content areas does a preburn plan for live fire training need to contain to comply with NFPA 1403?
2. Why is adherence to a preburn plan necessary? What should be done if there is a need to change the plan?
3. What is the purpose of a preburn inspection? What should the instructor be looking for?

CHAPTER 19

Conducting Burn Evolutions

NFPA 1041 JOB PERFORMANCE REQUIREMENTS

8.3.1 Identify and assign instructional tasks and duties in compliance with NFPA 1403, given staffing assignments, learning objectives, and instructor capabilities, so that safety officer(s), ignition officer, and crew/functional lead(s) are designated and rotated through duty assignments, instructor(s) implement participant accountability, proper instructor/student ratios are maintained, instructor(s) monitor and supervise all participants during evolutions, and awareness of changing conditions that impact training is maintained.

(A) Requisite Knowledge.

NFPA 1403, accountability procedures, supervisory techniques, and resource management.

(B) Requisite Skills.

Coaching and observation techniques.

8.3.2 Conduct a preburn briefing session, given the preburn plan, so that all facets of the evolution(s) are identified, training objectives are covered, a walkthrough of the structure or prop with all participants is performed, and established safeguards and emergency procedures are identified.

(A) Requisite Knowledge.

Preburn plan, safety rules, emergency procedures, and authority having jurisdiction (AHJ) policy and procedures.

(B) Requisite Skills.

Presentation and class management skills.

8.3.3 Maintain the training environment to safeguard participants, given participants in a live fire training evolution, so that signs and symptoms of fatigue and distress are recognized, action is taken to prevent injuries, and actions are documented.

(A) Requisite Knowledge.

Signs and symptoms of fatigue and distress, knowledge of environmental conditions; AHJ's safety, rehabilitation, and emergency procedures.

(B) Requisite Skills.

Evaluation of environmental conditions, class management, report completion, activation of the AHJ's emergency procedures.

KNOWLEDGE OBJECTIVES

After studying this chapter, participating in a structured learning environment, and completing assigned assessments, you will be able to:

- Describe the necessary prerequisite knowledge and skills for a fire fighter to become a live fire training instructor. (**NFPA 1041: 8.3.1**, pp 434–440)
- List and explain the instructional tasks and duties associated with a live burn. (**NFPA 1041: 8.3.1**, pp 431–434)
- Explain the purpose and procedures for a preburn briefing. (**NFPA 1041: 8.3.2**, pp 433–434)
- Identify safety hazards in live fire training and describe the steps to mitigate risk during the preparation and operation phases of live fire training. (**NFPA 1041: 8.3.3**, pp 440–441)
- Identify the need for staff and participant rotation during live fire training. (**NFPA 1041: 8.3.3**, pp 440–441)

SKILLS OBJECTIVES

After studying this chapter, participating in a structured learning environment, and completing assigned assessments, you will be able to:

- Assign instructional staff for a live burn evolution so that all positions required are staffed with qualified instructors. (**NFPA 1041: 8.3.1**, pp 431–440)
- Ensure the use of self-contained breathing apparatus (SCBA) by all personnel involved in an operation in which they could encounter immediately dangerous to life or health (IDLH) atmospheres. (**NFPA 1041: 8.3.3**, pp 438–441)
- Ensure a rapid intervention crew is provided during the live fire training evolution. (**NFPA 1041: 8.3.3**, pp 439–441)
- Conduct a preburn briefing session and walkthrough, ensuring all participants have knowledge of the layout of the structure, props, and training area details and that safeguards are in place. (**NFPA 1041: 8.3.2**, pp 433–434)
- Oversee a live burn evolution to maintain safety of all participants. (**NFPA 1041: 8.3.3**, pp 440–441)

You Are the Live Fire Instructor-in-Charge

You are one of eight instructors conducting a live fire training in-service at a non-gas-fired structure. During the pre-incident instructor meeting, the lead instructor makes the statement, "We are tight on time, so we are not going to do a walkthrough of the building. These guys have been in here before."

1. How do you address issues of unfamiliarity or disregard of NFPA 1403?

2. How do you avoid issues like this in the future?

3. Should this training be allowed to continue as planned?

 Access Navigate for more practice activities.

Introduction

The success, or failure, of any training event relies heavily on the instructors. The instructor-in-charge bears even greater responsibility. The National Fire Protection Association (NFPA) has recognized this and has established qualifications for an instructor and an instructor-in-charge.

Recall that, according to NFPA 1403: *Standard on Live Fire Training Evolutions*, an **instructor** is an individual who has the training and experience to supervise students during live fire training evolutions, who has met the requirements for Fire Instructor I, and has been qualified by the authority having jurisdiction (AHJ) to deliver fire fighter training. An **instructor-in-charge (IIC)**, according to NFPA 1403, is an individual qualified as an instructor and designated by the AHJ to be in charge of the live fire training evolution and who has met the requirements of an Instructor II in accordance with NFPA 1041. It is important to realize that, ultimately, a fire chief can name anyone whom he or she wishes to be considered instructors for a live burn. Choosing less-than-competent individuals to fill these roles can lead to accidents and the accompanying liability. The AHJ should consult NFPA 1041: *Standard for Fire and Emergency Services Instructor Professional Qualifications* to determine if an instructor has the training and experience necessary to supervise students in a live fire training evolution.

A main concern for any training event is the development of a lesson plan and individual action plans (IAPs). The IIC will be required to develop objectives and the lesson plan, schedule instructors, ensure logistical needs are met, and ensure that evaluations are conducted, as well as oversee the entire evolution(s). These skills are developed under the criteria for Fire Instructor II. There is a difference between the learning environment and an emergency incident. There should be no surprises while performing a live fire training evolution, and training should be the safest environment possible.

The scenario in "You Are the Live Fire Instructor-in-Charge" relates how a complacent instructor can play a role in potential mishaps in a live fire training session. Complacency, a lack of training as a live fire instructor, or just a lack of preparation contributes to injuries and even fatalities, as is seen in some of the incident reports in this text. NFPA 1403 is quite clear on the requirements to become a fire service instructor and that instructors must be qualified to deliver fire fighter training. It is also important to stress that these instructors have the training and expertise to supervise students during live fire training. Instructors acting as the safety officer also require training on the application of the requirements of the NFPA 1403 standard. The safety officer needs to have training to understand fire behavior, human physiology, water supply, and other factors covered in this standard. A comprehensive training program needs to be considered for instructors as well as for those acting as the IIC and/or safety officer. A number of regional or local training centers, as well as several states, have programs that regulate this process. In Florida and Pennsylvania, for example, a training program is legally required for all fire instructors participating in live fire training, and even requires state-issued fire instructor certification in addition to the training course.

> **LIVE FIRE TIP**
>
> Complacency is the cause of injury and even death during live fire training. Students are always observing their instructors, including how they wear their personal protective equipment (PPE). Instructor PPE and self-contained breathing apparatus (SCBA) usage sends a strong message without saying a word. An instructor cannot credibly espouse the danger of carcinogens and other toxic materials and wear dirty, damaged PPE. Instructors know their limits and the characteristics of the props they use. Students don't know what the instructor has trained on, so when instructors take short cuts and perform complacent acts, students believe these to be acceptable practices at all fires.

Instructor-in-Charge Responsibilities

The IIC has overall responsibility for all activities on the fire ground during live fire training. Every instructor has heard that the training ground is the safest environment possible as all aspects of the training are under the control of the instructors.

The IIC is responsible for full compliance with NFPA 1403, so the IIC and safety officer need to have a thorough understanding of all concepts of the standard.

The IIC must ensure that there are enough instructors available to meet the 1:5 instructor-to-student ratio and have one instructor for each functional crew and backup line, and one additional instructor for each additional functional assignment. Providing live fire training is a physically stressful event not only for students but for instructors as well. The provision for rehabilitation must extend to instructors as well. The ignition officer is subjected to additional heat especially as they need to add additional fuel during an evolution. NFPA 1403 states that instructors shall be rotated through assignments. An instructor should not serve as the ignition officer for more than one evolution in a row. Additionally, the standard now requires assignment rotation, rest, and rehabilitation for instructors. A good practice to follow is to have the instructors rehabilitate with their assigned groups, which also allows for further debriefing and discussion. Conditions and exposure to fire conditions vary tremendously; however, the instructor monitoring the fire (ignition officer) will still be required to rehabilitate. The IIC must ensure that instructors follow the standard. Many instructors want to "remain" in the burn room, and object to being placed on a backup line or outside. As an overheated and dehydrated instructor, you not only endanger yourself—it makes it more likely that you will lose mental acuity and physical dexterity, possibly not catching mistakes or being able to correct them as well as if you were in better condition.

It is necessary to ensure that enough instructors are onsite to safely conduct the evolution. This may require the IIC to have additional instructors due to weather extremes, a large number of participants being trained, or training that will take a long time. For example, there are academies that will train during the day with recruits, and then have additional outside burns in the nighttime for family members to attend. This event can be very taxing on resources, but the IIC must ensure that enough instructors who are able to conduct the training and deal with any emergency situation are available.

> **LIVE FIRE TIP**
>
> When learning objectives are clearly written and specific, it assists the instructor in setting up the training evolution.

Learning Objectives

Every live fire training evolution must have a clearly defined purpose. When conducting live fire training, clearly defined learning objectives for every training evolution must be established and communicated to all participants prior to beginning the drill.

Clear and measurable learning objectives help keep both students and instructors on track and ensure that the evolution has a learning purpose. An objective should be written for each specific task that is expected to be performed during the evolution. "Getting experience fighting interior fires" is not a specific learning objective. "Ladder crew will formulate and implement a plan for horizontal ventilation" is a specific learning objective.

A building block, or "crawl, walk, run approach," should be used to develop these training objectives. You can start with small-scale props such as the candle experiment or dollhouse prop to help students understand fire behavior before you move into a live fire environment. In the live fire environment, the first set of objectives regards understanding fire behavior with the instructors serving in the suppression roles. Then, as learning progresses and the students gain more experience, they can start stretching hose lines, performing search, and eventually running the type of scenario that requires a full-blown response.

When writing learning objectives, keep a picture in mind of what the tasks should look like when they are being performed correctly. For example, if a task were to include advancing a hose line and extinguishing a fire, a sample learning objective may be as follows:

> The engine crew of three will advance a 1¾″ (44-mm) hose line to the interior of the building and, while maintaining thermal balance, completely extinguish a room-and-contents fire.

This learning objective can be broken down into more detailed tasks, including the following:

- Check the hose stream before entering the structure.
- Check personal protective equipment (PPE) and self-contained breathing apparatus (SCBA) before entry.
- Identify and control the flow path.
- Ensure proper positioning of personnel on the hose line.
- Use a defined search pattern.
- Adapt and correct movement for conditions present.
- Position attack for best shielding and protection of personnel.
- Control proper stream application.

Each task expected to be completed at the live fire training should be written out in detail and included in the preburn plan. Task completion is measurable, making it possible to identify when the learning objective is met.

> **LIVE FIRE TIP**
>
> The importance of adherence to local standard operating procedures (SOPs) cannot be overemphasized. The most effective live fire training evolutions replicate the SOPs that are used during actual fire-ground operations.

For many live fire training evolutions, crew assignments of three are ideal. The three fire fighter crew should be made up of two fire fighters and one live fire training instructor, but we always have to maintain the student-to-instructor ratio of 5:1. The instructor may communicate the crew's progress and the completion of assigned learning objectives during the live fire training evolution.

Be careful not to over-assign fire fighters and create crews that are unrealistic to standard firefighting operations. Many complaints arise during the postevolution debriefing when crews are overstaffed or too many crews are assigned to work in one area. A common example would be when too many students are rotated through the nozzle position at the same time. Keep in mind that you need to be able to view and interact with all students at all times. This is not possible if the crews are too large. Having an interior safety officer or second instructor available may alleviate this.

Another factor to determine is if you have enough instructors to complete the rotations with smaller groups. Having an interior safety officer or backup instructors reduces the need for the entire crew to exit when a student has an SCBA issue, becomes overheated, or panics.

New Recruits and Live Fire Training Evolutions

When conducting live fire training with new recruits, the difficulty level of the planned evolutions should match the students' abilities. Some common training evolutions conducted with new recruits include the following:

- Fire behavior and fire growth patterns
- Simple search patterns
- Hose line advancement and stream application
- Thermal imager training
- Room-and-contents fire on ground-level floors
- Cold smoke drills
- Simulated, full-scale responses (multi-company scenarios)

Experienced Students

As much as a recruit can be a challenge for instructors during live fire training, the veteran or experienced fire fighter poses a different set of concerns. The veteran fire fighter comes to training with significant practice and experience in the trade, which can be a double-edged sword for instructors. Modifying or changing existing behaviors of a veteran fire fighter can be just as difficult as teaching new recruits, if not more so. Stay focused and follow every part of NFPA 1403, even though students may have performed live fire training many times throughout their careers.

Some particular concerns that experienced fire fighters pose include a possible reluctance to follow orders, complacency in wearing PPE, and a belief that live fire trainings are controlled and therefore pose no danger. This is a recipe for a tragedy on the training ground.

Another challenge is this: 20 experienced fire fighters may have received their initial training from 20 different sources. In addition, some veteran fire fighters may have been taught practices and protocols that differ from the SOPs of the instructor's fire department.

Unauthorized "shortcuts," such as the inappropriate use of tools or equipment or the use of damaged PPE, may be used on the fire ground by some experienced fire fighters. This will pose a problem when trying to teach recruit fire fighters how to perform tasks according to local SOPs. Many of these practices violate safety protocols and should not be allowed when conducting live fire training. Examples range from not using the throat closure on the turnout coat, to not wearing the waist strap on a SCBA, to many different "shortcuts" that may not provide for safety or backup should conditions change or something unexpected happen.

Experienced fire fighters may want to be challenged during their training. This desire to "show their stuff" may lead to variances from the standard. As an instructor, do not be pressured to modify the training evolution in order to provide a more "exciting" evolution for the experienced personnel. When the instructor-in-charge has set the evolution and determined the scope of the training, it shall not be altered. Deviating from the plan encourages bad things to happen on the training ground, up to and including line-of-duty deaths. During training, participants should do the following:

- Conduct the training exercise according to established fire department SOPs and applicable NFPA standards.
- Maintain personal discipline and accountability for actions during training exercises (National Institute for Occupational Safety and Health [NIOSH], 2016).

One advantage of having a class of experienced fire fighters is being able to include training evolutions that require more skill and work to complete. Evolutions that require multiple companies to work simultaneously are as realistic as it gets when conducting live fire training. Some examples of these evolutions include a room fire with simulated victims, content fires that require vertical ventilation, attic or cockloft fires, fires in an attached garage requiring an attack from the interior of the house, or fires that require forcible entry to be performed prior to entering the building.

Preburn Briefing

The preburn plan should be used prior to actual live fire training evolutions. The perfect time to review the preburn plan is during the preburn briefing session, during which all facets of each evolution are discussed and assignments for all crews are given.

Prior to the evolutions, all participants are required to conduct a walkthrough of the acquired structure to have a knowledge of and familiarity with the layout of the structure and to facilitate any necessary evacuation. Emergency procedures in the plan will be reviewed, as will the emergency egress routes.

Prior to conducting the live fire training evolution, all participants and instructors must be briefed and inspected by the safety officer. A simple rule to remember is: There can be no surprises when conducting actual training. This requires that all participants walk through the building and are made aware of all facets of the structure. During this walkthrough, paths of egress and windows, doors, and stairways that can be used to evacuate or retreat if necessary should be pointed out. The preburn briefing and inspection is the last opportunity to ensure the safety of the participants and the success of the drill.

During the preburn briefing, it is not necessary to disclose the actual location of simulated victims, but it is required to discuss the use of them if they will be included as part of the evolution. Participants need to be advised that there will not be any "live" victims. *No participant shall play the role of a victim under any circumstance.* Realizing the confusion that may occur between a fire fighter down and a manikin in turnout gear, NFPA 1403 has specified that rescue manikins outfitted in firefighting gear and used as victims must be uniquely colored or distinctively marked for easy identification. It is very important that participants be able to clearly identify any body they encounter as a manikin or actual fire fighter down.

Before beginning each training evolution, all participants shall be briefed on their roles during the evolution during the preburn briefing. NFPA 1403 does not require the disclosure of items such as where the fire will be located or where to find victims during the drill, provided that the possibility of victims is known. When the evolution begins, there should not be any confusion about what each fire fighter's task will be and what the learning objectives of the training evolution are.

Selecting Instructors for Live Burn Evolutions

NFPA 1403 now requires that the IIC has received the training to meet the minimum job performance requirements for Fire Instructor II (NFPA 1041). However, training centers and fire departments can set their own requirements for their live fire training instructors (LFTIs), unless there are state or provincial requirements. In those states, the department can specify requirements above those required by the state. Florida requires live fire instructors to also be certified as a Fire Instructor I, which requires 6 years of fire service experience, completion of a 40-hour course, and passing a state certification test. Pennsylvania's requirements include 8 years of fire service experience in addition to letters of recommendation.

LFTI candidates must have the following qualities:

- They have operated extensively in the environment they will teach in, and have an excellent understanding of all potential nuances and dangers the situation can present (**FIGURE 19-1**).
- They have proven, practical teaching skills, not just in the classroom but also in turnout gear and in a dynamic learning environment.
- They possess extensive knowledge regarding NFPA 1403 and other related NFPA standards, Occupational Safety and Health Administration

FIGURE 19-1 Live fire training instructors need to be experienced and well trained in fighting fires.
Courtesy of Chris Dilley.

(OSHA) standards, fire behavior, building construction, smoke reading, and hostile fire events.
- They are physically fit enough to perform their duties, including emergency actions.
- They are willing to lead a cultural change regarding the importance of participant safety, in both the training and emergency environments.

All of the listed qualities may seem straightforward—until you try and quantify and add detail to them. It should go without saying, but the LFTI candidate needs to be an excellent fire fighter. A good start in achieving this goal is for the candidate to meet the requirements of NFPA 1001: *Standard for Fire Fighter Professional Qualifications.*

The instructor candidate must have a solid foundation with the required knowledge, skills, and ability in the following topics:

- The history and orientation of the fire service
  - Fire fighter qualifications
  - Tactical priorities and fire department organization
  - Policies, procedures, and standard operating guidelines
  - Roles and responsibilities included in Fire Fighter I and II training

- Fire fighter safety
  - Causes of fire fighter injuries and deaths
  - Safety standards (NFPA 1500)
  - Safety and health programs
  - Safety during training
  - Fire fighter PPE
  - Protective clothing
  - Respiratory protection
  - Donning and doffing breathing apparatus
  - Inspection and maintenance of breathing apparatus
  - Use of breathing apparatus
  - Proper gear and equipment decontamination
  - Personnel monitoring for signs of medical concerns
- Accountability
- Fire service communications
  - Radio use
- Incident command system (ICS)
  - History and characteristics of the ICS
  - ICS organization
- Fire behavior
  - Chemistry of fire
  - Fire development and control
  - Smoke reading
  - Fuel loading
  - Heat saturation
- Building construction
  - Construction terminology, materials, and classifications
  - Hazards related to building construction
- Portable fire extinguishers
- Firefighting tools and equipment
  - Function and use
- Ropes and knots
- Response and size-up
  - Safety during emergency and nonemergency driving
  - Emergency operations
  - Scene safety
- Forcible entry
  - Breaching walls, floors, and ceilings
  - Forcing doors and windows
- Ground ladder functions and types
  - Inspection and maintenance of ladders
  - Handling, positioning, and safe use of ladders
- Search and rescue operations
  - Rescue and extrication
- Ventilation
  - Reasons for fire-ground ventilation
  - Considerations
  - Tactical priorities
- Water supply
  - Rural and municipal systems
  - Fire hydrant types, location, and operation
- Fire hose operations, nozzles, and streams
  - Fire hydraulics
  - Care and maintenance
  - Couplings, appliances, and tools
  - Fire stream patterns and nozzles
- Fire fighter survival
  - Safe operation procedures
  - Self-extrication
  - Mayday and emergency procedures
- Salvage and overhaul
- Fire fighter rehabilitation
  - Function, causes, and need
- Fire suppression
  - Handling hose lines while advancing and operating
  - Basic fire tactics at the company level
- Preincident planning
- Emergency medical care
  - Infection control
  - Emotional stress
  - Cardiopulmonary resuscitation (CPR)
  - Bleeding and shock
  - Burns
- Hazardous materials
  - Properties and effects
  - Recognizing and identifying
  - Proper protective equipment and decontamination

Not only does the AHJ need to make sure that the instructor candidates have the skills, ability, and knowledge to succeed, but they must also have the right attitude. This is one of the reasons why some states require written letters of recommendation.

Fire fighter training is all about the students, not the instructors. It is not the time or place for the instructors to show off their skills. It is time for the instructors to show the students the correct method for hose placement, advancement, and stream application and to discuss what is occurring around them, while in a hostile environment.

It is imperative that the instructor candidate understands that accountability must be continually maintained for all personnel operating on the scene. In order for the training process to run safely, the instructor candidates must understand the fundamentals while making safety their ultimate priority. If they do not operate in such a manner during emergency responses, how do you expect them to act during a live fire training evolution? Those individuals have no business being LFTIs. Ideally, if this is the case, it is discovered before the training has started.

IICs must ensure that instructors working under them are selected for their demonstrated proficiency at performing standard live fire training evolutions. The following are skills that IICs must be certain an instructional team member possesses before assigning him or her to a live fire exercise:

- Instructor is attentive to positioning in relation to the crew.
- Instructor can rotate student crews in the nozzle position while maintaining crew accountability and protection.
- Instructor can withdraw student crews while maintaining accountability and monitoring fire conditions.
- Instructor adjusts positioning and mobility in order to be able to assist the student crews in case of problems.
  - Instructor is able to reach and touch students for communication, reassurance, or to provide direction, and ultimately, remove students if needed (**FIGURE 19-2**).
  - Instructor is able to reach the nozzle to adjust the pattern, open or close it, or take control (**FIGURE 19-3**).
  - While monitoring and encouraging students, instructor is prepared to take action to guard students from harm (**FIGURE 19-4**).
  - Instructor remains on the side of the hose away from the wall, so as to not get pinned against it (**FIGURE 19-5**).
- Instructor is able to address evacuations and missing students.
- Instructor is able to use thermal imagers for monitoring students' location and performance, reading heat signatures and fire spread indicators.
- Under live conditions, instructors must understand how to shield themselves from radiant heat and how to protect themselves during interior training (**FIGURE 19-6**).

FIGURE 19-3 The instructor should be able to reach the nozzle to adjust the pattern, open or close it, or take control of it.
Courtesy of Joshua Bauer, Seminole Tribe of Florida Fire Rescue.

FIGURE 19-4 It may be necessary for the instructor to take action to keep the student out of harm's way.
Courtesy of Joshua Bauer, Seminole Tribe of Florida Fire Rescue.

FIGURE 19-2 The instructor needs to be able to reach and touch the students for communication or reassurance, or to provide assistance and direction.
Courtesy of Joshua Bauer, Seminole Tribe of Florida Fire Rescue.

FIGURE 19-5 The instructor needs to remain on the side of the hose away from the wall so as to not get pinned against it.
Courtesy of Joshua Bauer, Seminole Tribe of Florida Fire Rescue.

Instructors must be able to demonstrate the proper methods of training students to use features to protect themselves from hostile fires. These methods include using doors, corners, and furniture, and staying low on staircases.

- Instructor can address heat saturation inside PPE. This type of heat saturation most often occurs in the upper areas of the body where the PPE is pressed against the body, such as the upper arms, shoulders, and upper back. Moving about to reinstitute the air pockets helps.

FIGURE 19-6 Participants need to be shown how to protect themselves from radiant heat, and instructors need to use protection when necessary while crews enter and exit.
Courtesy of Dave Casey.

Staffing and Organization

Instructor positions must be filled by qualified and competent people. The IIC needs to assign instructors that meet the criteria to the following roles:

- One instructor to each functional crew, which shall not exceed five students
- One instructor to each backup hose line
- Additional personnel to backup lines to provide mobility
- One additional instructor for each additional functional assignment

It is important that this selection process follows an organized structure that is known and understood by all participants. Instructors and staff need to be assigned to one of three functional groups: command staff, attack group, and support group. Instructors may swap assignments within their group at any time, as long as the IIC is advised of the change. The accountability system must then be updated with these changes.

Groups are to be rotated at the discretion of the IIC. It is important to note that out of concern for divided responsibility, it is generally best for the IIC to also be the incident commander. The command structure should follow the normal National Incident Management System (NIMS) model. Following a strict incident command structure will help maintain a higher level of safety and accountability (**FIGURE 19-7**).

The IIC has the overall responsibility to coordinate the drill-ground activities to ensure correct levels of

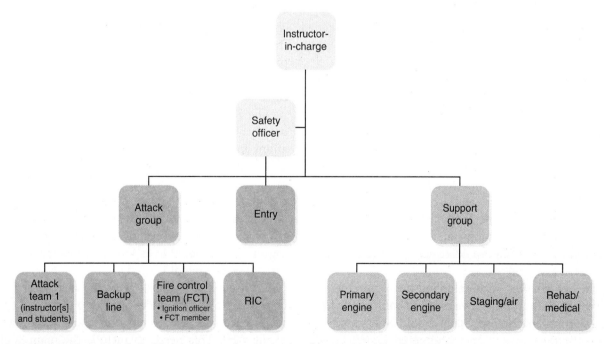

FIGURE 19-7 A sample incident command organization for structural live fire exercise.
© Jones & Bartlett Learning.

safety, whether the exercise involves a permanent prop or an acquired structure. The IIC must work closely with the safety officer (SO) to be sure that operations are conducted properly. By using the Go/No Go sequence, *all operational positions are empowered and given responsibility to watch for safety issues.*

Along with other considerations, personnel assignments will be based on the number of evolutions, the number of participants involved, the experience of the participants, and the desired objectives.

Safety Officer

Just as in fire-ground operations, the role of the safety officer is of the highest importance and the selection of the individual should be based on their abilities and knowledge of live fire training and the roles and duties of the incident SO. The assignment of an SO to all live fire training evolutions is required by NFPA 1403.

Unlike the incident SO on the fire ground, the SO for training has the ability to see and assess the structure before the fire starts. The SO should be involved in the planning and preparation of the structure.

Like emergency fire-ground operations, the SO will have the authority, regardless of rank, to intervene and control any aspect of the operations when, in his or her judgment, a potential or actual danger, accident, or unsafe condition exists.

The SO should provide for the safety of all persons on the scene, including students, instructors, visitors, and spectators. It is important to note that the scope of the SO's duties includes ensuring the safety of nonparticipants. The responsibilities of the SO shall include, but not be limited to, the following:

1. Prevention of unsafe acts
2. Elimination of unsafe conditions

The SO shall not be assigned other tasks that interfere with safety responsibilities. The role of the SO requires constant vigilance and considerable mobility to be able to observe operations and to maintain constant scrutiny of structural conditions. The SO must know about available safety features. There are responsibilities that may be deemed necessary by the SO that may require additional safety personnel. These personnel can be located strategically within the structure to react to any unplanned or threatening situation or condition, or to review geographical or functional assignments.

The SO is tasked with enforcing the requirements to wear SCBA and protective clothing appropriate for the environment. No individual should be exposed to smoke, toxic vapors or fumes, products of combustion, or other contaminated atmospheres or be exposed to an oxygen-deficient atmosphere. The use of SCBA and a personal alert safety system (PASS) device must be enforced for all participants, and the instructors are often the most difficult to influence to comply. SCBA with PASS must be used any time participants, or anyone for that matter, are operating in an atmosphere that might be any of the following:

- In an atmosphere that is oxygen deficient or contaminated by products of combustion, or both
- In an atmosphere that is suspected of being oxygen deficient or contaminated by products of combustion, or both
- In any atmosphere that can become oxygen deficient, contaminated, or both
- Below ground level

It is one of the major duties of the SO to ensure that these rules are strictly followed to avoid unnecessary injuries or fatalities.

Fire Control Team

It is the role of the IIC to appoint a **fire control team**. The fire control team comprises a minimum of two personnel: the ignition officer and at least one other individual. The ignition officer cannot be a student or the SO and is responsible for igniting, maintaining, and controlling the materials being burned. The second member of the team stays in the area of ignition with the ignition officer to observe the ignition and help to maintain the fire as well as recognize, report, and respond to any adverse conditions.

Ignition of the fire is initiated only when the fire control team is in full PPE and SCBA and there is a charged line in place. An instructor must visually confirm that the flame area is clear of training personnel prior to ignition. The decision to ignite is made by the IIC in coordination with the SO and then communicated to the ignition officer.

The SO will often need to be assisted by several deputy SOs to help maintain visual contact with operations and the structure, which is a distance from him or her. In permanent props, deputy SOs may be necessary for entry checks and to monitor student progress through the prop. In Class A props, deputy SOs may be needed to monitor the fire conditions. The SO has many important responsibilities during live fire training, and it is prudent to assign additional personnel to assist the SO when it is needed.

With gas-fired props, all of the instructors must be trained in the complete operation of the system in accordance with the manufacturer and the direction of the AHJ. This needs to include the main control station, the local controls for each prop, emergency shutdowns, and emergency procedures for the gas-fired prop. The safety officer must also be trained to operate all emergency shutdown controls and valves. The command staff may be smaller, but it follows the same organizational structure. As in acquired structures, nobody operates alone, instructor or student.

The following lists the sample incident command organization for live fire evolutions:

- The **command staff** shall consist of the IIC and SO.
- The **attack group** shall consist of the attack instructor, backup instructor, RIC, and the entry officer.
- The **support group** shall consist of the primary engine, secondary engine, rehab/medical, and staging/air supply.
- The **instructor-in-charge** has overall responsibility and authority for managing and coordinating the drill, ensuring that all procedures are completed in accordance with NFPA 1403. Generally, the IIC will also be the incident commander.
- The **safety officer** shall have the authority, regardless of rank, to intervene and control any aspect of the operations when, in his or her judgment, a potential or actual danger, accident, or unsafe condition exists.
- The **attack instructor** supervises the attack team and directs the fire attack. The attack team shall not exceed five students.
- The **backup instructor** supervises the backup team, which monitors conditions during the live burn. The backup team shall not exceed five students. As the term infers, this line is ready to be deployed, and is normally outside so it can be rapidly relocated to backup interior crews in case of unforeseen fire spread, water supply problems to the attack lines, or other issues.
- The **rapid intervention crew** should monitor the interior and exterior conditions and should not be made up of students, but rather instructors. This crew must be sufficiently staffed, equipped, and prepared to make entry at any point to remove a lost or entrapped fire fighter or crew. This works the same way in real emergencies, except that during training, the crew is familiar with the structure, so they can plan the rescues and be better prepared.
- The **entry officer** controls the entry point to the structure for accountability. The entry officer must maintain an entry board to hold the personnel accountability tags (PAT), PASSPORT, or other accountability system used, of students and instructors working inside

FIGURE 19-8 An accountability board.

to keep an accurate head count inside and out (**FIGURE 19-8**). The entry officer also monitors exterior conditions and performs a last check on all PPE. The entry officer reports to the SO and is part of that division.

- The fire control team includes the **ignition officer**. The ignition officer ignites the fire after the decision to ignite has been made by the IIC and SO. The ignition officer should not be a student. The ignition officer ignites, maintains, and controls the fire and fuel with a hose line available after assuring the area is clear of personnel. Ignition can be accomplished using propane lighters, fusees, kitchen-type matches, and similar devices that must be removed from the area after ignition. The fire control team is utilized any time a fire is ignited and consists of a minimum of two members, one of whom is the ignition officer or assigned ignition person. A member other than the ignition officer observes the ignition officer ignite and maintain the fire and watches for adverse conditions. Members must be in full PPE and SCBA when performing their functions.
- The **primary engine/pumper** establishes and maintains a continuous water supply from the primary water source. It operates to provide required water flow and discharge pressures to the attack line and all other lines as needed.
- The **secondary engine/pumper** establishes and maintains a continuous water supply from the secondary water source. It operates to provide required water flow and discharge pressures to the backup line and all other lines as needed.
- The **rehab/medical officer** assesses the established location for conditions such as positions upwind, uphill, and with easy access

to emergency medical services (EMS), and sets the rehabilitation plan into effect. The rehab officer ensures that baseline vital signs are taken on all students and instructors before the drill. The rehab officer also ensures that all students arrive at rehab from attack positions for vitals, hydration, cooling, and rest. After rehabilitation, the students are sent to staging for reassignment. The rehab officer manages patient care and is the contact person with local EMS.

- The **staging/air supply officer** assesses the established location for inhibiting conditions (upwind, uphill). The staging officer also maintains adequate air supply reserve, monitors all PPE, and maintains at least one crew in a state of readiness if assistance is needed in addition to the RIC and other team assignments.

While NFPA 1403 places the overall responsibility for the rehabilitation of personnel with the IIC, the SO oversees this process more directly. Appropriate health monitoring, time for rest and rehabilitation, food and fluid replenishment, and relief from climatic conditions must be provided to participants.

Staff and Participant Rotation

Live fire training is physically demanding on the instructors as well as the students. The NFPA requires that instructors rotate through assignments, keeping the instructor in condition to conduct live fire training. There are a number of methods for instructor and student rotation within the structure. Due to the interior crews' close proximity to the fire, the idea is to rotate instructor and student positions to keep them from overheating. This is especially true for Class A burn buildings and acquired structures, where the ignition officer and personnel assigned to monitor and restock the fire will be exposed to greater heat over a longer period of time. The ignition officer must be rotated after every Class A ignition. Rotations need to be planned for and "choreographed" to prevent delays between entries. Any delays that occur while a fire is burning can lead to an increase in the heat of a permanent prop or can make the entire acquired structure dangerous.

Crews of recruit students should be kept as small as possible. Four students to a crew is generally the most that can be supervised at one time, but three is more desirable. Routine crew staffing can be used to demonstrate what crews may actually experience. NFPA 1403 states there shall be no more than five students per instructor. Many agencies use one instructor and one SO for each attack hose line team for recruit student training. This is done so that the instructor or SO is always within reach of all of the students and they are able to closely monitor their activities.

When working with nonrecruit fire fighters, the rotation is different. The instructor has no more than five students at a time and he or she can maintain integrity with that crew throughout the entire evolution. The instructor and crew will move as one team from one assignment to the next. Remember, many communities will only have two to three fire fighters on an attack line, and some may only have one fire fighter and a company officer. When planning the evolutions, it is beneficial to try and reflect the actual staffing that will be present at the burn.

Evolution Safety

The IIC is ultimately responsible for the safety of everyone on the training ground. The safety rules pertinent to the training shall be reviewed prior to starting. These may include trip hazards, projections from the wall such as standpipe connections, emergency lights, or other drill-ground or prop hazards. This is also a good time to explain the building evacuation procedure and demonstrate the evacuation signal that will be used. Emphasize that the evacuation signal will only be used in real emergencies and shall not be used as a prop for any training evolution. An alternate method to simulate an emergency can be used with the knowledge of all instructors and support staff and should be preceded by the notice, "This is a drill." This still allows practice evacuation under emergency conditions and ensures accountability, RIC activation, and so on.

All hose lines used for the evolution shall be charged and checked prior to beginning. Ensure that you have adequate flow and water supply. This exercise of testing all hose lines is also a good way to confirm that the pump operator is competent, and there are no mechanical difficulties. It is recommended that all of the hose lines to be used are flowed simultaneously. This includes hose lines to be used in the evolution, the ignition hose line, and any exposure or safety hose lines. This will ensure that sufficient flow capability is available. At this time, assure that you have adequate personnel to safely handle all of the hose lines that will be used.

Determine if the tools necessary to complete the training evolution are in place and ready prior to starting the scenario. This includes ensuring that portable radios are working and on the proper channel. Check components such as fuel levels in saws and other tools.

Also, ensure that the RIC has all of the tools necessary to complete a rescue, if needed.

Each participant, including instructors, shall be checked before entering a live fire evolution to ensure that PPE is being worn correctly and that there are no defective pieces of equipment. There should be no exposed skin prior to personnel entering into the IDLH environment. Also, prior to beginning the evolution, the safety officer should inspect everyone's SCBA to confirm that they are beginning the training evolution with an adequate air supply. Inspection of PPE and SCBA must be done prior to every training evolution. It is not sufficient or acceptable to check all of these items at the beginning of the training day and never reevaluate them.

Emergency Plans

A training fire is unlike an emergency response. We can plan, gather personnel, lay out equipment and hose lines, and tour the building before a planned fire starts. However, when working with an acquired structure, unexpected fire spread and other problems can occur. This is why a solid understanding of fire behavior, dynamics, and flow path is so important; it will help instructors recognize potential problems. Unlike a fire in a permanent live fire training structure, in this case, the structure itself is actually on fire. The role of the RIC (which must now meet NFPA 1407) and the safety personnel, and the purpose of the walkthrough, all become more critical.

Despite the weeks of planning leading up to a training exercise, fire is unpredictable and can still spread into voids, concealed spaces, and exit paths. Like a working structure fire, all personnel must be vigilant for such fire spread. When fire is discovered in any of these areas, training operations should cease, as the fire should now be treated as an uncontrolled fire. Planning must determine how emergencies like this will be handled. The backup hose line may take over suppression while recruits are relocated to the exterior. However, personnel need to know that operations have shifted from training to suppression. The instructor-in-charge needs to advise all personnel of the shift from training to suppression and get an acknowledgment of that shift, just like when shifting from offensive to defensive operations at a structure fire. NFPA 1403 requires that any time the instructor-in-charge determines that any condition represents a potential hazard, the training evolution shall be stopped immediately. If the fire acts in an unexpected manner for whatever reason, or problems are encountered with students, hose lines, water supply, or anything that threatens the safety of the participants, withdraw the personnel and resolve the situation. In certain situations, this will require careful analysis of the building in order to ensure it is safe to continue. Crews on the outside may need to be reassigned with new orders to extinguish the fire, until the building is assessed for safety. If the fire can be extinguished and the area made safe, sometimes ceilings can be secured and operations started again. Be very cautious with fires getting into attic and large void spaces, as this is a recurring problem in acquired structures.

If an evolution is stopped, it should only be restarted once the hazard identified has been resolved. The safety officer and any assistants need to confer with the instructor-in-charge and, if in agreement, the evolution should continue only after the Go/No Go sequence is initiated.

Developing the Preburn Plan

The first preburn plan developed by a department or an instructor will always be the most difficult. An SOP needs to be put in place to ensure the plan is a policy that must be followed. Because this policy involves a hazardous activity with risks, it should be specific in nature. Once the SOP and a preburn plan are developed, they will serve as guides for future exercises. It is strongly recommended to use clear language that all participants can understand when developing the preburn plan. This is a critical piece of the planning stage—everyone needs to understand every element of the preburn plan.

Injuries and fatalities have occurred in the past when there was a deviation from the preburn plan and when leadership was not fully aware of the plan. It is crucial that once a preburn plan is in place, it be followed strictly to ensure the safety of all involved.

Items that should be a part of any preburn plan are as follows:

- Learning objectives
- Participants
- Water supply needs
- Apparatus needs
- Building plan
- Fuel needs
- Site plan
- Parking and areas of operation
- Emergency plans
- Weather
- List of training evolutions

- Order of operations
- Emergency medical plan
- Communications plan
- Staffing and organization
- Safety officer
- Fire control team
- Staff and participant rotation
- Personal protective equipment use
- Rehabilitation plan
- Agency notification checklist
- Demobilization plan

Live fire training records must be kept in accordance with NFPA 1403. Regardless of the type of structure used in training, an accounting of the activities conducted; a list of instructors, their assignments, and all other participants; documentation of unusual conditions encountered; a list of injuries incurred and treatment rendered; any changes or deterioration of the structure; and documentation of the condition of the structure and surrounding areas after the training exercise must be kept on record. Documenting these items and saving a record of each will only help you, the instructor, in the end. If something goes wrong, even after the fact, you will have these records to go back to.

An essential post-fire training task is thoroughly decontaminating and cleaning all PPE. Arrangements should also be made for personnel involved in training evolutions with products of combustion to shower and change uniforms.

After-Action REVIEW

IN SUMMARY

- The success, or failure, of any training event relies heavily on the instructors. The instructor-in-charge bears even greater responsibility. NFPA has recognized this and has established qualifications for an instructor and an instructor-in-charge.
- The IIC has overall responsibility for all activities on the fire ground during live fire training. Every instructor has heard that the training ground is the safest environment possible as all aspects of the training are under the control of the instructors.
- The IIC is responsible for full compliance with NFPA 1403, so the IIC and safety officer need to have a thorough understanding of all concepts of the standard.
- When conducting live fire training, clearly defined learning objectives for every training evolution must be established and communicated to all participants prior to beginning the drill.
- Clear and measurable learning objectives help keep both students and instructors on track and ensure that the evolution has a learning purpose. An objective should be written for each specific task that is expected to be performed during the evolution.
- NFPA 1403 now requires that the IIC has received the training to meet the minimum job performance requirements for Fire Instructor II (NFPA 1041). However, training centers and fire departments can set their own requirements for their live fire training instructors, unless there are state or provincial requirements. In those states, the department can specify requirements above those required by the state.
- It is imperative that the instructor candidate understands that accountability must be continually maintained for all personnel operating on the scene. In order for the training process to run safely, the instructor candidates must understand the fundamentals while making safety their ultimate priority.
- The first preburn plan developed by a department or an instructor will always be the most difficult. An SOP needs to be put in place to ensure the plan is a policy. Because this policy involves a hazardous activity with risks, it should be specific in nature.
- Once the SOP and a preburn plan are developed, they will serve as guides for future exercises. It is strongly recommended to use clear language that all participants can understand when developing the preburn plan.

KEY TERMS

Access Navigate for flashcards to test your key term knowledge.

Instructor An individual qualified by the authority having jurisdiction to deliver fire fighter training, who has the training and experience to supervise students during live fire training evolutions, and who has met the requirements of an Instructor I in accordance with NFPA 1041 (NFPA 1403).

Instructor-in-charge (IIC) An individual qualified as an instructor and designated by the authority having jurisdiction to be in charge of the live fire training evolution and who has met the requirements of an Instructor II in accordance with NFPA 1041 (NFPA 1403).

Fire control team A team consisting of at least two members including the ignition officer. The other fire fighter(s) can staff the charged line, observe the ignition, help maintain the fire, and watch for and report adverse conditions.

REFERENCES

National Fire Protection Association. 2018. NFPA 1403: *Standard on Live Fire Training Evolutions.* Quincy, MA: National Fire Protection Association.

National Fire Protection Association. 2019. NFPA 1041: *Standard for Fire and Emergency Services Instructor Professional Qualifications.* Quincy, MA: National Fire Protection Association.

REVIEW QUESTIONS

1. When selecting instructors to assign to a live burn evolution, what should an IIC consider?
2. Why is an understanding of fire dynamics one of the most important components that the instructor must have a complete understanding of when working in a live fire environment?
3. How can an IIC create a training climate that empowers all participants to speak up when they have a safety concern?

CHAPTER 20

Post Live Fire Evolution

NFPA 1041 JOB PERFORMANCE REQUIREMENTS

8.4.1 Conduct a postburn briefing session, given the learning objectives of the evolution, so that feedback on each learning objective is provided to each participant, and any needed corrective actions are identified.

(A) Requisite Knowledge.
Preburn plan, learning objectives, supervisory techniques, and authority having jurisdiction (AHJ) policy and procedures.

(B) Requisite Skills.
Presentation skills, evaluation skills, class management skills.

8.4.2 Conduct a postburn inspection of the structure or prop, given a structure or prop for live fire training, so that structural damage is identified, safety concerns are identified, and necessary corrective actions are taken.

(A) Requisite Knowledge.
AHJ policies, facility requirements, safety practices.

(B) Requisite Skills.
Observation techniques, inspection skills.

8.4.3 Complete records and reports in accordance with NFPA 1403, given a live fire evolution, so that all required reports are completed.

(A) Requisite Knowledge.
NFPA 1403, AHJ policy on records.

(B) Requisite Skills.
Technical writing and records completion.

KNOWLEDGE OBJECTIVES

After studying this chapter, participating in a structured learning environment, and completing assigned assessments, you will be able to:

- Describe the benefits of a postevolution debriefing. (**NFPA 1041: 8.4.1**, pp 451–452)
- Identify topics that should be covered in a postevolution debriefing. (**NFPA 1041: 8.4.1**, pp 451–452)
- Inspect and document any damage to exterior props and notify the owner or authority having jurisdiction. (**NFPA 1041: 8.4.2**, pp 451–452)
- Describe tasks involved in the overhaul of an acquired structure. (**NFPA 1041: 8.4.2**, pp 451–452)
- Describe methods to verify student prerequisite training has been met. (**NFPA 1041: 8.4.3**, pp 447–448)
- Identify the documentation that will be required for any live fire training evolution according to NFPA 1403. (**NFPA 1041: 8.4.3**, pp 447–449)
- Describe the additional documentation required to be maintained for acquired structures according to NFPA 1403. (**NFPA 1041: 8.4.3**, pp 447–449)

- Describe the documentation specific to gas-fired and non-gas-fired structures that must be maintained according to NFPA 1403. (**NFPA 1041: 8.4.3**, pp 447–450)
- Identify documentation that may be required when using an exterior prop for live fire training according to NFPA 1403. (**NFPA 1041: 8.4.3**, pp 447–450)
- Describe the process for turning over control of the structure to the property owner, including the completion of a standard form. (**NFPA 1041: 8.4.3**, p 450)
- Describe the process used to conduct a postevolution debriefing. (**NFPA 1041: 8.4.3**, pp 450–452)
- Describe the documentation that must be maintained after conducting a live fire training evolution. (**NFPA 1041: 8.4.3**, pp 450–452)

SKILLS OBJECTIVES

After studying this chapter, participating in a structured learning environment, and completing assigned assessments, you will be able to:

- Conduct a postevolution debriefing to collect feedback and identify problems. (**NFPA 1041: 8.4.1**, pp 451–452)
- Complete a postburn inspection of a structure or prop so that all damage is identified and documented. (**NFPA 1041: 8.4.2**, pp 451–452)
- Create a documentation file with required paperwork of a live fire training evolution. (**NFPA 1041: 8.4.3**, pp 447–449)

You Are the Live Fire Training Instructor

You have completed live fire training in an acquired structure. During the training, one student received first- and third-degree burns to his left wrist. Also during the event, a student became separated from her teammates during search and rescue evolutions, and a mayday was called.

1. What are your responsibilities now that the training is complete?

2. What paperwork needs to be completed and filed?

 Access Navigate for more practice activities.

Introduction

You may have heard the saying, "If it's not written, it didn't happen." An obvious reason for documentation is for legal purposes in the case of an injury or incident; however, documentation can also be used to improve future training events. Valuable information can be gained during the postincident analysis on the benefits of training, how smoothly the event went, and how to improve the training. This information comes from both instructors and participants at the events. Checklists of required paperwork can be developed to expedite the planning process for future events, which can also be used to verify the skill levels of personnel and may be required for recertification hours. If an injury or a line-of-duty death (LODD) occurs, federal, state, provincial, and local agencies will conduct an investigation.

Besides the legal ramifications, there is another practical reason for thorough documentation. An investigation can be a stressful and emotional event. If there has been an injury or LODD, instructors and participants may experience a wide range of emotions. Some participants and citizens may try to place blame, members may feel guilt, and others will question whether something could have been done to prevent the outcome and whether the event followed proper procedures. Investigators will scrutinize the department's policies, training, and equipment, and question all personnel involved in the incident. The investigation could carry on for a long period of time, and it may seem the investigator is looking for someone to blame. However, a proper investigation must be conducted to ensure that the problem does not occur again. Thorough documentation will assist investigators and make the process go more smoothly for all involved. Documentation will identify that NFPA 1403 standards were followed and that all aspects of the event followed proper guidelines.

Reports and Documentation

Documentation Before the Event

Much of the documentation will be done prior to the actual live fire training event. There are steps that must be completed prior to training, such as confirming there is no insurance on the acquired structure, visually inspecting props, calculating fuel load and water supply, preparing asbestos removal reports, securing owner documentation including insurance cancellation, and notifying neighbors of the event. All of these steps need to be documented. As mentioned, the first time you plan a live fire training event will be the hardest part of the process. Once the training is complete, you will have a checklist of what needs to occur. You will know what permits are required by the authority having jurisdiction (AHJ) and how long it takes to get them, understand what notifications need to be made, and know the steps involved in conducting a burn. You will be able to develop a checklist of required permits, a to-do list of actions prior to the event, and a checklist for the day of operations. And, of course, the incident action plan (IAP) becomes a template for future live fire training.

You will need to create a file that contains all the documentation, reports, and permits. It should include both general documentation and documentation specific to acquired structures, gas-fired and non-gas-fired buildings, and props. This file should be retained according to the AHJ's record retention policy. If an injury or unexpected exposure occurs, you may need to maintain records for the life of the participant or longer.

General Reports and Documentation

To comply with NFPA 1403, the standard operating procedure (SOP), IAP, and permits must be kept on file. The standard requires that all records and reports of live fire training be maintained, including an accounting of the activities conducted; a list of instructors present and their assignments; participants; unusual conditions encountered; injuries and treatment rendered; changes or deterioration of the structure; and documentation of the condition of the premises and adjacent area at the end of training.

Another document to maintain is the personal protective equipment (PPE) inspection. In live fire training, the safety officer (SO) is to conduct a PPE check. Include documentation of the gear condition and how the gear was inspected. This may also include a re-inspection of the gear after the live fire training event to ensure that no damage has occurred, and if it has, that it is documented and students are informed to follow AHJ procedures on PPE maintenance.

As part of the live fire event documentation, an IAP must be created (**FIGURE 20-1**). The IAP should contain the learning objectives for each evolution during training. For example, the first evolution may include search and rescue objectives, while the second evolution may include a mayday scenario. These evolutions have very different objectives and should be created by the instructor-in-charge (IIC) and SO. The IAP should also list all instructors and participants, their crew assignments, and the evolutions in chronological order. The building plan and site plans should be retained as part of the documentation file because they will indicate water sources, apparatus placement, areas of operation, number of rooms, and burn information. Site plans will also identify parking zones, emergency medical services (EMS) locations, access and egress routes for emergency vehicles, and parking areas for visitors.

Other documentation to maintain includes the following:

- Student prerequisite training: Prior to allowing personnel to participate in live fire training, verify that the participants meet the prerequisite training required by NFPA 1403. (Remember that there are prerequisite training requirements in NFPA 1403 that are not in the current NFPA 1001 standard and that the standard requires incumbent fire fighters to meet those

FIGURE 20-1 The IAP is used during the evolution. Notes need to be kept on issues or changes to the plan.
Courtesy of Dave Casey.

prerequisites.) Regarding personnel from within the fire department, you will either know their qualifications or be able to access their records. Departments need to ensure current certifications of their own personnel. This may be a state-issued certificate or nationally accredited certificate. If there are participants from other agencies, it is necessary to obtain written evidence of successful completion of the prerequisite training prior to allowing participation in live fire training. The types of written documentation can vary based on familiarity with the participant's level of training from outside agencies. All students from outside agencies should be allowed to participate only as official representatives of an established organization, with proper documentation required to facilitate planning of the training.

- Apparatus: Document all apparatus on scene, how it was used, and where it was located.
- Preburn plan: The preburn plan includes objectives of the evolutions, fuel loading, apparatus and hose placement, and other information covered previously in this text.
- Emergency plan: An emergency plan covers evacuation procedures, mayday procedures, accountability concerns, and communications procedures and channels.
- Emergency medical plan: The emergency medical plan includes details such as ambulance placement and level of staffing, helicopter landing zones, and location of hospitals and trauma/burn centers.
- Communications plan: This includes routine and emergency traffic channels. Identify who is assigned radios; include procedures for contacting dispatch center if needed.
- Order of operations: Evolutions in chronological order include which room(s) will be ignited and in what order, instructors assigned to evolution, rotation of crews and instructors.
- Fuel load: Document the fuel load for each evolution. This should include initial loading and reloading if needed. Anticipated and actual fuel load should be documented.
- Weather: Document the conditions that could impact the evolution and safety of participants. This may include temperature, humidity, wind chill, rain, lightning, and wind speed and direction.
- Water supply: Document the calculations/determinations on required water supply, identification of separate water sources, and any shuttle operations.
- Temperature: Document the temperatures taken throughout an evolution, whether using thermal imagers or thermocouples.
- Rehabilitation records: Document the time in, vitals, and rehabilitation operations.

Documentation of an Emergency Incident

There are some general procedures to follow in the case of an emergency during a live fire training event. After providing care to the injured person, all persons involved should be asked to write a statement as to where they were, what they were doing prior to the event, and what they saw from their perspective. This should be done prior to any discussion among participants. PPE and self-contained breathing apparatus (SCBA) should be secured; gauges on the bottle and regulator should be noted prior to shutting the unit down. Note whether the personal alert safety system (PASS) device is activated. The PPE and SCBA should be placed in a location with restricted access and only official personnel should be granted access to this. Photographs should be taken of all aspects of the event. This includes apparatus and hose placement, location of the event, photos of all areas in and around the training area, and any tools that were being used. Other personnel will need to be taken care of as well, and depending on the severity of the incident, you may need to consider critical incident stress management (CISM) and counseling. Appropriate authorities must be notified early following AHJ guidelines.

This is only a brief overview of what should occur in the case of an emergency. As the instructor, you should be aware of the AHJ's procedures on injuries including whether a governmental body handles worker's compensation or investigations. The IIC will need to address the continuance of the event on a case-by-case basis when an injury occurs. Minor injuries may not require the event to be stopped, but other events may require the event to be canceled. It is also important to consider factors such as what caused the emergency, the emotional and mental fitness of participants, and the ability to conduct the event safely.

Photographs and Videos

Photographs and videos are valuable documentation devices; however, there are a few important points and guidelines to keep in mind. The individual taking photographs cannot interfere with the operations of preparation or the event itself. Entrance into the immediately dangerous to life or health (IDLH) area is restricted, so photographers may not be able to enter.

Another concern surrounds liability. If everything is done according to NFPA 1403, you should not have any issues; however, there have been occasions when fire fighters were freelancing and were photographed in the hot zone without proper PPE, improper fuels were used, acquired structures had egresses blocked, and personnel were conducting overhaul without SCBA being used. Photographs such as these are clear evidence that NFPA 1403 and other safety practices were violated. Live fire training is an exciting event, and your members will want to video and snap pictures and then place them on social media. It is strongly suggested that you have a solid SOP for social media, which should include the review by the AHJ of photos and videos before being released. Social media has given instructors numerous examples—both positive and negative—to review with students on live fire evolutions.

Documentation of Unusual Events

Documentation of the unusual events listed here should include, but not be limited to, the following:

- Evolution stopped by the IIC or SO: Documentation should include the reason(s) the event was stopped (e.g., fire behavior was unexpected, room temperatures were above expected levels, a student panicked), the steps taken to make the event safe again (e.g., ventilation, allowing the building to cool, removal of the student), and when the event began. It should also include the points at which the order of operations was suspended and then resumed.
- Water flow issues: In the case of a loss of pressure or if the attack lines or backup lines lose water supply, the evolution will be stopped immediately. Again, documentation should include steps taken to ensure participant safety, steps taken to restore water flow, and the point at which the evolution was resumed.
- Evacuation ordered: Unless the evacuation was planned as part of the training evolution and objectives written as such, any order for an evacuation is an unusual event and should be documented as to circumstances, personnel accountability reports (PARs), emergency actions, and whether the event was resumed or ended.
- Mayday: Any situation resulting in an unscheduled mayday must be carefully documented, as well as rapid intervention crew (RIC) activation, rescue activities, and any additional actions or medical treatment.
- Medical emergency: If an injury occurs during training, the first priority is to ensure treatment for the injured. Whether the treatment is simply a bandage, oxygen, or rehydration, a report should be filed. Even if an injured participant is not transported, there must be a report on file. Health Insurance Portability and Accountability Act (HIPAA) rules do apply, so files containing medical records must be retained and only personnel with a need to know should have access to them. If you are going to use the event as a template, you may want to note the run report number and where the report can be found rather than retaining the report in the file.
- Changes or deterioration of structure: In most acquired structures, the structure is burned to the ground after all training has been accomplished. These are expected changes and deterioration of the structure. The IAP should clearly identify that the final evolution is that of burn down. This time is not a training event, but rather a controlled burn that should follow established procedures and command structure. However, when the order of operations is developed, document an outline of how many times a room can be burned and the fuel load. The fire behavior is anticipated and the evolution controlled. However, if, during the event, the fire does not get extinguished as expected and extends to another room or ceiling (which was not planned), the event must be documented. Carefully document circumstances regarding the unplanned event. This will provide valuable information later on. Possible causes could be early ignition of the fire, participants not ready to attack, or more fuel load than needed. Whatever the cause, identifying this will help to make future training events run smoother.

In permanent structures, it is important to identify any changes or deterioration to the building. Spalling that was not present before will need to be documented and the owner notified. A visual inspection is required prior to use, and the same individual should conduct a visual inspection after evolutions to ensure the safety of the building, making sure it is ready for future use and identifying any changes to the structure. Rebar may be exposed during the burn, and the owner needs to be notified of this prior to using the building. Doors or windows that swung freely before the evolution may be difficult to operate, indicating that maintenance will need to be performed.

The importance of this documentation cannot be overstated. In some cases, liability for damage has become an issue when multiple agencies used a structure, damage resulted, and there was no documentation of what was found in the preinspection and postinspection reports.

The Conclusion of the Training Exercise

As part of the IAP, there is a conclusion to all training events. Regardless of the structure or prop, the environment should be returned to a safe and ready state. This may include extra pallets, hay, and straw being removed to a trash container, shutting off fuel sources and safety valves, securing buildings, ensuring that all fires have been extinguished, and other activities required by the AHJ. Documentation should include what was done, any cleanup activities, and, if there were any effects on the adjacent areas, how they were mitigated. All smoke that affects roadways should be gone prior to removing any barriers, and the area should be made safe before taking barriers down that cordoned off visitors and spectators.

Acquired Structures

Besides the general documentation that must be completed, there are certain documents that will need to be retained after using an acquired structure. These include the following:

- Asbestos removal documentation: Document that either there is no asbestos in the structure, or, if there was asbestos, subsequent proof of its removal.
- EPA documentation (if needed): Jurisdictions may require the Environmental Protection Agency (EPA) to review acquired structure plans. This review and any mitigation should be documented. EPA may require samples drawn of oil–water separators. Document the results and keep it on file at the training center.
- Runoff considerations: Document where the runoff went and whether it needed to be contained.
- Preparation records: Records should include any repairs, rooms that were barricaded (which should be on site plan), removed items, notification of neighbors, identification of areas affected by smoke, and designated visitor/spectator areas.
- Owner documentation: Document evidence of ownership and clear title, proof of insurance cancellation, written permission on what will be done with the structure, and the condition of the structure at the end of evolution.
- End of evolution documentation: A document should be signed by both the AHJ and the owner turning over control of the property. This should include documentation of any debris that may need to be removed or any structures remaining standing and their condition. For example, a house may be burned, but an outside shed left according to the owner's request. The document should include that the shed is undamaged, intact, and in a safe condition. This should be a formal process that includes a standard form transferring the authority.

Gas-Fired and Non-Gas-Fired Structures

Documentation for gas-fired and non-gas-fired structures should include the following:

- Preinspection condition: Note any damage that is found prior to the evolution. Include the condition of gas-fueled props and documentation of a test burn.
- Safety issues: Document whether emergency shut-offs, emergency stops, automatic ventilation systems, and so on, were tested and worked properly.
- Fuel usage: The AHJ may require that you document fuel usage.
- Thermocouple readings: Most AHJs have preset temperatures that are not to be exceeded. As such, an instructor should be assigned to ensure that the temperature is not exceeded in the burn building.
- Inspections: Before using a burn building, the IIC or SO should ensure that the AHJ has completed annual inspections and required engineer inspections. Consider keeping a record of these inspections in your files.
- Burn sequence chart: The burn sequence chart will document fuel load and the number of burns in the order in which they were performed.

Exterior Training Props

Documentation for exterior training props will depend heavily on the prop or props used. Manufactured props will require specific maintenance that will need to be documented. There should be a visual inspection based on the manufacturer's guidelines as well. Homemade props need to be visualized for structural integrity, possible rust or damage, damage that may affect proper operation, properly functioning fuel supply lines, debris that may have accumulated in or around the prop, and anything that may affect proper operation. This can be difficult if you are not familiar with the prop, so familiarization and training with the prop in advance of training is essential. Documentation should include the visual inspection, operation of the prop prior to the evolution, and ensuring the prop is left in a safe condition after each use. The AHJ may require the documentation of fuel usage. If you use Class A materials to ignite the prop, fuel load must be documented. Include documentation of extinguishing agents, the amounts needed, and reserves.

If you are using a homemade prop, preparation of the prop needs to be well documented. For example, if you use a vehicle, specify that drums and cylinders were vented, emptied, and made safe. A dumpster needs to be cleared of any debris that could cause unexpected fire behavior. Include the completion of dry runs and non-fire attacks to show familiarization and knowledge of the prop.

Postburn Inspection

Overhaul of Acquired Structures

The structure needs to be reevaluated after the final burn down, to determine what measures need to take place to render the site safe. Metal roofing, chimneys, large beams, and parts of walls that remain could be dangerous to children or scavengers. Arrangements need to be made for heavy equipment to knock over remaining walls or chimneys and to make the site safe.

It should be expected that scavengers will want to go through the debris to collect salvageable metals and materials, and the curious will want to explore the remains. Signs and barrier or scene tape can warn adults of the danger, but children may not understand the warning signs and some adults won't recognize the danger present. Known by the legal term *attractive nuisance*, many work sites and even some fire departments have been sued for injuries incurred by "trespassers." The added step of installing construction or barrier fencing and warning signs will help protect the property owner and the fire department until the debris is removed (Stittleburg, 1998). Often local government will have the fencing that they use on hand for large events and construction projects, or it can be included in the agreement with the property owner that they provide it. It is also available for rental.

Postburn Inspection of Structures and Props

A thorough postburn inspection of the structure/prop is necessary. If there is any damage or deficiencies, or any repairs are needed to the facility, they must be reported. The facility needs to be left in a safe condition. Be respectful of the next crew of fire fighters who will be using the facility. Carefully cleaning up your crew's mess will allow for an easier transition for the next crew.

> **LIVE FIRE TIP**
>
> Do not forget the importance of decontamination after a live fire event. Just as documentation of PPE condition should be included, so should documentation of the decontamination of gear. Field decontamination should be conducted. Use sanitation wipes to remove as much soot as possible from head, neck, jaw, throat, underarms, and hands immediately and while still at the training event. Instructors and students should be reminded to change clothes and wash them as soon as possible when they return to their station, to shower within an hour after training, and to clean PPE, gloves, hood, and helmet, and tools. Used PPE should not be taken back to fire fighters' homes until it is cleaned. Fire apparatus should also be decontaminated after training. These steps will help to reduce exposure to carcinogens.

The Postevolution Debriefing

Just as a critique should be conducted after every emergency call, so should a critique be done after every training evolution. The purpose of this critique, or **postevolution debriefing**, is to ensure there are no injuries, determine what was done correctly and went well, and identify any areas that need improvement in future training events. All participants should be encouraged to give input during the debriefing.

The postevolution debriefing should be held after the training event and cleanup. Even though personnel will be tired, the event is fresh in their mind. In some cases, the students may gain confidence from the postevolution debriefing and it should reinforce what they have learned and clarify proper procedures.

A common format should be used for the debriefing. It may begin by asking questions such as, "Do you think that we met the objectives of the training event?" or, "Do you feel this training was beneficial? Why or why not?" The debriefing can include a review of learning objectives, a description of events, and an assessment of any obstacles that may have been encountered. The postevolution debriefing should be held in a comfortable area away from environmental issues and distractions, such as foot traffic and excessive street noise.

Students and instructors need to be reminded to use constructive criticism and that the event is a learning environment. Remember to focus on the objectives and training evolution and to use positive reinforcement of any activities that went well. Limit the discussion when necessary to keep the critique on track and within allotted time frames. It is often tempting to debate tactics and policy following an evolution. If these subjects do not add to the constructive criticism of the evolution, limit the discussion on those matters.

Finally, before ending the postevolution debriefing, confirm that there were no injuries sustained during the training evolution.

Any comments made during this debriefing should be documented and become part of the file. Any injuries identified will require a medical evaluation report. In addition, make sure to document follow-up activities and completion of the IAP components, such as proper decontamination, cleanup considerations, and what to do if a student or instructor becomes ill or experiences medical problems after the event.

Postevolution Evaluation Forms

To make the postevolution debriefing a valuable tool, review the participants' comments. Objectives may need to be rewritten or evolutions changed to make the training more valuable. A critique is valuable only if the lessons learned are applied to future training events. In some cases, instructors and students may not feel comfortable critiquing aspects of the evolution in an open format. They may be more likely to honestly report their feedback in writing; therefore, evaluation forms should be provided to both students and instructors to be completed anonymously. The evaluations provided to the students should provide an opportunity to evaluate not only the training event but also the instructors who were involved. Participants should be given the opportunity to identify what they felt was a good learning experience, if they had any concerns with the training, and how they would change the training event to make it a better learning experience in the future. This documentation will give the AHJ the opportunity to receive honest feedback and identify possible concerns with the instructors or training.

Instructors may not share freely in the debriefing if other instructors are senior or higher ranking. As an instructor, you can use the written postevolution evaluation to identify what was done correctly, what could be improved, or any concerns or issues that need to be addressed.

These evaluations, as with any training evaluation, should be maintained in the file. They should also be reviewed by the AHJ, IIC, and SO, and documented so that future live fire training can be improved. If evaluations are done but nothing is considered or changed, the evaluation becomes useless.

After-Action REVIEW

IN SUMMARY

- An obvious reason for documentation is for legal purposes in the case of an injury or incident; however, documentation can also be used to improve future training events.
- Valuable information can be gained during the postincident analysis on the benefits of training, how smoothly the event went, and how to improve the training. This information comes from both instructors and participants at the events.
- You will need to create a file that contains all the documentation, reports, and permits. It should include documentation specific to acquired structures, gas-fired and non-gas-fired buildings, and props. This file should be retained according to the AHJ's record retention policy. If an injury or unexpected exposure occurs, you may need to maintain records for the life of the participant or longer.
- To comply with NFPA 1403, the SOP, IAP, and permits must be kept on file. The standard requires that all records and reports of live fire training be maintained, including an accounting of the activities conducted; a

- list of instructors present and their assignments; participants; unusual conditions encountered; injuries and treatment rendered; changes or deterioration of the structure; and documentation of the condition of the premises and adjacent area at the end of training.
- As part of the IAP, there is a conclusion to all training events. Regardless of the structure or prop, the environment should be returned to a safe and ready state.
- Documentation for exterior training props will depend heavily on the prop or props used. Manufactured props will require specific maintenance that will need to be documented. Documentation should include the visual inspection, operation of the prop prior to the evolution, and ensuring the prop is left in a safe condition after each use.
- The AHJ may require the documentation of fuel usage. If you use Class A materials to ignite the prop, fuel load must be documented. Include documentation of extinguishing agents, the amounts needed, and reserves.
- The structure needs to be reevaluated after the final burn down, to determine what measures need to take place to render the site safe. Arrangements need to be made for heavy equipment to knock over remaining walls or chimneys and to make the site safe.
- A thorough postburn inspection of the structure/prop is necessary. Any damage or needed repairs must be reported. Be respectful of the next crew of fire fighters who will be using the facility. Carefully cleaning up your crew's mess will allow for an easier transition for the next crew.
- A critique should be done after every training evolution. The purpose of this postevolution debriefing is to ensure there are no injuries, determine what was done correctly and went well, and identify any areas that need improvement in future training events. All participants should be encouraged to give input during the debriefing.
- To make the postevolution debriefing a valuable tool, review the participants' comments. Objectives may need to be rewritten or evolutions changed to make the training more valuable. A critique is valuable only if the lessons learned are applied to future training events.

KEY TERM

Access Navigate for flashcards to test your key term knowledge.

Postevolution debriefing A review of the training event. It is used as a critique to evaluate the objectives and training evolution to determine positive events as well as those that need improvement.

REFERENCES

National Fire Protection Association. 2018. NFPA 1403: *Standard on Live Fire Training Evolutions*. Quincy, MA: National Fire Protection Association.

National Fire Protection Association. 2019. NFPA 1041: *Standard for Fire and Emergency Services Instructor Professional Qualifications*. Quincy, MA: National Fire Protection Association.

Stittleburg, Phillip C. "When a live burn can come back to haunt you." *Fire Chief Magazine* February 1998: 18–20.

REVIEW QUESTIONS

1. At what stage of the live fire training must the following be completed: visually inspecting props, preparing asbestos removal reports, and securing owner documentation?

2. What is the purpose of the postevolution debriefing?

3. During the review of an incident, what document would include information on the participants, the instructors, and the live fire training activities they performed?

Appendix A
Resources for Fire and Emergency Services Instructors

Contents

- Training Record Attendance Report Form
- Training and Education Report
- Sample Training Policy
- Instructor Training and Experience Validation Form
- Fire Fighter Skill Performance Ratings Template
- Examples of Verbs Used to Write Learning Competencies
- Personal Improvement Agreement
- Outside Training Request Form
- Fire Fighter Performance Expectations
- JPR Skill Sheet Template
- Resources for Training Officers

Training Record

Training Attendance Form

Date: _____ Station(s): _____ Description: _____

Start Time: _____ End Time: _____ Credit Hours (Total Time): _____

Method of Training: ☐ Classroom   ☐ Practical   ☐ Self-Directed   ☐ Certification Credit: ☐ Yes ☐ No

Lead Instructor: _____ Additional Instructor(s): _____

Print Name	Dept ID #	Signature	Hours Attended

Objectives: _____

Description of Training (Notes): _____

_____

_____

_____   _____

Instructor Signature   Training Officer Approval

Type of Training (ISO Category)

| Company | Multi-Comp. | Officer | Mutual Aid | Night |
| Tower Burn | Classroom | Practical | Combo. | Driver |

Equipment Used in Training Session	Feet of 1¾" Hose Used	Feet of 2½" Hose Used	Supply Hose Used (ft)	Feet of Ladders	Number of Engines	Number of Trucks	Gallons of Water	Number of SCBA	Total Number of Fire Fighters

Training and Education Report

Fire Department	Department Logo	Training Information Sheet

Name: _____ ID#: _____ Date of Birth: _____

SS #: _____ Date Entered Fire Service: _____ Date of Hire: _____

Driver's License #: _____ Expiration Date: _____ D/L Classification: _____

Certification Record
Indicate date of certification in boxes below

Fire Fighter I (NFPA 1001)	Fire Fighter II (NFPA 1001)	Apparatus Operator (NFPA 1002)	HazMat Operations (NFPA 472)
HazMat Technician (NFPA 472)	HazMat Specialist (NFPA 472)	HazMat Specialist (NFPA 472)	HazMat Incident Command (NFPA 472)
NFPA 1006—Rope	NFPA 1006—Trench	NFPA 1006—Confined Space	NFPA 1006—Structural Collapse
NFPA 1006—Vehicle and Machinery	NFPA 1006—Wilderness Search	NFPA 1006—Water Rescue	NFPA 1006—Water Specialty
Instructor I	Instructor II	Instructor III	Training Program Manager
Fire Officer I (NFPA 1021)	Fire Officer II (NFPA 1021)	Fire Officer III (NFPA 1021)	Fire Officer IV (NFPA 1021)
Airport Fire Fighter (NFPA 1003)	Fire Inspector and Plan Examiner (NFPA 1031)	Fire Investigator (NFPA 1033)	Wildland Fire Fighter (NFPA 1051)
NIMS 100	NIMS 200	NIMS 300	NIMS 400
NIMS 700	NIMS 800	Command and General Staff	Incident Safety Officer
CPR/AED	EMS First Responder	EMT Basic	EMT Paramedic

Formal Education Record

High School Attended _____   Year Graduated _____ GED _____
College Attended _____   Credits Earned _____ Degree _____
Advanced Degree _____   Credits Earned _____ Degree _____
Advanced Degree _____   Credits Earned _____ Degree _____

Fire Department
Administrative Directive/Policy/SOP
Department Training Program

Purpose and Scope

This directive provides definition as to the training responsibilities of each rank within the department. This directive also establishes the basic format of the department's training program, including training requirements and documentation.

Responsibilities

Training Officer

The training officer is responsible for the overall administration and management of the department's Training Division. Through working with the staff and line officers, the training officer is to develop and implement a comprehensive yearly training plan. The training officer has overall responsibility and accountability to ensure that department training activities are current and consistent with applicable standards and practices.

Instructor-in-Charge

The instructor-in-charge is the designated person who is responsible for the overall delivery of a specific lesson plan or training objective. All resources assigned to the training session are under the responsibility of the instructor-in-charge, including personnel, apparatus, facilities, and equipment. The use of all required and best practice safety procedures throughout the training session will be the responsibility of the instructor-in-charge. All elements of documentation will also be overseen by the instructor-in-charge.

Instructor(s)

Instructors are responsible for assisting the instructor-in-charge at any high-risk training event or routine training session where it has been determined that additional instructors are needed. This may be for the purpose of safety, accountability, or reduction in the student-to-instructor ratio. Instructors are responsible for delivering lesson plans, evaluating performance, providing feedback, and documentation of the evolution or objectives as necessary. Instructors will be responsible for the safety of the members who are assigned to them during the evolution.

Shift Commanders

Shift commanders are responsible for administering and monitoring the department training plan within their assigned shift. This includes assisting in the coordination, presentation, and evaluation of specific department-level training sessions. Shift commanders should make every attempt to attend high-risk training events, training events based on procedural operations where their presence is needed to simulate operations, and any other training event where they can assist in the evaluation of member performance.

Through periodic evaluation of companies and/or individuals during training and emergency operations, the shift deputy commanders are responsible for identifying training deficiencies and providing recommendations to the training officer regarding the specific training needs of their shift.

Captains (or Other Mid-Level Shift Supervisors)

Captains are responsible for implementing and monitoring the department's training plan within their assigned area of authority (example—shift 2, west side). This includes reviewing monthly training reports to ensure company and individual compliance with training assignments. They also may be assigned by the training officer to assist in the coordination, presentation, and evaluation of specific department-level training sessions. Captains are responsible for completing the training responsibilities of a company officer.

Company Officers/Acting Company Officers

The company officer is the key to the department's training program. They are the individuals most responsible for the training and readiness of their personnel. Company officers are required to complete each month's training assignments and submit all necessary documentation, including training reports and skills checklists. Officers are to coordinate the various daily company activities so that training assignments are completed. They are responsible for coordinating company-level training so that all members receive the training regardless of time off, vacations, Kelly days, etc.

The monthly training assignments represent the minimum of what must be done.

The officer is not limited to this because each individual has strengths and weaknesses that must be addressed by the officer. It is the officer's responsibility to improve the performance of the personnel assigned to him or her. It is the officer's responsibility to foster an environment that encourages his or her company toward continuous improvement.

The monthly training calendar and company officer's packet lists the assigned training activities for the month. Some activities will have specific dates and/or time periods designated. For company-level training assignments, the company officer has the authority to vary from the published schedule, if necessary for valid reasons. The company officer shall be responsible for scheduling and completing the training. The objective is for all training assignments to be completed by the end of each month.

Company officers are responsible for the safety of their personnel while training.

Company officers are responsible for maintaining the licensure required of their position. This includes EMT or Paramedic and appropriate driver's license classification.

Fire Fighters

Department Fire fighters are expected to maintain a high level of preparedness through regular training and individual study. This includes keeping current on both fire service and departmental changes and notifying their company officer of their training needs.

Fire fighters are responsible to participate in an aggressive, safe, and positive manner in all classroom and practical training. Fire fighters are responsible for maintaining the licensure required of their position. This includes EMT or Paramedic and appropriate drivers license classification.

Monthly Training Assignments

The Training Division is responsible for providing monthly training assignments through the Training Bulletin, training calendar, and the Company Officer's Training Packet issued at the beginning of each month. This shall specify the assigned and make-up dates for training (if applicable), who is required to attend, and other necessary information.

Monthly training will be classified as follows:

Mandatory

This means training that must be accomplished by all members. This may also include 40-hour personnel. Mandatory training will include federally and state-mandated courses and courses deemed as mandatory by the Anytown Fire Department. These training sessions and make-up sessions will be scheduled by the Training Division and coordinated through the shift commanders.

Regular

Company level–Training that is to be completed by each company during the month. Company training designated as part of the Essential Skills program shall be completed by each company member. Other types of company-level training will specify whether make-up sessions are required for personnel who miss the training. Company officers are responsible for monitoring this and scheduling and conducting any necessary make-up training.

Department level–Training that is primarily scheduled and conducted through the Training Division. Department-level training that is designated as mandatory will have make-up sessions scheduled by the Training Division.

Documentation of Training

The training officer shall be responsible for providing to the shift commanders and captains a report as to the completion of training assignments from the previous month. The report shall be provided within the first 10 days of the next month. The shift commanders and captains are responsible for following up with their officers on the training that has not been completed.

The instructor of a specific training activity is responsible for completing and submitting the training report to the Training Division.

The company officer is responsible for ensuring that all skills checklists that may be required as part of a training assignment are completed and submitted to the Training Division.

For individual type training activities (i.e., independent study, reviewing fire service publications, physical fitness, etc.) the individual is responsible for completing the report and submitting it to his or her company officer for his or her signature.

Individuals attending classes outside of the fire department (i.e., fire officer classes, tactics seminars, etc.) are responsible for completing a training report upon returning from the course.

BY ORDER OF: _____

DATE: _____

Fire Department
Live Fire Training Instructor
Training and Experience Validation

Name: **Department:**

Years of Fire Service Experience: **Rank:**

Certification Levels

☐ Instructor I   ☐ Instructor II   ☐ Instructor III   ☐ Training Program Manager   ☐ TPM   ☐ Advanced Degree

☐ Fire Officer I   ☐ Fire Officer II   ☐ Fire Officer III   ☐ EFO   ☐ Other

➤ Copies of certifications must be available upon request

Previous Live Fire Training Educational Experience

☐ ISFSI Fixed Facility Instructor credential

☐ ISFSI Acquired Structure credential

☐ ISFSI Instructor-in-Charge credential

☐ NFPA 1403 course/class at state training facility, local academy, or agency

☐ Attended FDIC Live Fire Training programs

☐ Attended other national conference live fire training courses

➤ Copies of course/class completion certificates are available upon request

Previous Live Fire Training Practical Experience

Approximate Number of Live Fire Training Exercises Conducted at:

☐ Training Tower (Class A)   ☐ Training Tower (Gas Fired)

☐ Acquired Structures   ☐ Container/Simulator/Portable Unit

☐ Exterior Burn Prop   ☐ Class B/Gas Fuel

☐ Approximate Number of Years of Live Fire Training Experience

Validation/Attestation Statement

I have read and understand all components of NFPA 1403, Standard on Live Fire Training Evolutions, *and agree to abide by all requirements specified for the positions/roles that I am assigned for this training event. I understand and accept responsibility for the position requirements for the activities I am assigned to complete and do so knowing the hazards and dangers associated with these assignments.*

_____   _____

Signature   Date

Fire Department
Fire Fighter Skill Performance Ratings

- **Unskilled (0)**
  - Member failed evolution or skill (mandatory reevaluation will take place).
    - Exceeded time limit
    - Missed step in procedure
    - Created safety hazard to self or other member
    - Unable to perform task
    - Repeated failure of task attempt
  - Requires Personal Improvement Agreement and formal documentation on standard evaluation form.
  - No credit is given for purpose of progress reporting or evaluation toward applicable certification.

- **Moderately Skilled (1)**
  - Performance meets the minimum requirement for the task and is performed on first attempt. With additional practice, improved performance levels can be attained.
  - Members performance of skill may require supervision on fireground in order to perform.
  - Time performance near minimum requirement.
  - All appropriate safety precautions are taken.
  - Instructor/evaluator determines need for additional training or repeat of skill.

- **Skilled (2)**
  - Performance is above the minimum level because:
    - The time was above average.
    - Skill meets all performance levels and could be performed on fireground without supervision.
    - No serious/critical errors were committed.
    - All appropriate safety practices were observed and performed.

- **Highly Skilled (3)**
  - Performance is at a high level of competence because:
    - It was error free.
    - It was the fastest time.
    - Member could supervise others doing same task and identify errors or suggest improvements.
    - Member knows the role and importance of this task in relation to other fireground operations.

Example

Subject	Skill	Subset	Equipment	Performance Rating
Forcible Entry	Force an Inward Swinging Door	Hand Tools	Halligan Bar/Flat Head Axe	3

Examples of Verbs Used to Write Learning Competencies

AFFECTIVE DOMAIN

Levels of Learning

Receiving	Responding	Valuing	Organization	Value Complex
ask	answer	complete	adhere	act
choose	assist	describe	alter	discriminate
describe	comply	differentiate	arrange	display
follow	conform	explain	combine	influence
give	discuss	form	compare	listen
hold	greet	initiate	complete	modify
identify	help	invite	defend	perform
locate	label	join	explain	propose
name	perform	justify	identify	qualify
point to	practice	propose	integrate	question
select	present	read	modify	revise
set erect	read	report	order	serve
reply	recite	select	organize	solve
use	report	share	synthesize	use
	select	study	verify	
	tell	work		
	write			

Examples of Verbs Used to Write Learning Competencies (Cont'd)

COGNITIVE DOMAIN					
Levels of Learning					
Knowledge	Comprehension	Application	Analysis	Synthesis	Evaluation
define	acquire	apply	break down	categorize	appraise
describe	convert	change	correct	combine	compare
identify	defend	compute	diagram	compile	conclude
label	distinguish	create	differentiate	compose	contrast
list	estimate	demonstrate	discriminate	create	criticize
match	explain	develop	discuss	devise	diagnose
name	extend	discover	distinguish	design	discriminate
outline	generalize	manipulate	identify	explain	enhance
provide	give	modify	illustrate	generate	justify
reproduce	examine	operate	infer	modify	interpret
select	infer	predict	outline	organize	relate
state	paraphrase	prepare	point out	plan	research
	predict	produce	program	process	summarize
	rewrite	relate	relate	rearrange	support
	summarize	show	review	reconstruct	
		solve	select	relate	
		use	separate	reorganize	
			study	revise	
			subdivide	rewrite	
				sequence	
				summarize	
				tell	
				write	

Examples of Verbs Used to Write Learning Competencies (Cont'd)

PSYCHOMOTOR DOMAIN			
Levels of Learning			
Imitation	Manipulation	Precision	Articulation
adjust	arrange	administer	conduct
apply	code	book	document
assemble	control	clip	encircle
build	design	derive	graph
calibrate	dismantle	draw	pull
change	display	focus	push
clean	drill	handle	regulate
combine	encapsulate	identify	sculpt
compose	expand	introduce	set
compute	fasten	locate	sketch
connect	fix	manipulate	slide
construct	follow	mend	start
correct	frame	mix	stir
create	graph	modify	transfer
debug	grind	nail	use
display	hammer	paint	vend
insert	heat	preserve	vocalize
install	input	point	weigh
map	interface	sand	work
operate	loop	transport	
probe	maintain		
repair	organize		
shade	punch		
transform	support		
troubleshoot	switch		
	transmit		
	work		

Division of Training
Personal Improvement Agreement

Agreement Initiated by: **Today's Date:**

Member's Name: **Department ID#:**

I. **Concerns/Area Needing Improvement:** (Description of behavior, situation, or objective causing PIA)

II. **Standard Reference:** (What is the acceptable level of performance, attitude, conduct, or ability?)

III. **Shift Commander/Station Officer Action Plan:** (What the officer will do to help improve the performance.)

IV. **Personal Action Plan:** (Written by the member describing what they will do to improve themselves.)

V. **Document Action Plan Progress:** (How we are doing and progress check dates.)

Initial Follow-Up Date

VI. **Has the area of concern been corrected?** (Has improvement been seen in this area?)

Yes: ☐

No: ☐

Completion Date

VII. **Ongoing Monitoring:** (Describe how monitoring of issue will take place and for what duration.)

Fire Fighter Name (print): _____

Agreement Date: _____

Station Officer Name (print): _____

Shift Commander Name (print): _____

Training Division Review: _____

Appendix A   **467**

Outside Training Request Form
Out-of-Department School Request Form

Training Division Only	☐ Approved ☐ Denied ☐ Hold	Comments
	Training Officer Signature	

Attach Class Registration Form to Request

Payment Information/Registration Tracking (Training Division Use)

Department Purchase Order No.	Reimbursable (Y/N)	Method of Payment
Other Members Attending (Y/N)	Registration Completed	Member Notification Date/Method
Firehouse Class Code Created	Member Instructions Issued	Attendance Confirmation Received
Certificate Received	Certification Received	Monthly Report Entry

Member Information

Member Name	Date of Request	Shift
Station	Company Officer Name	Dates of Course
Course Title	Course Location	Course Sponsor
Class Fee	Hours of Program	Is this a certification course?
Member Signature	Signature	

Training School Request Acknowledgments

Company Officer Signature	Battalion Chief Signature	Special Teams Leader* (*if applicable)

Out-of-Town/Overnight Travel Information (Completed by Training Division)

Transportation

Department Vehicle Authorized	Personal Vehicle Use	Mileage Reimbursement Eligible* *Reimbursement for mileage at approval of Chief
Rental Car Agency	Payment Method	Airfare, Rail, Other
Reimbursements to Employee	Secondary Billing	

Equipment Issued

Helmet	Personal Flotation Device
Eye Protection	Buoyancy Compensator
Hearing Protection	Mask/Snorkel/Fins
Gloves (*Type*):	SCUBA
Bunker Coat	Other Resp. Protection (*Type*):
Hood	HazMat CPC (*Type*):
Bunker Pants	SCBA
Safety Boots	Spare Cylinders
Radio/Spare Batteries	

Out-of-Town Arrangements

Hotel Name	Confirmation Number	Direct Billing Arranged (Method)
Hotel Address	Hotel Phone	Check-in/Check-out Dates
Meals in Conference Fee	Per Diem Rate/Days/Total	TOTAL EXPENSES (Estimated)

Fire Fighter Performance Expectations

Work Ethic
- Actively seeks academic and technical knowledge for self-improvement
- Completes tasks assigned without shortcuts and without repeating of tasks
- Work is complete, thorough, and done in a professional manner
- Actively seeks out additional work as it improves the team's ability to thrive
- Accomplishes tasks or goals with a "safety-first" attitude
- Keeps commitments and meets deadlines
- Can be trusted with confidential information
- Can be trusted with the property of others
- Is committed to help at fire department events off duty, as necessary

Judgement and Problem Solving
- Makes reasonable and safe decisions when attempting to accomplish a task or solve a problem
- Approaches problems in a safe, logical, and well thought out fashion
- Seeks proactive solutions to problems
- Applies critical thinking skills to complex and varied situations

Time Management
- Consistently punctual and completes assignments on time
- Manages work so that quality of work is satisactory and not hurried, incomplete, or overwhelming to self and team

Teamwork/Interpersonal Skills
- Places the success of the team above self-interest
- Effectively works with others in order to accomplish tasks or solve problems
- Offers to help other company members
- Understands and follows chain-of-command
- Is courteous and respectful of peers and supervisors
- Does not undermine team
- Helps and supports other team members

Adaptability/Stress Management
- Remains calm in stressful situations
- Adapts behavior in order to deal with changing situations in a safe manner
- Adapts behavior in order to accomplish individual and department goals
- Recognizes symptoms of stress in self and others and seeks to deal with stress appropriately
- Communicates with others to resolve problems
- Remains flexible and open to change

Practical Competence/Physical Ability
- Demonstrates a desire to develop skills that are above minimal performance levels
- Can accomplish multiple tasks in succession
- Can retain and recall previously mastered skills
- Strives to improve practical abilities
- Knows all applicable safety behaviors and actions related to practical skills
- Maintains a high level of physical fitness, dexterity, flexibilty, and strength through ongoing fitness program participation

Communication
- Uses appropriate tone of voice
- Articulates in a clear, logical, and understandable manner
- Displays confidence in message
- Is persuasive and makes a positive impression
- Demonstrates appropriate nonverbal communication techniques
- Avoids letting stress control a communication process or method
- Writes legibly using correct grammar and punctuation
- Listens actively

Initiative/Motivation/Decisiveness
- Accomplishes tasks or goals without being ordered, coerced, or motivated by others
- Demonstrates desire for personal and professional development
- Makes decisions definitively and consistently

Empathy
- Shows compassion for others and responds appropriately to heightened emotional responses
- Demonstrates a calm, compassionate, and helpful demeanor towards those in need
- Mindful of the impact of his/her demeanor on those in need, family, bystanders, and other members of the public

Community Awareness
- Exercises compassion and willingness to help persons in varied situations with varied backgrounds
- Is sensitive to individual and cultural differences
- Knows the role a member of department and represents to the community

Appearance and Personal Hygiene
- Always clean, neat, well-groomed, and in good personal hygiene
- Always wears appropriate uniforms in excellent condition

XYZ Fire Department/Academy
In-Service Training Program
JPR Skill Sheet

DESCRIPTION: This JPR Training Guideline references the format identified in NFPA Professional Qualification Standard series. Knowledge, skill, performance, and topic description are referenced from the applicable state and local certification objectives to meet the JPR. Local application of each training guideline should be made by instructors utilizing department procedures or guidelines.

Duty Area	Level	Subject	JPR Number
Text Reference		Training Module	

Job Performance Requirement:

Objective Number	Skill/Knowledge/Performance/ Topic Description	Reference	Standard	Validated
			Pass/Fail	
			Pass/Fail	
			Pass/Fail	
			Pass/Fail	
			Pass/Fail	
			Pass/Fail	
			Pass/Fail	

Critical Safety Points
-
-
-

Required Equipment
-
-
-

Instructor Notes:

No.	TASK STEP	FIRST TEST		RETEST	
		Pass	Fail	Pass	Fail
1.					
2.					
3.					
4.					
5.					
6.					
7.					
8.					
9.					
10.					

Resources for Training Officers

TABLE AA-1 Resources for Training Officers

SCBA Training

FF Near-miss incidents involving SCBA	www.firefighternearmiss.com
OSHA Standard 1910.134	www.osha.gov
SCBA training ideas	www.firehouse.com

Driver Training

USFA Emergency Vehicle Driver Training program	www.usfa.fema.gov—publications section of website
Driver/Operator lesson plans	www.firehouse.com—training section of website
Fire apparatus and traffic safety related information	www.respondersafety.com
Driver safety information	www.drivetosurvive.org
Sample apparatus driving SOGs	www.vfis.com

Company Officer Development

IAFC Officer Development Handbook	http://www.iafc.org/associations/4685/files/OffrsHdbkFINAL3.pdf
Wildland Fire Leadership Development	http://www.fireleadership.gov
Long Beach (CA) Fire Department	http://www.lbfdtraining.com/index.html

Miscellaneous Resources

National Fire Academy library	www.lrc.fema.gov 1-800-638-1821
Fire Fighter Close Calls	www.firefighterclosecalls.com
National Fire Fighter Near Miss Reporting System	www.firefighternearmiss.com
Everyone Goes Home	www.everyonegoeshome.org
With the Command	www.withthecommand.com
Training Resources and Data Exchange (TRADE)	https://www.lsu.edu/feti/municipal/nfstrade.php

Regulatory Agencies

OSHA	www.osha.gov
U.S. Environmental Protection Agency	www.epa.gov
NIOSH Fire Fighter Fatality Reports	www.cdc.gov/niosh/fire

NFPA Standards

National Fire Protection Association (NFPA)	www.nfpa.org

Organizations

Illinois Society of Fire Service Instructors	www.ill-fireinstructors.org
International Association of Fire Chiefs (IAFC)	www.iafc.org
International Association of Fire Fighters (IAFF)	www.iaff.org
International Society of Fire Service Instructors (ISFSI)	www.isfsi.org
National Volunteer Fire Council	www.nvfc.org

Fire Service Publications

Fire Engineering Magazine	www.fireengineering.com
Firehouse Magazine	www.firehouse.com
FireRescue Magazine	www.firefighternation.com/

Online Magazines

Vincent Dunn	www.vincentdunn.com

Appendix B

An Extract from: NFPA 1041, Standard for Fire and Emergency Services Instructor Professional Qualifications, 2019 Edition

Chapter 4 Fire and Emergency Services Instructor I

4.1 General.

4.1.1 The Fire and Emergency Services Instructor I shall meet the JPRs defined in Sections 4.2 through 4.5 of this standard.

4.2 Program Management.

4.2.1 Definition of Duty. The management of basic resources, records, and reports essential to the instructional process.

4.2.2 Assemble course materials, given a specific topic, so that the lesson plan and all materials, resources, and equipment needed to deliver the lesson are obtained.

4.2.2(A) Requisite Knowledge. Components of a lesson plan, policies and procedures for the procurement of materials and equipment, and resource availability.

4.2.2(B) Requisite Skills. None required.

4.2.3 Prepare requests for resources, given training goals and current resources, so that the resources required to meet training goals are identified and documented.

4.2.3(A) Requisite Knowledge. Resource management, sources of instructional resources and equipment.

4.2.3(B) Requisite Skills. Oral and written communication, forms completion.

4.2.4 *Schedule single instructional sessions, given a training assignment, AHJ scheduling procedures, instructional resources, facilities and timeline for delivery, so that the specified sessions are delivered according to AHJ procedure.

4.2.4(A) Requisite Knowledge. AHJ scheduling procedures and resource management.

4.2.4(B) Requisite Skills. Training schedule completion.

4.2.5 Complete training records and reports, given policies and procedures, so that required reports are accurate and submitted in accordance with the procedures.

4.2.5(A) Requisite Knowledge. Types of records and reports required, and policies and procedures for processing records and reports.

4.2.5(B) Requisite Skills. Report writing and record completion.

4.3 Instructional Development.

4.3.1 *Definition of Duty. The review and adaptation of prepared instructional materials.

4.3.2 *Review instructional materials, given the materials for a specific topic, target audience, learner characteristics, and learning environment, so that elements of the lesson plan, learning environment, and resources that need adaptation are identified.

4.3.2(A) Requisite Knowledge. Recognition of student learner characteristics and diversity, methods of instruction, types of resource materials, organization of the learning environment, and policies and procedures.

4.3.2(B) Requisite Skills. Analysis of resources, facilities, and materials.

4.3.3 *Adapt a prepared lesson plan, given course materials and an assignment, so that the needs of the student and the objectives of the lesson plan are achieved.

4.3.3(A) *Requisite Knowledge. Elements of a lesson plan, selection of instructional aids and methods, and organization of the learning environment.

4.3.3(B) Requisite Skills. Instructor preparation and organization techniques.

4.4 Instructional Delivery.

4.4.1 Definition of Duty. The delivery of instructional sessions utilizing prepared course materials.

4.4.2 Organize the learning environment, given a facility and an assignment, so that lighting, distractions, climate control or weather, noise control, seating, audiovisual equipment, teaching aids, and safety are addressed.

4.4.2(A) Requisite Knowledge. Learning environment management and safety, advantages and limitations of audiovisual equipment and teaching aids, classroom arrangement, and methods and techniques of instruction.

4.4.2(B) Requisite Skills. Use of instructional media and teaching aids.

4.4.3 Present and adjust prepared lessons, given a prepared lesson plan that specifies the presentation method(s), so that the method(s) indicated in the plan are used and the stated objectives or learning outcomes are achieved, applicable safety standards and practices are followed, and risks are addressed.

4.4.3(A) *Requisite Knowledge. The laws and principles of learning, methods and techniques of instruction, lesson plan components and elements of the communication process, and lesson plan terminology and definitions; learner characteristics; student-centered learning principles; instructional technology tools; the impact of cultural differences on instructional delivery; safety rules, regulations, and practices; identification of training hazards; elements and limitations of distance learning; distance learning delivery methods; and the instructor's role in distance learning.

4.4.3(B) Requisite Skills. Oral communication techniques, methods and techniques of instruction, ability to adapt to changing circumstances, and utilization of lesson plans in an instructional setting.

4.4.4 *Adjust to differences in learner characteristics, abilities, cultures, and behaviors, given the instructional environment, so that lesson objectives are accomplished, disruptive behavior is addressed, and a safe and positive learning environment is maintained.

4.4.4(A) *Requisite Knowledge. Motivation techniques, learner characteristics, types of learning disabilities and methods for dealing with them, and methods of dealing with disruptive and unsafe behavior.

4.4.4(B) Requisite Skills. Basic coaching and motivational techniques, correction of disruptive behaviors, and adaptation of lesson plans or materials to specific instructional situations.

4.4.5 Operate instructional technology tools and demonstration devices, given a learning environment and equipment, so that the equipment functions, the intended objectives are presented, and transitions between media and other parts of the presentation are accomplished.

4.4.5(A) Requisite Knowledge. Instructional technology tools, demonstration devices, and selection criteria.

4.4.5(B) Requisite Skills. Use of instructional technology tools, demonstration devices, transition techniques, cleaning, and field level maintenance.

4.5 Evaluation and Testing.

4.5.1 *Definition of Duty. The administration and grading of student evaluation instruments.

4.5.2 Administer oral, written, and performance tests, given the lesson plan, evaluation instruments, and evaluation procedures of the AHJ, so that bias or discrimination is eliminated, the testing is conducted according to procedures, and the security of the materials is maintained.

4.5.2(A) Requisite Knowledge. Test administration, laws and policies pertaining to discrimination during training and testing, methods for eliminating testing bias, laws affecting records and disclosure of training information, purposes of evaluation and testing, and performance skills evaluation.

4.5.2(B) Requisite Skills. Use of skills checklists and assessment techniques.

4.5.3 Grade student oral, written, or performance tests, given class answer sheets or skills checklists and appropriate answer keys, so the examinations are accurately graded and properly secured.

4.5.3(A) Requisite Knowledge. Grading methods, methods for eliminating bias during grading, and maintaining confidentiality of scores.

4.5.3(B) Requisite Skills. None required.

4.5.4 Report test results, given a set of test answer sheets or skills checklists, a report form, and policies and procedures for reporting, so that the results are accurately recorded, the forms are forwarded according to procedure, and unusual circumstances are reported.

4.5.4(A) Requisite Knowledge. Reporting procedures and the interpretation of test results.

4.5.4(B) Requisite Skills. Communication skills and basic coaching.

4.5.5 *Provide evaluation feedback to students, given evaluation data, so that the feedback is timely; specific enough for the student to make efforts to modify behavior; and objective, clear, and relevant; also include suggestions based on the data.

4.5.5(A) Requisite Knowledge. Reporting procedures and the interpretation of test results.

4.5.5(B) Requisite Skills. Communication skills and basic coaching.

Chapter 5 Fire and Emergency Services Instructor II

5.1 General. The Fire and Emergency Services Instructor II shall meet the requirements for Fire and Emergency Services Instructor I and the JPRs defined in Sections 5.2 through 5.5 of this standard.

5.2 Program Management.

5.2.1 Definition of Duty. The management of instructional resources, staff, facilities, records, and reports.

5.2.2 Assign instructional sessions, given AHJ scheduling policy, instructional resources, staff, facilities, and timeline for delivery, so that the specified sessions are delivered according to AHJ policy.

5.2.2(A) Requisite Knowledge. AHJ policy, scheduling processes, supervision techniques, and resource management.

5.2.2(B) Requisite Skills. Select resources, staff, and facilities for specified instructional sessions.

5.2.3 Recommend budget needs, given training goals, AHJ budget policy, and current resources, so that the resources required to meet training goals are identified and documented.

5.2.3(A) Requisite Knowledge. AHJ budget policy, resource management, needs analysis, sources of instructional materials, and equipment.

5.2.3(B) Requisite Skills. Resource analysis and preparation of supporting documentation.

5.2.4 Gather training resources, given an identified need, so that the resources are obtained within established timelines, budget constraints, and according to AHJ policy.

5.2.4(A) *Requisite Knowledge. AHJ policies, purchasing procedures, and budget.

5.2.4(B) Requisite Skills. Records completion.

5.2.5 Manage training record-keeping, given training records, AHJ policy, and training activity, so that all AHJ and legal requirements are met.

5.2.5(A) Requisite Knowledge. Record-keeping processes, AHJ policies, laws affecting records and disclosure of training information, professional standards applicable to training records, and systems used for record-keeping.

5.2.5(B) Requisite Skills. Records management.

5.2.6 Evaluate instructors, given an evaluation tool, AHJ policy, and objectives, so that the evaluation identifies areas of strengths and weaknesses, recommends changes in instructional style and communication methods, and provides opportunity for instructor feedback to the evaluator.

5.2.6(A) Requisite Knowledge. Personnel evaluation methods, supervision techniques, AHJ policy, and effective instructional methods and techniques.

5.2.6(B) Requisite Skills. Coaching, observation techniques, and completion of evaluation records.

5.3 Instructional Development.

5.3.1 Definition of Duty. The development of instructional materials for specific topics.

5.3.2 *Create a lesson plan, given a topic, learner characteristics, and a lesson plan format, so that learning objectives, a lesson outline, course materials, instructional technology tools, an evaluation plan, and learning objectives for the topic are addressed.

5.3.2(A) Requisite Knowledge. Elements of a lesson plan, components of learning objectives, instructional methodology, student-centered learning, methods for eliminating bias, types and application of instructional technology tools and techniques, copyright law, and references and materials.

5.3.2(B) Requisite Skills. Conduct research, develop behavioral objectives, assess student needs, and develop instructional technology tools; lesson outline techniques, evaluation techniques, and resource needs analysis.

5.4 Instructional Delivery.

5.4.1 Definition of Duty. Conducting classes using a lesson plan.

5.4.2 Conduct a class using a lesson plan that the instructor has prepared and that involves the utilization of multiple teaching methods and techniques, given a topic and a target audience, so that the lesson is delivered in a safe and effective manner and the objectives are achieved.

5.4.2(A) Requisite Knowledge. Student-centered learning methods, discussion methods, facilitation methods, problem-solving techniques, methods for eliminating bias, types and application of instructional technology tools, and evaluation tools and techniques.

5.4.2(B) *Requisite Skills. Facilitate instructional session, apply student-centered learning, evaluate instructional delivery; use and evaluate instructional technology tools, evaluation techniques, and resources.

5.4.3 *Supervise other instructors and students during training, given a specialized training scenario so that applicable safety standards and practices are followed and instructional goals are met.

5.4.3(A) Requisite Knowledge. Safety rules, regulations, and practices; the incident management system; and leadership techniques.

5.4.3(B) Requisite Skills. Conduct a safety briefing, ability to communicate, and implement an incident management system.

5.5 Evaluation and Testing.

5.5.1 Definition of Duty. The development of student evaluation instruments to support instruction and the evaluation of test results.

5.5.2 Develop student evaluation instruments, given learning objectives, learner characteristics, and training goals, so that the evaluation instrument measures whether the student has achieved the learning objectives.

5.5.2(A) Requisite Knowledge. Evaluation methods, evaluation instrument development, and assessment of validity and reliability.

5.5.2(B) Requisite Skills. Evaluation item construction and assembly of evaluation instruments.

5.5.3 *Develop a class evaluation instrument, given AHJ policy and evaluation goals, so that students have the ability to provide feedback on instructional methods, communication techniques, learning environment, course content, and student materials.

5.5.3(A) Requisite Knowledge. Training evaluation methods.

5.5.3(B) Requisite Skills. Development of training evaluation instruments.

Chapter 6 Fire and Emergency Services Instructor III

6.1 General. The Fire and Emergency Services Instructor III shall meet the requirements for Fire and Emergency Services Instructor II and the JPRs defined in Sections 6.2 through 6.5 of this standard.

6.2 Program Management.

6.2.1 Definition of Duty. The administration of AHJ policies and procedures for the management of instructional resources, staff, facilities, records, and reports.

6.2.2 *Administer a training record system, given AHJ policy and type of training activity to be documented, so that the information captured is concise, meets all AHJ and legal requirements, and can be accessed.

6.2.2(A) Requisite Knowledge. AHJ policy, record-keeping systems, professional standards addressing training records, legal requirements affecting record-keeping, and disclosure of information.

6.2.2(B) Requisite Skills. Development of records and report generation.

6.2.3 Develop recommendations for policies to support the training program, given AHJ policies and procedures and the training program goals, so that the goals are achieved.

6.2.3(A) Requisite Knowledge. AHJ procedures and training program goals, and format for AHJ policies.

6.2.3(B) Requisite Skills. Technical writing and decision making.

6.2.4 Select instructional staff, given personnel qualifications, instructional requirements, and AHJ policies and procedures, so that staff selection meets AHJ policies and achievement of AHJ and instructional goals.

6.2.4(A) Requisite Knowledge. AHJ policies regarding staff selection, instructional requirements, the capabilities of instructional staff, employment laws, and AHJ goals.

6.2.4(B) Requisite Skills. Evaluation techniques and interview methods.

6.2.5 Construct a performance-based instructor evaluation plan, given AHJ policies and procedures and job requirements, so that instructors are evaluated at regular intervals, following AHJ policies.

6.2.5(A) Requisite Knowledge. Evaluation methods, employment laws, AHJ policies, staff schedules, and job requirements.

6.2.5(B) Requisite Skills. Evaluation techniques, scheduling, technical writing.

6.2.6 Formulate budget needs, given training goals, AHJ budget policy, and current resources, so that the resources required to meet training goals are identified and documented.

6.2.6(A) Requisite Knowledge. AHJ budget policy, resource management, needs analysis, sources of instructional materials, and equipment.

6.2.6(B) Requisite Skills. Resource analysis and required documentation.

6.2.7 Write equipment purchasing specifications, given curriculum information, training goals, and AHJ guidelines, so that the equipment is appropriate and supports the curriculum.

6.2.7(A) Requisite Knowledge. Equipment purchasing procedures, available AHJ resources, and curriculum needs.

6.2.7(B) Requisite Skills. Preparation of procurement documents, technical writing.

6.2.8 Present evaluation findings, conclusions, and recommendations to AHJ administrator, given data summaries and target audience, so that recommendations are unbiased, supported, and reflect AHJ goals, policies, and procedures.

6.2.8(A) Requisite Knowledge. Statistical analysis and AHJ goals.

6.2.8(B) Requisite Skills. Presentation skills and report preparation following AHJ guidelines.

6.3 Instructional Development.

6.3.1 Definition of Duty. Plans, develops, and implements comprehensive programs and curricula.

6.3.2 Conduct an AHJ needs analysis, given AHJ goals, so that instructional needs are identified and solutions are recommended.

6.3.2(A) Requisite Knowledge. Needs analysis, gap analysis, instructional design process, instructional methodology, learner characteristics, instructional technologies, curriculum development, facilities, and development of evaluation instruments.

6.3.2(B) Requisite Skills. Conducting research and needs and gap analysis, forecasting, and organizing information.

6.3.3 Design programs or curricula, given needs analysis and AHJ goals, so that the goals are supported, learner characteristics are identified, audience-based instructional methodologies are utilized, and the program meets time and budget constraints.

6.3.3(A) Requisite Knowledge. Instructional design, instructional methodologies, learner characteristics, principles of student-centered learning and research methods.

6.3.3(B) Requisite Skills. Technical writing and selecting course reference materials.

6.3.4 Write program and course outcomes, given needs analysis information, so that the outcomes are clear, concise, measurable, and correlate to AHJ goals.

6.3.4(A) Requisite Knowledge. Components and characteristics of outcomes, and correlation of outcomes to AHJ goals.

6.3.4(B) Requisite Skills. Technical writing.

6.3.5 Write course objectives, given course outcomes, so that objectives are clear, concise, measurable, and reflect specific tasks.

6.3.5(A) Requisite Knowledge. Components of objectives and correlation between outcomes and objectives.

6.3.5(B) Requisite Skills. Technical writing.

6.3.6 Construct a course content outline, given course objectives, and reference sources, so that the content outline supports course objectives.

6.3.6(A) Requisite Knowledge. Correlation between course objectives, instructor lesson plans, and instructional methodology.

6.3.6(B) Requisite Skills. Technical writing.

6.4 Instructional Delivery. No JPRs at the Instructor III Level.

6.5 Evaluation and Testing.

6.5.1 Definition of Duty. Develops an evaluation plan; collects, analyzes, and reports data; and utilizes data for program validation and student feedback.

6.5.2 Develop a system for the acquisition, storage, and dissemination of evaluation results, given AHJ goals and policies, so that the goals are supported and so that those affected by the information receive feedback consistent with AHJ policies and federal, state, and local laws.

6.5.2(A) Requisite Knowledge. Record-keeping systems, AHJ goals, data acquisition techniques, applicable laws, and methods of providing feedback.

6.5.2(B) Requisite Skills. The evaluation, development, and use of information systems.

6.5.3 *Develop a course evaluation plan, given course objectives and AHJ policies, so that objectives are measured and AHJ policies are followed.

6.5.3(A) Requisite Knowledge. Evaluation techniques, AHJ constraints, and resources.

6.5.3(B) Requisite Skills. Decision making and technical writing.

6.5.4 Develop a program evaluation plan, given AHJ policies and procedures, so that instructors, course components, program goals, and facilities are evaluated, student input is obtained, and needed improvements are identified.

6.5.4(A) Requisite Knowledge. Evaluation methods and AHJ goals.

6.5.4(B) Requisite Skills. Construction of evaluation instruments, technical writing.

6.5.5 Analyze student evaluation instruments, given test data, objectives, and AHJ policies, so that validity and reliability are determined and necessary changes are made.

6.5.5(A) Requisite Knowledge. AHJ policies and applicable laws, test validity and reliability, and item analysis methods.

6.5.5(B) Requisite Skills. Item analysis.

Chapter 7 Live Fire Instructor

7.1 General.

7.1.1 For qualification at Live Fire Instructor, the candidate shall meet the requirements of Fire Fighter II as defined in NFPA 1001 or Interior Structural Fire Brigade Member as defined by NFPA 1081, the requirements of Fire and Emergency Services Instructor I as defined in Chapter 4, and the job performance requirements defined in 7.2 through 7.3 of this standard.

7.1.2 A Live Fire Instructor shall demonstrate competency in knowledge and skills in all subjects, methods, and equipment being taught and the objectives contained in NFPA 1403 and identified for the live fire evolution.

7.2 Pre-Live Fire Evolution.

7.2.1 Inspect live fire participants' PPE and SCBA, given participants and PPE and SCBA, so that equipment is determined to be serviceable and worn in accordance with manufacturer's instructions.

7.2.1(A) Requisite Knowledge. Manufacturers' instructions.

7.2.1(B) Requisite Skills. Visual inspection, using an inspection checklist.

7.3 Live Fire Evolution.

7.3.1 Predict stages of fire growth in a compartment, flow path, flashover, rollover, and backdraft, given a live fire evolution, so that a safe environment is maintained.

7.3.1(A) Requisite Knowledge. Fire dynamics, including fuel load, fire growth, flow path, flashover, rollover, and backdraft.

7.3.1(B) Requisite Skills. Configure fuel loads to meet the objectives of the live fire evolution, recognize changing conditions of the live fire environment.

7.3.2 Supervise a group during a live fire evolution, given a live fire structure or prop and a group of participants, so that instructional objectives are met, crew integrity is maintained, the instructor maintains a position to supervise the crew, fire conditions are monitored, and emergency actions are taken as necessary.

7.3.2(A) Requisite Knowledge. Group dynamics, instructor positioning, egress routes, fire dynamics, including fuel load, fire growth, flow path, flashover, rollover, and backdraft.

7.3.2(B) Requisite Skills. Supervisory skills, fire suppression operations.

7.3.3 Conduct a personnel accountability report (PAR) upon entering and exiting a live fire structure or prop, given a group of participants in a live fire evolution, so that all participants are accounted for and safety is ensured and maintained.

7.3.3(A) Requisite Knowledge. Knowledge of incident management system, AHJ personnel accountability procedures.

7.3.3(B) Requisite Skills. Use of AHJ's accountability system, ability to recognize inadequacies in the use of the accountability system.

7.3.4 Monitor live fire participants to safeguard participants, given a live fire evolution, so that signs and symptoms of fatigue and distress are recognized and action is taken to prevent injury.

7.3.4(A) Requisite Knowledge. Signs and symptoms of fatigue and distress, knowledge of environmental conditions, AHJ safety, rehabilitation, and emergency services procedures.

7.3.4(B) Requisite Skills. Evaluation of environmental conditions, class management, activation of AHJ emergency procedures.

Chapter 8 Live Fire Instructor in Charge

8.1 General.

8.1.1 For qualification at Live Fire Instructor in Charge, the candidate shall meet the requirements of Fire Emergency Services Instructor II as defined in Chapter 5, the requirements of Live Fire Instructor as defined in Chapter 7, and the job performance requirements defined in 8.2 through 8.4 of this standard.

8.1.2 The Live Fire Instructor in Charge shall demonstrate competency in knowledge and skills in all subjects, methods, and equipment being taught, and in the objectives contained in NFPA 1403 and identified for the live fire evolution.

8.2 Pre-Live Fire Evolution.

8.2.1 Prepare a pre-burn plan in compliance with NFPA 1403, given the AHJ policy and procedures for live fire training evolutions, the facility policies applicable to evolutions, learning objectives, and all conditions affecting the evolution, so that learning objectives are developed, the plan meets all AHJ requirements, existing conditions are identified, and the plan meets the developed learning objectives.

8.2.1(A) Requisite Knowledge. NFPA 1403, components of learning objectives, AHJ and facility policies and procedures, hazards associated with live fire training, fuel packages, burn room size, ventilation strategies, time between sequential burn evolutions, evidence-based practices for fire control, and training procedures.

8.2.1(B) Requisite Skills. Learning objective development, technical writing, pre-burn plan development.

8.2.2 Conduct a pre-burn inspection of the structure or prop, given a structure or prop for live fire training, so that structural damage is identified, structural preparation is determined, and safety concerns are identified and addressed prior to the live fire evolution.

8.2.2(A) Requisite Knowledge. Facility requirements, structure or prop considerations.

8.2.2(B) Requisite Skills. Observation techniques, inspection and evaluation skills.

8.2.3 Calculate the minimum water supply required for a live fire evolution in compliance with NFPA 1403, Section 4.12, given a structure or prop so that the required minimum water supply is determined.

8.2.3(A) Requisite Knowledge. NFPA 1403, fire flow calculations.

8.2.3(B) Requisite Skills. Calculation of water supply requirements, development of water supply documentation

8.2.4 Calculate the minimum water flow application rate for a live fire evolution in compliance with NFPA 1403, Section 4.12, given a structure or prop so that the required minimum water flow application rate is determined.

8.2.4(A) Requisite Knowledge. NFPA 1403, fire flow calculations, capacity of hose lines, fireground hydraulics.

8.2.4(B) Requisite Skills. Calculation of minimum water flow application rate.

8.3 Live Fire Evolution.

8.3.1 Identify and assign instructional tasks and duties in compliance with NFPA 1403, given staffing assignments, learning objectives, and instructor capabilities, so that safety officer(s), ignition officer, and crew/functional lead(s) are designated and rotated through duty assignments, instructor(s) implement participant accountability, proper instructor/student ratios are maintained, instructor(s) monitor and supervise all participants during evolutions, and awareness of changing conditions that impact training is maintained.

8.3.1(A) Requisite Knowledge. NFPA 1403, accountability procedures, supervisory techniques, and resource management.

8.3.1(B) Requisite Skills. Coaching and observation techniques.

8.3.2 Conduct a pre-burn briefing session, given the pre-burn plan, so that all facets of the evolution(s) are identified, training objectives are covered, a walk-through of the structure or prop with all participants is performed and established safeguards and emergency procedures are identified.

8.3.2(A) Requisite Knowledge. Pre-burn plan, safety rules, emergency procedures, and AHJ policy and procedures.

8.3.2(B) Requisite Skills. Presentation and class management skills.

8.3.3 Maintain the training environment to safeguard participants, given participants in a live fire training evolution, so that signs and symptoms of fatigue and distress are recognized, action is taken to prevent injuries, and actions are documented.

8.3.3(A) Requisite Knowledge. Signs and symptoms of fatigue and distress, knowledge of environmental conditions; AHJ's safety, rehabilitation, and emergency procedures.

8.3.3(B) Evaluation of environmental conditions, class management, report completion, activation of the AHJ's emergency procedures.

8.4 Post Live Fire Evolution.

8.4.1 Conduct a post-burn briefing session, given the learning objectives of the evolution, so that feedback on each learning objective is provided to each participant, and any needed corrective actions are identified.

8.4.1(A) Requisite Knowledge. Pre-burn plan, learning objectives, supervisory techniques and AHJ policy and procedures.

8.4.1(B) Requisite Skills. Presentation skills, evaluation skills, class management skills.

8.4.2 Conduct a post-burn inspection of the structure or prop, given a structure or prop for live fire training, so that structural damage is identified, safety concerns are identified, and necessary corrective actions are taken.

8.4.2(A) Requisite Knowledge. AHJ policies, facility requirements, safety practices.

8.4.2(B) Requisite Skills. Observation techniques, inspection skills.

8.4.3 Complete records and reports in accordance with NFPA 1403, given a live fire evolution, so that all required reports are completed.

8.4.3(A) Requisite Knowledge. NFPA 1403, AHJ policy on records.

8.4.3(B) Requisite Skills. Technical writing and records completion.

Appendix C

Correlation to NFPA 1041: Standard for Fire and Emergency Services Instructor Professional Qualifications, 2019 Edition

NFPA 1041, Fire and Emergency Services Instructor I	Corresponding Chapters	Corresponding Pages
4.2.2	5	126–131
4.2.2 (A)	5	126–141
4.2.3	5	130–131
4.2.3 (A)	5	130–131
4.2.3 (B)	4, 5	105–113; 131
4.2.4	5	130–131
4.2.4 (A)	5	130–131
4.2.4 (B)	5	130–131
4.2.5	1	16–21
4.2.5 (A)	1	20–21
4.2.5 (B)	1	16–21
4.3.1	5	131–134
4.3.2	2, 3, 5	39; 53–62; 85–94; 130–134
4.3.2 (A)	2, 3, 5	41–46; 53–54; 85–92; 130–134
4.3.2 (B)	3, 5	91–92; 130–134
4.3.3	3, 5, 6	85–94; 131–134; 145–150
4.3.3 (A)	3, 5, 6	78–94; 131–134; 145–150
4.3.3 (B)	3, 5	77–82; 131–134
4.4.2	1, 3, 6, 7	11–12; 81–92; 145–151; 162–172

NFPA 1041, Fire and Emergency Services Instructor I	Corresponding Chapters	Corresponding Pages
4.4.2 (A)	1, 3, 6, 7	11–12; 81–92; 145–151; 162–172
4.4.2 (B)	1, 3, 6	11–12; 75–79; 145–151
4.4.3	2, 3, 4, 5, 7	41–45; 75–84; 93–94; 103–112; 162–175
4.4.3 (A)	2, 3, 4, 5, 7	41–45; 75–84; 93–94; 103–112; 162–175
4.4.3 (B)	2, 3, 4, 5	41–45; 75–84; 93–94; 103–112
4.4.4	2, 3	39–63; 82–84
4.4.4 (A)	2, 3	39–63; 82–84
4.4.4 (B)	2, 3	39–63; 82–84
4.4.5	6, 7	142–151; 162
4.4.5 (A)	6	142–150
4.4.5 (B)	6	151
4.5.1	8	185–192
4.5.2	8	185–187; 192–193
4.5.2 (A)	8	185–193
4.5.2 (B)	8	187–192
4.5.3	8	192–193
4.5.3 (A)	8	192
4.5.4	8	192–193
4.5.4 (A)	8	186; 192–193
4.5.4 (B)	8	192–193
4.5.5	8	192–193
4.5.5 (A)	8	192–193
4.5.5 (B)	8	192–193

NFPA 1041, Fire and Emergency Services Instructor II	Corresponding Chapters	Corresponding Pages
5.2.2	12	267–276
5.2.2 (A)	12	267–276
5.2.2 (B)	12	267–276
5.2.3	12	278–285
5.2.3 (A)	12	278–285
5.2.3 (B)	12	278–285
5.2.4	12	282–287
5.2.4 (A)	12	282–287
5.2.4 (B)	12	282–287
5.2.5	12	276–277
5.2.5 (A)	12	276–277
5.2.5 (B)	12	276–277
5.2.6	10	229–232
5.2.6 (A)	10	230–232
5.2.6 (B)	10	229–232
5.3.2	9	203–215
5.3.2 (A)	9	203–215
5.3.2 (B)	9	203–215
5.4.2	10	223
5.4.2 (A)	10	223
5.4.2 (B)	10	223
5.4.3	10	223–236
5.4.3 (A)	10	225; 230; 233–234
5.4.3 (B)	10	233–234
5.5.2	11	240–256

NFPA 1041, Fire and Emergency Services Instructor II	Corresponding Chapters	Corresponding Pages
5.5.2 (A)	11	240–256
5.5.2 (B)	11	240–256
5.5.3	11	256–258
5.5.3 (A)	11	256–258
5.5.3 (B)	11	256–258

NFPA 1041, Fire and Emergency Services Instructor III	Corresponding Chapters	Corresponding Pages
6.2.2	15	341–342
6.2.2 (A)	15	341–342
6.2.2 (B)	15	341–342
6.2.3	15	350–354
6.2.3 (A)	15	350–354
6.2.3 (B)	15	350–354
6.2.4	15	354–357
6.2.4 (A)	15	354–357
6.2.4 (B)	15	354–357
6.2.5	15	356–357
6.2.5 (A)	15	356–357
6.2.5 (B)	15	356–357
6.2.6	15	358–364
6.2.6 (A)	15	358–364
6.2.6 (B)	15	358–364
6.2.7	15	361–362
6.2.7 (A)	15	361–362
6.2.7 (B)	15	361–362

NFPA 1041, Fire and Emergency Services Instructor III	Corresponding Chapters	Corresponding Pages
6.2.8	13	299, 301, 216
6.2.8 (A)	13	299, 301, 316
6.2.8 (B)	13	299, 301, 316
6.3.2	13	297–301
6.3.2 (A)	13	297–301
6.3.2 (B)	13	297–301
6.3.3	13	299–307
6.3.3 (A)	13	301–307
6.3.3 (B)	13	299–307
6.3.4	13	299–307
6.3.4 (A)	13	301–307
6.3.4 (B)	13	299–307
6.3.5	13	301–306
6.3.5 (A)	13	301–306
6.3.5 (B)	13	301–306
6.3.6	13	299–306
6.3.6 (A)	13	301–306
6.3.6 (B)	13	299–306
6.5.2	14	330–331
6.5.2 (A)	14	330–331
6.5.2 (B)	14	330–331
6.5.3	13	314–317
6.5.3 (A)	13	314–317
6.5.3 (B)	13	314–317

NFPA 1041, Fire and Emergency Services Instructor III	Corresponding Chapters	Corresponding Pages
6.5.4	13	299, 310–315
6.5.4 (A)	13	299, 310–315
6.5.4 (B)	13	299, 310–315
6.5.5	14	327–334
6.5.5 (A)	14	327–334
6.5.5 (B)	14	327–334

NFPA 1041, Live Fire Instructor	Corresponding Chapters	Corresponding Pages
7.2.1	17	383–385
7.2.1 (A)	17	383–385
7.2.1 (B)	17	383–385
7.3.1	17	386–388
7.3.1 (A)	17	386–388
7.3.1 (B)	17	386–388
7.3.2	17	388–390
7.3.2 (A)	17	388–390
7.3.2 (B)	17	388–390
7.3.3	17	385–386; 390
7.3.3 (A)	17	385–386; 390
7.3.3. (B)	17	385–386; 390
7.3.4	17	393–400
7.3.4 (A)	17	393–400
7.3.4 (B)	17	393–400

NFPA 1041, Live Fire Instructor in Charge	Corresponding Chapters	Corresponding Pages
8.2.1	18	405–409; 411–427
8.2.1 (A)	18	405–409; 411–427
8.2.1 (B)	18	405–409; 411–427
8.2.2	18	416–426
8.2.2 (A)	18	416–426
8.2.2 (B)	18	416–426
8.2.3	18	406–409
8.2.3 (A)	18	406–409
8.2.3 (B)	18	406–409
8.2.4	18	406–409
8.2.4 (A)	18	406–409
8.2.4 (B)	18	406–409
8.3.1	19	431–440
8.3.1 (A)	19	431–440
8.3.1 (B)	19	431–440
8.3.2	19	433–434
8.3.2 (A)	19	433–434
8.3.2 (B)	19	433–434
8.3.3	19	438–441
8.3.3 (A)	19	438–441
8.3.3 (B)	19	438–441
8.4.1	20	450–452
8.4.1 (A)	20	450–452
8.4.1 (B)	20	450–452

NFPA 1041, Live Fire Instructor in Charge	Corresponding Chapters	Corresponding Pages
8.4.2	20	451–452
8.4.2 (A)	20	451–452
8.4.2 (B)	20	451–452
8.4.3	20	447–452
8.4.3 (A)	20	447–452
8.4.3 (B)	20	447–452

Appendix D

An Extract from: NFPA 1401, Recommended Practice for Fire Service Training Reports and Records, 2017 Edition

Chapter 4 Elements of Training Documents

4.1 General.

4.1.1 Training records and reports should be utilized by the training officer and line officers for analysis of the effectiveness of the training program in terms of time, staffing, individual performance rating, and financing.

4.1.2* Training records and reports should be utilized to develop specific training objectives and to evaluate compliance with, or deficiencies in, the training program.

4.1.3 Compliance with mandated training requirements should be documented.

4.1.4 The management of training functions should be performed in a closed-feedback loop.

4.1.5 The training functions should not operate as an open-ended cycle.

4.1.6 The closed-feedback loop should consist of the following:
  (a) Planning
  (b) Organization
  (c) Implementation
  (d) Operation
  (e) Review
  (f) Feedback/alteration

4.1.7* In each phase of the cycle, information should be provided for management to perform effectively.

4.1.7.1 The information is provided through various types of records, reports, and studies; therefore, records should be designed to fit into the overall training management cycle.

4.1.8 In order to be most effective, these records should contribute to the overall organization information cycle.

4.2 Elements of Information.

4.2.1* Training documents, regardless of their intent or level of sophistication, should focus on content, accuracy, and clarity.

4.2.2* These documents should relay to the reader at least five specific elements of information as follows *(see Annex B for examples of training record forms)*:
  (1) Who
      (a) Who was the instructor?
      (b) Who participated?
      (c) Who was in attendance?
      (d) Who is affected by the documents?
      (e) Who was included in the training (individuals, company, multi-company, or organization)?
  (2) What
      (a) What was the subject covered?
      (b) What equipment was utilized?
      (c) What operation was evaluated or affected?
      (d) What was the stated objective, and was it met?
  (3) When
      (a) When will the training take place? or
      (b) When did the training take place?
  (4) Where
      (a) Where will the training take place? or
      (b) Where did the training take place?
  (5) Why
      (a) Why is the training necessary? or
      (b) Why did the training occur?

4.3 Additional Information.
Additional information or detail, which should include but not be limited to the following, should be included to explain or clarify the document as necessary:
  (1) Source of the information used as a basis for the training
      (a) Textbook title and edition
      (b) Lesson plan title and edition
      (c) Policy name and reference number
      (d) Videotapes, CDs, and DVDs
      (e) Distance learning sources
      (f) Internet address
      (g) Industry best practices
      (h) Post-incident analysis (PIA)
      (i) Other
  (2) Method of training used for delivery
      (a) Lecture
      (b) Demonstration
      (c) Skills training
      (d) Self-study
      (f) Video presentation
      (g) Mentoring
      (h) Drill(s)
      (i) Other
  (3) Evaluation of training objectives
      (a) Written test
      (b) Skills examination
      (c) Other

Chapter 5 Types of Training Documents

5.1 Training Schedules.

5.1.1 Need for Training Schedules.

5.1.1.1 All members of a fire department should receive standardized instruction and training.

5.1.1.2 Standardized training should include considerable planning; however, standardization can be improved through the preparation of training schedules for use by department personnel.

5.1.1.3 Standardized training schedules should be prepared and published for both short-term scheduling (considerable detail), intermediate-term scheduling (less detail), and long-term scheduling (little detail) to facilitate long-term planning by the training staff, instructional staff, company officers, and personnel.

5.1.2* Types of Training Schedules. Training schedules should be prepared for all training ground and classroom sessions.

5.1.2.1 Periodic Training Schedule — Station Training. The station training schedule, which is prepared by the training officer, should designate specific subjects that are to be covered by company or station officers in conducting their station training.

5.1.2.1.1 The company officers should use this schedule to set their own in-station training schedule.

5.1.2.1.2 A balance between manipulative skills training and classroom sessions should be considered in the preparation of training schedules.

5.1.2.1.3 Such training schedules should include all of the topics necessary to satisfy job knowledge requirements and to maintain skills already learned.

5.1.2.2* Periodic Training Schedule — Training Facility Activities. The training facility activities schedule details when companies should report to the training facility for evolutions or classes.

5.1.2.2.1 Days also should be set aside for make-up sessions.

5.1.2.2.2 Training activities conducted outside the training facility or by outside agencies also should be shown on this schedule.

5.1.3* All Other Training. Schedules should be prepared for all training, including, but not limited to, the following:
 (a) Recruit or entry-level training
 (b) In-service training
 (c) Special training
 (d) Officer training
 (e) Advanced training
 (f) Mandated training
 (g) Medical training
 (h) Safety training

5.2* Training Reports.

5.2.1* Logical Sequence. A training report should be complete and should follow a logical sequence.

5.2.1.1 A report should clearly and concisely present the essentials so those conclusions can be grasped with a minimum of effort and delay.

5.2.1.2 Furthermore, a report should provide sufficient discussion to ensure the correct interpretation of the findings, which should indicate the nature of the analysis and the process of reasoning that leads to those findings.

5.2.2 Purpose. Each item of a report should serve a definite purpose.

5.2.2.1 Each table and chart in a report should be within the scope of the report.

5.2.2.2 The tables and charts should enhance the information stated or shown elsewhere, and they should be accurate and free of the possibility of misunderstanding, within reason.

5.2.3 Organization.

5.2.3.1* The process of writing reports should include five steps that are generally used in identifying, investigating, evaluating, and solving a problem.

5.2.3.2 These five steps, which should be accomplished before the report is written, are as follows:
 (1) The purpose and scope of the report should be obtained.
 (2) The method or procedure should be outlined.
 (3) The essential facts should be collected.
 (4) These facts should be analyzed and categorized.
 (5) The correct conclusions should be arrived at and the proper recommendations should be made.

5.2.4* While there are differing needs among fire departments, certain reports should be common to most departments.

5.2.4.1 Typical recommended training reports should include the following:
 (1) A complete inventory of apparatus and equipment assigned to the training division
 (2) Detailed plans for training improvements that include all equipment and facility needs and cost figures
 (3) A detailed periodic report on and evaluation of the training of all probationary fire fighters
 (4) A monthly summary of all activities of the training division
 (5) An annual report of all activities of the training division
 (6) A complete inventory of training aids and reference materials available to be used for department training

5.2.4.2 The annual report should describe the accomplishments during the year, restate the goals and objectives of the training division, and describe the projected plans for the upcoming year.

5.2.5 Narrative Report. There are times when a narrative report should be necessary.

5.2.5.1 Before writing a narrative report, the writer should consider the audience for the report.

5.2.5.2 The comprehensiveness of the report should be determined by the recipients' knowledge of the subject.

5.3* Training Records.

5.3.1 Training records should be kept to document department training and should assist in determining the program's effectiveness. Information derived from such records should, for example, provide the supporting data needed to justify additional training personnel and equipment.

5.3.1.1 Training records can include paper or electronic media.

5.3.2 Performance tests, examinations, and personnel evaluations should contribute to the development of the training program if the results are analyzed, filed, and properly applied.

5.3.2.1 Training records should be kept current and should provide the status and progress of all personnel receiving training.

5.3.2.2 Frequent review of training records should provide a clear picture of the success of the training program and document lessons learned.

5.3.3 Properly designed training records should be developed to meet the specific needs of each fire department.

5.3.3.1 Training records should be detailed enough to enable factual reporting while remaining as simple as possible.

5.3.3.2 The number of records should be kept at a minimum to avoid confusion and duplication of effort.

5.3.4 Typical training records should include an evaluation of the competency of the student, as well as hours attended.

5.4* State Certification Records.
5.4.1 Minimum Information.

5.4.1.1 Information and documentation that should serve as a foundation for submission to state certification programs should include, as a minimum, the following:
(1) A single file that includes all training accomplished by the individual fire fighter during his/her career
(2) Dates, hours, locations, and instructors of all special courses or seminars attended
(3) Monthly summaries of all departmental training

5.4.1.2 These records should require the signatures of the instructor and the person instructed as a valid record of an individual's participation in the training.

5.4.2 The format used for state certification should be different from that utilized by an individual department. Otherwise, this is likely to cause considerable problems with accurate record submission and should be addressed on the state level by all parties concerned. Various state certification forms are contained in Annex B.

Chapter 6 Evaluating the Effectiveness of Training Records Systems

6.1 Evaluating Records of Individuals.

6.1.1 The evaluation of training records should be done at specified intervals by the local department training officer or training committee.

6.1.2 Each training record should be evaluated to determine the following:
(1) Has the individual taken all the required training?
(2) If not, has the individual been scheduled for missed classes?
(3) Do performance deficiencies show up on the individual's training record?
(4) If performance deficiencies exist, what kind of program is being developed to overcome them?
(5) Have companies met all the required job performance standards established by the department?
(6) If job performance standards have not been met, have the problems been identified and a program developed to overcome them?
(7) Are there areas of training that are being overlooked completely?
(8) Is the cycle of training sufficient to maintain skill levels?

6.2 Evaluating the Record-Keeping System.

6.2.1 All training records and the record-keeping system should be evaluated at least annually.

6.2.2 During the evaluation process, the following questions should be applied to each record:
(1) What is the purpose of the record?
(2) Who uses the information compiled?
(3) Is the record providing the necessary information?
(4) Do other records duplicate the material being compiled?
(5) How long should records be retained?
(6) Can training trends be determined from a compilation of the records?
(7) Is there a simpler and more efficient way of recording the information?

Chapter 7 Legal Aspects of Record Keeping

7.1 Privacy of Personal Information.

7.1.1* Employee training and educational records and other examination data included in an individual's training file should be disclosed only with written permission of the employee, unless required by law or statute, or by a court order.

7.1.2 The fire chief and the training officer should verify with legal counsel the federal, state, provincial, and local laws and ordinances regulating the disclosure of confidential information, and ensure adequate control measures are in place for the privacy of personal information.

7.1.2.1 Training records should not use the student's Social Security number for identification purposes.

7.1.2.2 The fire chief or training officer should ensure that training records do not include any confidential medical information.

7.1.2.3 All medical records should be kept in a completely separate file and not mixed with any other records or personnel files.

7.1.2.4 Access to any personally identifiable or proprietary information should be restricted.

7.1.3 Length of Time for Keeping Records or Reports.

7.1.3.1* Legal counsel should be contacted concerning the length of time records or reports, or both, need to be kept available and documented in a records retention schedule. *[See Figure B.1(m) for a sample schedule.]*

7.1.3.2 Documents should be maintained for a period of time as specified by law or as required by certain agencies and organizations.

7.1.4 Most training records should be maintained in their entirety in a computerized form, thus greatly reducing the amount of paper that needs to be stored.

7.1.4.1* Some training records should be maintained in their original hard-copy form, as required by certain agencies and organizations.

7.1.4.2* Computerized records should be backed up periodically and stored in an off-site location to avoid destruction.

7.2* Record Keeping and Risk Management. Agencies that conduct multijurisdictional training should have a signed release form for those individuals who participate in certain training activities.

Appendix E

An Extract from: NFPA 1500, Standard on Fire Department Occupational Safety, Health, and Wellness Program, 2018 Edition

Chapter 5 Training, Education, and Professional Development

5.1 General Requirements.

5.1.1* The fire department shall establish and maintain a training, education, and professional development program with a goal of preventing occupational deaths, injuries, and illnesses.

5.1.2 The fire department shall provide training, education, and professional development for all department members commensurate with the duties and functions that they are expected to perform.

5.1.3 The fire department shall establish training and education programs that provide new members initial training, proficiency opportunities, and a method of skill and knowledge evaluation for duties assigned to the member prior to engaging in emergency operations.

5.1.4* The fire department shall restrict the activities of new members during emergency operations until the member has demonstrated the skills and abilities to complete the tasks expected.

5.1.5 The fire department shall provide all members with training and education on the department's risk management plan.

5.1.6 The fire department shall provide all members with training and education on the department's written procedures.

5.1.7 The fire department shall provide all members with a training, education, and professional development program commensurate with the emergency medical services that are provided by the department.

Δ 5.1.8* The fire department shall provide all members with a documented training and education program that covers all assigned personal protective equipment (PPE).

5.1.8.1 Training shall comply with applicable governing standards and follow the manufacturer's instructions and guidelines to include the following topics:
(1) The organization's overall program for the selection and use of protective ensembles, ensemble elements, and SCBAs
(2) Technical data package (TDP) where applicable
(3) Proper overlap and fit
(4) Proper donning and doffing (including emergency doffing)
(5) Construction features and function
(6) Usage and performance limitations (including physiological effects on user and effects of heat transfer on the protective ensemble)
(7) Recognizing and responding to indications of protective ensemble and SCBA failure
(8) Routine inspection cleaning, maintenance, and retirement
(9) Special incident procedure operation
(10) Proper storage

Δ 5.1.8.2 For maintenance of structural and proximity protective ensembles and ensemble elements, refer to NFPA 1851.

5.1.8.3 For maintenance of SCBA, refer to NFPA 1852.

5.1.8.4 For maintenance of protective ensemble for technical rescue incidents, refer to NFPA 1855.

5.1.9 As a duty function, members shall be responsible to maintain proficiency in their skills and knowledge, and to avail themselves of the professional development provided to the members through department training and education programs.

5.1.10 Training programs for all members engaged in emergency operations shall include procedures for the safe exit and accountability of members during rapid evacuation, equipment failure, or other dangerous situations and events.

5.1.11 All members who are likely to be involved in emergency operations shall be trained in the incident management and accountability system used by the fire department.

5.2 Member Qualifications.

5.2.1 All members who engage in structural fire fighting shall meet the requirements of NFPA 1001.

5.2.2* All driver/operators shall meet the requirements of NFPA 1002.

5.2.3 All aircraft rescue fire fighters (ARFF) shall meet the requirements of NFPA 1003.

5.2.4 All fire officers shall meet the requirements of NFPA 1021.

5.2.5 All wildland fire fighters shall meet the requirements of NFPA 1051.

Δ 5.2.6* All members responding to hazardous materials incidents shall meet the operations level as required in NFPA 472.

N 5.2.7 All members who engage in fire investigations shall meet the requirements of NFPA 1033.

N 5.2.8 All members who engage in fire inspections shall meet the requirements of NFPA 1031.

5.3 Training Requirements.

5.3.1* The fire department shall adopt or develop training and education curriculums that meet the minimum requirements outlined in professional qualification standards covering a member's assigned function.

5.3.2 The fire department shall provide training, education, and professional development programs as required to support the minimum qualifications and certifications expected of its members.

5.3.3 Members shall practice assigned skill sets on a regular basis but not less than annually.

5.3.4 The fire department shall provide specific training to members when written policies, practices, procedures, or guidelines are changed and/or updated.

5.3.5* The respiratory protection training program shall meet the requirements of NFPA 1404.

5.3.6 Members who perform wildland fire fighting shall be trained at least annually in the proper deployment of an approved fire shelter.

5.3.7 All live fire training and exercises shall be conducted in accordance with NFPA 1403.

5.3.8 All training and exercises shall be conducted under the direct supervision of a qualified instructor.

5.3.9* All members who are likely to be involved in emergency medical services shall meet the training requirements of the AHJ.

5.3.10* Members shall be fully trained in the use, limitations, care, and maintenance of the protective ensembles and ensemble elements assigned to them or available for their use.

5.3.11 All members shall meet the training requirements as outlined in NFPA 1561.

5.3.12 All members shall meet the training requirements as outlined in NFPA 1581.

***N* 5.3.13** All members shall be trained in the risks associated with workplace exposure to products of combustion, carcinogens, fireground contaminants, and other incident-related health hazards.

N* 5.3.13.1 Members shall be trained to recognize when a workplace exposure has occurred and to know the control methods for personal decontamination, decontamination of protective clothing and equipment, and the risks of cross-contamination.

5.4 Special Operations Training.

5.4.1 The fire department shall provide specific and advanced training to members who engage in special operations as a technician.

5.4.2 The fire department shall provide specific training to members who are likely to respond to special operations incidents in a support role to special operations technicians.

5.4.3 Members expected to perform hazardous materials mitigation activities shall meet the training requirements of a technician as outlined in NFPA 472.

Δ 5.4.4 Members expected to perform technical operations at the technician level as defined in NFPA 1670 shall meet the training requirements specified in NFPA 1006.

5.5 Member Proficiency.

5.5.1 The fire department shall develop a recurring proficiency cycle with the goal of preventing skill degradation and potential for injury and death of members.

5.5.2 The fire department shall develop and maintain a system to monitor and measure training progress and activities of its members.

5.5.3* The fire department shall provide an annual skills check to verify minimum professional qualifications of its members.

N 5.6 Training Activities.

N* 5.6.1 All training and exercises shall be conducted under the direct supervision of a qualified instructor.

N* 5.6.2 All live fire training and exercises shall be conducted in accordance with NFPA 1403.

***N* 5.6.2.1** Emergency medical services shall be provided for live fire training exercises in accordance with Section 4.10 of NFPA 1403.

N* 5.6.3 For non–live fire training exercises, fire departments shall conduct a risk assessment to determine the appropriate emergency medical capabilities to be available at the training site.

Shaded text = Revisions. Δ = Text deletions and figure/table revisions. • = Section deletions. *N* = New material.

Appendix F

An Extract from: NFPA 1403, Standard on Live Fire Training Evolutions, 2018 Edition

Chapter 5 Acquired Structures

5.1 Structures and Facilities.

5.1.1* Any acquired structure that is considered for a structural fire training exercise shall be prepared for the live fire training evolution.

5.1.1.1 Buildings that cannot be made safe as required by this chapter shall not be utilized for interior live fire training evolutions.

5.1.2 Adjacent buildings or property that might become involved shall be protected or removed.

5.1.3* Preparation shall include application for and receipt of required permits and permissions.

5.1.4* Ownership of the acquired structure shall be determined prior to its acceptance by the AHJ.

5.1.5 Evidence of clear title shall be required for all structures acquired for live fire training evolutions.

5.1.6* Written permission shall be secured from the owner of the structure in order for the fire department to conduct live fire training evolutions within the acquired structure.

5.1.7* A clear description of the anticipated condition of the acquired structure at the completion of the evolution(s) and the method of returning the property to the owner shall be put in writing and shall be acknowledged by the owner of the structure.

5.1.8* Proof of insurance cancellation or a signed statement of nonexistence of insurance shall be provided by the owner of the structure prior to acceptance for use of the acquired structure by the AHJ.

5.1.9 The permits specified in this chapter shall be provided to outside, contract, or other separate training agencies by the AHJ upon the request of those agencies.

5.1.10 A search of the acquired structure shall be conducted to ensure that no unauthorized persons, animals, or objects are in the acquired structure immediately prior to ignition.

5.1.11 No person(s) shall play the role of a victim inside the acquired structure.

5.1.12 Only one fire at a time shall be permitted within an acquired structure.

5.2 Hazards.

5.2.1 In preparation for live fire training, an inspection of the structure shall be made to determine that the floors, walls, stairs, and other structural components are capable of withstanding the weight of contents, participants, and accumulated water.

5.2.2* All hazardous storage conditions shall be removed from the structure or neutralized in such a manner as to not present a safety problem during use of the structure for live fire training evolutions.

5.2.3 Closed containers and highly combustible materials shall be removed from the structure.

5.2.3.1 Oil tanks and similar closed vessels that cannot be removed shall be vented to prevent an explosion or overpressure rupture.

5.2.3.2 Any hazardous or combustible atmosphere within the tank or vessel shall be rendered inert.

5.2.4 All hazardous structural conditions shall be removed or repaired so as to not present a safety problem during use of the structure for live fire training evolutions.

5.2.4.1 Floor openings shall be covered to be made structurally sound.

N 5.2.4.2 Fires shall not be ignited under exposed structural members.

5.2.4.3 Missing stair treads and rails shall be repaired or replaced.

5.2.4.4 Dangerous portions of any chimney shall be removed.

5.2.4.5 Holes in walls and ceilings shall be patched.

5.2.4.6* Roof ventilation openings that are normally closed but can be opened in the event of an emergency shall be permitted to be utilized.

5.2.4.7* Low-density combustible fiberboard and other highly combustible interior finishes shall be removed.

5.2.4.8* Extraordinary weight above the training area shall be removed.

5.2.5* All hazardous environmental conditions shall be removed before live fire training evolutions are conducted in the structure.

5.2.5.1 All forms of asbestos deemed hazardous shall be removed by an approved manner and documentation provided to the AHJ.

5.2.6 Debris creating or contributing to unsafe conditions shall be removed.

5.2.7 Any toxic weeds, insect hives, or vermin that could present a potential hazard shall be removed.

5.2.8 Trees, brush, and surrounding vegetation that create a hazard to participants shall be removed.

5.2.9 Combustible materials, other than those intended for the live fire training evolution, shall be removed or stored in a protected area to preclude accidental ignition.

5.3 Utilities.

5.3.1 Utilities shall be disconnected.

5.3.2 Utility services adjacent to the live burn site shall be removed or protected.

5.4 Exits.

5.4.1 Exits from the acquired structure shall be identified and evaluated prior to each training burn.

5.4.2 Participants of the live fire training shall be made aware of exits from the acquired structure prior to each training burn.

5.4.3 Fires shall not be located in any designated exit paths.

Δ **5.5 Rapid Intervention Crew (RIC).** A RIC trained in accordance with NFPA 1407 shall be provided during a live fire training evolution.

N **5.6 Water Supply.** For acquired structures the minimum water supply and delivery for the live fire training evolutions shall meet the criteria identified in NFPA 1142.

Chapter 6 Gas-Fired Live Fire Training Structures and Mobile Enclosed Live Fire Training Props

6.1 Structures and Facilities.

6.1.1 This section pertains to all interior spaces where gas-fired live fire training exercises occur.

6.1.2 Live fire training structures shall be left in a safe condition upon completion of live fire training evolutions.

6.1.3 Debris hindering the access or egress of fire fighters shall be removed prior to the beginning of the training exercises.

6.1.4 Flammable gas fires shall not be ignited manually.

6.2 Inspection and Testing.

6.2.1* Live fire training structures shall be inspected visually for damage prior to live fire training evolutions.

6.2.1.1* Damage shall be documented and the building owner or AHJ shall be notified.

6.2.2 Where the live fire training structure damage is severe enough to affect the safety of the participants, training shall not be permitted.

6.2.3 All doors, windows and window shutters, railings, roof scuttles and automatic ventilators, mechanical equipment, lighting, manual or automatic sprinklers, and standpipes necessary for the live fire training evolution shall be checked and operated prior to any live fire training evolution to ensure they operate correctly.

6.2.4 All safety devices, such as thermal sensors, combustible gas monitors, evacuation alarms, and emergency shutdown switches, shall be checked prior to any live fire training evolutions to ensure they operate correctly.

6.2.5 The instructors shall run the training system prior to exposing students to live flames in order to ensure the correct operation of devices such as the gas valves, flame safeguard units, agent sensors, combustion fans, and ventilation fans.

6.2.6* The structural integrity of the live fire training structure shall be evaluated and documented annually by the building owner or AHJ.

6.2.6.1 If visible structural defects are found, such as cracks, rust, spalls, or warps in structural floors, columns, beams, walls, or metal panels, the building owner shall have a follow-up evaluation conducted by a licensed professional engineer with live fire training structure experience and expertise, or by another competent professional as determined by the building owner or AHJ.

6.2.7* The structural integrity of the live fire training structure shall be evaluated and documented by a licensed professional engineer with live fire training structure experience and expertise, or by another competent professional as determined by the AHJ, at least once every 10 years, or more frequently if determined to be required by the evaluator.

6.2.8* All structures constructed with calcium aluminate refractory structural concrete shall be inspected by a structural engineer with expertise in live fire training structures every 3 years.

6.2.8.1 The structural inspection shall include removal of concrete core samples from the structure to check for delaminations within the concrete.

6.2.9* Part of the live fire training structure evaluation shall include, at least once every 10 years, the removal and reinstallation of a representative area of thermal linings (if any) to inspect the hidden conditions behind the linings.

Chapter 7 Non-Gas-Fired Live Fire Training Structures and Mobile Enclosed Live Fire Training Props

7.1 Structures and Facilities.

7.1.1 This section pertains to all interior spaces where non-gas-fired live fire training exercises occur.

7.1.2 Live fire training structures shall be left in a safe condition upon completion of live fire training evolutions.

7.1.3 Debris hindering the access or egress of fire fighters shall be removed prior to the beginning of the training exercises.

7.2 Inspection and Testing.

7.2.1* Live fire training structures shall be inspected visually for damage prior to live fire training evolutions.

7.2.1.1* Damage shall be documented, and the building owner or AHJ shall be notified.

7.2.2 Where the live fire training structure damage is severe enough to affect the safety of the participants, training shall not be permitted.

7.2.3 All doors, windows and window shutters, railings, roof scuttles and automatic ventilators, mechanical equipment, lighting, manual or automatic sprinklers, and standpipes necessary for the live fire training evolution shall be checked and operated prior to any live fire training evolution to ensure they operate correctly.

7.2.4 All safety devices, such as thermal sensors, oxygen and toxic and combustible gas monitors, evacuation alarms, and emergency shutdown switches, shall be checked prior to any live fire training evolutions to ensure they operate correctly.

7.2.5* The structural integrity of the live fire training structure shall be evaluated and documented annually by the building owner or AHJ.

7.2.5.1 If visible structural defects are found, such as cracks, rust, spalls, or warps in structural floors, columns, beams, walls, or metal panels, the building owner shall have a follow-up evaluation conducted by a licensed professional engineer with live fire training structure experience and expertise or by another competent professional as determined by the AHJ.

7.2.6* The structural integrity of the live fire training structure shall be evaluated and documented by a licensed professional engineer with live fire training structure experience and expertise or by another competent professional as determined by the AHJ at least once every 5 years or more frequently if determined to be required by the evaluator.

7.2.7* All structures constructed with calcium aluminate refractory structural concrete shall be inspected by a structural engineer with expertise in live fire training structures every 3 years.

7.2.7.1 The structural inspection shall include removal of concrete core samples from the structure to check for delaminations within the concrete.

7.2.8* Part of the live fire training structure evaluation shall include, once every five years, the removal and reinstallation of a representative area of thermal linings (if any) to allow inspections of the conditions hidden behind the linings.

7.3 Sequential Live Fire Burn Evolutions.

7.3.1 The AHJ shall develop and utilize a safe live fire training action plan when multiple sequential burn evolutions are to be conducted per day in each burn room.

7.3.2 A burn sequence matrix chart shall be developed for the burn rooms in a live fire training structure.

7.3.2.1 The burn sequence matrix chart shall include the maximum fuel loading per evolution and maximum number of sequential live fire evolutions that can be conducted per day in each burn room.

7.3.3* The burn sequence for each room shall define the maximum fuel load that can be used for the first burn and each successive burn.

7.3.4* The burn sequence matrix for each room shall also specify the maximum number of evolutions that can be safely conducted during a given training period before the room is allowed to cool.

7.3.5 The fuel loads per evolution and the maximum number of sequential evolutions in each burn room shall not be exceeded under any circumstances.

Shaded text = Revisions. Δ = Text deletions and figure/table revisions. • = Section deletions. *N* = New material.

Glossary

ABCD method A process for writing lesson plan objectives; it includes four components: audience, behavior, condition, and degree.

Acquired structure A building or structure acquired by the authority having jurisdiction from a property owner for the purpose of conducting live fire training evolutions.

Active listening The process of hearing and understanding the communication sent; demonstrating that you are listening and have understood the message.

Adapt To make fit (as for a specific use or situation).

Affective domain The domain of learning that affects attitudes, emotions, or values. It may be associated with a student's perspective or belief being changed as a result of training in this domain.

Agency training needs assessment A needs assessment performed at the direction of the department administration, which helps to identify any regulatory compliance matters that must be included in the training schedule.

Ambient noise The general level of background sound.

Andragogy The art and science of teaching adults.

Assignment The part of the lesson plan that provides the student with opportunities for additional application or exploration of the lesson topic, often in the form of homework that is completed outside of the classroom.

Asynchronous learning An online course format in which the instructor provides material, lectures, tests, and assignments that can be accessed at any time. Students are given a time frame during which they need to connect to the course and complete assignments.

Attention-deficit/hyperactivity disorder (ADHD) A disorder in which a person has a chronic level of inattention and an impulsive hyperactivity that affects daily functions.

Audience Who the students are.

Audience analysis The determination of characteristics common to a group of people; it can be used to choose the best instructional approach.

Baby boomers The generation born after World War II (i.e., 1946–1964).

Backdraft A deflagration (explosion) resulting from the sudden introduction of air into a confined space containing oxygen-deficient products of incomplete combustion (NFPA 1403).

Base budget The level of funding that would be required to maintain all services at the currently authorized levels including adjustments for inflation and salary increases.

Behavior An observable and measurable action for the student to complete.

Behaviorist perspective The theory that learning is a relatively permanent change in behavior that arises from experience.

Blended learning An instruction method that combines online/independent study with face-to-face meetings with the instructor.

Bloom's Taxonomy A classification of the different objectives and skills that educators set for students (learning objectives).

Budget An itemized summary of estimated or intended revenues and expenditures.

Capital budget A budget that is used to purchase an item that would last for several years such as a fire engine.

Certification Document awarded for the successful completion of a testing process based on a standard.

Coaching The process of helping individuals develop skills and talents.

Cognitive domain The domain of learning that effects a change in knowledge. It is most often associated with learning new information.

Cognitive perspective An intellectual process by which experience contributes to relatively permanent changes (learning). It may be associated by learning by experience.

Communication process The process of conveying an intended message from the sender to the receiver and getting feedback to ensure accuracy.

Competency-based learning Learning that is intended to create or improve professional competencies.

Condition The situation in which the student will perform the behavior.

Conduction Heat transfer to another body or within a body by direct contact (NFPA 1403).

Confidentiality The requirement that, with very limited exceptions, employers must keep medical and other personal information about employees and applicants private.

Continuing education Education or training obtained to maintain skills, proficiency, or certification in a specific position.

Convection Heat transfer by circulation within a medium such as a gas or liquid (NFPA 1403).

Core temperature The body's internal temperature.

Course A unit of instruction. In college or high school, a course usually lasts one term or semester. In the fire service, a course may be a single-topic course, for example a confined space course that consists of four classes of 4 hours each.

Courseware Any educational content that is delivered via a computer.

Curriculum The complete body of study offered. A college will have a curriculum for each major it offers. (*Curricula* is the plural form of curriculum.)

Degree The last part of a learning objective, which indicates how well the student is expected to perform the behavior in the listed conditions.

Delegation Transfer of authority and responsibility to another person for the purpose of teaching new job skills or as a means of time management.

Demographics Characteristics of a given population, possibly including such information as age, race, gender, education, marital status, family structure, and location of agency.

Direct threat A situation in which an individual's disability presents a serious risk to his or her own safety or the safety of his or her co-workers.

Disability A physical or mental condition that interferes with a major life activity.

Discrimination index The value given to an assessment that differentiates between high and low scorers.

Dyscalculia A learning disability in which students have difficulty with math and related subjects.

Dyslexia A learning disability in which students have difficulty reading due to an inability to interpret spatial relationships and integrate visual information.

Dysphasia A learning disability in which students lack the ability to write, spell, or place words together to complete a sentence.

Dyspraxia A lack of physical coordination with motor skills.

Enabling objective An intermediate learning objective, usually part of a series that directs the instructor on what he or she needs to instruct and what the learner will learn in order to accomplish the terminal objective.

Essay test A test that requires students to form a structured argument using materials presented in class or from required reading.

Ethics Principles used to define behavior that is not specifically governed by rules of law but rather in many cases by public perceptions of right and wrong. They are often defined on a regional or local level within the community.

Expenditures Money spent for goods or services that are considered appropriations.

Face validity A type of validity achieved when a test item has been derived from an area of technical information by an experienced subject-matter expert who can attest to its technical accuracy.

Feedback The fifth and final link of the communication chain. It allows the sender (the instructor) to determine whether the receiver (the student) understood the message.

Fire control team A team consisting of at least two members including the ignition officer. The other fire fighter(s) can staff the charged line, observe the ignition, help maintain the fire, and watch for and report adverse conditions.

Fire flow rate The amount of water pumped per minute (gallons per minute or liters per minute) for a fire. There are several different formulas that are commonly used to calculate this.

Flameover (or rollover) The condition in which unburned fuel (pyrolysate) from the originating fire has accumulated in the ceiling layer to a sufficient concentration (i.e., at or above the lower flammable limit) that it ignites and burns. Flameover can occur without ignition of or prior to the ignition of other fuels separate from the origin (NFPA 1403).

Flashover A transition phase in the development of a compartment fire in which surfaces exposed to thermal radiation reach ignition temperature more or less simultaneously and fire spreads rapidly throughout the space, resulting in full room involvement or total involvement of the compartment or enclosed space (NFPA 1403).

Flipped classroom An instructional approach that reverses or "flips" the standard learning environment by delivering online content for self-study and reserving the physical classroom for mentored practice that traditionally would have been assigned as homework.

Flow path The movement of heat and smoke from the higher pressure within the fire area towards the lower pressure areas accessible via doors, window openings, and roof structures (NFPA 1410).

Formal communication An official fire department communication. The letter or report is presented on stationery with the fire department letterhead and generally is signed by a chief officer or headquarters staff member.

Formative evaluation Process conducted to improve the instructor's performance by identifying his or her strengths and weaknesses.

Four-step method of instruction The primary process used to relate the material contained in a lesson plan to the students; the steps are preparation, presentation, application, evaluation.

Fuel load The total quantity of combustible contents of a building, space, or fire area, including interior finish and trim, expressed in heat units or the equivalent weight in wood (NFPA 1403).

General orders Short-term documents signed by the fire chief that remain in force for a period of days to 1 year or more.

Generation X People born after the baby boomers.

Generation Y People born immediately after Generation X.

Generation Z People born immediately after Generation Y, starting in the late 1990s.

Go/No Go sequence A verbal confirmation via radio communication that each and every participant is ready for action in the live fire environment.

Gross negligence An act, or a failure to act, that is so reckless that it shows a conscious, voluntary disregard for the safety of others.

Heat release rate The rate at which heat energy is generated by burning (NFPA 921).

Hold harmless An agreement or contract wherein one party holds the other party free from responsibility for liability or damage that could arise from the transaction between the two parties.

Hybrid learning Learning environment in which some content is available for completion online and other learning requires face-to-face instruction or demonstration; sometimes referred to as "blended" or "distributed" learning.

Identifying The process of selecting those persons whom the instructor would like to mentor, coach, and develop.

Ignition officer An individual who is responsible for igniting and controlling the material being burned at a live fire training.

Immediately dangerous to life or health (IDLH) Any condition that would pose an immediate or delayed threat to life, cause adverse health effects, or interfere with an individual's ability to escape unaided from a hazardous environment.

Incident scene rehabilitation A function on the fireground that cares for the well-being of the fire fighters. It includes physical assessment, revitalization, medical evaluation and treatment, and regular monitoring of vital signs.

Indemnification agreement An agreement or contract wherein one party assumes liability from another party in the event of a claim or loss.

Independent study A method of distance learning in which students order the course materials, complete them, and then return them to the instructor.

Informal communications Internal memos, e-mails, instant messages, and computer-aided dispatch/mobile data terminal messages. Informal reports have a short life and are not archived as permanent records.

In-service drill A training session scheduled as part of a regular shift schedule.

Instructor An individual qualified by the authority having jurisdiction to deliver fire fighter training, who has the training and experience to supervise students during live fire training evolutions, and who has met the requirements of an Instructor I in accordance with NFPA 1041 (NFPA 1403).

Instructor-centered learning Instructional technique where the student is a passive listener and the instructor is the sole source of knowledge, such as a lecture with the use of presentation material.

Instructor-in-charge (IIC) An individual qualified as an instructor and designated by the authority having jurisdiction to be in charge of the live fire training evolution and who has met the requirements of an Instructor II in accordance with NFPA 1041 (NFPA 1403).

Instructor-led training (ILT) A learning environment in which a human instructor facilitates both the in-person sections and the online components of coursework. Instructors usually set deadlines for submission, create quizzes and assignments, and track student progress.

Interrogatory A series of formal written questions sent to the opposing side of a legal argument. The opposition must provide written answers under oath.

Item analysis A listing of each student's answer to a particular question used for evaluation.

Job-content/criterion-referenced validity A type of validity obtained through the use of a technical committee of job incumbents who certify the knowledge being measured is required on the job and referenced to known standards.

Kinesthetic learning Learning that is based on doing or experiencing the information that is being taught.

Learner characteristics Designating a target group of learners and defining those aspects of their personal, academic, social, or cognitive self that may influence how and what they learn (NFPA 1041).

Learning A relatively permanent change in behavior potential that is traceable to experience and practice.

Learning domains Categories that describe how learning takes place—specifically, the cognitive, psychomotor, and affective domains.

Learning environment A combination of a physical location (classroom or training ground) and the proper emotional elements of both an instructor and a student.

Learning management system (LMS) A web-based software application that allows online courseware and content to be delivered to learners, that also includes classroom administration tools such as activity tracking, grades, communications, and calendars.

Learning objective A goal that is achieved through the attainment of a skill, knowledge, or both, and that can be measured or observed.

Learning outcomes Statements of what a student can do as a result of completion of a program or course of study.

Lesson outline The main body of the lesson plan; a chronological listing of the information presented in the lesson plan.

Lesson plan A detailed guide used by an instructor for preparing and delivering instruction.

Lesson summary The part of the lesson plan that briefly reviews the information from the presentation and application sections.

Lesson title The part of the lesson plan that indicates the name or main subject of the lesson plan. Also called the lesson topic.

Level of instruction The part of the lesson plan that indicates the difficulty or appropriateness of the lesson for students.

Liability Responsibility; the assignment of blame. It often occurs after a breach of duty.

Live fire Any unconfined open flame or device that can propagate fire to the building, structure, or other combustible materials (NFPA 1403).

Major life activity Basic functions of an individual's daily life, including, but not limited to, caring for oneself, performing manual tasks, breathing, walking, learning, seeing, working, and hearing.

Malfeasance Dishonest, intentionally illegal, or immoral actions.

Master training schedule Form used to identify and arrange training topics by the number of times they must be trained on or by the type of regulatory authority that requires the training to be completed.

Mean A value calculated by adding up all the scores from an examination and dividing by the total number of students who took the examination.

Median The score in the middle of the score distribution for an examination.

Medium The third link of the communication chain, which describes how you convey the message.

Mentoring A relationship of trust between an experienced person and a person with less experience for the purpose of growth and career development.

Message The second and most complex link of the communication chain; it describes what you are trying to convey to your students.

Method of instruction Various ways in which information is delivered to student, both in a classroom and on the training ground (NFPA 1041).

Minimum water supply The quantity of water required for fire control and extinguishment (NFPA 1142).

Misfeasance Mistaken, careless, or inadvertent actions that result in a violation of law.

Mode The most commonly occurring value in a set of values (e.g., scores on an examination).

Motivation The activator or energizer for an activity or behavior.

Motivational factors States of the person that are relatively temporary and reversible and that tend to energize or activate the behavior of the individual.

Negligence An unintentional breach of duty that is the proximate cause of harm.

Networking An activity or process of like-minded individuals meeting and developing relationships for professional and personal growth through sharing of ideas and beliefs.

Operating budget A budget that is used to pay for the day-to-day expenditures such as fuel, salaries, or fire prevention flyers.

Oral test A test in which the answers are spoken in essay form in response to direct or open-ended questions. They may accompany a presentation or demonstration.

Order of operations The sequence of steps to conduct a procedure. In this context, the steps are in proper sequence to conduct the live fire evolution, but order of operations could also refer to any emergency scene operation. Most commonly refers to the sequence a mathematical equation is solved.

Organizational chart A graphic display of the fire department's chain of command and operational functions.

P+ value Number of correct responses to the test item. Example: P = 67 means that 67 percent of test takers answered correctly.

Participant Any student, instructor, safety officer, visitor, or other person who is involved in the live fire training evolution within the operations area (NFPA 1403).

Passive listening Listening with your eyes and other senses without reacting to the message verbally.

Pedagogy The art and science of teaching children.

Performance test A test that measures a student's ability to do a task under specified conditions and to a specific level of competence. Also known as a skills evaluation.

Personnel accountability report (PAR) A verification by the person in charge of each crew or team that all of their assigned personnel are accounted for.

Pilot course The first offering of a new course, which is designed to allow the developers to test models, applications, evaluations, and course content.

Postevolution debriefing A review of the training event. It is used as a critique to evaluate the objectives and training evolution to determine positive events as well as those that need improvement.

Preburn plan A briefing session conducted for all participants of live fire training in which all facets of each evolution to be conducted are discussed and assignments for all crews participating in the training sessions are given.

Prerequisite A condition that must be met before a student is allowed to receive the instruction contained within a lesson plan—often a certification, rank, or attendance of another class.

Program A program of instruction comprises the fully body of teaching in a particular area. A fire academy may divide their curricula into several program areas such as firefighting, officer development, and technical rescue.

Program budget A system in which expenditures are allocated for specific activities such as training.

Program outcomes The measurable outcomes of an assessment process intended to determine whether program goals are being met.

Psychomotor domain The domain of learning that requires the physical use of knowledge. It represents the ability to physically manipulate an object or move the body to accomplish a task or use a skill.

Pyrolysis A process in which material is decomposed, or broken down, into simpler molecular compounds by the effects of heat alone; pyrolysis often precedes combustion (NFPA 921).

Qualitative analysis An in-depth research study performed to categorize data into patterns to help determine which test items are acceptable.

Quantitative analysis Use of statistics to determine the acceptability of a test item.

Radiation Heat transfer by way of electromagnetic energy (NFPA 1403).

Reasonable accommodation An employer's attempt to make its facilities, programs, policies, and other aspects of the work environment more accessible and usable for a person with a disability.

Receiver The fourth link of the communication chain. In the fire service classroom, the receiver is the student.

Recommendation report A decision document prepared by a fire officer for the senior staff. The goal is support for a decision or an action.

Reference blank Where the current job-relevant source of the test-item content is identified.

Reliability The characteristic that a test measures what it is intended to measure on a consistent basis.

Reliability index Value that refers to the reliability of a test as a whole in terms of consistently measuring the intended material.

Revenues The income of a government from all sources, which is appropriated for the payment of public expenses and is stated as estimates.

Safety officer An individual appointed by the authority having jurisdiction as qualified to maintain a safe working environment at all live fire training evolutions (NFPA 1403).

Self-directed study A learning environment in which learners can stop and start as they desire, progress to completion of a module, and submit a final score to determine whether they passed or failed the module, all without an instructor's involvement.

Sender The first link of the communication chain. In the fire service classroom, the sender is the instructor.

Sharing The basic concept of giving to others with nothing expected in return.

Skills evaluation A test that measures a student's ability to do a task under specified conditions and to a specific level of competence. Also known as a performance test.

Social learning An informal method of learning using technologies that enable collaborative content creation. Examples include blogs, social media, collaborative discussions, and wikis.

Standard deviation The value to which data should be expected to vary from the average.

Standards A set of guidelines outlining behaviors or qualifications of positions or specifications for equipment or processes. Often developed by individuals within the regulated profession, they may be applied voluntarily or referenced within a rule or law.

Stroke volume The volume of blood pumped by the left ventricle with each contraction of the heart.

Student-centered learning Educational methodologies that focus on student engagement and require students to be active, responsible participants in the learning experience (NFPA 1041).

Subject-matter expert (SME) An individual who is technically competent and who works in the field for which test items are being developed.

Succession planning The act of ensuring the continuity of the organization by preparing its future leaders.

Summative evaluation Process that measures the students' achievements to determine the instructor's strengths and weaknesses.

Supplemental training schedule Form used to identify and arrange training topics available in case of a change in the original training schedule.

Syllabus An outline of a course that identifies course objectives, course material, assignments, and the assessment process.

Synchronous learning An online class format that requires students and instructors to be online at the same time. Lectures, discussions, and presentations occur at a specific hour and students must be online at that time to participate and receive credit for attendance.

Systems approach to training (SAT) process A training process that relies on learning objectives and outcome-based learning.

Tailboard chat An informal gathering where fire fighters discuss various issues.

Technical-content validity A type of validity that occurs when a test item is developed by a subject-matter expert and is documented in current, job-relevant technical resources and training materials.

Technology-based instruction (TBI) Training and education that uses the internet or a multimedia tool.

Terminal objective A broader outcome that requires the learner to have a specific set of skills or knowledge after a learning process.

Thermal gradient The rate of temperature change with distance.

Thermal tolerance The body's ability to cope with high heat conditions.

Thermoregulation The process by which the body regulates temperature.

Training needs analysis (TNA) A process of identifying organizational training and development needs.

Validity The documentation and evidence that supports the test item's relationship to a standard of performance in the learning objective and/or performance required on the job.

Ventilation-controlled fire A fire in which the heat release rate or growth is controlled by the amount of air available to the fire (NFPA 1403).

Vent point ignition Smoke is at or above its ignition temperature and is lacking oxygen. The smoke will ignite as it exits the opening and falls within the flammable range.

Virtual classroom A digital environment where content can be posted and shared, and where instructors can create quizzes and assignments to be completed and tracked online via an internet-connected computer.

Webinar An online meeting, occurring in real time, which usually allows for participants to share their desktop screens, web browsers, and documents live as the meeting is occurring.

Willful and wanton conduct An act that shows utter indifference or conscious disregard for the safety of others.

Written test A test that may be made up of several types of test items, such as multiple choice, true/false, matching, essay, and identification questions.

Index

Note: Page numbers followed by 'f' and 't' refers to figures and tables, respectively.

A

AAR. *See* after-action review
Abbotsford, Wisconsin (incident report), 366
ABCD learning objective method, 124–126, 202, 303–304
    audience, 204–205
    behavior, 205–208, 205–206t, 207f
    condition, 208
    degree, 208, 208t, 210
accountability, 385–386
acquired structures, 405
    deterioration of, 449
    documentation for, 450
    inspection and preparation of
        access, 416–417, 417t
        entry and egress routes, 417, 418f
    live fire training in, 168, 374, 411–412
    overhaul of, 451
    water supply requirements for, 375
active learning, 54, 56
active listening, 104–105
activity-oriented learners, 40, 61
ADA. *See* Americans with Disabilities Act
adaptation, lesson plan, 123
    accommodating instructor style, 134
    evaluating
        facilities for appropriateness, 133
        local conditions, 133
        student limitations, 134
    local SOPs, 133
    method of instruction, 134
    reviewing instructional materials, 132–133
    student needs, 134
    vs. creation, 131–134
ADDIE (analyze, design, develop, implement, and evaluate) model, 301
ADEA. *See* Age Discrimination in Employment Act
ADHD. *See* attention-deficit/hyperactivity disorder
adult learners
    defined, 39
    generational characteristics, 41–44
        baby boomers, 42, 42f
        Generation X, 42–43
        Generation Y, 43–44, 44f
        Generation Z, 44
    influences on, 40–41
    learning skills, 47–48
        classroom study tips, 47
        personal study time, 47–48
        test preparation, 48
    motivation of, 61–62
    psychological state, 61
    senses, use of, 45, 45f
    types of, 40
adult learning, 39–40. *See also* adult learners
affective domain objectives, 207–208
affective learning
    examples of, 53
    levels of, 52
after-action review (AAR), 345
Age Discrimination in Employment Act (ADEA), 17–18, 355–356
agency notification checklist, 416
agency training needs assessment, 267
AHJ. *See* authority having jurisdiction
ambient noise, 106
American Psychological Association, 241
Americans with Disabilities Act (ADA), 17–20, 355, 356
    learning disabilities classification, 62–63
analysis and report, data, 299, 301
andragogy. *See* adult learning
announcements, formal communication, 112
Apple Keynote, 142, 145, 147
applied research project (ARP), 27
ARP. *See* applied research project
arrangement test item, 189–190, 190f, 248, 249f
    disadvantage of, 248
    preparation of, 248–249
asbestos, preburn inspection, 418
asphalt, preburn inspection, 418
assignment, 79, 130
asynchronous learning, 79, 143
attack group, 439
attack instructor, 439
attention-deficit/hyperactivity disorder (ADHD), 62–63
attics, preburn inspection, 422–423
audience
    acronym, 42
    learning objective, 124, 204–205
    methods of instruction, 85
audience analysis, 41–42, 107–108
audio systems, 150–151
authority having jurisdiction (AHJ), 16, 447
average score, examination
    mean score, 330, 330f
    median score, 330–331, 331t
    mode score, 330, 330f

B

baby boomers, 42, 42f
backdraft, 378
backup instructor, 439
Baltimore, Maryland (incident report), 66–68
base budget, 278
Beaumont, Texas (incident report), 260–261

behavior
  distracting, 224–225
  learning objective, 124, 205–208
behavioral outcomes. *See* learning objective
behaviorist perspective, learning, 45–46
blended learning, 224. *See also* hybrid learning
Bloom's Taxonomy, 51, 205–206, 205–206t, 304
  revised, 206, 207f
Blue Card Hazard Zone Management, 150
body language, 80–81
brainstorming technique, 60
bubble-form format, 257, 258f
budget, 357, 359. *See also* budget process
  categories of, 359
  constraints, 79
  justifications, 361
  organization of, 359, 360t
  purchasing guidelines and policies, 361–362
    procedures, 362, 363t
    request for proposal, 362
    writing specifications, 362
  purchasing process, 362, 364
  skills application, 361
  spreadsheet, 361
budget process, 278, 359–361, 361f
  capital expenditures, 279, 281
  cycle, 279, 280–281t
  preparation, 278–279, 279f
  terminologies, 278
  training, 281
    expenses, 281–282
building, marking, 419, 419f
building plan, for preburn plan, 409, 409f
burn evolutions
  experienced students, 433
  instructor-in-charge responsibilities, 431
  learning objectives, 432
  live, selecting instructors for, 434–441, 434f, 436–437f, 439f
  new recruits and live fire training, 432–433
  preburn plan
    briefing, 433–434
    developing, 441–442

C

cadets. *See* audience
Cambridge, Minnesota (incident report), 289
capital budget, 278, 359
capital expenditures, 279, 281
cardiac emergencies, 398–399
  prevention of, 399
cardiovascular strain, 391–392, 392f
  environmental conditions, 393, 394f
  fitness level, 396
  hydration status, 396–397, 397f
  individual characteristics, 394–395, 395f
  medical conditions, 395–396, 395t
  personal protective equipment, 394
  work performed, 393
case studies, for enhanced instructional method, 79, 224
CBT. *See* computer-based training
certificate, 14
certification, 14
certification record, 457
certification training record (fire fighter), 347
challenging process, Instructor II skill, 226
CHD. *See* coronary heart disease
cheating, during exam, 192
chief officers' periodic training summary, 347
chimneys, preburn inspection, 419–420
CISM. *See* critical incident stress management

Civil Rights Act, 18, 355
class clown, disruptive behavior, 83
class evaluation forms, 256–258, 256f, 258f
classroom study tips, adult learners, 47
clean wooden pallets, 421
closed-ended questions, 82
coach, fire and emergency services instructor as, 10, 10f
coaching
  defined, 10
  Instructor II skill, 226
  succession planning, 29–30
cognition, defined, 46
cognitive domain objectives, 205–206
cognitive learning
  examples of, 51
  levels of, 51
cognitive objectives, 124, 125t
cognitive perspective, learning, 45–46
command staff, 439
communication, effective, 81
communication process
  audience analysis, 107–108
  cone of learning, 108
  elements, 103–104, 103f
  environment, 104
  feedback, 103–104
  medium, 103
  nonverbal, 104–105
  receiver, 103
  sender, 103
  verbal, 105–107
  written, 108–113, 109f
communications plan, 415
competency-based learning, 46, 46f
complacency, 431
completion test item, 190
  types, 190–191
complex true/false test item, 191, 191f
computer-based testing, 254–256, 255f
computer-based training (CBT), 143–145, 145f
condition, learning objective, 125, 208
conduction, 397
conferences, professional development, 25–26
confidence building, 15
confidentiality, 17, 20–21, 277
  in evaluation results, 325
construction classification numbers, 407
contingency plans, outdoor classroom, 92
convection, 397
Copyright Act, 21, 215
core temperature, 391
coronary heart disease (CHD), 395
courses, designing, 303
  content outlines, 306–307
    conducting research and identifying resources, 307, 308f
    evaluating instructional resources, 307–309, 309t
    modifying courses, 309–310
  evaluation plans, 310
    feedback, 310–313, 310f
    monitored performance evaluations, 313–316, 314–315f
    interpreting evaluation results, 316
  objectives, writing, 303–304
    converting JPRs into learning objectives, 305–306, 305f
    level of achievement, 304–305
courseware, 144
crawl–walk–run sequence, 75–76
credentials and qualifications, of instructor
  confidence building, 15
  continuing education, 15
  laying groundwork, 14
  standards, 14–15

criterion-related validity. *See* job-content/criterion-referenced validity
critical incident stress management (CISM), 448
critical reflection, 41
critique. *See* postevolution debriefing
curriculum, 301

D

data collection, 328–329
    average score, 330–331, 330*f*, 331*t*
    isolation of data, 328*t*
    score distribution, 329*f*, 329–330, 329*t*
    test-item analysis tally sheet, 328*t*
data gathering, 298–299
data projectors, 148, 148*f*
daydreamer, disruptive behavior, 83
decision document, 113
decontamination
    importance of, 451
    of props, 421
degree
    defined, 14, 125
    learning objective, 208, 208*t*, 210
delegation, 23–24, 226
delivery, method of, 128
demobilization plan, 416
demographics, 93
    adapting lesson plan based on, 94, 94*f*
    dress, 93–94
    gender, 93
    gestures, 93–94
    offensive language, 93–94
demonstration/skill drill method, 57–58, 75–76, 76*f*, 223, 223*f*
    classroom setup, 88*t*
department culture, 85
departmental training record, 346–347
digital audio players, 148–149, 148*f*
digital light processor (DLP) projectors, 148
direct questions, 82
direct threat, 19
disabilities, learning
    attention-deficit/hyperactivity disorder, 62–63
    color blindness, 63
    dyscalculia, 62
    dyslexia, 62
    dysphasia, 62
    dyspraxia, 62
    hearing loss, 63
    instructing students with, 63–64
    poor vision, 63
disability. *See also* disabilities, learning
discrimination index, 331–333, 331*t*, 332*t*
discussion, 56–57, 76–77, 76*f*, 223
    classroom setup, 88*t*
disruptive behavior, students
    management, 84
    type of, 82–83
disruptive students, learning, 64, 64*f*
distance learning. *See also* technology-based instruction (TBI)
distractions, indoor classroom, 89
distributed learning. *See* hybrid learning
diversity, learning environment, 16
DLP projectors. *See* digital light processor projectors
documentation, 21
    for acquired structures, 450
    of emergency incident, 448
    before event, 447
    for exterior training props, 450–451
    for gas-fired and non-gas-fired structures, 450
    general reports and, 447–448, 447*f*
    photographs and videos, 448–449
    reasons for, 446–447
    of unusual events, 449–450
domain objective
    affective, 207–208
    cognitive, 205–206
    psychomotor, 206–207
doors, preburn inspection, 422, 422*f*
dress code, 79
drill assignment (incident report), 318
dry-erase boards, 149–151
dyscalculia, 62
dyslexia, 62
dysphasia, 62
dyspraxia, 62

E

e-instruction. *See* technology-based instruction (TBI)
easel pad, 149–150
education, continuing, 15
education, defined, 7
educational courses, training record, 347
EEOC. *See* Equal Employment Opportunity Commission
EFOP. *See* Executive Fire Officer Program
electronic learning (e-learning), 143–144
electronic results, 192
electronic training records, 341–342, 343*f*
emergency incident, documentation of, 448
emergency medical services (EMS), 142, 185, 412
emergency plans, 412–413, 413*t*, 441
    medical, 415
    nonexercise emergencies, 413–414
emergency shutdown controls, for fire control team, 438–439
employee selection, guidelines for, 242
EMS. *See* emergency medical services
enabling objective, 124, 211
English as a second language (ESL), 63
enhanced instructional methods, 78–79, 78*f*
enthusiasm, 106
entry and egress routes, preburn inspection, 417, 418*f*
entry officer, 439, 439*f*
environment, communication process, 104
Equal Employment Opportunity Act, 355
Equal Employment Opportunity Commission (EEOC), 241–242, 355
ESL. *See* English as a second language
essay test items, 191–192, 250
    formatting, 250–251
    grading, 251–253, 252*f*
    key, preparation of, 253, 253*f*
    objective, preparation of, 251
ethics, in training environment, 15–17, 16*f*
    leading by example, 16
    recordkeeping, 16
    student accountability, 16
    trust, and confidentiality, 17
evaluation plan
    for instructors performance, 356–357
    writing, 215
evaluation process, 226, 229
    formative, 226–227
    forms and tools, 230, 231*f*
    lesson plan, 230
    observation, 229–230
    preparation, 229, 229*f*
    student, 227
    summative, 227
evaluation results
    acquiring, 324
    analyzing tools, 327
    dissemination of, 325, 325*f*
    storing, 325
evaluator, fire and emergency services instructor, 10–11

evaporative cooling, 391
"everyone goes home", 374
evolution safety, 440–441
exam, cheating during, 192
Executive Fire Officer Program (EFOP), 25, 27
exit routes, preburn inspection, 423, 423f
expenditures, 278, 359
expense budget. See operating budget
expert, disruptive behavior, 83
exterior training props, documentation for, 450–451
eye contact, 80

F

face-to-face communication, 104
face validity, 242
fatigue, 392–393
FDIC. See Fire Department Instructors Conference
Federal Emergency Management Agency (FEMA), 27
federal law, 17–18
feedback, 310–311, 310f
   in communication process, 103–104
   instructor, 231–232, 232f
      evaluation review, 232, 232f
   surveys, 311
      format, 313
      questions, 311, 313
      rating scales, 311, 312f
   on testing results, 193
FEMA. See Federal Emergency Management Agency
FESHE. See Fire and Emergency Services Higher Education
FFR. See fire flow rate
final training test, 241
Fire and Emergency Services Higher Education (FESHE), 7, 27
fire and emergency services instructor. See also specific instructor
   audience analysis, 107–108
   as coach, 10, 10f
   credentials and qualifications, 14–15
   disruptive behavior management, 82–84, 83f
   effective, roles of, 8–11
   ethics in training environment, 15–17, 16f
   as evaluator, 10–11
   law applies to, 17–21, 17f
   as leader, 8–9, 8f, 9f
   learning environment, 11
      physical elements affecting, 11–12
   levels of, 5–6, 5f
   managing multiple priorities, 21–24
   as mentor, 9–10, 9f
   next generation, 28–30
   organizational charts, 7–8
   presentation skills and techniques, 79–82
   professional development, 24–27
   qualities of, 4–5
   role in
      department's future, 13
      succession planning, 13–14
   as teacher, 11
      emotional elements affecting, 12
      physical elements affecting, 12f
Fire and Emergency Services Instructor I
   communication skills. See communication process
   evaluation. See testing, process
   learning process. See learning
   lesson plan. See lesson plan
   methods of instruction. See instruction methods
   safety. See training safety
   training. See training process
Fire and Emergency Services Instructor II
   evaluation process, 226–230, 229f, 231f. See also testing
   feedback, 231–232, 232f

   instructional development. See lesson plan
   method of instruction, 223–225, 223f
   skills, 256
   supervision during high-risk training, 233–234, 233f
   as supervisor, 225–226
   training program management. See training program management
Fire and Emergency Services Instructor III
   program administration. See training program administration
   program development. See training program development
   program evaluation. See training program evaluation
fire and EMS organizations, 254–256
fire control team, 438–440, 439f
Fire Department Instructors Conference (FDIC), 26
fire dynamics, 386
fire fighter skill performance ratings template, 462
fire fighter trainees. See audience
fire fighters
   demographics, 93–94
   departmental culture, 85
   hands-on training, 76
   learning environment, 85–92
   risk management, 85
   safety, 91
fire flow rate (FFR), 407
fire-ground operations, 432, 438
fire officers. See audience
fire training, physiological aspect of
   cardiac emergencies, 398–399
      prevention of, 399
   cardiovascular and thermal strain, 391–393, 392f
      factors affecting, 393–397, 393–395f, 395t, 397f
   heat illness, 398, 398–399t
   thermal balance, 397–398
fire fighter performance expectations, 469–470
five Ws, of writing, 110
flameover, 386
flashover, 378, 386
Flesch–Kincaid readability index, 109
flipped classroom, 57, 144
floor, preburn inspection, 421, 421f
flow path, 386
FOIA. See Freedom of Information Act
forced learning, 46–47
formal communication, 111
   announcements, 112
   general orders, 112
   legal correspondence, 112
   standard operating procedure, 111–112
formal education record, 458
formal learning, 38, 86–87
formal training course, 270–271
formative assessment, 241
formative evaluation process, 226–227
four-step method of instruction, 77t
   application, 77–78
   evaluation, 78
   preparation/motivation, 77
   presentation, 77
Freedom of Information Act (FOIA), 20, 277
front-end analysis, 298
FSRI. See UL Firefighter Safety Research Institute
fuel loads, 386
fuel needs, in live fire training, 409–411
furniture, preburn inspection, 420–421

G

gas-fired structure, 438–439
   documentation for, 450
   preparation and inspection of, 425
gender, demographics, 93

general fund budget. *See* operating budget
general orders, 112
Generation Next. *See* Generation Z
Generation X, 42–43, 43f, 84–85
Generation Y, 43–44, 44f
Generation Z, 44, 84
gestures, demographics, 93–94
gifted learner, disruptive behavior, 83
global positioning satellite (GPS), 415
goal-oriented learners, 40, 61
Go/No Go sequence, 413, 413t, 437–438, 441
good facilitator, qualities, 47
Google Slides, 147
GPS. *See* global positioning satellite
grading essay test items, 251–253, 252f
    key, preparation of, 253, 253f
grading scales, 193
Greenwood, Delaware (incident report), 32–33
gross negligence, 173
group training records, 347

H

hands-on training method, 76, 86
    personal protective equipment, 163
    rehabilitation practices, 163, 163f
hazards, preburn inspection, 420
Health Insurance Portability and Accountability Act (HIPAA), 449
heat emergencies
    heat illness, 398, 398–399t
    thermal balance, 397–398
heat release rate, 386–387, 387t
hidden hazards, during training, 169
high-risk training, 168, 168f
    supervision during, 233, 233f
        live fire training, 233, 233f
        safety briefings, 233–234
high-stakes tests, 241
higher education, professional development, 25
HIPAA. *See* Health Insurance Portability and Accountability Act
historian, disruptive behavior, 83
hold harmless agreement, 174, 277
Hollandale, Minnesota (incident report), 96–97
Homewood, Illinois (incident report), 336
hybrid learning, 143–144

I

IAP. *See* incident action plan
identification test item, 190, 190f, 249–250, 249f
identifying, succession planning, 28–29
IDLH. *See* immediately dangerous to life or health
iGen. *See* Generation Z
ignition officer, 170, 233, 439
IIC. *See* instructor-in-charge
ILT. *See* instructor-led training
immediately dangerous to life or health (IDLH), 384, 448
in-service drills, 268
    skill and knowledge development, 268, 270f
    skill and knowledge improvement, 268, 270, 270t
    skill and knowledge maintenance, 268, 270f
incident action plan (IAP), 447, 447f
incident report
    Abbotsford, Wisconsin, 366
    Baltimore, Maryland, 66–68
    Beaumont, Texas, 260–261
    Cambridge, Minnesota, 289
    drill assignment, 318
    Greenwood, Delaware, 32–33
    Hollandale, Minnesota, 96–97
    Homewood, Illinois, 336
    Lairdsville, New York, 154–155

    Milford, Michigan, 136–138
    Pennsylvania State Fire Academy, 177–179
    Pinckneyville, Illinois, 217
    Poinciana, Florida, 115–117
    Port Everglades, Florida, 195–196
    Tarrytown, New York, 236
incident scene rehabilitation, 400
indemnification agreement, 174, 277
independent study, 79, 224
individual action plans (IAPs), 431
individual special course record, 347
individual training record, 347
indoor classroom, learning environment
    arrangement for testing, 89
    distractions, 89
    internet use, 87
    lighting, 89
    safety, 89
    setup, 87
    temperature, 87, 88f, 88t
informal communications, 111
informal learning environment, 38, 86
information, currency of, 242
infrequent reports, 113
inspection, preburn
    acquired structures, 416–418, 417–418f
    asphalt and asbestos, 418
    attics, 422–423
    chimneys, 419–420
    decontamination of props, 421
    exits, 423, 423f
    flooring, 421, 421f
    furniture, 420–421
    hazards, 420
    kitchens, 422, 423f
    marking the building, 419, 419f
    oil tanks, 422
    of props and facilities, 423–425, 424f
        gas-fired live fire training structures, 425
        non-gas-fired live fire training structures, 425–427, 426–427f
    utilities, 418–419, 419f
    ventilation, 419, 420f
    walls and ceilings, 421–422
    windows and doors, 422, 422f
instruction methods
    audience, 85
    blended learning, 79
    case studies, 79, 224
    definition, 75
    demonstration/skill drill, 75–76, 223, 223f
    department culture, 85
    discussion, 76–77, 76f
    enhanced, 78–79
    four-step, 77–78, 77t
    independent study, 79
    labs/simulations, 78–79, 78f, 224
    learning environment, 85–92
    lecture, 75, 76f
    lesson presentation skills and techniques, 79–82, 80f
    managing disruptive behavior, 82–84, 83f
    online/distance learning, 79
    out-of-class assignments, 224
    pre-course survey, 78
    prerequisites, 78
    role-play, 224
    self-actualization, 92
    teamwork, 92
        demographics in, 93–94
    techniques and transitioning, 224–225
    transitioning between, 84–85

instructional design, 301
instructional materials, lesson plan, 130
instructional preparation, lesson plan, 130
    organizational techniques, 130–131
    procuring materials/equipment, 131
    rehearse, 131
    scheduling, 131
    student, 130
instructional resources, evaluation, 307–309, 309t
instructor-centered learning, 203–204, 204f
instructor, defined, 430
Instructor I, 123, 131–132, 209. *See also* Fire and Emergency Services Instructor I
    defined, 6
    demonstration/skill drill, 75–76
    lecture method, 75, 76f
    roles and responsibilities of, 6–8
    student-centered teaching methods/strategies, 56–60
Instructor II, 6, 123, 128, 132, 209. *See also* Fire and Emergency Services Instructor II
Instructor III, 6, 209. *See also* Fire and Emergency Services Instructor III
instructor-in-charge (IIC), 170–172, 233, 374, 376–378, 439
    defined, 430
    evolution safety, 440–441
    fire control team, 438–440, 439f
    instructional team member skills, 436–437, 436–437f
    instructor and student rotation, 440
    responsibilities, 431
    staffing and organization, 437–438, 437f
instructor-led training (ILT), 144
"instructor-made tests", 254
instrument. *See* class evaluation forms
Insurance Service Office (ISO), 267, 341
interactive boards, 149, 149f
interest, areas of, 47
International Society of Fire Service Instructors (ISFSI), 27, 374
internet-based instruction. *See* technology-based instruction (TBI)
interrogatory, 112
investigation, accident and injury, 172
iPod, 148
ISFSI. *See* International Society of Fire Service Instructors
ISO. *See* Insurance Service Office
item analysis, evaluation tool, 327–328
item-difficulty score, 327–328

J

job-content/criterion-referenced validity, 242
job performance requirements (JPRs), 124
    in evaluation, 228, 243
    into learning objectives, converting, 210–211, 305–306, 305f
    in lesson plan, 209
    NFPA 1001, importance of, 377–378
    program evaluation, 326
    skill sheet template, 470
    in training program
        administration, 358
        development, 300
        resources, 273
JPRs. *See* job performance requirements

K

K-W-L approach, 60
Keynote. *See* Apple Keynote
kinesthetic learning. *See* psychomotor learning
kitchens, preburn inspection, 422, 423f
know-it-all, disruptive behavior, 83
knowledge, skills, and abilities (KSAs), 298
KSAs. *See* knowledge, skills, and abilities
Kuder–Richardson reliability formula, 333

L

labs/simulations, 78–79, 78f, 224
Lairdsville, New York (incident report), 154–155
laminar smoke, 387, 387f
language, verbal communication, 106–107
law, instructor's conduct, 17–21, 17f
    Americans with Disabilities Act, 18–20
    confidentiality, 20–21
    copyright, 21
    federal employment laws, 18–20
    Freedom of Information Act, 20
    of learning, 45
    public domain, 21
    records, 20–21
    reports, 20–21
law of effect, 45
law of exercise, 45
law of intensity, 45
law of primacy, 45
law of readiness, 45
law of recency, 45
LCD projectors. *See* liquid crystal display projectors
leader, fire and emergency services instructor as, 8–9
learner-centered teaching. *See* student-centered teaching
learner characteristics, 39
    effective teaching, 54–55, 54f
    individual, 53
learner-oriented adults, 40
learning
    adult, 39–44
    affective, 52–53
    areas of interest, 47
    behaviorist perspective, 45–46
    cognitive, 51
    cognitive perspective, 45–46
    competency-based, 46, 46f
        examples of verbs used to write, 463–465
    defined, 38
    disabilities, 62–64
    disruptive students, 64
    domains, 50–53, 51f
    forced, 46–47
    instructor-centered, 203–204, 204f
    interactive process of, 39
    laws/principles of, 45–47
    lifelong, 26–27
    Maslow's hierarchy of needs, 48–50
    motivation and, 60–62
    psychomotor, 51–52, 52f
    questioning techniques, 82
    skills, for adult learners, 47–48
        classroom study tips, 47
        personal study time, 47–48
        test preparation, 48
    student-centered, 203–204
learning course management system (LCMS). *See* learning management system (LMS)
learning environment, 11, 86, 86f
    audience, 85
    emotional elements affecting, 12
    indoor classroom, 87–89, 88f, 88t
    outdoor classroom, 89–92, 90f
    physical elements affecting, 11–12, 12f
    physical environment, 86–87
    safety in, 162–163
learning management system (LMS), 143–145, 145f
learning objective, 406
    ABCD method, 125–126, 202, 204t
        audience, 124, 204–205
        behavior, 124, 205–208, 205–206t, 207f

condition, 125, 208
    degree, 125, 208, 208t, 210
components of, 124–126
converting JPRs into, 210–211, 210f
defined, 124
developing preburn plan, 406
examples of, 125t
and lesson plan, 128
live fire training evolution, 432
learning-oriented learner, 61
learning outcomes, 302
learning science, 54–55
lecture, 223
lecture method, 75, 76f
    classroom setup, 88t
legal considerations
    testing, 185
    training safety, 172–175
legal correspondence, 112
lesson outline, 130
    copyright and public domain, 215
    creation of, 211–214
    identifying instructional materials, 214–215
    sample, 211–214f
lesson plan
    adapting vs. creating, 131–134
    based on demographics, 94
    components of, 127–128f, 202–203
        assignment, 130
        delivery, method of, 128
        instruction, level of, 126
        learning objectives, 128
        lesson outline, 130
        lesson summary, 130
        lesson title, 126, 129f
        materials needed, 130
        references/resources, 129
    creation of, 203
        determining course purpose, 204
        evaluation plan, 215
        instructor-centered learning, 203–204, 204f
        learning objectives, 204–211, 204t
        lesson outline, 211–215, 211–214f
        student-centered learning, 203–204
    defined, 123
    instructional preparation, 130–131
    learning objective, 124–126
    uses of, 123–124
lesson presentation skills, and techniques, 79–81, 80f
    effective communication, 81
    questioning techniques, 82
lesson summary, 130
lesson title/topic, 126, 129f
level of instruction, 126
LFI. See Live Fire Instructor
LFTI. See live fire training instructor
liability, 173
lighting
    indoor classroom, 89
    learning environment, 12
line-of-duty death (LODD), 161–162, 446
liquid crystal display (LCD) projectors, 148
live burn evolutions, selecting instructors for, 434–437, 434f, 436–437f
    emergency plans, 441
    evolution safety, 440–441
    fire control team, 438–440, 439f
    instructor and student rotation, 440
    safety officer, 438
    staffing and organization, 437–438, 437f

live fire evolution
    accountability, 385–386
    fire behavior and structural fire dynamics, 386
    heat release rate, 386–387, 387t
    personal protective equipment, 385, 385f
    protective clothing, 383–385
    sample incident command organization for, 439–440, 439f
    self-contained breathing apparatus, 383–385
    smoke, 387–388, 387–388f
live fire exercise, 436–437, 437f
live fire instructor (LFI)
    establishing clear objectives, 376–377
    history of, 373–374, 373f
    incident scene rehabilitation, 400
    live fire evolutions, 385–388, 385f, 387–388f
    live fire training environment
        preparing for, 383–385
        students in, 388–390, 389f, 390f
    live fire training evolutions, history of, 373–374, 373f
    NFPA 1001 prerequisite JPRs, importance of, 377–378, 377f
    NFPA 1403
        impact in live fire training, 374
        importance of, 373
        referenced standards, 374–376
        uses of, 376
    participant safety and training effectiveness, 372–373, 373f
    physiological aspect, of fire training, 390–399
    vs. live fire instructor-in-charge, 376
live fire instructor-in-charge. See also instructor-in-charge (IIC)
    vs. live fire instructor, 376
live fire training, 170–171, 233, 233f, 372, 383
    evolutions, new recruits and, 432–433
    fire fighter style, 389–390, 390f
    recruit students, 388, 389f
live fire training instructor (LFTI), 389, 431, 432, 434, 434f
    knowledge, skills, and ability, 435–436
    preburn inspection and planning. See preburn plan
    qualities of, 434
    training and experience validation form, 461
LMS. See learning management system
LODD. See line-of-duty death
logistics, outdoor classroom, 91–92
low-risk training, 168, 168f

M

Mager, Dr. Robert F., 244
major life activities, 19
malfeasance, 173–175
Maslow's hierarchy of needs, 48–50, 49f
    esteem and status, 50, 50f
    physiological needs, 49, 49f
    safety, security, and order, 49
    self-actualization, 50
    social needs and affection, 49–50, 50f
master training schedule, 274–275, 274f
matching test item, 189, 189f, 247–248, 248f
mayday, documentation, 449
mean score, 330, 330f
median score, 330–331, 331t
medical emergency
    documentation, 449
    plans, 415
medium, communication process, 103
medium-risk training, 168, 168f
memorandum (memo), 109–110, 110f
mentor, fire and emergency services instructor, 9–10, 9f
mentoring, 29, 226
method of instruction, 75, 223. See also instruction methods
Milford, Michigan (incident report), 136–138
Millennials. See Generation Y

minimum water supply (MWS), 407
misfeasance, 173–174
mode score, 330, 330f
monopolizer, disruptive behavior, 82–83
monthly activity report, 112–113
motivation, and learning, 39
    adult learners, 61–62
    factors, 60–61
        in class design, 62
multimedia tools, training, 145, 145f
    advantages, 145–146
    audio systems, 150–151
    best practices, 146–147
    data projectors, 148
    disadvantages, 146
    dry-erase boards, 149–150
    easel pads, 149–150
    interactive boards, 149, 149f
    presentation programs, 147
    presentation-ready programs, 147–148
    simulations, 150
    tablet computers, 149, 149f
    virtual reality devices, 150
multiple choice exam, 185
multiple choice test item, 189, 189f, 247, 247f
MWS. See minimum water supply

N

National Board of Fire Underwriters, 418
National Emergency Training Center, 27
National Fallen Firefighters Foundation (NFFF), 161
National Fire Academy (NFA), 24, 27, 132, 226
    five-step model, 301
National Fire Protection Association (NFPA) standards, 126, 210
    30, flammable and combustible liquids code, 375
    58, liquefied petroleum gas code, 375
    59, utility LP-gas plant code, 375
    1001, fire fighter professional qualifications, 375, 434–435
    1021, fire officer professional qualifications, 226, 270–271
    1041, fire and emergency services instructor professional qualifications, 5–6, 131–132, 374, 376, 430, 473–486
    1142, water supplies for suburban and rural fire fighting, 375
    1401, recommended practice for fire service training reports and records, 277, 325, 341, 346–347, 487–489
    1403, live fire training evolutions, 18, 164, 170–171, 174, 233, 350, 372–373, 430, 492–494
        impact in live fire training, 374
        importance of, 373
        referenced standards, 374–376
        uses of, 376
    1406, outside live fire training evolutions, 374
    1407, training fire service rapid intervention crews, 376
    1451, fire and emergency service vehicle operations training program, 174
    1500, fire department occupational safety, health, and wellness program, 167, 174, 490–491
    1582, comprehensive occupational medical program for fire departments, 164
    1583, health-related fitness programs for fire department members, 396
    1584, rehabilitation process for members during emergency operations and training exercises, 91, 163, 167
    1971, protective ensembles for structural fire fighting and proximity fire fighting, 375
    1975, emergency services work clothing elements, 375
    1981, open-circuit self-contained breathing apparatus for emergency services, 375
    1982, personal alert safety systems, 376
National Incident Management System (NIMS), 437
National Institute for Occupational Safety and Health (NIOSH), 79, 268, 373, 373f
National Oceanic and Atmospheric Administration (NOAA), 414
National Professional Development Model, 27
National Volunteer Fire Council, 376
needs analysis, 298
needs assessment, 298
negligence, 173–174
networking, 26
NFA. See National Fire Academy
NFFF. See National Fallen Firefighters Foundation
NFPA Professional Qualifications standard, 14, 305
NFPA standards. See National Fire Protection Association standards
NIMS. See National Incident Management System
NIOSH. See National Institute for Occupational Safety and Health
NOAA. See National Oceanic and Atmospheric Administration
noise, outdoor classroom, 91
non-gas-fired structure
    documentation for, 450
    preparation and inspection of, 425–426
        fuel load, 426
        safety, 426–427, 426–427f
nonverbal communication, 104, 105f
    active listening, 104–105
    passive listening, 105, 105f

O

objective essay test items, 251
objectives, defined, 124
occupancy hazard classification numbers, 407
Occupational Safety and Health Administration (OSHA), 174, 267, 414
off-site training, 174
offensive language, demographics, 93–94
Office of the State Fire Marshal (OSFM), 342
oil tanks, preburn inspection, 422
online learning, 79, 86, 224
open-ended questions, 82
operating budget, 359
operations budget, 278
oral communication. See verbal communication
oral tests, 185–186, 186f
order of operations, 414–415
organizational charts, 7–8, 8f
organizational techniques, in instructional preparation, 130–131
OSFM. See Office of the State Fire Marshal
OSHA. See Occupational Safety and Health Administration
out-of-class assignments, 224
outdoor classroom, 89–90, 90f
    communication, 90–91
    contingency plans, 92
    logistics, 91–92
    noise, 91
    resources, 91–92
    safety, 91
    weather, 91
outside training request form, 467–468
overhaul, of acquired structures, 451
oxygen-deficient atmosphere, 438

P

P+ value, 331
paper training records, 341–342, 343f
PAR. See personnel accountability report
participant
    defined, 376
    safety and training effectiveness, 372–373, 373f
participating student-to-instructor ratio, 406
PASS. See personal alert safety system
passive listening, 105, 105f

Index

pedagogy, defined, 40
peer tutoring approach, 59
Pennsylvania State Fire Academy, incident report, 177–179
performance-based training program, 244
performance evaluation, 313
    checklists, 313–316, 314–315f
performance outcomes. See learning objective
performance/skill testing, 185, 253–254, 254f
periodic company summary, training record, 347
permits, training site evaluation, 405–406
personal alert safety system (PASS), 438, 448
personal improvement agreement, 466
personal protective equipment (PPE), 76, 103, 163, 209, 377, 377f, 378, 383, 383f, 394, 431, 433, 438, 440, 442
personal study time, adult learners, 47–48
personnel accountability report (PAR), 385–386, 413, 449
photographs, documentation, 448–449
Piaget's theory, 53
pilot course, 306
Pinckneyville, Illinois (incident report), 217
PIO. See public information officer
Poinciana, Florida (incident report), 115–117
policies and procedures
    fire department, 17–18
    training program, 350, 352–353, 352f
        adoption and implementation, 353–354
        writing, 353
    training safety, 164, 165–167f
Port Everglades, Florida (incident report), 195–196
portable media devices, 148–149
post-incident analysis, 345
post live fire evolution
    postburn inspection, 451
    postevolution debriefing, 451–452
    reports and documentation, 447–450, 447f
    training exercise, conclusion of, 450–451
Post-Millennials. See Generation Z
postburn inspection, of structures and props, 451
postevolution debriefing, 451–452
postevolution evaluation forms, 452
posttest, evaluation tool, 240–241
posttest item analysis, 328
    revising test items based on, 334
PowerPoint (PPT), 142, 145, 147
PPE. See personal protective equipment
PPT. See PowerPoint
pre-course survey, 78
pre-test, evaluation tool, 240
preburn plan, 405
    briefing, 433–434
    developing, 441–442
        agency notification checklist, 416
        apparatus needs, 409
        building plan, 409, 409f
        communications plan, 415
        demobilization plan, 416
        emergency medical plan, 415
        emergency plans, 412–414, 413t
        fuel needs, 409–411
        learning objectives, 406
        list of training evolutions, 414
        order of operations, 414–415
        parking, staging, and areas of operations, 412, 412f
        participants, 406
        protective clothing and self-contained breathing apparatus, 416
        rehabilitation plan, 416
        site plan, 411–412
        staffing and organization, 416
        visitors and spectators, 412
        water supply, 406–409, 409f
        weather, 414
    inspection and preparation
        of acquired structures, 416–418, 417–418f
        exterior, 418–420, 419–420f
        interior, 420–423, 421–423f
        of props and facilities, 423–425, 424f, 426–427f
    training site, initial evaluation of, 405–406
prerequisite, 78, 126
presentation-ready programs, 147–148
Prezi, 142, 147
primary engine/pumper, 439
proctoring tests
    oral, 187
    performance, 187, 188t
    written, 186–187
professional development, 24–27
    conferences, 25–26
    higher education, 25
    lifelong learning, 26–27
    professional organizations, 27
    staying current, 24–25
program, 301
    evaluation, 302
    goals and structure, 301–302
    outcomes, 302
    planning, 22–23
program budget system, 359
progress chart, 347
promotional test, 241
psychomotor domain objectives, 206–207
psychomotor learning
    examples of, 52, 52f
    levels of, 51–52
psychomotor lesson, method of instruction, 76
psychomotor objectives, 124, 125t
psychomotor skills evaluations, 334
public domain, 21
public information officer (PIO), 412
pyrolysis, 386

Q

qualitative analysis, 244
quantitative analysis, 244
questioning techniques, 58, 59t, 82

R

radiation, 397
rapid intervention crew (RIC), 411, 413, 439, 449
reasonable accommodation, 19, 19f
"reasonable person" concept, 172
receiver, communication process, 103
recommendation report, 113
recordkeeping, 16, 20–21, 276–277
records, 20–21
    management of, 276–277
reference blank, 246
references, lesson plan, 129
rehabilitation plan, in fire training program, 416
rehabilitation practices, and hands-on training, 163
rehab/medical officer, 439–440
reliability test, 189, 190, 245
reports, 20–21, 112
    infrequent, 113
    monthly activity, 112–113
    recommendation, 113
    special, 113
    training, 112–113
request for proposal (RFP), 362

resources, outdoor classroom, 91–92
revenues, 278
RFP. *See* request for proposal
rhetorical questions, 82
RIC. *See* rapid intervention crew
risk management, fire fighters, 85
role-play, 58–59, 78, 224
room temperature, learning environment, 12

S

safety. *See also* training safety
    fire fighters, 377–378, 377f
    indoor classroom, 89
    for instructors and participants, 223–224
    non-gas-fired live fire training structures, 426–427, 426–427f
    outdoor classroom, 91
safety officer (SO), 170, 233, 438, 439
SAT process. *See* systems approach to training process
scaffolding, 63–64
SCBA. *See* self-contained breathing apparatus
scheduling, training session, 131
secondary engine/pumper, 439
selection test, 241
selection-type objective test items
    arrangement, 248, 249f
    essay, 251–253, 252f, 253f
    identification, 249–250, 249f
    matching, 247–248, 248f
    multiple choice, 247, 247f
    true/false, 250, 250f
self-actualization, 50, 92
self-contained breathing apparatus (SCBA), 90, 163, 303, 431–433, 438, 440, 441, 448
self-directed learning, 41
self-directed study, 144
seminars, training record, 347
sender, communication process, 103
senses, use of, 45, 45f
sharing, succession planning, 30
signs, in live fire evolutions, 412, 412f
simulators, 150
site plan, 411–412
skill sheet format, 253, 254f
skills evaluation, 245. *See also* performance test
SlideDog, 147
slides, usage suggestions, 147
small-group discussion method, 57
SMART Board®, 149f
SME. *See* subject-matter expert
smoke, 387–388, 387–388f
SO. *See* safety officer
social learning, 144
SOG. *See* standard operating guideline
SOP. *See* standard operating procedure
Spearman–Brown reliability formula, 333
special reports, 113
spectators, 412
split-half method. *See* Spearman–Brown reliability formula
staff and participant rotation, live fire training, 440
staff selection, for training program, 354
    candidate evaluation, 354–355
    performance-based evaluation plans, 356–357
    selection policy and procedures, 355–356
staffing and organization, 437–438
staging/air supply officer, 440
standard deviation, 333
standard operating guideline (SOG), 186, 374
standard operating procedure (SOP), 111–112, 133, 163, 186, 225, 271, 374
    importance of, 432

standard testing procedures, 186
standards, 14–15. *See also* National Fire Protection Association (NFPA) standards
state law, 17–18
storytelling, learning environment, 59–60
student accountability, ethics, 16
student-centered learning, 53, 55–56, 203–204. *See also* student-centered teaching
student-centered teaching, methods/strategies
    brainstorming, 60
    demonstration, 57–58
    discussion, 56–57
    peer tutoring, 59
    questioning, 58
    role playing, 58–59
    small-group discussion, 57
    storytelling, 59–60
student evaluation instruments
    data collection, 328–329
        average score, 330–331, 330f, 331t
        isolation of data, 328t
        score distribution, 329f, 329–330, 329t
        test-item analysis tally sheet, 328t
    discrimination index, 331–333, 331t, 332t
    identifying items as acceptable/needing review, 333–334
    item analysis, 327–328
    P+ value, 331
    posttest analysis, revising test items based on, 334
    psychomotor skills, 334
    reliability index, 333
student prerequisite training, documentation, 447–448
student-to-instructor ratio, 170
students. *See also* audience
    disruptive behavior, 64, 64f
        management, 82–84, 83f
        type of, 82–83
    exercise's safety implications, 167–168
    experienced, 433
    grading oral, written, and performance tests, 192–193
    information handling, 192
    instructional preparation, 130
    learning environment. *See* learning environment
    limitations in lesson plan, 134
    in live fire training environment, 388
        fire fighter style, 389–390, 390f
        recruitment, 388, 389f
    reading level, 109
    training safety responsibilities, 172
subject-matter expert (SME), 185, 242
succession planning, 28, 28f
    coaching, 29–30
    identifying, 28–29
    instructor's role in, 13–14
    mentoring, 29
    sharing, 30
summative assessment, 241
summative evaluation process, 227
supervision, during high-risk training, 233, 233f
    live fire training, 233, 233f
    safety briefings, 233–234
supervisor, Instructor II as, 225–226
    skills and techniques, 226
supplemental training schedule, 270, 270t
support group, 439
surveys, evaluation strategy, 311
    format, 313
    questions, 311, 313
    rating scales, 311, 312f
syllabus, 306

synchronous learning, 79, 143
systems approach to training (SAT) process, 244

T

table-top fire-ground lab, 78
table arrangements, learning environment, 12, 12f
tablet computers, 149, 149f
tailboard chat, 87, 107, 108f
Tarrytown, New York (incident report), 236
TBI. *See* technology-based instruction
teacher-centered passive learning, 55–56
teacher, fire and emergency services instructor as, 11
teaching aids, 81
teamwork, 92
technical-content validity, 242
technology-based instruction (TBI)
   computer-based training, 143–145
   defined, 142
   distance learning, 143
   learning management systems, 143–145
terminal objective, 124, 211
test analysis, 244
test generation strategies and tactics, 254, 255f
   computer/web-based testing, 254–256, 255f
test item, 244
   components of, 187, 246–247, 247f
   development of, 245
      and documentation forms, 246
      specifications, 245–246, 246f
   validity, 242
      face, 242
      job-content/criterion-referenced, 242
      technical-content, 242
test preparation skills, adult learners, 48
test results, reporting
   feedback to students, 193
   scores, confidentiality of, 192
testing
   class evaluation forms, 256–258, 256f, 258f
   performance, 253, 254f
   problems in, 240
   process
      classroom arrangement for, 89
      exam, cheating during, 192
      grading student oral, written, and performance tests, 192–193
      legal considerations, 185
      problems in, 184–185
      proctoring tests, 186–187
      purposes and types of, 185–186
      standard procedures, 186
      web-based training and, 193
      written test items, 187–192, 246–247, 247f
   selection-type objective test items, 247–253, 247–250f, 252f, 253f
   in systems approach to training process, role of, 244
   test generation strategies and tactics, 254–256, 255f
   test item
      development, 245–246, 246f
      and test analysis, 244
   types and purpose of, 240–243
theatrical smoke, training, 169
thermal gradient, 397
thermal imager (TI), 388
thermal strain, 391
   environmental conditions, 393, 394f
   fitness level, 396
   hydration status, 396–397, 397f
   individual characteristics, 394–395, 395f
   medical conditions, 395–396, 395t
   personal protective equipment, 394
   work performed, 393

thermal tolerance, 396
thermoregulation, 397
think-pair-share technique, 57
TI. *See* thermal imager
time management, instructor, 21–22
Title VII of Civil Rights Act, 18
TNA. *See* training needs analysis
TO. *See* training officer
TRADE. *See* Training Resource and Data Exchange
training attendance form, 456
training budget, 281
   expenses, 281–282
training evolutions, list of, 414
training exercise, conclusion of
   acquired structures, 450
   exterior training props, 450–451
   gas-fired and non-gas-fired structures, 450
training needs analysis (TNA), 297–298
   analysis and report, 299, 301
   data gathering, 298–299
   parts of, 298
training officer (TO), 267
   resources for, 471–472t
training policy, 459–460. *See* standard operating procedure (SOP)
training priorities, 22, 23f
training process, 142
   classification, 168, 168f
   defined, 7
   maintaining technology, 151
   multimedia applications, 145–151
      troubleshooting procedure, 151, 152t
   technology-based instruction, 142–145
training program administration
   budgets, 357, 359–364, 360t, 361f, 363t
   policies, 350, 352–354, 352f
   reports for, 347–350, 348f, 349f, 351f
   staff selection, 354–357
   training records, 341–347, 342t, 343–346f, 347t
training program development, 297, 298f
   course design, 303
      content outlines, 306–310, 308f, 309t
      evaluation plans, 310–316, 310f, 312f, 314–315f
      interpreting evaluation results, 316
      objectives, writing, 303–306, 305f
   instructional design, 301
      evaluation, 302
      goals and structure, 301–302
      outcomes, 302
   training needs analysis, 297–301
training program evaluation
   results
      acquiring, 324
      analyzing tools, 327
      dissemination of, 325, 325f
      storing, 325
   student evaluation instruments, 327–334, 328t, 329t, 329f, 330f, 331t, 332t
training program management
   budget development, and administration, 278–282, 279t, 280–281t
   instruction, scheduling of, 267–275, 269f, 270t, 270f, 272f, 274f, 275f
   instructors, selection of, 276
   record management, 276–277
   resources, acquiring and evaluating, 282–284, 282f, 283f, 285t, 286–287f
training records, 341
   additional information on, 345–346
   elements of, 345, 346f, 347t
   examples per regulatory agency, 342t

training records (*Continued*)
    required signatures, 344–345
    types of, 341
        certification (fire fighter), 347
        chief officers' periodic training summary, 347
        departmental, 346–347
        educational courses, 347
        electronic and paper records, 341–342, 343f
        exams, 342
        group, 347
        individual, 347
        individual special course, 347
        performance tests, 342
        periodic company summary, 347
        personnel evaluations, 342
        progress chart, 347
        progress reports, 343, 344f
        seminars, 347
        state certification records, 343, 345f
        vocational courses, 347
    typical training file, 343
    uses of, 341
training report, 112–113, 347–348
    format for, 348–349, 348f
    legal aspects of, 350
    storage and retention, 350, 351f
    uses of, 349–350, 349f
Training Resource and Data Exchange (TRADE), 307
training resources, 282, 282f
    defined, 282
    evaluating, 284
        purchase request form, 286–287f
        resource evaluation form, 285t
    Hand-Me-Down, 282–283, 283f
    lesson plan example, 283f
    purchasing, 283–284
training safety
    developing
        anticipating problems, 171–172
        investigation, 172
        student responsibilities, 172
    general procedures, 175t
    hands-on, 163
    hidden hazards, 169
    influencing, 164, 167
    leading by example, 162, 162f
    in learning environment, 162–163
    legal considerations, 172–175
    live fire training, 170–171
    overcoming obstacles, 169–170, 169f
    planning, 167–169, 168f
    policies and procedures, 164, 165–167f
    sixteen fire fighter life safety initiatives, 161–162, 161t
training schedule, 267
    creation, 267–268
    master, 274–275, 275f
    for success, 271–274, 272f
    supplemental, 270, 270t
    types of, 268, 269f
        formal training course, 270–271
        in-service drill, 268–270
training site, initial evaluation of, 405
    required permits, 405–406
transition techniques, presentation, 81

troubleshooting, multimedia problems, 151, 152t
true/false test item, 191, 191f, 250, 250f
trust, 17
turbulent smoke, 387, 387f

U

UL Firefighter Safety Research Institute (FSRI), 27
utilities, preburn inspection, 418–419, 419f

V

validity test, 189, 190, 192, 245
vent point ignition, 387, 388f
ventilation-controlled fire, 386
ventilation, preburn inspection, 419, 420f
verbal communication, 105–106
    factors affecting, 106
    language, 106–107
videos, documentation, 448–449
virtual classroom, 144
virtual learning environment (VLE). *See* learning management system (LMS)
virtual reality devices, 150
visitors, 412
vocational courses, training record, 347

W

walls and ceilings, preburn inspection, 421–422
water flow issues, documentation, 449
water supply, for live fire training, 406–407
    protecting exposures from radiant heat, 408, 409f
    requirements to determine, 407
    source, 407–408
    technique, 409
weather
    in live fire training, 414
    outdoor classroom, 91
web-based training and testing, 193
webinar, 144
willful and wanton conduct, 173
windows, preburn inspection, 422
writing course objectives, 303–304
    converting JPRs into learning objectives, 305–306, 305f
    level of achievement, 304–305
written communication skill, 108–109
    decision document, 113
    reading level, 109
    reports, 112–113
    rules, 110–111
    style, 111
        formal, 111–112
        informal, 111
    test items, 185
        arrangement, 189–190, 190f
        completion, 190–191
        components, 187
        essay, 191–192
        identification, 190, 190f
        matching, 189, 189f
        multiple choice, 189, 189f
        true/false, 191, 191t
    writing format, 109
        five Ws of, 110
written test, 246
    development and documentation forms, 246–247, 247f